AMERICAN CULINARY FEDERATION FOUNDATION eCULINARY PROFESSIONAL DEVELOPMENT INSTITUTE

About Us...

THOMSON
—✦—
TM
DELMAR LEARNING

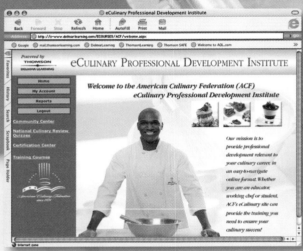

AMERICAN CULINARY FEDERATION FOUNDATION

The American Culinary Federation (ACF), established in 1929, is the premier professional organization for culinarians in America. Its primary mission was to promote professionalism among cooks and chefs in America with an emphasis on education, apprenticeship and professional networking. The ACF Foundation was formed later to develop a variety of educational products to aid in ACF's original mission. In addition, ACF currently holds the presidium for the World Association of Chefs Societies (WACS), the international network of over seventy countries' chef associations representing an estimated 8 million cooks and chefs. For more information, please visit www.acfchefs.org.

THOMSON DELMAR LEARNING

The Thomson Corporation, with 2004 revenues of $8.10 billion, is a global leader in providing integrated information solutions to business and professional customers. Thomson Learning Career & Professional Group is a leading provider of tailored learning solutions for customers in the educational, computer and professional markets. It publishes under the brands of Thomson Delmar Learning and Thomson Course Technology. For more information on our culinary offerings, please visit www.culinary.delmar.com.

“ *Whether you are an educator, working chef, or student, ACF's eCulinary Institute can provide the training you need to ensure your culinary success!* ”

Visit www.acfchefs.org/eculinary • Call (800) 624-9458 or (904) 824-4468

Modern
Garde Manger

Join us on the web at

culinary.delmar.com

Modern
Garde Manger
A GLOBAL PERSPECTIVE

Robert Garlough, MS, FMP, AAC

Angus Campbell, C & G

THOMSON

DELMAR LEARNING

Australia Canada Mexico Singapore Spain United Kingdom United States

THOMSON

DELMAR LEARNING

Modern Garde Manger
by Robert Garlough and Angus Campbell

**Vice President, Career Education
Strategic Business Unit:**
Dawn Gerrain

Director of Learning Solutions:
Sherry Dickinson

Acquisitions Editor:
Matthew Hart

Managing Editor:
Robert L. Serenka, Jr.

Product Manager:
Patricia M. Osborn

Editorial Assistant:
Patrick B. Horn

Director of Production:
Wendy A. Troeger

Production Editor:
Matthew J. Williams

Project Editor:
Maureen M.E. Grealish

Technology Project Manager:
Sandy Charette

Director of Marketing:
Wendy E. Mapstone

Marketing Channel Manager:
Kristin B. McNary

Marketing Coordinator:
Scott A. Chrysler

Cover and Text Design:
Potter Publishing Studio

Cover Photography:
Randy Van Dam

For permission to use material from this text or product,
contact us by
Tel   (800) 730-2214
Fax   (800) 730-2215
www.thomsonrights.com

Library of Congress Cataloging-in-Publication Data
Garlough, Robert. 1954-
Modern garde manger / Robert Garlough, Angus Campbell.
    p. cm.
Includes bibliographical references and index.
ISBN-13: 978-1-4018-5009-8
ISBN 1-4018-5009-X (alk. paper)
1. Quantity cookery. 2. Cookery (Cold dishes) 3. Garnishes (Cookery) 4. Buffets (Cookery) I. Campbell, Angus, 1956- II. Title.

TX820.G37 2005
641.7'9--dc22

2005032270

NOTICE TO THE READER

Publisher does not warrant or guarantee any of the products described herein or perform any independent analysis in connection with any of the product information contained herein. Publisher does not assume, and expressly disclaims, any obligation to obtain and include information other than that provided to it by the manufacturer.

The reader is notified that this text is an educational tool, not a practice book. Since the law is in constant change, no rule or statement of law in this bok should be relied upon for any service to any client. The reader should always refer to standard legal sources for the current rule or law. If legal advice or other expert assistance is required, the services of the appropriate professional should be sought.

The Publisher makes no representation or warranties of any kind, including but not limited to, the warranties of fitness for particular purpose or merchantability, nor are any such representations implied with respect to the material set forth herein, and the publisher takes no responsibility with respect to such material. The Publisher shall not be liable for any special, consequential, or exemplary damages resulting, in whole or part, from the readers' use of, or reliance upon, this material.

We would like to both dedicate this book to our parents in deep appreciation for the love and guidance they've shown over our lifetimes. Our passion for food and appreciation for family were nurtured from the beginning. We dedicate this book to:

William and Charlotte Garlough
Holland, Michigan USA

Alexander and Murdina Campbell
Isle of Lewis, Scotland

Thomson Delmar Learning is excited to announce the *About Series,* the first installment in a robust line of culinary arts textbooks from a leader in educational publishing. You'll soon discover why it's all *About* baking, garde manger, and wine! The first three publications in this series are outlined below and will be available in early 2006. These essential textbooks for culinary arts students present the tools and techniques necessary to ensure success as a culinary professional in a highly visual, accessible, and motivating format. It is truly the first culinary arts series written for today's culinary arts students.

The following principles represent the vision behind the *About* series:

- A highly visual, accessible, and motivating format
- A comprehensive instructor support package that provides the tools necessary to make your life easier and your in-class time as effective and stimulating as possible
- A thorough and complete review by a team of academic and industry professionals ensuring that the *About* series is the most up-to-date and accessible culinary arts series ever published

About Professional Baking by Gail Sokol. With over 700 full-color photographs demonstrating best practices and key techniques, your students will be motivated and prepared for each and every baking laboratory exercise. Features include profiles of professional bakers; an entire chapter on *mise en place;* fully kitchen-tested recipes written in an easy-to-comprehend format; clearly stated objectives and key terms presented at the beginning of each chapter; and hundreds of detailed step-by-step procedural photographs.

About Wine by J. Patrick Henderson and Dellie Rex. This introductory wine textbook presents culinary arts and hospitality students with practical and detailed knowledge necessary to manage wine and wine sales. The five distinct sections of the text cover the basics of wine, the wine regions of the world, types of wine, and the business of wine. Special features of *About Wine* include detailed color diagrams, maps, and photographs throughout to keep the text interesting and engaging. Useful appendices designed for use as a quick reference or as a basis for more research are also included, making this text a valuable resource even after formal training has ended.

Modern Garde Manger by Robert Garlough and Angus Campbell. This innovative and comprehensive text is designed to meet the educational needs of both culinary arts students and experienced culinary professionals. Carefully researched content and fully tested recipes span the broad international spectrum of the modern garde manger station. Seventeen chapters are divided between five areas of instruction, each focusing on a different aspect of the garde manger chef's required knowledge and responsibilities. With nearly 600 color photographs, more than 250 recipes, and 75 beautifully illustrated graphs and charts, *Modern Garde Manger* is the most comprehensive text of its kind available for today's culinary arts student and the professional chef.

We look forward to providing you with the highest quality educational products available. Please contact us at (800) 477-3692 to order your desk copies of this exciting new series.

All the Best!

Matthew Hart
Culinary Arts Acquisitions Editor
matthew.hart@thomson.com

Instructor Resources

INSTRUCTOR'S MANUAL

The *Instructor's Manual* provides chapter outlines, answers to end-of-chapter review questions, and additional assessment questions and answers. The *Instructor's Manual* is available at no charge to adopters of the text in print and on-line.

ELECTRONIC CLASSROOM MANAGER (ECM)

The *Electronic Classroom Manager* is a CD-ROM designed as a complete teaching tool for *Modern Garde Manger*. It assists instructors in creating lectures, developing presentations, constructing quizzes and tests, and offers additional lesson plans. This valuable resource simplifies the planning and implementation of the instructional program. This complimentary resource package is available upon adoption of the text and consists of the following components:

Electronic Support Slides, divided by chapter, offer a visually-appealing way to extract key points of the textbook and enhance class lectures.

Computerized Test Bank consists of a variety of test questions including multiple choice and true/false.

Lesson Plans are available as an additional resource to aid the instructor in preparing lessons.

Instructor's Manual

Online Resources

Available to both instructor and student, the *Online Companion to Accompany Modern Garde Manger* is a valuable resource providing additional study materials (activities, web links, and other features.) The Online Companion is available on Thomson Delmar Learning's Web site at http://www.culinary.delmar.com.

Key Features

OBJECTIVES: Answering the question "What am I about to learn?" will best describe this chapter-opening feature. These learning objectives are used to help students understand that by the end of the chapter they will have a working knowledge of the material presented.

After reading this chapter, you should be able to

- discuss the functional origins of salad in modern gastronomy.
- list the classification of salads.
- explain how salads are composed.
- evaluate the appropriate selection and use of salads.
- prepare a variety of international cold sauces and dressings.
- demonstrate techniques in the preparation of salad ingredients.

KEY TERMS: Key terms are listed at the front of each chapter enforcing the importance of new terminology presented in each chapter.

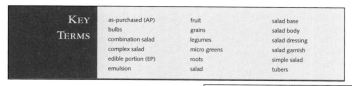

KEY TERMS	as-purchased (AP)	fruit	salad base
	bulbs	grains	salad body
	combination salad	legumes	salad dressing
	complex salad	micro greens	salad garnish
	edible portion (EP)	roots	simple salad
	emulsion	salad	tubers

Key terms are also bolded at first use within the chapters for easy identification.

SIMPLE

Simple salads are those that are basic in nature and composition. They include light salads made from a variety of one or more greens, fruits, pastas, or grains, but not in combination with each other. They are generally dressed with mild seasonings, such as vinaigrette, and are used as a course served around the entrée. Examples include coleslaw, macaroni salad, spinach salad, gelatins, and fruit cup.

COMPLEX OR MIXED

Complex salads, also known as mixed salads, are heartier in character than simple salads, and are composed of raw or cooked vegetables, fruits, meats, seafood, game, or poultry. **Complex salads,** are seasoned with flavorful dressings or marinades and usually contain multiple ingredients from

ACTIVITIES AND APPLICATIONS EXERCISES: Exercises include Small Group Discussions, Research Projects, Group Activities, and Individual Activities for the classroom to aid in the student's use of critical thinking skills.

ACTIVITIES AND APPLICATIONS

A. Group Discussions

Discuss the role played by grains in salad preparation; report on new ways of presenting them.

B. Research Project

As a group, research the use of **micro greens**, reporting on the quality and how functional they are as salad greens.

C. Group Activity

Compose three new combination salads using drawings to illustrate their construction.

D. Individual Activity

Create two recipes for salad dressing using fresh fruits oils and vinegars suitable for dressing bitter greens.

REVIEW QUESTIONS: Each chapter ends with a series of assessment questions for the student for further consideration of content to encourage application and synthesis of material presented.

GLOSSARY: Key terms and definitions are provided at the back of the book.

Special Features

PHOTOGRAPHS: This beautifully illustrated textbook contains nearly 600 full-color photographs including culinary equipment, tools, and ingredients.

PREPARATION STEPS: Colorful photographs represent step-by-step procedures for recipes throughout the text to help the student learn both cognitively and visually including the latest in kitchen equipment, and plated finished products.

DETAILED ILLUSTRATIONS: Colorful line art is displayed throughout the text emphasizing and simplifying different processes with ease.

CHAPTER TIPS AND NOTES: Special chapter tips and notes elaborate on essential techniques, giving students insight into the minds of culinary experts.

PROFESSIONAL PROFILES: Interviews with leading culinary personalities offering a special recipe with step-by-step preparation instructions.

Professional Profile

BIOGRAPHICAL INFORMATION

Name: Muaynee (Marnee) Siriyarn   **Recipe Provided:** Green Papaya Salad
Place of Birth: Bangkok, Thailand

CULINARY EDUCATION AND TRAINING HIGHLIGHTS

As the third child in a family of eight children, Muaynee had to help her parents with the family restaurant at an early age. By the age of seven, she had begun doing odd jobs in the kitchen and dining room, and by the time she was 12 years old, she was cooking. Muaynee's father died when she was 13, making it necessary for her to have her own food court in the local market. As a result of working to help her family,

---

TABLES AND CHARTS offer visual representation of important relevant information for students to comprehend more easily.

Table 5–2   Comparison of Salad Green Yields

INGREDIENT NAME	AS PURCHASED WEIGHT (U.S.)	AS PURCHASED WEIGHT (METRIC)	EDIBLE PORTION BY WEIGHT (U.S.)	EDIBLE PORTION BY WEIGHT (METRIC)	EDIBLE PORTION BY VOLUME (U.S.)	EDIBLE PORTION BY VOLUME (METRIC)
Arugula	6 ounces	168 g	3.5 ounces	98 g	2.5 cup	591 mL
Boston	5.25 ounces	147 g	4 ounces	112 g	2.5 cup	591 mL
Green Cabbage	2 lb	896 g	1 lb, 14 ounces	840 g	8 cup	1.89 L
Napa Cabbage	2 lb	896 g	1 lb, 12 ounces	784 g	7 cup	1.66 L
Dandelion Greens	6.5 ounces	182 g	3.25 ounces	91 g	3 cup	709 mL
Escarole	12.5 ounces	350 g	9 ounces	252 g	6 cup	1.42 L
Frisée	12 ounces	336 g	7.5 ounces	210 g	6 cup	1.42 L
Green Leaf Lettuce	11 ounces	308 g	7 ounces	196 g	4 cup	946 mL
Radicchio	9.5 ounces	266 g	4.5 ounces	126 g	2 cup	473 mL
Romaine	14 ounces	392 g	9.25 ounces	259 g	4.5 cup	1.06 L
Spinach, flat leaf	11 ounces	308 g	7.5 ounces	210 g	7 cup	
Watercress	5 ounces	140 g	3 ounces	84 g	2.5 cup	

MEASUREMENTS: The book is written for universal application, as all recipes are printed using both U.S. customary and metric measurements.

RECIPE 5–6

LEMON AIOLI

Recipe Yield: 2½ cups

Note The aioli may be kept in the refrigerator for up to a day in a covered container.

MEASUREMENTS		INGREDIENTS
U.S.	METRIC	
4 each	4 each	Large egg yolks, warmed
½ cup	118 mL	Lemon juice, fresh
8 each	8 each	Garlic cloves, minced
1 teaspoon	5 mL	Kosher salt
2 tablespoons	30 mL	Lemon zest, fresh grated
⅛ teaspoon	½ mL	White pepper
1½ cups	355 mL	Olive oil, extra virgin

PREPARATION STEPS:

1. Combine egg yolks, lemon juice, garlic, salt, zest, and pepper in a blender.

2. With blender running, slowly drizzle in olive oil and blend until an emulsion forms.

3. Adjust seasoning, as needed.

4. Refrigerate in a tightly covered container.

A **RECIPE INDEX** is provided at the back of the book for quick identification and location.

CONTENTS

RECIPES

CHAPTER 13

Curing, Sausage Making, and Smoking

CHAPTER 14

Pâtés, Terrines, and Mousselines

CHAPTER 15
Kitchen-Made Cheeses and Creams

CHAPTER 16
Sculpting, Carving, and Modeling

CHAPTER 17
Food Decoration, Platter Presentation, and Culinary Competition

In the classic French kitchen the term *mise en place* could well be the absolute foundation not only of its justifiably famous cuisine but also the essential starting point of every excellent meal ever made by anyone.

A culinary instruction usually comprises two parts: the *recipe,* which is the list of measured ingredients, followed by the *method* by which they are assembled.

Mise en place means to select and ready those recipe ingredients ahead of time, allowing the cook's absolute focus to be upon the cooking process because everything that is needed is "at hand."

The garde manger is the classic description of the department of a kitchen that does this "prep work" and therefore it must be argued that, when done with all due diligence, the output of the garde manger will have an enormous impact upon the quality of the eventual dining experience and as a result, the financial success of the entire operation.

I began my food career in the garde manger of the hotel my father was managing at the time. I learned to handle a knife, came to understand the reason why all manner of foods were cut in certain ways, and watched as the freshest of ingredients were inspected before I was allowed to "participate" in the daily quest for perfection.

All of this effort and training served me well in my ongoing career in both the hotel business and eventually in television.

Television is impossible without complex stages of mise en place and excellent raw materials for those close up "beauty" shots. I have been brilliantly served by my own team of garde manger associates over the years and freely admit to being in their debt.

How, I wonder, will it be for you? Will you go on to fame and fortune, delighting future diners? If this is your purpose you will do well to read every word of this book . . . it is an essential mise en place for your future. The authors have invested much hard work and the fruits of their extensive experience into this book; now you have only to read, learn, digest, and apply.

Much success in all you set your hearts to achieve.

Graham Kerr

". . . instead of copying servilely, we ourselves should seek new approaches so that we too may leave behind us methods of working that have been adapted to the customs and needs of our time."

GEORGES-AUGUSTE ESCOFFIER

Modern Garde Manger is written for both the working chef and the serious student engaged in the practice and study of culinary arts. It is intended to be useful in several capacities as a teaching tool; it seeks to explain the scientific reasons behind techniques, while also demonstrating methods of preparation. It is well suited to serve as a primary course book for beginning and advanced courses in garde manger, as well as a hands-on course in banquets and catering.

This book is designed to improve upon existing expertise, as well as for those wishing to gain a solid foundation in a variety of required cooking-related skills. It is written with the intent of reaching culinarians at various stages of their own career development. Its lessons and methods cut across most disciplines within the professional kitchen and it serves as an easy study companion for both the novice and expert chef who continue to enhance their skills.

As the foodservice industry continues to evolve, chefs must be ever vigilant to the trends and demands of the trade. Customers are more informed today and they have an eye for fanciful presentations and a taste for complex flavors. And as cooking methods, ingredients, and flavor profiles continue to blend, foodservice operators continually seek to refine their menus. The adept garde manger chef must be able to meet higher expectations of his or her patrons, and increased responsibilities of challenging positions. A broad understanding of modern global cuisine is paramount to the success of a garde manger chef.

Purpose of This Textbook

During the last three decades of our culinary-related travels to many diverse regions of the planet, we have had the opportunity to study the culinary arts with some very talented chefs, farmers, winemakers, fishmongers, cheese makers, butchers, and other associated craftsmen. We have also had the incredible good fortune to work among their staff, using the ingredients, equipment, and methods as they use them. This book is based on our travels and personal experiences in the culinary world and is reflected in our recipes by converging flavors and ingredients from across the world. To the chefs and consumers who reside in these various countries of origin, their cuisine is equally noble and worthy of inclusion.

Organization of the Text

Recognizing the broad range of responsibilities and interests a garde manger chef carries, *Modern Garde Manger* was written to address a varied selection of skill sets. It serves as a handy recipe guide, providing countless proven formulas that have been tested and used by the authors.

There are five distinct sections covering seventeen chapters of text and recipes:

Part I: Foundations in Garde Manger, opens with an overview of the position of garde manger chef. It reviews some of the more significant historical events and milestones in culinary history related to this ancient responsibility, and then brings the position forward to the modern kitchen. The section then discusses the global nature of the position, and how chefs may use indigenous foods and styling from other cultures to enhance their menus. The mathematical skills required by the garde manger are discussed, with emphasis on recipe costing and performing financial analysis on the menu. Part 1 concludes with a chapter on kitchen and banquet organization, including useful information on workflow design. It also approaches the planning and presentation of buffets and considerations that can be made to enhance the dining experience.

Part II: Preparation Skills of the Garde Manger is a storehouse of recipes with their preparation methods for the working chef. Gathered from years of working with many talented foodservice professionals from around the world, this section addresses the basic areas of food preparation that commonly fall to the garde manger department of a large hotel or club operation. It does so from a distinctly global viewpoint because as the role of the garde manger chef has evolved over the years, so too has the nature of the garde manger's recipes. Part 2 also provides the reader with insightful information on the ingredients commonly used in the preparation of appetizers, salads, and sandwiches. We believe this section will aid the chef in meeting his or her modern needs.

Part III: Fabrication Skills of the Garde Manger provides an elaborate selection of photographs and drawings to illustrate the many technical skills required of the garde manger chef. Special attention is paid to the butchery of poultry, game, and various meats, plus the fabrication of fish and shellfish are carefully detailed. Many cuts not often used in the traditional Western kitchen are discussed and illustrated, again addressing the global nature of the food business. In an effort to refrain from duplication of technique, we have chosen to demonstrate different techniques that may be used on most of the same products in the chapter.

Part IV: Preserved Foods of the Garde Manger includes several important chapters on food preservation, including the use of food chemicals, dehydration, and canning. Each technique of preservation provides plentiful benefits to the kitchen. Methods of smoking and curing foods are explained in detail, while incorporating numerous recipes on sausage making. The section also includes a comprehensive chapter on the use and production of pâtés, terrines, and mousselines, with numerous step-by-step photographs detailing their production. Part 4 ends with a careful treatment of cheese making, and includes several recipes for kitchen-made cheeses and creams.

Part V: Displayed Arts of the Garde Manger illustrates the imaginative side of the garde manger chef. The various media used for sculpting, carving, and modeling are discussed, and techniques learned from some of the leading foodservice sculptors are revealed. Ideas for using individual plate sculptures along with buffet centerpieces are explored. Techniques in coating with aspic and classical chaud-froid are demonstrated for use in buffet presentations and culinary competitions.

Pedagogical and special features are throughout the chapters providing structure and guidance to your learning. Features include clear Chapter Objectives, Key Terms, Review Questions, Activities and Applications, and a Glossary and Appendix located at the back of the book include additional resources.

Supplements

Instructor's Manual The *Instructor's Manual* provides chapter outlines, answers to end of chapter review questions, additional assessment questions and answers, and definitions of key terms. The Instructor's Manual is available at no charge to adopters of the text.

Electronic Classroom Manager (ECM) The *Electronic Classroom Manager* is a CD-ROM designed as a complete teaching tool to accompany *Modern Garde Manger*. This valuable resource simplifies the planning and implementation of the instructional program. Included on the ECM are chapter-specific PowerPoint® presentations, a computerized test bank, additional lesson plans, and a PDF file of the Instructor's Manual.

Online Companion Available to both instructors and students, the *Online Companion* is a valuable resource providing activities, Web links, and much more. Access the Online Companion through Thomson Delmar Learning's Web site at www.culinary.delmar.com.

ACKNOWLEDGMENTS

The development and production of a text manifests itself from conception through completion. All during the process, transformations occur within the manuscript that result in, what the editors and authors' hope will be, a better-finished product. Such was the case with this book.

We would like to acknowledge the many people who took interest in this project. Without question, their involvement helped us focus our efforts on the essence of the modern garde manger. In particular, we would like to thank the following individuals; we believe the result is a superior book because of their contributions:

The entire staff, faculty and student population of the Hospitality Education Department at Grand Rapids Community College for their interest in and support of the project; and in particular Randy Sahajdack, Bill Jacoby, Mike Kidder, Bob Monaldo, Luba Petrash, Katie Nickels and Dale Vandenburg for their involvement. It is a pleasure to come to work every day with such a talented and committed group of colleagues.

The management at My Chef, Inc. of Naperville, Illinois, Bill and Karen Garlough, for use of their relevant photos and documents. Bob Sullivan at Plitt Seafood of Chicago, Illinois for use of his fishery and quality products. Thad Lyman, Executive Chef/Partner of the Napa Valley Grille, for use of his menu and Market Platter. 20th Century Market for their assistance with sausage photography and Kent Butcher Supply for the use of their facility, both of Grand Rapids. Bob Tansing and Athens Foods for lending their directions and drawings for the use of phyllo dough. Our gratitude to Ron Stein, sculptor and Visual Arts professor, for his assistance with the armature and stands used in chapter 16.

Our photographer, Randy Van Dam of Grand Rapids, Michigan and Chef Joseph George of Midland Country Club, our food stylist, whose talented work together brought our textbook to life. GRCC students Chris Ball, Jonathan Bartelson, Patrick Cummisky, Katie Nickels, and James Taylor for their invaluable assistance to our food stylist.

Our celebrated food professionals who we were honored to include at the end of each chapter; our book, and indeed our entire food industry, is elevated by these people every day. Their dedication, tenacity, raw talent, and artistry are to be admired and respected.

Graham Kerr, who first touched our lives via television, and later as a featured speaker to our culinary program, has been an important influence throughout our careers. Graham's ceaseless energy, hunger for knowledge, commitment to healthy living, passion for food, and sincere friendship made him the singular choice to write our Foreword. We are truly honored that he accepted.

The team at Thomson Delmar Learning of Clifton Park, New York for their care, guidance and production of the text; we are indebted to them for their professional contributions. We particularly would like to thank Joan Gill, the Acquisitions Editor who initially contracted us for

The food styling team for Modern Garde Manger

BACK ROW (L TO R): Chris Ball, Patrick Cummisky, Jonathan Bartelson, Katherine Nickels, James Taylor, Mike Kidder, Dale Vanden Berg, Randy Sahajdack

FRONT ROW (L TO R): Joe George, Robert Garlough, Angus Campbell

Randy Van Dam (not pictured)

the project and Matthew Hart, the Acquisitions Editor who enthusiastically adopted our project; Sherry Dickinson, the Director of Learning Solutions who willingly championed our plan with Thomson Delmar Learning; Pat Gillivan and Patricia Osborn, our Product Managers who steadily shepherded our manuscript through development; Maureen Grealish, our Project Editor whose proofreading skills and wordsmith abilities proved invaluable; Matt Williams, Production Editor and Wendy Troeger, Director of Production for their yeoman efforts in the layout and design of the finished product; Kristin McNary, our Channel Manager and Wendy Mapstone, Director of Marketing for their comprehensive promotion of the text; Lisa Flatley and Pat Horn, our Editorial Assistants who kept the communications going and the details straight; and all of the other staff at Delmar Learning. It takes an army of talented individuals to bring a book to fruition; they were all vitally important to the process.

Lastly, we would like to thank our families who allowed us the enormous measure of time to devote to this project. They are generous and caring, and our most treasured gift.

Our thanks also to the following reviewers, who both applauded our efforts and challenged our thinking, providing many excellent suggestions and ideas for improving the text. Their comments were insightful and respected:

ROBERT ANDERSON
Withlacoochee Technical Institute

DAVID BEARL
The Southeast Institute of the Culinary Arts

WILFRED BERIAU
Southern Maine Technical College

JULIAN DARWIN
Cascade Culinary Institute

JONATHAN DEUTSCH
Kingsborough Community College

JILL DOEDERLEIN
Lansing Community College

GREG FORTE
Utah Valley State College

ROBERT HUDSON
Pikes Peak Community College

HELMUT KAHLERT
Newbury College

KEVIN KEATING
Capital Culinary Institute

JOHN KINSELLA
Cincinnati State Technical & Community
      College

DR. KEITH MANDABACH
New Mexico State University

DAVID PANTONE
Florida Culinary Institute

MICHAEL PICCININO
Shasta College

TINA POWERS
Metropolitan Community College

HEATH STONE
Johnson & Wales University

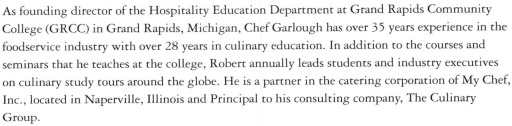

Although they originally hailed from different countries, Chefs Garlough and Campbell have worked together for over fifteen years as culinary educators at one of America's leading culinary colleges. A common passion for exploration of distant cultures, and an intense interest in global cuisine has resulted in their leading annual international cuisine and culture study tours to over 30 countries on six continents. Many hundreds of culinary students have traversed the planet with them, as they have continued to study world gastronomy.

During their travels, they have taken seminars in various culinary schools in Italy, France, Greece, Australia, South Africa, Scotland, Brazil, and Malta. Chefs Garlough and Campbell have shopped the world's food markets, including Barcelona, Nassau, Zurich, Istanbul, Auckland, New York, Paris, Puerto Vallarta, Lucerne, Glasgow, Athens, Rome, Valetta, Hong Kong, Frankfurt, London, Cairo, Beijing, and Lisbon. They have successfully competed in culinary competitions in the Bahamas, England, Scotland, United States, Malta, Germany, and France.

Their joint quest for learning the art and science of cookery has culminated in this textbook, a reflection of their observations and understanding of the modern garde manger.

Robert Garlough, MS, FMP, AAC
Chef-Emeritus, Grand Rapids Community College

As founding director of the Hospitality Education Department at Grand Rapids Community College (GRCC) in Grand Rapids, Michigan, Chef Garlough has over 35 years experience in the foodservice industry with over 28 years in culinary education. In addition to the courses and seminars that he teaches at the college, Robert annually leads students and industry executives on culinary study tours around the globe. He is a partner in the catering corporation of My Chef, Inc., located in Naperville, Illinois and Principal to his consulting company, The Culinary Group.

Chef Garlough is certified as a Foodservice Management Professional by the Educational Foundation of the National Restaurant Association, and is a member of the American Academy of Chefs, The Honorable Order of the Golden Toque, and the Craft Guild of Chefs. He holds an Associate in Occupational Studies degree in Culinary Arts from the Culinary Institute of America, a Bachelor of Business Administration degree in Restaurant and Lodging Management from Davenport University, and a Master of Science in Occupational Education degree from Ferris State University.

A recipient of numerous culinary salon competition awards, Robert personally earned silver and bronze medals at the 1988 Internationale Kochkunst Ausstellung (considered the "Culinary Olympics") in Frankfurt, Germany while managing a six-member team of GRCC faculty and graduates. He served as Manager for the 1993 Pastry Team USA that represented the United States at the 1993 Coupe du Monde de la Patisserie (World Pastry Cup) in Lyons, France. Chef Garlough also served as Manager for Team USA 1998, a culinary student team representing America at the Malta International Students Culinary Salon in St. Julian's Bay, Malta. In 1984, Chef Garlough was awarded the Chef Herman Breithaupt Memorial Award by CHRIE, naming him their national chef-instructor of the year. The American Culinary Federation Educational Institute honored him as their National Educator of the Year in 1992.

His professional affiliations included serving as President of the Michigan Council on Hotel, Restaurant and Institutional Education, President of the American Culinary Federation Greater Grand Rapids Chapter, Chairman of the American Culinary Federation Educational Institute Accrediting Commission, and Executive Director of the International Consortium of Hospitality and Tourism Institutes.

Chef Garlough is a co-author of *Ice Sculpting the Modern Way: For Beginning and Advanced Ice Artists,* also published by Thomson Delmar Learning.

He and his wife Nancy reside in Grand Rapids, Michigan where they have raised three children: Jonathan, Jeremy, and Kristen.

Angus Campbell, C & G
Chef-Instructor, Grand Rapids Community College

Chef Campbell is from the Island of Lewis in the Outer Hebrides of Scotland, where as a young lad his passion for the culinary arts was nurtured by his constant dabbling in his mothers excellent home cooking. His formal European-style training brought him to the mainland of Scotland to the cities of Elgin, Aberdeen, and Glasgow and finally to Troon on the Ayrshire coast and the Marine Highland Hotel where he ran the unique French à la carte restaurant, l'Auberge de Cuisine. While there, the restaurant was awarded an AA Rosette for fine cuisine. His certification is with the City and Guilds of London Institute, and he achieved master craftsman level with The Craft Guild of Chefs.

He then turned his culinary talents to teaching at Glasgow College of Food Technology, in Glasgow, Scotland where he became a senior lecturer in food production. After eight years, and having earned his teaching qualification from Jordanhill College of Education, it was time for Chef Campbell to continue his travels. He became the Departmental Chair of Food and Beverage at the Bahamas Hotel Training College, where he was instrumental in planning the building of the new Hotel Training College. While in the Bahamas, he also held the position of apprenticeship chairman for the Bahamas Culinary Association. After three years in the Caribbean, he moved to Grand Rapids Community College where he now teaches food production for the Hospitality Education Department at their public restaurant, The Heritage. Chef Campbell also serves as an adjunct faculty member at Western Michigan University in Kalamazoo, Michigan, in the department of Family and Consumer Sciences.

He has won medals at many levels of food competition and is an accomplished junior culinary team coach, taking his students to the national finals of junior competition in 1996, and coaching them to a 15 gold medal performance at an International competition in Malta in 1999. His production of a monthly cable TV show has won a national bronze medal at the Paragon awards, and he has two multi-part video series featured on his personal Web page.

After 30 years in the culinary field he still loves nothing more than going into the kitchen to start a busy day of food production and passing on his knowledge to each and every student he meets.

He and his Scottish-born wife, Katie, live in Grand Rapids, Michigan with their two daughters, Fiona who was born in the Bahamas and Cara who was born in the United States.

PART I

Foundations in Garde Manger

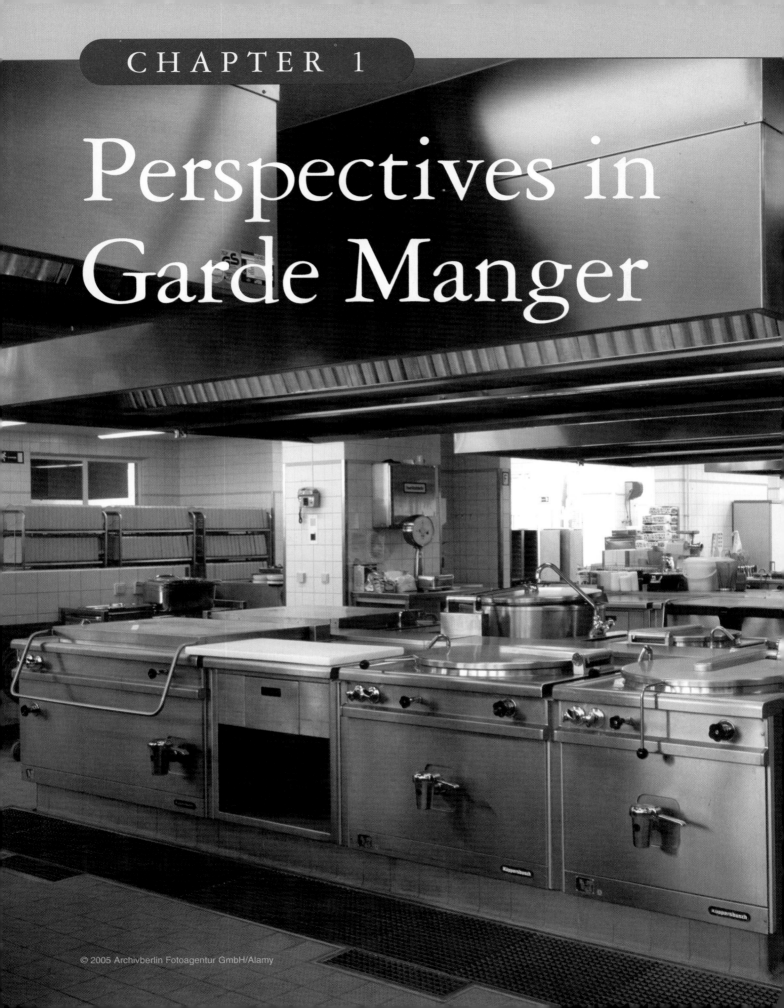

CHAPTER 1

Perspectives in Garde Manger

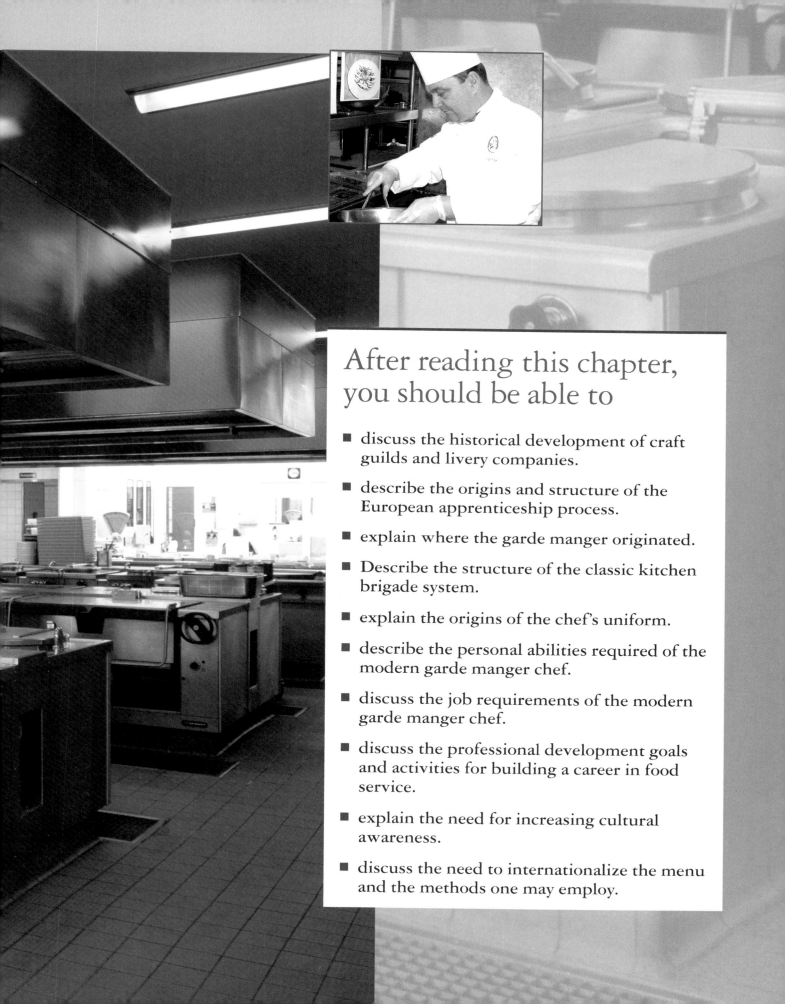

After reading this chapter, you should be able to

- discuss the historical development of craft guilds and livery companies.

- describe the origins and structure of the European apprenticeship process.

- explain where the garde manger originated.

- Describe the structure of the classic kitchen brigade system.

- explain the origins of the chef's uniform.

- describe the personal abilities required of the modern garde manger chef.

- discuss the job requirements of the modern garde manger chef.

- discuss the professional development goals and activities for building a career in food service.

- explain the need for increasing cultural awareness.

- discuss the need to internationalize the menu and the methods one may employ.

KEY TERMS

chef de partie

garde manger

guilds

job descriptions

kitchen brigade system

la grande cuisine

nonplanner's cycle

nouvelle cuisine

performance standards

Siamese twins of management

trinity of business

unity of command

The art of garde manger is a blend of the aesthetic and the practical, for while being one of the most artful presentations of the creative chef, it is also one of the most lucrative merchandising tools of the creative manager.

HENRY OGDEN BARBOUR

Craft Guilds and Livery Companies

Although much of the origins are lost behind the curtains of antiquity, the tendency of workers and townsfolk to organize themselves into fraternities, **guilds**, and mysteries is one of the great social accomplishments of the Middle Ages. Some of these organizations were religious; others were engaged in trades, crafts, or occupations.

The livery companies—similar to the fraternities, guilds, and mysteries that flourished throughout Europe for many centuries—probably had their origins in England before the end of the eleventh century. The development of guilds was not confined to London. Many cities throughout Britain had craft guilds, some of which still exist. Both Scotland and Ireland maintain strong, active guild traditions, as do many countries in continental Europe, particularly Switzerland, France, and Germany. Across the Atlantic, the United States and Canada have several fraternal and federated organizations. Although present-day functions vary considerably, the organizations are generally based on trade and craft support.

In London, names such as Milk Street, Bread Street, and Poultry Street mark the sites where these guilds began. People following the same craft tended to live and work near each other, making informal arrangements amongst themselves for the regulation of sanitation and quality standards, commerce, and competition. These early guilds greatly controlled the provision of services and the production and selling of foods. The guilds protected their customers, employers, and employees by checking for inferior work, underweight portions, and the use of poor-quality product. In London, separate guilds were formed for each trade, but in smaller towns these groups tended to be federated guilds. In general, guilds were the forerunners of the modern unions and served to regulate occupations and preserve the craft.

There were two general classifications of food-related guilds: those that sold raw goods and those that sold prepared goods. Each guild was given its own charter, granting it specific rights and jurisdiction. Each guild was to stay within the constraints of its charter when doing business. But as businesses sought to expand their trade and services, they would often test the limits of their charter. Entrepreneurial spirit gave momentum to product development, and new goods were constantly developed for sale. For example, when prohibited by the baker's guild to sell pâté en croûte (liver paste encased in bread dough), the charcuterie guild created a new product by selling their pâtés in a terrine.

Training with Craft Guilds

Merchant guilds, which controlled the trade within a town, were separate from the craft guilds, which regulated the quality, working hours, and conditions of their members. There were three levels of craftsmen: apprentices, journeymen, and masters. Parents paid a fee to place a boy with a master craftsman to work and learn as an apprentice. The boy received food, lodging, clothes, and instruction in the craft. Often, the conditions were poor and the hours long.

The period of apprenticeship lasted 2 to 7 years, after which time the apprentice became a journeyman. (The term originates from the French word *journée*, meaning day, indicating he was paid by the day for his work.) After several years, the journeyman could submit a sample of his work to the guild's masters, and if his "masterpiece" was judged acceptable, he could be called a master craftsman and have his own shop.

As time progressed, the guild system became increasingly rigid. This inflexibility led to the development of new trade and industry by the capitalists who readily adapted themselves to the needs of commerce.

In France, the strict enforcement of the guild system was officially ended in 1791 during the French Revolution. After that, food service workers were free to engage in any employment without restriction to training, certification, or charter membership. Although this relaxation of the once-enforced charters opened up broader opportunities for all guild members, the initial lack of structure within the hotel and restaurant kitchens made it difficult for those with only specialized skills.

In England, the power of the guilds had withered by the seventeenth century and their privileges were officially abolished by the Crown in 1835. The German, Italian, and Austrian guilds were abolished later, in the nineteenth century.

In the Larder: Keeping to Eat

The **garde manger** is a hybrid position within the kitchen. Historically, the position was adopted to put a responsible person in charge of keeping food fit to eat. That meant a certain amount of understanding of sanitation and food preservation, coupled with training in meat butchery and food preparation (Figure 1–1).

As our ancestors moved from a culture of hunters and gatherers to an agrarian society, they developed the practical skills necessary to preserve their foodstuffs, ensuring a reliable food supply. Settling in the fertile valley between the Tigris and Euphrates rivers in around 3500 BC, the Sumerians learned to domesticate animals and raise crops. The annual deposit of rich silt from the rivers ensured a steady harvest of grain, allowing them to raise feed for their livestock and make porridge for themselves. The Sumerians became skilled in the science of food preservation by pickling, salting, brining, dry curing, and smoking their stored goods.

Over the centuries, seasonings used in food storage such as sugar, oils, spices, and salt were in high demand. Trade routes to East Asia and Africa were established to secure spices. Inland

FIGURE 1–1 Craftsmen working in a medieval kitchen. Courtesy of the Historical Picture Archive/CORBIS

cities like Rome, Wieliczka, and Salzburg were founded near salt mines, while coastal cities like Sfax, Athens, Ephesus, and Genoa took their salt from the sea. As Kurlansky (2002) writes:

> Most Italian cities were founded proximate to saltworks, starting with Rome in the hills behind the saltworks at the mouth of the Tiber. Those saltworks, along the northern bank, were controlled by the Etruscans. In 640 BC, the Romans, not wanting to be dependent on Etruscan salt, founded their own saltworks across the river in Ostia. They built a single, shallow pond to hold seawater until the sun evaporated it into salt crystals. The first of the great Roman roads, the Via Salaria, Salt Road, was built to bring this salt not only to Rome but across the interior of the peninsula.

In 252 BC, the Governor of Shu ordered the drilling of the world's first brine wells in the Sichuan province of China. However, natural brine pools had already existed in what is now Southern Poland as early as 3500 BC, where it was gathered and boiled in clay pots.

After the foods were suitably preserved, they were stored in earthenware jars and kept in cool stone storage houses and cold cellars below ground. When ice was available from the frozen lakes or mountaintops, it was covered with hay and used to preserve the meats, fruits, and vegetables. Manor houses and castles each had these cold rooms known as the *garde manger*, a French term meaning "keeping to eat." A trusted member of the household staff was given the responsibility of managing the garde manger and issuing foods upon demand.

THE CHEF'S UNIFORM

The classic double-breasted chef's jacket (styled after the Turkish army coat) originated when chefs were servants of the king and could, upon a moment's notice, be called upon to serve in battle as well as in the noble households. By the early nineteenth century, chefs were wearing uniforms reportedly based on those worn by soldiers in the Turkish army. White eventually became the standard chef's uniform color to emphasize cleanliness and good sanitation.

Numerous legends are attributed to the origin of the chef's toque blanche—the tall, pleated white hat characteristically worn by chefs. One version attributes it to the tubular black hats worn by Greek Orthodox priests. It is believed that men from all walks of life fled to Greek Orthodox monasteries for protection at the time the Byzantine Empire was under siege by the Barbarians. The refugees adopted the same dress as the priests so as not to be recognized. After the threat of persecution lessened, the chefs wore white hats to differentiate themselves. Antonin Carême, the eighteenth-century chef to Tallyrand and various Rothschilds, is also credited with bringing the cylinder-shaped toque into the kitchen. As legend holds, it was inspired by a woman's hat. Still other stories suggest the taller white hat originated as a practical means of identifying the chef-in-charge, among a sea of kitchen employees, from across the expanse of the medieval kitchens.

CLASSIC KITCHEN BRIGADE SYSTEM

The larger kitchens of the nobility had for centuries been loosely divided into different sections responsible for the preparation of various foodstuffs. The kitchens were sometimes chaotic and generally wasteful and inefficient. However frenetic they were, they functioned with little reorganization for several hundred years.

The traditional order of modern kitchen structure, known as the **kitchen brigade system** and led by the chef, has its roots in European military organizations dating back to the fourteenth century. Traveling armies had to be fed, and their cooks were selected from among the ranks. During times of peace, various rulers organized tournaments to keep their armies prepared for future conflicts. The military cooks who accompanied the knights to the tournaments, held on castle grounds, ultimately became cooks to the kings and nobility. Reasonably skilled in their

FIGURE 1–2 Georges-Auguste Escoffier. Courtesy of Hulton Archive

ability to nourish large armies of soldiers, they transferred those organizational skills to feed the multitude of guests who were part of the king's court.

Centuries later, the modern kitchen brigade system of **chef de partie** was devised and implemented by Georges-Auguste Escoffier (Figure 1–2) in London's Savoy Hotel in the late nineteenth century. Following a French army career, Escoffier devised an organized system to ensure against duplication of work, and to further increase communication and efficiency among the staff.

For maximum efficiency, Chef Escoffier organized the kitchen into a strict hierarchy of authority, responsibility, and function. The positions of the classic kitchen brigade system persist in many larger Western hotel and club kitchens (Figure 1–3). The following lists their order of rank and authority:

- *Chef de Cuisine* The executive chef is in charge of the whole kitchen operation, similar to the commanding general.
- *Sous Chef* The "under-chef" is second in command and coordinates the responsibilities of the various station chefs. In charge during the executive chef's absence, the sous chef is in training to become the executive chef. The sous chef also acts as the *aboyeur* and "expedites" the orders during meal times.
- *Chef de Partie* The station chef has production responsibility for a selected part of the menu, traditionally divided by the method of cooking or the ingredients used. The chef de partie usually oversees the work of several *demi-chefs* and *commis* assigned to the station.
- *Demi-chef* The assistant station chef prepares much of the food in the particular station, according to the standards of the station chef, and may be in charge of the station during the station chef's absence.
- *Commis* The attendant is given lower skilled work to perform in training to become an assistant station chef.

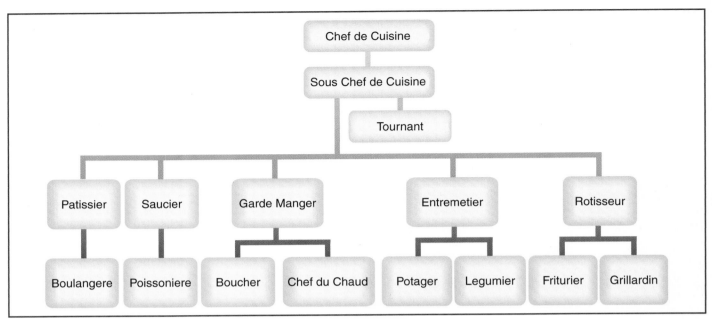

FIGURE 1–3 Classic kitchen brigade system

- *Apprentice* The student chef is tasked with the most laborious work, studying the art and science of the culinary trade while also being trained for the commis position.

  The number of chefs de partie (otherwise known as *station chefs*) can vary, but generally include some the following areas:
- *Saucier* One of the most difficult stations, the saucier makes all of the sauces and stews, and is responsible for all sautéed items.
- *Poissonier* This station is responsible for all fish and shellfish dishes.
- *Grillardin* This station is responsible for all grilled and broiled foods.
- *Friturier* This station prepares all fried items done in vats.
- *Rotisseur* The rotisseur is responsible for all roasted and braised foods, and prepared stuffing as needed for dressed meats.
- *Potager* This lower skilled station is generally used as a starting point in one's training. The potager prepares all stocks and soups, and assists the *saucier.*
- *Legumier* This position is responsible for all vegetables and starches.
- *Entremetier* This is often a combined position of the *legumier* and *potager.*
- *Patissier* The pastry chef, responsible for pastries and desserts, is also assisted by the baker for the production of breads and rolls.
- *Garde Manger* A very skilled position, the garde manger is responsible for cold foods, including salads, cold meats, pâtes and terrines, sausages, hors d'oeuvres, decorative carvings, and buffet items. The garde manger sometimes oversees the butchery of meats and fish.
- *Tournant* Generally a skilled person in all areas of the kitchen, the tournant helps out the various stations as needed. The tournant is generally in training for the *sous chef's* position

Note: Depending on the production needs of the establishment and the size of the staff, several of the stations are often combined under one *chef de partie.*

MODERN KITCHEN BRIGADE SYSTEM

Many large, modern kitchens are organized into a modified version of Georges-Auguste Escoffier's classic kitchen brigade system (Figure 1–4). The invention and implementation of modern food preparation equipment has radically changed how food is fabricated and prepared. This efficiency in food production has resulted in the consolidation of employee responsibilities and even in staff reductions.

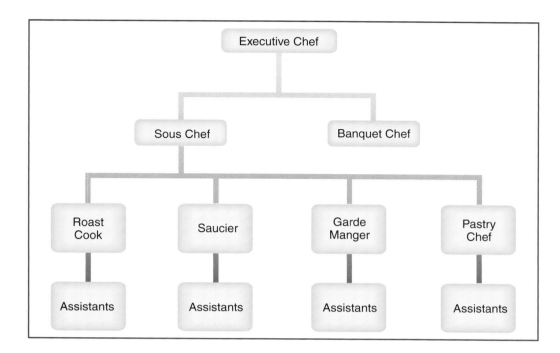

- *Executive Chef* The executive chef coordinates all kitchen activities, including directing the training and work of the staff, managing all related costs, and establishing the professional standards of the establishment.
- *Sous Chef/Assistant Chef* The sous chef, or assistant chef, is also second in command of the modern kitchen. The primary responsibility of the sous chef is to oversee the preparation, portioning, and presentation of the menu items in accordance with the standards established by the executive chef.
- *Area Chefs* Most large hotels and clubs with multiple dining facilities have one or more area chefs who are responsible for a particular production area. The garde manger and pastry chef are two such area chefs.
- *Line Cooks/Station Chefs* Under the guidance and instruction of the area chefs, the line cooks act as station chefs in assigned positions within an area. They are often involved in the finished production of the area's menu items.
- *Preparation Cooks* The "prep" cooks are responsible for doing the lowest skill level of food pre-preparation in an area. Their work is often done to assist the line cooks with simple, often cumbersome, tasks.

Role of the Modern-Day Garde Manger Chef

Tastes in food and presentation have evolved with the dining public over the decades, and demand for the products of the garde manger have never been stronger in modern times. The role of the modern-day garde manger has changed little over the last century, although technology has had an important impact on the processes used. The modern garde manger is still responsible for the production and preservation of cold soups and salads, appetizers, plate garnishes, charcuterie products, cheeses, condiments, and buffet platters (Figure 1–5). The garde manger chef is also likely to be tasked with the creation of centerpiece displays made from sculpted ice, pastry margarine, and tallow, plus carved fruits and vegetables. Additionally, the garde manger chef is commonly involved in poultry, game, meat, and seafood fabrication. The skills required of the *chef de partie* in the garde manger station are as technical as any position in the organization.

The sensibilities of a skilled garde manger are often equated to those of a skilled pastry chef. The talents are more visible to the consumer and are dependent upon a fundamental understand-

FIGURE 1–5 Modern garde manger chef

ing necessary for presenting a pleasing appearance, while requiring a substantial foundation in science and taste. It is the role of the garde manger, among other areas of responsibility, to add form to the function of the hospitality operation. Additionally, a capable garde manger is expected to safely convert salvaged scraps and lesser cuts into sufficient and profitable foodstuffs for their business. The opportunities are endless for a talented garde manger, and both retail and commercial food service operations seek the skills offered by these culinary artisans.

PERSONAL CHARACTERISTICS

The responsibilities of the modern garde manger demand skill sets in a variety of diverse areas. A competent garde manger chef must possess qualities that are both imaginative and logical; clearly, such attributes are not common among all artists and business-minded people. But, to truly excel in the position, a person must have several such dissimilar gifts—*organizational, technical, artistic, and business.*

Organizational Leadership Abilities

Christian theologian Reinhold Niebuhr once remarked, "Order is the first desideratum for the simple reason that chaos means nonexistence." Although he was speaking of societal needs, the same holds true for business. Organized departments only exist when the individual in charge is personally organized. We have a colloquialism that we use in our supervisory management class to reinforce the notion of organizational responsibility. "The body rots from the head down," we tell the students, graphically describing the effect poor organizational leadership has on an operation. Disorganization can be the ruin of a business.

There are many assets to organize and variables to consider when managing a garde manger department. Equipment, food products, and staff are balanced against time and budgetary constraints. The deft garde manger must be able to plan and organize assets according to the needs of the operation and, ultimately, the customer. Every day, time must be allocated away from production to allow for planning and communication. Failure to do so usually results in what is referred to as the **non-planner's cycle** (Figure 1–6), wherein the manager is constantly busy dealing with emergencies caused by lack of organizational forethought.

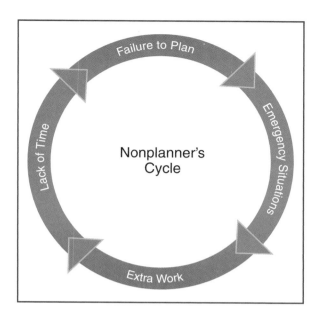

FIGURE 1–6 Nonplanner's cycle

Technical Abilities

Although we have seen the half-century mark of our lives, it is inconceivable to us that we will ever possess all the technical abilities one might hold in this business. Having spent the better part of our professional careers in pursuit of technical skills and cognitive understanding, we know that greater knowledge is ever possible, and there are always skills to master. Having said that, there are some specific fundamental skills that can be identified as crucial to the position of garde manger chef. These skills are discussed and demonstrated in detail in Part 3, Fabrication Skills of the Garde Manger, and Part 5, Displayed Arts of the Garde Manger.

Furthermore, it is our recommendation to the serious culinarian to commit to a regimen of working alongside other professionals in areas beyond his or her expertise. Forming rinds with a cheese maker, drying herbs with a gardener, salting roe with a fishmonger, and curing hams with a sausage maker are experiences not to be missed when seeking to develop broad skill sets. The dedicated culinary professional must create a long-term training plan that encompasses all major areas of technical responsibility. If possible, these plans should include opportunities for international short-term training and work experiences. The skills are career changing; the benefits are broad and immeasurable.

Artistic Abilities

Artistic ability is a gift to be constantly nurtured and refined. As modern French artist and sculptor Jean Cocteau observed, "Art is not a pastime, but a priesthood." The serious culinarian must commit to a lifetime of continuing education and professional development in order to fully develop the craft. Whether studying sculpture at the local community college, visiting a museum of art, or reading books on design and architecture, the garde manger must learn to transcend the limiting walls of a kitchen. Useful knowledge can be gained from other media and other fields of art. The challenge is to find the proper blend of the aesthetic and the practical (Figure 1–7).

Business Management Abilities

Among the many skills one must possess to be a garde manger, and certainly to be an executive chef, is the ability to create pleasing food in a profitable manner (Figure 1–8). The food service industry is in the business of providing food and service to its customers. With little exception, this must be done at a profit. To satisfy the operational expenses associated with running a food service company, the garde manger chef must be a faithful steward of the company's assets as well as an adroit manager of its staff.

FIGURE 1–7 Napa Grill's Market Platter, featuring California cured olives, smoked salmon, fresh honeycomb, artisanal goat's cheese, fresh figs, air-dried beef, grain mustard, foie-gras pâté, prosciutto, and raisins on the vine

FIGURE 1–8 A highly-decorative buffet presentation. Courtesy of Jackson Vereen/Foodpix

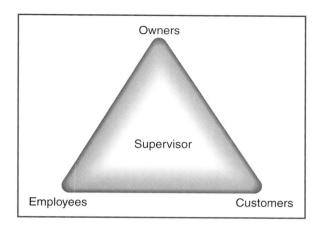

FIGURE 1–9 Trinity of business

This relationship is referred to as the **trinity of business** (Figure 1–9). The garde manger's ability as a manager is challenged by the needs of top management, the staff, and the customer. Each has needs, which often conflict financially. The skill lies in the ability to meet the expectations of each, without sacrificing quality or ethical behavior. This is usually accomplished by having a strong system of planning and controls, known as the **Siamese twins of management**. Each activity must be well planned in advance and then evaluated upon completion to measure the success of the plan. This system of planning and controlling generally results in an efficient and well-trained staff producing food within budget.

Accuracy is principal to achieving plans and obtaining financial success. The ability to order, prepare, and price food in an accurate manner is crucial to the financial health of any food service business. The garde manger chef can be particularly helpful to an operation when he or she approaches the craft with a business mentality for being accurate.

BUSINESS ORGANIZATIONAL STRUCTURES

Beyond the walls of the kitchen, there lies a labyrinth of halls and corridors leading to other departments of the food service operation. The modern garde manger must possess a clear conceptual understanding of the organization, and where the garde manger department fits within the plan. Organizational structures are developed both vertically and horizontally, tying various positions and units together. Each services many; each is dependent on another.

The principle of **unity of command** states that each employee reports to only one supervisor, that each job fits within a hierarchal order of the overall business. The organizational chart reflects the chain of command within the operation. The garde manger chef de partie generally directly reports to the executive sous chef or the executive chef de cuisine. In return, several commis and assistant chefs report to the garde manger chef (Figure 1–10).

JOB DESCRIPTIONS AND STANDARDS

It is incumbent upon all well-run food service operations that **job descriptions** be written and revised regularly for all jobs within the business (Figure 1–11). People holding positions at the property should be provided current job descriptions detailing their job objectives and requirements. The job descriptions should reflect management's expectation of the skills, abilities, experience, and education required to hold the position, as well as a summary of all of the job objectives inherent with the job.

Based on the actual job requirements, **performance standards** are developed to identify to what extent each job requirement is to be met (Figure 1–12). These standards will become the foundation for training within the department, as well as the standards by which the employee is to be evaluated by the supervisor.

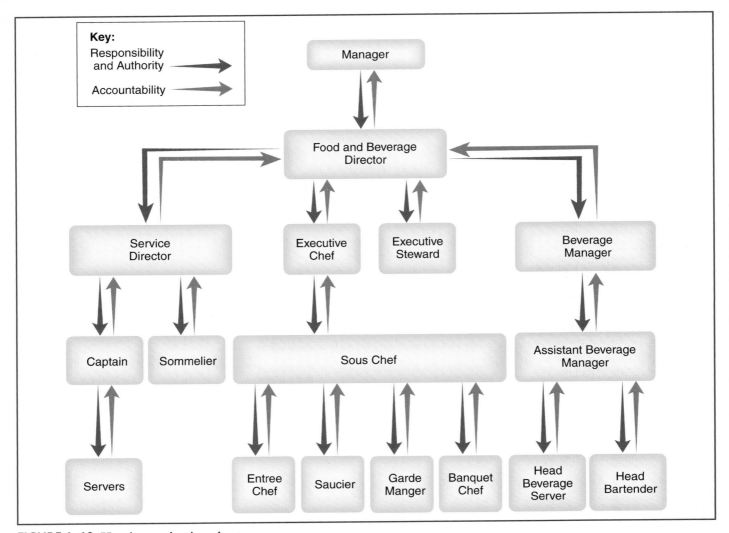

FIGURE 1–10 Hotel organization chart

All employees should be given these materials upon hiring so they are fully aware of management's expectations for their positions. Professional garde mangers should, in turn, develop these materials for their own staff if they do not exist already.

Focusing on Professional Development

The craft of the chef is global. Every country prepares food items that fall under the realm of the garde manger. And, as the dining public becomes more traveled and sophisticated in its tastes, the demand on chefs to be creative and diverse in their preparations continues to grow. To meet this ever-increasing demand for world food, the garde manger must undertake the responsibility of researching new menu items. Television networks devoted solely to 24/7 food programming, local cable shows, countless trade magazines, infinite numbers of seminars and hands-on cooking classes, and the Internet are only the beginning of the resources available to the serious culinarian.

With the demands of work and personal responsibilities always vying for the chef's time, the garde manger must be selective and focused in order to maximize time. There are four primary areas where the chef should focus energy: (1) nutrition and sanitation awareness, (2) product knowledge, (3) preparation techniques, and (4) trends and classics.

FIGURE 1–11 Job description of a garde manger chef

Position:
Garde Manger

Essential Functions:
Supervises cold kitchen and prepares products according to club recipes and standards.

Additional Responsibilities:
1. Slices and proportions cold meat, fish and poultry; garnishes them in an appetizing and tasteful manner
2. Prepares appetizers, hors d'oeuvres, center pieces and relishes in an attractive manner
3. Prepares cold sauces, jellies, stuffing, salad dressings and sandwiches using club recipes and standards
4. Supervises pantry crew in the preparation of salads, dressings, etc.
5. Schedules and supervises cold kitchen staff; responsible for their performance standards, evaluation, discipline and motivation
6. Requisitions food supplies necessary to produce the items on the menu
7. Adheres to local health and safety regulations.
8. Maintains the highest sanitary standards
9. Notifies sous chef in advance of expected shortages
10. Maintains security and safety in work area
11. Maintains neat professional appearance and observes personal cleanliness rules at all times
12. Ensures that work area and equipment are clean and sanitary
13. Covers, dates and properly stores all leftover products that are re-usable.
14. Assists with other duties as assigned by the sous chef

Reports to:
Sous Chef

Supervises:
Cold food personnel

Key Objectives/Goals:
Key objectives and goals are established with the Executive Chef and reviewed with the annual performance evaluation.

NUTRITION AND SANITATION AWARENESS

The professional garde manger chef must be an advocate for the health and well-being of the consumer. Graham Kerr, the legendary "Galloping Gourmet" who stormed the airwaves in 1969 and inspired millions of viewers to joyfully experiment in the kitchen, penned a cookbook entitled *The Gathering Place* after taking a world cruise in 1996. Enthused by the diversity of ingredients and foods available from across the globe, he and his wife Treena decided to wed their personal philosophy of healthy eating and balanced lifestyles with the rich flavors and culinary heritage of many lands. Kerr (1997) wrote:

> As I set about developing a book based on our journey, I tried to create dishes that were not pale reflections of their classic namesakes, but new recipes that pay homage to their history and branch out into the future. I was looking for new tastes from other lands that could marry with more familiar tastes, textures, and aromas that we already love in order to create new favorites. In many cases I took recipes for famous traditional dishes and "springboarded" them into formats using less fats, salt, and sugar. (p. 14)

Modern garde manger chefs must always be mindful of the impact their food and practices have on the guest. Grateful and faithful consumers will reward proper sanitary procedures, nutri-

Garde Manger Performance Standards

Position Purpose:
To ensure the garde manger department is run in an efficient manner, and that the finest quality food is served to our guests; to organize and supervise subordinates and staff; to control food cost and labor cost in order to achieve and/or surpass budgeted goals and profits as well as establishing hotel cold food quality and artistry as the finest in the midwest.

Major Responsibilities:
As chef garde manger it is of the utmost importance that you set an example for all other employees through your behavior, your adherence to company policy and your relations with managers and staff members, to a large extent, you will determine the personality of the garde manger department.

General Responsibilities:
1. Comply with company policies and regarding scheduling, uniforms, training, absenteeism and tardiness.
2. Follow and implement all company rules and regulations.
3. Follow and implement all personal appearance and grooming standards.
4. Conduct yourself in a professional manner at all times as a "Goodwill Ambassador" of the hotel to ensure customer satisfaction.

Specific Responsibilities:
1. Responsible for reparation of all cold food products including sauces, dressings, salads, aspic, cold soups, meats, seafood, canapes, terrines, pates, Cornucopia, platter layout, sculptures, ice carving, show pieces, chaud-froid and all other cold food items required by the hotel.
2. Advises on and follows thru the set-up for every buffet table with the banquet supervisor, catering director, following standard specs set by the executive chef.
3. Follow thru completion and execution of every item leaving garde manger department to assure *each* item is labeled with (time, room, function) information.
4. Maintain and enhance photo specs and standard recipe card file. Assure food is prepared and presented in accordance with these approved cards at all times.
5. Ensure a smooth running operation, high quality food and maximum production from all employees during shift.
6. Check stock food and other supplies daily and make required requisitions daily.
7. Check for proper utilization of left over food as well as checking on waste.

8. Prepare daily prep list for production, anticipate advance production, and place necessary orders with Bake Shop, etc.
9. Checks all refrigerators at different times during the day for cleanliness and proper rotation of food.
10. Follows strongly the sanitary standards set by the health department.
11. Conduct daily menu meeting to: review day before discrepancies, next day's production, and any changes in menu or forecasts.
12. Make sure that all stations are clean, food is properly stored and reach-in refrigerators are cleaned.
13. Supervise and train all garde manger staff according to hotel standards and policies.
14. Responsible for employees proper uniform appearance and personal hygiene.
15. Know all job descriptions of all employees reporting to you.
16. Make daily assignments in the garde manger department and monitor payroll to assure the balance cost goals are being accomplished.
17. Enforce "clean-as-you-go" policy.
18. Practice safety standards and report all unsafe conditions.
19. Know all food and beverage outlet locations, hours of operation, and all other facilities within the hotel.
20. Handles personnel problems to the best of his knowledge. Does not allow employees who are ill or suffer from infections to take part in the preparation or handling of food.
21. Ensure that no food leaves the kitchen without proper function order or transfer sheet.
22. Attend scheduled meetings as required.
23. Responsible for month end inventory.
24. Perform any related duties as assigned by the executive chef or executive sous chef. Assist all other employees in their duties whenever required.

Measurable Criteria:
1. Ability to supervise, train, organize and assist staff.
2. Ability to understand and perform assigned tasks.
3. Attitude to maintain high culinary standards and ability to get along with co-workers and superiors.
4. Attitude and willingness to perform other duties as assigned by management.
5. Ability to control food and labor costs.

FIGURE 1–12 Sample hotel performance standards for a garde manger chef

tional sensibility, and imaginative, attractive, and flavorful food. An effort to keep abreast of proper sanitation techniques, coupled with an ongoing interest in nutrition, is time well spent when building a career and customer following.

PRODUCT KNOWLEDGE

The commitment to continuously seek new product knowledge is most important to a serious chef. Over the decades, modern systems of preservation and transportation have created a global market for foodstuffs. Additionally, numerous products, such as different varieties of fish, cheese, and produce, are being introduced annually into the world market. Most have existed in their local areas, but never received greater exposure on the world market.

It is important for professional chefs to constantly seek out these foodstuffs, whether from abroad or from within their own lands, as a means of providing innovative menus to their customers. Communicating directly with the fishmongers, cheese makers, and growers can bring a wealth of knowledge to the interested chef. Additionally, suppliers are always interested in what chefs are looking for in new products; it serves as a mutually beneficial relationship.

PREPARATION TECHNIQUES

Too often, chefs think they are so familiar with certain food items that they assume all methods of preparation have been exhausted. This idea may be justified in some cases, but we believe, based on personal experiences, that preparation techniques are infinite. A case in point: It was only within the last few years that we learned several methods of filleting, slicing, and presenting fish that we never knew before. Having both been raised close to water, and having both lived on islands where we featured local and imported seafood in our restaurants, we were astonished by this revelation. Humility in the face of knowledge is the chef's ally.

TRENDS AND CLASSICS

The spectrum is broad when it comes to food preparation and presentation. Chefs, particularly in garde manger, benefit greatly by having a foundation in classical food. Whether it is the adherence to a disciplined style of food preparation and presentation or the historical reference, being versed in the classics of food has merit.

The great chef and scholar Marie-Antoine Carême founded **la grande cuisine**, "Classical French Cookery," in the early nineteenth century. Although his style of *service à la Française* was somewhat based on the heavy and ostentatious banquets of the eighteenth century, he systematized techniques, standardized recipes, and began to streamline menus. Georges-Auguste Escoffier embraced the styles of *service à la Russe* and *service à la Carte,* in which dishes were served individually instead of all at once. At the turn of the twentieth century, Escoffier created the French School of Culinary Arts, *La Cuisine Classique,* further shaping and simplifying the modern menu.

The classics have, over time, given way to modern preparation styles. In the mid-twentieth century, chef Fernand Point went further than Escoffier to bring about changes in cooking styles, laying the foundation for **nouvelle cuisine**. Chefs Paul Bocuse, Jean Troisgros, Roger Vergé, and Michel Guerard—known as the "Bande à Bocuse"—helped popularize Point's concept of nouvelle cuisine.

There are very few places in the culinary world where food is still prepared as it was 100 years ago. The availability of ingredients, many in a user-friendlier manner (not processed, per se, but plucked and eviscerated), has eliminated some of the drudgery of pre-preparation that existed. Some chefs, however, would argue, "At what cost?" Has freshness been supplanted by convenience? As in all matters in life, the choice remains a question of cost versus benefit.

Today's modern garde manger has the benefits of a world market and a dining public interested in global cuisine. It is important to the career of the chef, and the reputation of the chef's employer, to constantly evaluate products, techniques, and trends that will be of interest to customers. Some styles are merely fads, destined to fade quickly from the minds of the consumer, whereas others will take their rightful place on menus as trends with staying power. For example, it wasn't too long ago that sushi was thought to appeal only to those of Japanese descent or to jet setters and the sophisticated city crowd. Today, sushi is common fare, even in the supermarkets of midwestern America.

Lifelong Learning

To be a professional, in any field, means a commitment to lifelong learning. Every occupation requires that its leaders continue their training, because inherent in leadership is a vision for the future. As the world turns at a feverish pace, all people must continuously stay abreast of the factors that will affect their ability to live and work. Still, to ignore the traditions and achievements of the past is to build a house upon sand. While making a mark in food service, the modern chef must responsibly embrace change, but with a respect for the art and science that is garde manger.

Increasing Cultural Awareness

It is no revelation that our lives continue to evolve as the world continues to change. Not in the literal sense, of course, but certainly in most other ways. Advancements in science and technology have allowed the inhabitants of this planet to learn more about each other. Whether through global telecommunications or transcontinental travel, we are quickly realizing that we are citizens of one world. What happens in one country is discussed in, and often affects, another. Products grown or manufactured on one continent are distributed and sold to inhabitants on other shores. The demographics are changing from the world we knew, to the world that will be.

Patricia Heyman supports this view in her comprehensive text, *International Cooking: A Culinary Journey* (2003). She writes:

> Immigration and birthrates continue to change the demographics throughout the world. Predictions released from the United Nations estimate that about 87 percent of the world's population will consist of people from Asia, Africa and Latin America by 2050. The remaining 13 percent will reside in other regions, including North America and Europe. People from densely populated, developing countries will continue to seek opportunities in more prosperous nations. As a result, many immigrate to the more affluent countries. So although the world is not shrinking, it certainly is changing, and that change results in people being exposed to more countries, more cultures, and more cuisines. (p. vii)

Because of this reality, the profile of the consumer has also changed. Today's consumer has a more worldly and adventuresome appetite. Comparing the list of offerings from a menu printed 30 years ago to today's typical bill of fare, the ethnic and cultural influences are very noticeable. The older menus offered a narrow selection and rarely ventured into cultural diversity beyond Western Europe (Figure 1–13). As the United Kingdom, France, Germany, and Italy claimed to be the vanguard of world cuisine (ignoring the monumental culinary contributions of countries like China and Turkey), most American and continental menus reflected their cuisine for generations.

Contemporary menus generally offer selections that have been strongly influenced by many divergent and distinct cultures (Figure 1–14). In addition to a Western European persuasion, they generally feature dishes founded in the cuisines of Asia and the Far East, Africa, Latin America and the Caribbean, and the Middle East. Typical consumers are not solely from Western Europe and North America anymore, and if they are, they probably have acquired a much broader taste through travel. Menus continue to move with the times, as does the consumer. In an article in the magazine *Flavor & The Menu* (2004):

> An earlier study by DLG Market Research found that Mongolian barbecue was the number one cuisine high school students wanted to try, followed by Japanese at 43 percent, Vietnamese at 40 percent, Thai at 40 percent, Japanese (sushi) at 38 percent, Greek at 38

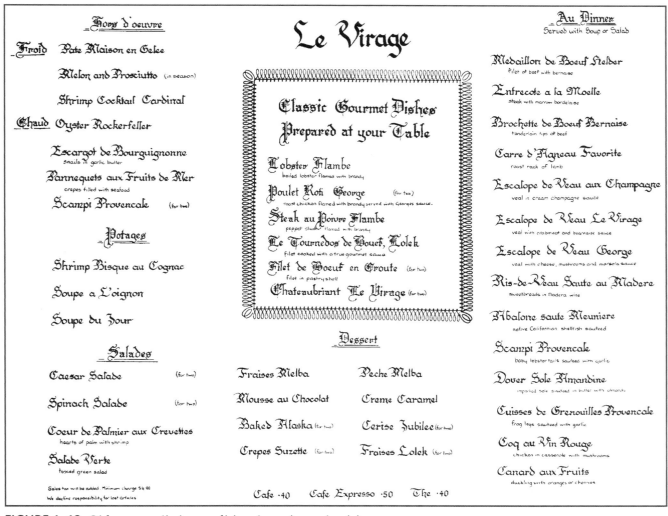

FIGURE 1–13 Older menu listing traditional continental cuisine

percent, Indian at 45 percent, Latin at 31 percent, Middle Eastern at 29 percent and Spanish 23 percent. (p. 20)

To meet the demand for change, chefs must seek awareness of the differences in cultures and tastes.

Internationalizing the Menu

"No pleasure endures unseasoned by variety," wrote the Roman philosopher Publius Syrus as his Maxim number 406. He noted that change keeps life interesting, and variety is a means to change. The modern garde manger must embrace change and use variety in a menu to attract new customers and retain existing clients.

In addition to local and regional specialties, the chef must include a broad range of menu items from around the world. The goal is to create a garde manger menu that reflects the global nature of food and culture, and to prepare it in a way that is representative of those cultures. A country's culinary identity is closely linked to its cultural history. To be aware of a country's cuisine, one must have an awareness of many other contributing factors. Heyman (2003) writes:

many of a cuisine's culinary traits result from conditions that naturally exist in the region or country—factors such as geography, topography, climate, what grows/is raised there, and historical influences from settlers, invaders, and bordering countries. (p. viii)

FIGURE 1–14
Contemporary menu
featuring globally diverse
selections

In promoting their annual international *Worlds of Flavor* conference and festival, the Culinary Institute of America (2003) writes:

And yet even today American menus just scratch the surface of these culinary traditions. As American chefs, operators, and foodservice managers look to the future, enormous opportunities exist to dig deeper into these marvelous, centuries-old food cultures to find a new generation of menu ideas that will capture the imagination of tomorrow's customer.

Recognizing the diversity in cultural backgrounds and consumer preferences for food, the contemporary chef must seek a means of internationalizing the menu. This requires both effort and confidence on the part of the chef. Research, whether through various media or by personal experience, is always necessary.

And although the product may be new to the chef, its preparation has been refined over centuries by the people of its origin. Local methods for preparing indigenous ingredients generally prove to be well researched and dependable. The consumer will appreciate nearly all dishes that have been authentically prepared. To be authentic, the chef must consider and incorporate three variables: (1) the use of indigenous ingredients and seasonings, (2) the use of various regional cooking styles, and (3) the use of traditional presentation styles.

TRADITIONAL INGREDIENTS AND SEASONINGS

One of the most common means of internationalizing the menu is to offer food and beverage items that are traditionally associated with other countries. For example, sake and sushi are representative of the Japanese culture. Their inclusion on a menu reflects an attempt to internationalize the selections. Certain spices and flavor profiles are often associated with specific countries or regions of the world, such as the prevalent use of sesame oil in Asian cooking, primarily in Chinese dishes.

However, one growing challenge in this area lies in the strength of the world market. Improved efficiency of transportation and distribution has made most ingredients appear universal. Pineapples, mangoes, and papaya are no longer considered exotic fruits, although their production is still restricted to tropical climates. The dramatically increased availability of most foodstuffs has even caused the buying public, in some cases, to forget the origins of many products.

This recognition by some growers and food producers has caused them to pursue "brand identification" for their products. Not unlike the "Appellation Contrôllée," which monitors and certifies the origins of French wines, brand identification is an attempt to prevent lesser quality or competing products from being substituted for leading products or from using their name recognition. Copper River Salmon, Kobe Beef, and Roquefort Cheese have designations that ensure quality and bring specific geographic locations to mind.

REGIONAL COOKING STYLES

Although the known methods of cookery have not changed in centuries (with a few notable exceptions, such as halogen heating, microwave cooking, and induction cooking), we often associate cooking styles with certain regions or countries. Skewered meats cooked over charcoals, hearth breads and pizzas made in wood-fired stone ovens, and with ladles of ingredients cooked in fiery woks remind us of certain flavors and nationalities. Whether steamed in a bundled banana leaf, roasted in a dried cornhusk, or baked inside a mound of damp salt, methods of cookery raise the interest of the consumer and heighten their dining experience.

TRADITIONAL PRESENTATION STYLES

Food must satisfy the eyes before it can enter the mouth. Visual presentation is important to both the garde manger and the consumer. On a buffet, smoked fish presented with its head and tail attached appears more fresh and exciting than merely a fillet. Paella, portioned from its traditional round pan, appears much more authentic to the consumer. As Henry Barbour said, "The art of garde manger is a blend of the aesthetic and the practical."

Professional Profile

BIOGRAPHICAL INFORMATION

Name: Robert William Kendrew
Place of Birth: West Hendon, London

Recipe Provided: Compote of Fruit
with Elderflower Cream

CULINARY EDUCATION AND TRAINING HIGHLIGHTS

Chef Kendrew started his culinary career as an apprentice in the bakery and confectionery department of London's South East Durham Co-op in 1950. During this time, he attended the local Durham College to study with the City & Guilds of London Institute. The courses were then the C&G 151, a basic cookery qualification, and C&G 152, the professional qualification for chefs employed in the industry. At 18 years old, he fulfilled his Military National Service and was conscripted to serve in the Army Catering Corps. He served in Aldershot, England, in Cyprus, and in Egypt where he attained the rank of cook sergeant. He additionally acquired Army qualifications B1 and B2, which equated to the equivalent of the advanced City Guilds 153, then the top qualification attainable in the United Kingdom. In 1956, Chef Kendrew joined the prestigious House of Fraser as sous chef to one of their restaurants, later rising to the position of executive chef for the corporation in 1970. After 35 years with House of Fraser, he joined Edinburgh's Telford College in 1990 as a lecturer in food studies, teaching practical cookery and related theory to HNC (Higher National Certificate), HND (Higher National Diploma), Scotvec, and school programs. During his career, Chef Kendrew was instrumental in the creation and development of numerous Scottish culinary heritage activities, including the Taste of Scotland and the Scottish Catering Training Board. He served as the governor and past national chairman of the United Kingdom's Cookery and Food Association and as chairman of the judges for numerous international culinary competitions located in the United Kingdom, Malta, and the United States. A crowning achievement in his career was being chosen to attend Queen Elizabeth's (the Queen Mother) 100th Birthday Parade as one of 100 selected chefs from the United Kingdom.

ADVICE TO A JOURNEYMAN CHEF

The standards expected of a chef engaged in the preparation of food are now so great, that it takes a dedicated person to fulfill their ambition. This may be a daunting task, but with passion and belief in one's self, it becomes attainable. Never accept second-hand performances given to you, as you will have to give your best at all times. It is easy to be a great chef for a short time, but to be a great chef all the time requires the support of friends, colleagues and, above all, your family. The respect of your peers and colleagues is, in itself, the highest honor life can bestow on you.

Robert William Kendrew

RECIPE 1-1

CHEF ROBERT KENDREW'S COMPOTE OF FRUIT WITH ELDERFLOWER CREAM

Recipe Yield: 8 portions

MEASUREMENTS		INGREDIENTS
U.S.	**METRIC**	
		Elderflower Cream
2 ounces	50 g	Sugar
1 cup	300 mL	Water
2 ounces	50 g	Elderflower heads, dried (or 12 fresh heads)
1 cup	300 mL	Double cream
		Compote
½ pound	225 g	Rhubarb
½ pound	225 g	Gooseberries
½ pound	225 g	Cherries
½ pound	225 g	Strawberries
1 pint	600 mL	Apple or orange juice, unsweetened
2 each	2 each	Cinnamon stick
¼ teaspoon	1 mL	Lemon zest
To taste	To taste	Clear honey

PREPARATION STEPS:

1. To make the elderflower cream, put the sugar and water in a pan and heat gently until the sugar has dissolved, then boil rapidly until the liquid is reduced by half. Take off the heat and submerge the elderflowers in the syrup.

2. Leave to infuse for at least 2 hours, then press the syrup through a sieve, discarding the elderflowers. Whip the cream until it is just beginning to hold its shape, and then fold in the elderflower syrup. Chill until ready to serve.

3. Put the fruit, fruit juice, cinnamon, and lemon zest in a large saucepan and simmer gently for 3 to 5 minutes until the fruits are softened, but still retain their shape.

4. Serve the compote, warm or cold, with the elderflower cream.

REVIEW QUESTIONS

1. Explain the term *craft guild.*

2. Briefly describe the term *keeping to eat.*

3. Why was the brigade system so important to the success of the kitchen?

4. Describe a chef de partie.

5. Explain the term *sous chef.*

6. What is the garde manger section responsible for in the kitchen?

7. Explain the term *tournant.*

8. Explain the term *commis.*

9. Discuss the origin of the modern chef's uniform.

10. What is a saucier?

A. **Group Discussion**

Discuss the role of the modern garde manger with emphasis on the change in technology.

B. **Research Project**

Research the impact that the garde manger chef can have on the profitability of the kitchen, as a whole, considering that the majority of the raw food arrives at his or her department.

C. **Group Activity**

Design a brigade system for a large hotel that feeds 1,000 guests breakfast, lunch, afternoon tea, and dinner daily.

D. **Individual Activity**

Interview a garde manger chef in a local restaurant to establish what that chef considers to be his or her most important role within the kitchen brigade.

CHAPTER 2

Calculations and Controls for the Garde Manger

After reading this chapter, you should be able to

- understand the process of converting U.S. customary measurements to metric measurements.

- explain the concept of performing make-or-buy decisions.

- perform a recipe costing.

- explain the value in using standardized recipes.

- demonstrate the process used to calculate quantities of food.

- understand how to determine prime costs.

- list the many activities involved in production management.

- list the garde manger's responsibilities in recipe development, cross-utilization, waste control, and product usage.

- understand the value and purpose in using production reports.

- explain how to conduct a menu analysis using food cost percentage, contribution margin, and goal value decision making.

- explain how to price events using cost-based pricing, multiplier-based pricing, competitive-based pricing, and client-based pricing.

Obvious is the most dangerous word in mathematics.

ERIC TEMPLE BELL

Calculations: A Job Requirement

Inherent in the position of chef is the responsibility of making a variety of calculations. Accuracy is vital to success in business. Whether determining ingredient quantities in recipes, food to order, recipes to cost, or the number of portions to prepare, the garde manger chef must be capable of making precise calculations. Profitability and superior performance are the expectations placed on contemporary chefs.

Chefs should embrace the modern technology now available in this area. Whether using a calculator or menu and recipe software, there are many tools on the market to aid the garde manger in calculating figures correctly.

Converting Metric and U.S. Customary System Measures

In order for the modern garde manger to use recipes and packaged ingredients from distant lands, he or she must be versed in both metric and U.S. customary systems of measurement. Even though the Department of Commerce submitted a report in 1971 to Congress recommending that the United States convert to metric and in 1975 President Ford signed into law the Metric Conversion Act, the United States has long resisted making the change to metric.

Over the years, however, U.S. governmental agencies have switched some of their standards to metric. And, several industries (e.g., medical, science, engineering, and automotive) have converted because of their need to participate in world trade.

Today, more than 90 percent of the world's population and 90 percent of its nations use the metric system. Yet, U.S. food service operations still operate on a system based on U.S. customary measures. As a result, in order for chefs to be able to work in the world market, they must be familiar with both systems of measurement.

The **metric** system is a decimal system based on the number 10. All measurements are multiples of 10. The system provides standard rules for amounts of its units through the use of prefixes. For example:

1 meter = 10 *deci*meters = 100 *centi*meters = 1,000 *milli*meters

1 decimeter = 0.1 meter
1 centimeter = 0.01 meter
1 millimeter = 0.001 meter

Also, units of measure have specific names, such as:

Meters	representing length
Grams	representing mass or weight
Liters or cubic meters	representing volume
Degrees Celsius	representing temperature

WEIGHTS AND MEASURES EQUIVALENTS

In order for chefs to work with recipes and food containers from foreign lands, they must be able to convert the measurements from one system to their own standard measuring system. Measuring cups and scales now commonly feature both metric and U.S. units of measure, making it easier on the chef to use unfamiliar recipes. Table 2–1 lists the most commonly used forms of measurements. (*Note:* In many cases, these equivalent numbers are approximate.)

Many culinary textbooks and professional recipe books, including this one, are published using both systems of measurement. As chefs from around the world hunt for the means to expand their repertoire of recipes and skill sets, publishing houses seek to meet this demand. The arrangement is mutually beneficial, and it points to the trend of global education.

Converting Temperatures

In the metric system, temperatures are measured in degrees Celsius (°C), whereas degrees Fahrenheit (°F) are used in U.S. customary measurements. On the Celsius scale, water boils at 100°C and freezes at 0°C. On the Fahrenheit scale, water boils at 212°F and freezes at 32°F.

To convert Fahrenheit temperatures to degrees Celsius, subtract 32 from the given Fahrenheit temperature and multiply the result by 5/9 (or 0.56).

For example:　　　$212°F - 32° = 180°F$

$180°F \times 5/9 = 100°C$

To convert degrees Celsius to Fahrenheit temperatures, multiply the Celsius temperature by 9/5 (or 1.8) and add 32 to the result.

For example:　　　$100°C \times 9/5 = 180°F$

$180°F + 32° = 212°F$

Table 2–2 shows common temperature conversions. A greatly expanded chart of measurement equivalents is included in the Appendix for additional reference. It is recommended that the reader adopt the habit of looking at both columns of measurements while following the recipes in this textbook. Repeated exposure to dual systems of measurements will eventually train the reader to think in dual terms, thereby creating a skill that is useful and practical.

Table 2–1　Weights and Measures Equivalents

U.S. CUSTOMARY SYSTEM	METRIC SYSTEM
1 gram	= 0.035 ounces
1 kilogram	= 2.2 pounds
28 grams	= 1 ounce
454 grams (0.45 kg)	= 1 pound
5 milliliters	= 1 teaspoon
15 milliliters	= 1 tablespoon
240 milliliters (0.24 liters)	= 1 cup
0.47 liters	= 1 pint
0.95 liters	= 1 quart
1 liter	= 1.06 quarts

Table 2–2   Common Temperature Conversions

FARENHEIT	CELSIUS
32°F	0°C
50°F	10°C
68°F	20°C
86°F	30°C
100°F	40°C
115°F	45°C
120°F	50°C
130°F	55°C
140°F	60°C
160°F	70°C
170°F	75°C
180°F	80°C
185°F	85°C
195°F	90°C
200°F	95°C
212°F	100°C
230°F	110°C
250°F	120°C
265°F	130°C
285°F	140°C
300°F	150°C
325°F	165°C
350°F	175°C
360°F	180°C
375°F	190°C
400°F	200°C
425°F	220°C
450°F	230°C
485°F	250°C
500°F	260°C
575°F	300°C

Recipe Analysis

Profitability and customer acceptance should not be achieved by chance. The purpose of recipe and menu analysis is to determine the related costs and sales (profit) contributions for each menu item. It is important to the profitability of the operation to periodically measure the performance of each menu item. This includes raw costs, contribution to profit, and sales history. Chefs and restaurateurs traditionally have the flexibility to control how recipes are made and what items are offered on the menu. A proactive chef manager can engineer a menu that is both profitable and consumer driven when proper evaluation is given.

BUTCHER'S YIELD TEST: AN EXAMPLE OF MAKE-OR-BUY

One of the fundamental considerations that chef managers give to the selection of menu items, and the ingredients necessary to create them, is the **make-or-buy decision**. Basic to the development of the menu is the acquisition of the menu's ingredients. Chefs always have to consider the sourcing of their ingredients and in what finished state of preparedness should they purchase these items. Sometimes it is better to fabricate the ingredients; other times it is more practical to buy items premade.

In the past, most large clubs and hotels had their own butcher shops where sides and quarters of hanging meat were brought in on rails and taken to walk-in coolers to **dry-age** before cutting. Staff butchers would cut up the chickens, veal, lamb, beef, and pork to the specifications of the chef de cuisine and chef garde manger. Menus were painstakingly written to use all of the primary cuts, lesser cuts, by-products, and trimmings. Bones were used for stock and fat was turned to tallow or roux.

Nowadays, chefs usually buy primal cuts, boxed meat, or precut portions that are **wet-aged**. The decision whether to purchase precut portioned veal scallops versus whole legs of veal to be broken down and portioned by the staff is an important one that is usually well considered by the chef. To decide which is most cost efficient, chefs perform a **butcher's yield test** (Figure 2–1). The test is used to evaluate the difference in the cost of cutting their own portions (including the resultant by-products of trim, bones, and fat) or buying them pre-portioned. It is a classic make-or-buy decision. The garde manger is often called upon to turn the trimmings into profitable appetizers, sausages, pâtés, and other forcemeats when the decision is to cut their own meat.

CALCULATING RECIPE COSTS

The process of analyzing recipes for their food costs has become a much easier task with the advent of computers and specialty software (Figures 2–2 and 2–3). Even without the use of customized software packages that evaluate each ingredient's cost and nutritional value, standard spreadsheet software can be used to itemize the impact each ingredient has on the total cost of a recipe. These evaluations are important for understanding the true costs involved in each recipe.

Whatever method is used, it is important for the chef to periodically cost out the recipes being used, analyzing the impact of each ingredient. Also, beyond maintaining an understanding of current costs, an awareness of seasonal fluctuations in pricing, dependable sources (or less than dependable sources), health considerations, and the latest list of endangered species must also be regularly addressed.

Standardized Recipes

As Judy Lieberman (1991, p. 52) writes in her text on catering, "In order to make any food-pricing system valid and to improve the efficiency of a catering kitchen, it is important to use standardized recipes. . . . The use of standardized recipes ensures that menu items are properly prepared, of consistent quality, and of predictable yield and cost each time it is made." To meet the projected revenues and profits earned by the sales of each menu item, a standard recipe must be followed each time the menu item is prepared.

The **standardized recipe** must be developed before the menu is priced, because as menu prices are based on the actual costs attributed to the standardized recipe (Figure 2–4). Additionally, the taste, appearance, and resultant quantity of product produced is greatly affected by the recipe. Using a standardized recipe guarantees consistency, which is best for both the customer and the operation.

Portion Control

In addition to using a standardized recipe for yielding consistent appearance and taste, the use of **portion control** is vital for yielding the anticipated number of portions each time the recipe is prepared. Portion control is the conscious act of alloting the same quantity of product each time

FIGURE 2–1 Butcher's yield test

BUTCHERS' YIELD TEST

Item _Veal Leg_   Specification _334 B_   Condition _____

Other Remarks _Fresh_____

Date Received _____   Weight at Receipt _42 lb.____   No. of Pcs. _1___

Purchased Price per Pound _$ 3.45/lb._____   Total as Purchased Cost _$ 144.90_____

Date of Test _____   Weight at Time of Test _42 lb.____   No. of Pieces _1_____

Supplier _____   Fabricating or Cutting Instructions _____

Results of test:

Credits	Weight	% of Total	Value Per Pound	Extension
Usable Trim	4 lb.	9.5%	$ 4.49	$ 17.96
Bones	7 lb.	16.6%	1.05	7.35
Fat	3.5 lb.	8.3%	.20	.70
Shrinkage	0	0	0	0
Cutting Loss	2 lb.	4.8%	3.45	6.90
TOTAL CREDITS	16.5 lb.	39.2%	/ / / / / / / / / /	$ 32.91

A.P. Cost	$ 144.90
Less Credit	$ 32.91
Cost of Fabricated Product	$ 111.99

Yield of Fabricated Product	42 – 16.5 = 25.5
% of Total	60.8%
Cost per Pound	$ 4.39
Cost Factor (new/old)	$^{4.39}$/$_{3.45}$ = 1.27

Additional Remarks: _____

_____

_____

_____
Signature

NOTE: Show percentages of all products to total. Indicate reasons for loss or gain in total weight, such as shrinkage through storage or cutting loss or added material as may occur in ground beef, etc.

it is served. In this way the anticipated profit, based on recipe costing and the resultant menu pricing, will be realized.

Many portion control utensils are available to assist the garde manger in maintaining proper control in recipe preparation and service. Portion scales, measuring spoons and cups, scoops, ladles, and "spoodles" are all means by which the chef can meet their financial and production goals. In the United States, portion scoops are used frequently to measure and portion food. The size of the scoop, often stamped either on the thumb guard or baler of the scoop, indicates the number of servings that will come from one quart (32 ounces) of product. For example, a number 16 scoop will yield sixteen ¼ cup (2 ounce) portions, and a number 8 scoop will yield eight ½ cup (4 ounce) portions. See the Appendix for a *Size and Capacity of Scoops* chart.

FOOD COST FORM

Menu Item: Date:

Number of Portions: Size:

Cost per Portion: Selling Price: Food Cost %:

Ingredients	Recipe Quantity (EP)			Cost			Total Cost
	Weight	Volume	Count	APC/Unit	Yield %	EPC/Unit	
						TOTAL RECIPE COST	

FIGURE 2–2 Sample recipe cost sheet

CALCULATING QUANTITIES TO ORDER

Controlling the amounts to order is an important function of the garde manger. Ordering as closely as possible to the proper amount needed to prepare a recipe will reduce inventories and help reduce waste from spoilage and overproduction.

When ordering food for a specific number of customers, the chef multiplies the serving size by the number of guests. The result indicates the total number of ounces or grams required. The standard unit of issue then divides this number, with the quotient equaling the total number of units to order. The chef must round up the quotient when placing the order, because partial units generally cannot be ordered.

To illustrate, assume the Brittany Hotel is serving an hors d'oeuvre banquet for 200 guests. The party planner has contracted for two Beef Satés per guest, among other appetizers. Each saté is made with $1\frac{1}{2}$ ounces (as purchased) of beef bottom round. Beef bottom round is ordered by the pound. The calculations are as follows:

200 guests $\times$ 2 satés each = 400 satés

400 satés $\times$ $1\frac{1}{2}$ ounces = 600 ounces of beef bottom round

600 ounces $\div$ 16 ounces per pound = 37.5 pounds

Rounding up, the chef needs to order 38 pounds of beef bottom round

My Chef Holiday Hors D'Oeuvre Packages

Roman Holiday – Serves 36 Guests	PP Qty	Retail	36		
Caeser Salad – Salad for 24 Guests – 2 1/2 lb. Romaine	1	$ 2.75	24	$ 66.00	
Caesar Dressing – QUART			1.5		
Grilled Medallions of Beef Tenderloin with Aus Jus – LB	3 oz	$ 28.95	7	$ 202.65	
French Dinner Rolls – EACH	1.5	$ 0.65	50	$ 32.50	
Sherried Horseradish Sauce – PINT			1		
Portabello Mushroom Ravioli, Olive Oil/Herbs or Mornay – LB	5 pc	$ 12.98	6	$ 77.88	
Shredded Parmesan on the Side – PINT	0		1		
Shrimp – 23 Count – LB	4	$ 29.95	6	$ 179.70	
Cocktail Sauce – QUART			1		
Bruschetta Bar – TRAY OF 48	1.25	$ 59.95	1	$ 59.95	
Antipasti Tray – Italian Meats, Cheeses Crudités – LG	1.25	$ 54.95	1	$ 54.95	
	11			$ 607.63	**$599.95**

FIGURE 2–3 Recipe evaluation showing cost contribution of ingredients. Courtesy of My Chef, Inc. Naperville, Illinois

FIGURE 2-4 Sample standardized recipe card format

Item _____ Code_____
No. Portions _____
Portion Size _____
Garnish _____
Serving Piece_____

Ingredients	Quantities	Cost		Method
		Unit	Total	
(Ingredients are listed by groups or item, as used, separated by spaces.)				(The methods of preparation and/or procedure for each group or item are separated by spaces. This form shows the method for each group of ingredients or single ingredient and then the method for combining with the following ingredients.)

Total Cost _____
Cost per Portion _____

Prime Variable Costs: Calculating Food, Beverage, and Labor Costs

There are two categories of cost in most business situations. The first category is known as the **fixed cost (expenses)**. It is defined in the following manner:

- It is known in advance.
- It is expressed in dollars ($).
- It is not affected by sales.

Examples of fixed costs include rent or mortgage, bank loans, insurance, salaries, and equipment rental or leases.

The second category of cost is known as **variable (controllable) costs (expenses)**. It is defined in the following manner:

- It is not known in advance.
- It is expressed as a percentage of sales (%).
- It is affected by sales.

Examples of variable costs include supplies, utilities, food, beverage, and labor.

Food, beverage, and labor are considered **prime costs (expenses)** in that they collectively represent the greatest costs to an operation. Although there is no absolute rule of thumb for comparing these costs to sales, many operations aim for a combined prime cost of between 50 and 70 percent of total sales. The following example illustrates how a prime cost may be divided:

- 15 to 30 percent food cost
- 10 to 30 percent beer, wine, and liquor cost
- 15 to 30 percent labor cost

Note: Cost percentages are of total sales and represent what percentages of total sales are needed to satisfy these costs.

Each restaurant, hotel, country club, and catering company is different. For example, a steak house could have a high food cost and low payroll cost. On the other hand, a country club could have a high food cost and high payroll cost. A hotel could have a low food cost and low payroll cost. The prime cost ratio is a very important tool for monitoring a food service operation.

CALCULATING FOOD AND BEVERAGE COSTS

Food and beverage costs are often jointly referred to as the **cost of goods sold**. They can be calculated together as the cost of goods or calculated individually to reveal separate food or beverage costs. The calculations for cost of goods, food cost, or beverage cost are made in a similar manner. The following example is for calculating food cost.

$$
\begin{array}{r}
\textbf{Beginning food inventory} \\
+ \textbf{ Food purchases} \\
\hline
= \textbf{Foods available for sale} \\
- \textbf{ Ending food inventory} \\
\hline
= \textbf{Cost of food}
\end{array}
$$

Calculating Food Cost Percent

The food cost percentage represents the relationship of food costs to sales. Specifically, it represents the percentage of sales used to prepare an item for retailing. The lower the food cost percentage, the less food expense involved in making the sale. The **food cost percentage** (FC%) is calculated by dividing the cost of food consumed (FC) by the sales (S).

$$
\frac{\textbf{Total food cost (FC)}}{\textbf{Total sales (S)}} = \textbf{Food cost \% (FC\%)}
$$

To illustrate, assume a small cheese tray costs the Benham Hotel kitchen $38 in **food costs (FC)** to prepare. The cheese tray sells on their banquet menu for $125 (S). The food cost percentage (FC%) is calculated by dividing the food cost (FC) by sales (S): $38 ÷ $125 = 0.304, or 30.4%.

If the hotel could reduce its food cost by $2 to make the cheese tray, the net result to the food cost percentage would be: $36 ÷ $125 = 0.288, or 28.8%.

Thus, the food cost percentage is also lowered when the cost to make an item is reduced. However, even though the calculation of food cost percentage illustrates the important ratio between costs and sales, it does not report the volume of sales and profits earned. Rather, it is used as a control tool by management to measure performance.

CALCULATING LABOR COSTS

Labor costs are often more difficult to identify initially for individual menu items without the benefit of conducting time studies. However, after a few days of preparing these items, most chefs can streamline their work and organize their production schedules for optimum efficiency. **Labor costs** are generally calculated by identifying the wages (and sometimes related benefits) of employees for a specific period of time. Upper management generally dictates the desired labor cost percentage and assigns the responsibility to the executive chef.

Calculating Labor Cost Percentage

Similar to calculating food cost percentages, the labor cost percentage represents the relationship of labor costs to sales. Specifically, it represents the percentage of sales used to prepare an item for retailing. The lower the labor cost percentage, the less expense involved in making the sale. The **labor cost percentage (LC%)** is calculated by dividing the cost of labor expended (LC) by the sales (S).

$$\frac{\textbf{Total labor cost (LC)}}{\textbf{Total sales (S)}} = \textbf{Labor cost \% (LC\%)}$$

To illustrate the calculation of labor costs, assume the same small cheese tray costs the Benham Hotel kitchen $24 in labor costs (LC) to prepare. The cheese tray sells on their banquet menu for $125 (S). The labor cost percentage (LC%) is calculated by dividing the labor cost (LC) by sales (S): $24 ÷ $125 = 0.192, or 19.2%.

If the hotel could reduce its labor cost by $4 to make the cheese tray, the net result to the labor cost percentage would be: $20 ÷ $125 = 0.16, or 16%.

Thus, the labor cost percentage is also lowered when the time or wages to make an item are reduced.

In most hotels, banquet, and catering situations, and often in some restaurants, management is able to forecast sales for a given period. They will predict their future sales based on historical data and contracted events. To control labor costs, they will provide specific labor budgets to their managing chefs and department heads, and instruct them not to exceed the prescribed limits. The skill is in balancing fixed labor costs with variable labor costs, while meeting the production needs of the operation and payroll needs of the staff. The scenario shown in Table 2–3 reflects how the fluctuation of sales affects available money for hourly labor.

In most cases, chefs are able to average their labor costs over a month's time. Note in the scenario depicted in Table 2–3, the total monthly sales equal $228,000 with $45,600 available for the monthly labor budget. As long as the chef spends no more than $11,400 per week in labor, he or she will meet the desired labor cost budget. However, the chef's ability to manage the staff's productivity is paramount in achieving labor cost goals.

Production Management

The contemporary concept of **production management** in food service parallels the modern systems used by the manufacturing industry. In reality, the similarities between the two industries are greater than the differences. Both are involved in taking raw materials and fabricating them to the specifications and needs of their customers. Although the shelf life of food products

Table 2–3   Available Monthly Labor Cost Budget Based on Projected Food Sales

	WEEK 1	WEEK 2	WEEK 3	WEEK 4
Projected food sales	$ 54,000	$ 60,000	$ 44,000	$ 70,000
Desired labor cost percentage	X 20%	X 20%	X 20%	X 20%
Available budget for labor	$ 10,800	$ 12,000	$ 8,800	$ 14,000
Less (fixed) labor costs of salaried chefs	$ 3,500	$ 3,500	$ 3,500	$ 3,500
Available budget for (variable) hourly labor	$ 7,200	$ 8,500	$ 5,300	$ 10,500

is vastly shorter than that of most other goods, the stages of production and the overall management concerns are comparable.

A quality product can be produced by most operations, if the owners and management staff are willing to provide the environment and leadership to make it happen. Among the many responsibilities inherent in production management are the following activities:

- Accounting
- Product development
- Estimating
- Procurement
- Staffing and scheduling
- Production efficiency
- Waste management
- Quality control
- Storage
- Distribution

Among other areas of responsibility, the garde manger is chiefly responsible for recipe (product) development, cross-utilization (production efficiency), waste control (scrap rate control), and turning leftovers into profit.

RECIPE DEVELOPMENT

Recipes are the foundation of production management, because they are the blueprints for food manufacturing. Better sets of directions are more likely to yield superior results. The garde manger must use a careful strategy when creating recipes so that other issues are addressed in addition to customer appeal and satisfaction. When developing recipes, it is an obvious consideration that they must meet the menu expectations of the customers while also meeting the profitability demands of the operation. However, they should also be designed for other less apparent, but still important, needs of the business entity.

It is within the control of a skilled garde manger to design recipes and menus that use the ingredients in advantageous ways. Because the ingredients themselves are the biggest variables in any recipe, thoughtful and clever use of these ingredients can be a major benefit to the operation.

CROSS-UTILIZATION

The benefits of **cross-utilization**, wherein both raw and prepared products are used in multiple fashions, affect many areas of the kitchen operation, including procurement, inventory management, cash flow, and food production. The ability to use the same product specifications for items like sausage, chicken breast, shrimp, and cut vegetables helps reduce inventory while streamlining garde manger production.

However, one of the toughest challenges when cross-utilizing products is to still offer variety to the consumer in both taste and appearance. The skillful garde manger must always be vigilant in offering a broad menu, one that offers range and depth. One certain way of addressing this concern is to internationalize the menu with flavors and presentations that are representative of vastly different cultures and cuisines. Eastern Indian, German, Italian, Korean, and Cajun spices and prepared foods are very different from each other, yet they use many common base ingredients.

SCRAP RATE CONTROL

In manufacturing, unusable by-products generated by standard production methods (some even caused by error) are called scrap. The amount of loss is measured in ratio to the overall quantity used. The objective is to reduce the **scrap rate**, or amount of unusable material, to a minimal amount. So it is with food production; the less waste, the better the food cost. As Michael Baskette (2001) writes:

> Every buying decision should be based on the following factors:
> - Amount of trim needed to obtain proper ingredient preparation
> - Possible use of the usable trim from foods
> - Amount of expertise needed to obtain properly trimmed foods
> - Cost of buying fully or partially trimmed and pre-portioned foods

A skilled chef, in any area of production, is concerned with minimizing waste. As menus have been continuously developed over centuries, much effort has been given to using the trimmings and lesser cuts. Soups, satés, sausages, feijoadas, gumbos, and forcemeats have all benefited from the parcels off the butcher's table. This is where some of the greatest skills of the garde manger are put to the test: to create first quality products out of lesser cuts that command as much money as do the better cuts.

That is not to say that the garde manger works only with lesser cuts and trimmings. The point is that an accomplished garde manger is capable of producing superior products from almost any wholesome ingredient. The garde manger can reduce the scrap rate to a negligible amount. As we tell our students, "We don't make any money by putting good food in the garbage can!"

LEFTOVERS INTO PROFIT

As that lesson preaches, money is made only on products that are sold. And more money can be made from products that are sold twice. It is a rare opportunity to resell products, and usually only occurs when the number of banquet guests fails to meet the guaranteed count for an event. However, in all cases, *the well-being of the guests and the reputation of the operation takes precedence over any decision to use leftovers.*

The alimentary health of the guest is always the major consideration along with an absolute dedication to excellent food and service. Useful leftovers are items that are "reapplied," not reused. Reapplied foods are those that were previously available and guaranteed for the guest. They were safely stored under refrigeration while waiting to be heated or displayed. When not used for one event, there exists a limited window of opportunity when they may be safely used for another, resulting in a lowered food cost for that event.

That being said, planning for leftovers must be part of the strategy of production management. Sometimes referred to as *peripheral food*, in that it was previously designated for a different purpose, food that is left over can be well suited for use in pâtés, terrines, soups, and mousses. Without a plan, food becomes "old and tired," and often ends up being discarded after losing its luster from freezer-burn. A varied menu must allow for the broad application of the products being used, and profitable strategies must be designed to use the leftovers from production.

The Three Cs of Production Reports

Production reports (Figure 2–5) serve three primary purposes, including (1) control, (2) communication, and (3) calculation. As an instrument of *control*, the production report is used to record the activity surrounding all prepared menu items. Both the quantity of each item prepared and the amount sold, and their difference, is recorded daily. The food is accounted for on a regular basis, thus reducing the incidence of theft or waste.

Sandwiches	FRENCH	7 GRAIN	FOCC	WRAP	TOM WRAP	CROIS	WHITE	MARBLE	PITA	PANNINI	CIABATTA
TURKEY & CHEDDAR											
HAM & SWISS											
ROASTBEEF & PROV											
TURKEY CARVER											
FRSH MOZZ & TOMATO											
RST BEEF & MOREL											
GRILLED CHIX											
CHIX CAESAR											
TUNA SALAD											
CHIX ALMOND SALAD											
TURKEY CLUB SALAD											
VEGETARIAN											
TURKEY BACON CLUB											
TURKEY PESTO											
SESAME STEAK											
SESAME CHIX											
ROAST BEEF/RED ONION											
PORTABELLO MUSH											
MEATLOAF											
SPICY ITALIANA											
FAJITA CHIX											
TUSCAN CHIX											
TURKEY/HAVARTI											

SALAD

CHEF SALAD	
CAESAR SALAD	
COBB	
MANDARIN ORANGE	

CUPS

FRUIT SALAD	
ITALIAN VEG	
GREEK ISLE	
DILL POT SALAD	
BROCC FLORET	
HONEY MUSTARD	
SA. COLESLAW	
OVEN RSTED POT	
AMERICAN POT SALAD	

Slicing Sheet : _____

FIGURE 2–5 Sample production report

As a means of *communication* to the staff, the report gives them direction as to how much food to prepare. The chef does not need to remain at the staff's side throughout the day; they are trained to reference the production reports as they use standardized recipes to prepare the menu items.

As a means of *calculating* the value of food sold, the report also predicts the expected profit to be realized by the sale. The chef can roughly estimate his or her own periodic food cost percentage using the value of storeroom goods received compared to the value of finished goods sold during the same time period.

Calculating Quantities to Make

One of the most important responsibilities for any chef is to produce the appropriate quantities of food for the customers. Making too little often results in disappointed customers and their guests who may speak ill of the establishment. Conversely, making too much will most certainly negatively affect food costs and profit, and may likely lead to termination of the chef.

The ideal situation is to have either one of two scenarios: (1) sell the event on a cost-plus basis in which food is constantly supplied as needed, and a reasonable profit is added on at the end over real costs; or (2) make the customer responsible for all decisions regarding quantities of food. The problem is that there are few customers who will agree to the first scenario, and customers will still blame the establishment for not providing better guidance if they have too much or too little food in the second situation.

ESTIMATING HORS D'OEUVRES

Chefs and party planners need to use their skills in planning for and preparing the correct amount of food. Our experience as chefs, caterers, and instructors has shown us that there are three main variables that must be considered when planning for an hors d'oeuvres party. In order to closely estimate the quantity and type of appetizers that need to be prepared and served, the chef must know at least the following information: (1) the length of time for the party, (2) the time of day, and (3) the makeup of the attendees.

■ *Length of Time for the Party* Generally speaking, the longer the guests are present, the more food is required to feed them. However, over time, the attendees slow their consumption as their hunger is satisfied.

■ *Time of Day* The time of day has much to do with both the type and the quantity of food required to satisfy a group of guests. Appetizers served just before the evening meal need not be hearty nor in massive amounts, but hors d'oeuvres offered during the dinner hour will be considered dinner by the guest, and substantial quantities and portions would be appropriate. Again, the amount required to satiate the guests will quickly diminish later in the second hour.

■ *Makeup of the Attendees* The demographics of the group affects the quantity and type of food required to meet their needs. A team of college football players will consume a different menu than those attending a monthly meeting of the Gardener's Guild. Generally, men eat more than women, and guests younger than 50 years of age eat more than those older than 50.

Table 2–4 reflects estimations by the authors of the quantities of 1½- to 2-ounce (42- to 57-g) appetizers required to adequately serve a group of average customers, given variables of time of day and length of event. The chef must also consider the makeup of the group and make adjustments as needed.

CONVERTING STANDARDIZED RECIPES

As mentioned earlier, food service operations should use standardized recipes for the production of their foodstuffs. However, because they are written for specific yields, the chef would need to vary the quantities to prepare for different party sizes. For this reason, it is common practice for a

Table 2–4 Quantity of Appetizers Required Per Person

LENGTH OF EVENT	DURING MID-DAY	DURING PREDINNER	AS DINNER	DURING POST-DINNER
1 Hour	8 total appetizers	6 total appetizers	10 total appetizers	6 total appetizers
2 Hours	14 total appetizers		16 total appetizers	12 total appetizers
3 Hours	16 total appetizers		18 total appetizers	14 total appetizers

club, hotel, resort, or catering business to create hors d'oeuvre party menus that sell their appetizers by preestablished quantities (Figure 2–6).

The menus will list the various selections available while also indicating the actual number of units or volume measure (in the case of liquids) provided for the selling price. The standardized recipes are then written to produce the exact number of units or volume measure as listed on the menu. The customer is required to choose a variety of menu selections, and sometimes multiples of some favorite selections, to arrive at the total number of appetizers required to meet the expected needs of the party. The chef then follows the standard recipe, or simple multiples of the recipe, to prepare the exact amount required.

On the other hand, there are many circumstances in which menu items are sold according to the exact customer count. This is common practice with pre-plated and buffet functions that are complete meals. The garde manger department may be responsible for preparing appetizer, sor-

Elegant Hot Hors D'Oeuvres
Min Order 2 Dozen ~ Order Additional in Dozens
Complete Heating Instructions Included

	Price Per Dozen
Asparagus Roll-Ups with Asiago	17.95
Mini Chicken Dijon in Puff Pastry	21.95
Mushroom Pillows	15.95
Shrimp Puffs	14.95
Spinach Pesto Puffs	14.95
Feta and Sundried Tomato Fillo Rolls	14.95
Mini Brie Puffs with Spicy Walnuts	17.95
Crab Cakes with Cajun Remoulade	19.95
Spanikopita	13.95
Petite Quiche Lorraine	14.95
Mini Tomato and Sausage Tartlet	15.95
Steak and Cheese Tartlet	17.95
Mini Chicken Pot Pie	17.95
Italian Sausage Stuffed Mushrooms	15.95
With Spinach, Walnuts and Cheese	
Mushrooms Lorraine-Swiss Cheese, Bacon	14.95
Beef and Boursin Stuffed Mushrooms	15.95
Grilled Chicken Skewers	18.95
With Thai Peanut or Sesame Ginger	
Buffalo Bleu BBQ Skewers	19.95
With Blue Cheese Dressing	
Coconut Chicken with Plum Sauce	15.95
Bacon Wrapped Dates	19.95
With Red Pepper Coulis	
Beef Sate - Sesame Teriyaki Glaze	21.95
Bacon Wrapped Scallops	21.95
Peking Duck Rolls with Plum Sauce	19.95
Crab Rangoon	14.95
With Hot Mustard or Plum Sauce	
Vegetable Spring Rolls with Plum Sauce	13.95

Gourmet Trays
Beautifully Garnished on a Tray – Serves 36
Crackers are Not Included with Trays

Imported & Domestic Cheese Tray	52.95
Gourmet Cracker Tray - Approx 175 pc	21.95
Antipasto Tray	54.95
Fresh Vegetable Tray - *With Choice of Dip*	46.95
Fresh Fruit Tray - With Amaretto Dip	49.95
Fresh Fruit Kabobs (24) – Amaretto Dip	44.95

© My Chef, Inc. - October 2004
Hors D'Oeuvre Menu - Page 2 of 2

Elegant Cold Hors D'Oeuvres
Min Order 2 Dozen ~ Order Additional in Dozens
Beautifully Garnished on a Tray

	Price Per Dozen
Tenderloin Wraps – *Beef Tenderloin,*	28.95
Herbed Cheese Wrapped Around Scallion Onions	
Petite New Zealand Lamb Chops	129.95
with Minted Vinegar Dipping Sauce - Tray of 40	
Fresh Raspberry and Brie Tartlet	15.95
Grilled Vegetable Tartlets	14.95
Chicken Almond Tartlets	15.95
Chicken Tartlet with Apricot and Curry	15.95
Tuscan Skewers - *Tortellini, Salami, Olive*	14.95
Artichoke and Sundried Tomato Crostini	17.95
Roasted Red Pepper Crostini	17.95
Shrimp Crostini - *Garlic Remoulade*	19.95
Roast Beef Crostini	18.95
Mild Goat Cheese Crostini	17.95
with Roasted Tomato and Basil Relish	

Cold Seafood Trays & Displays

Shrimp with Zesty Cocktail Sauce	29.95 lb
23 per lb; 2 lb minimum	
Grilled Peppered Ahi Tuna Skewers	89.95
Presented Rare with Wasabi Mayo - Tray of 48	
Smoked Trout Platter - Serves 36	69.95
Smoked Trout Spread with Apple and Horseradish	
plus 1 lb of Trout Fillets and Toast Points	
Smoked Salmon Tray - Serves 40	99.95
2 ½ lb Side with Capers, Chopped Egg, Red	
Onion, Tomato, Dill Cream Cheese and Crackers	
Poached Salmon Display - Serves 100	295.95
With Capers, Chopped Egg, Red Onion, Tomato,	
Dill Cream Cheese and Crackers	
Smoked Salmon Canapé Tray	47.95
Smoked Salmon Pinwheels with Dilled Cream	
Cheese, Salmon Mousse in Pastry Shell	

Specialty Brie – Serves 36

Pinwheel Brie – *Baked with Preserves & Nuts*	40.00
Brie with Fresh Fruit	45.00
Tray For Brie – *French Bread, Crackers,*	24.00
Clusters of Grapes, Apple Slices	

My Chef Catering
630/717-1167 ~ www.mychef.com

Cold Hors D'Oeuvre Trays
Beautifully Garnished ~ Trays Serve 48

Assorted Canapé Tray	46.95
Artichoke and Sundried Tomato Canapes,	
Bruschetta Bouche, Chutney Chicken Tartlets	
Cucumber Canapés	38.95
Salami and Cheese Cornucopias	42.95
Lavosh Pinwheel Tray	46.95
Roast Beef and Blue Cheese, Italian,	
Vegetable and Cheese	
Eggplant & Prosciutto Skewers	49.95
Bruschetta Bar	62.95
Roasted Tomato and Basil, Tomato and	
Fresh Herbs, Spicy Olive Tapenade,	
Toasted Sliced Bread Brushed with Olive Oil	
Seasoned Pita Chips - **Serves 24**	29.95
With Artichoke Romano Dip or Prosciutto and	
Cheddar Spread	

Gourmet Pizza Triangles
Ready to Serve ~ 48 Pieces Per Order

Tuscany Pizza	39.95
Spinach, Goat Cheese, Sundried Tomatoes,	
Toasted Pine Nuts	
Fresh Vegetable Pizza	34.95
Herbed Cream Cheese and Fresh Vegetables	
North of the Border Pizza	36.95
Grilled Chicken, Provolone and BBQ Sauce	

Casual Hot Hors D'Oeuvres
Min Order 2 Dozen ~ Order Additional in Dozens

Meatballs	7.95 dz
Red Wine, BBQ or Brandied Mushroom	
Italian Sausage Bites - *Approx 80 Pieces*	49.95
With Red and Green Peppers, French Bread	
Spicy Chicken Quesadillas	19.95 dz
Vegetable Quesadilla Cornucopias	15.95 dz
Hearty Pork Skewers (4 oz) *Apricot Glaze*	39.95 dz
Beef Franks in Puff Pastry	11.95 dz
Chicken Strips of Fire - *Ranch Dressing*	11.98 lb
3 lb minimum - 10 pieces per lb	

How Many? Plan on serving 4 to 5 hors d'oeuvres (pieces) per guest if dinner follows; 10 hors d'oeuvres for a cocktail party. For "heavy" appetizers served in place of dinner, plan on 10 - 15 pieces per guest.

FIGURE 2-6 Sample of catering sales menu. Courtesy of My Chef, Inc. Naperville, Illinois

bet, salad, and cheese courses for a wide range of customer counts. To accomplish this feat, while still using standardized recipes, the chef must know how to make accurate adjustments.

In order to revise the standard recipe for the proper yield, the chef needs to create a working factor. A **working factor** is a multiplier used to increase or decrease the quantities of each ingredient while maintaining a consistent ratio between the ingredients, in order to create the new yield desired by the chef. It is a simple matter for most people to double, triple, or halve a recipe without making difficult calculations on paper. It can be more challenging to exactly alter a recipe's yield from 50 portions to 165 portions. The use of a working factor makes the process easy and accurate.

The first step in converting a recipe is to find the working factor. The process is as follows:

Step 1:
Divide the new yield desired by the yield of the original (old) recipe: Working factor = New yield ÷ Old yield.

Step 2:
Multiply each ingredient in the original recipe by the working factor to determine the new quantity of ingredients: Ingredient old quantity × Working factor = Ingredient new quantity.

For ease of computation when using ounces and pounds, it is best to convert all ingredient weights to ounces before multiplying by the working factor. Then convert back to whole pounds and ounces after multiplying. This step is not necessary when using metric or decimal-pounds.

The process of using the working factor remains the same whether the chef wishes to increase or decrease a recipe's yield. In either case, the working factor is always determined by dividing the new yield by the old yield.

For example, assume the chef wishes to make 30 portions of crème fraîche from a batch recipe that normally yields 40 portions. First, the working factor is determined by dividing the new yield by the old yield, or 30 ÷ 40. The fraction is reduced from 30/40 to ¾ (or 0.75) for simplicity, and then each ingredient is multiplied by the working factor of ¾ (or 0.75) to reveal the new quantities needed for the recipe.

Now, assume the same chef wishes to make 225 pounds of chorizo from a batch recipe that normally yields 75 pounds. First, the working factor is determined by dividing the new yield by the old yield, or 225 ÷ 75. The fraction is reduced from 225/75 to 3/1 (or 3) for simplicity, and then each ingredient is multiplied by the working factor of 3 to reveal the new quantities needed for the recipe.

Example: Convert the following recipe for Chicken and Crayfish Terrine that yields 2 pounds (12 portions) to the desired new yield of 5 pounds (30 portions). The working factor (WF) is 5/2 (or 2.5).

INGREDIENTS	OLD QTY	×	WF =	NEW QTY
Chicken breast, ground	16 ounces		2.5	40
Egg whites	2 each			5
Salt	2 teaspoons			5
Ground black pepper	.5 teaspoon			1.25
Shellfish essence	6 ounces			15
Heavy cream	2 ounces			5
Crayfish tails, cooked	16 ounces			40
Chipoltes in adobo sauce, minced	4 each			10
Chicken liver, cooked diced	2 ounces			5
Shitake mushrooms, medium diced	4 each			10
Cilantro, chopped	2 tablespoons			5
Dill, chopped	1 tablespoon			2.5

Menu Analysis

Menu analysis is the disciplined act of recording the sales history of all menu items sold and evaluating both the item's contribution to profit and its customer appeal. Because the menu is the primary means of soliciting revenue through customer sales, it is crucial to the operation that the menu be developed with deliberate and considered forethought. The crucial question becomes "How does the selling of this menu item help us achieve success?"

Menu engineering is the science of menu analysis and modification. It is used to create a menu that is both profitable to the operation and acceptable to the customer. It considers both financial gain and customer demand. Several variables are generally considered when analyzing a menu, including the following:

- Food cost percentage
- Contribution margin
- Goal value
- Popularity
- Selling price
- Variable expenses
- Fixed expenses

Different food service operators favor various methods according to their own management philosophy. However, the three methods used most widely are (1) food cost percentage, (2) contribution margin, and (3) goal value. With each of these methods, popularity of the menu item is also factored in the analysis.

FOOD COST PERCENTAGE

When analyzing the menu by food cost percentage, the garde manger is seeking to offer menu items that have the effect of minimizing overall food costs. However, before the food cost percentage of an item can be determined, selling price and cost figures must be identified.

$$\frac{\text{Cost of the menu item}}{\text{Menu selling price}} = \text{Food cost \% (FC\%)}$$

A review of recipe costs, pricing policies, portioning, purchasing strategies, and waste management procedures can all benefit the chef in helping lower food cost percentages.

CONTRIBUTION MARGIN

The calculation of **contribution margin** is used to determine the actual profit (before operating expenses are deducted) earned on the sale of a menu item. This is an important calculation to make for food service operations, in that it provides important information regarding the contribution to profit that each item listed on the menu makes. The garde manger seeks to produce a menu that maximizes the overall contribution margin.

The selling price (SP) minus its cost to prepare the food (FC) equals the dollar contribution to profit (CM). It is mathematically calculated as:

$$\text{Selling price (SP)} - \text{Food cost (FC)} = \text{Contribution margin (CM)}$$

To illustrate, assume the following scenario regarding three other menu items offered by the Benham Hotel:

- Whole Smoked Salmon Platter sells for $225, with a food cost of $80
  $225 (SP) − $80 (FC) = $145 (CM)
- Whole Smoked Whitefish Platter sells for $200, with a food cost of $40
  $200 (SP) − $40 (FC) = $160 (CM)
- Jumbo Shrimp Cocktail Platter sells for $200, with food cost of $70
  $200 (SP) − $70 (FC) = $ 130 (CM)

In this scenario, the Smoked Whitefish Platter has the largest contribution margin, making it more profitable to the operation than the sale of Smoked Salmon Platters. However, Smoked Salmon Platters may be more desired by the clientele. The Shrimp Cocktail Platter may have the lowest contribution margin of the three, but its popularity may be many times greater than the other platters.

All menu items are evaluated as to their contribution margin (contribution to profit) and unit sales generated. In this instance, food cost percentages are not evaluated or considered, because they do not reflect actual dollars earned after expenses. Although an important barometer for measuring prescribed performance and efficiency, the food cost percentage does not adequately reflect the volume of business that a company does.

GOAL VALUE

The **goal value** concept involves setting a target number that is neither a dollar figure nor a percentage of sales. It is determined by multiplying a series of numbers together to reveal a score. The garde manger then establishes a minimum threshold score known as the *average goal value*, above which items remain on the menu, below which items are removed or adjusted. To determine the goal value for each menu item, the garde manger must determine the following figures: food cost percentage, variable cost percentage, and selling price. Item popularity is determined by calculating the actual number of portions sold for each item.

The values for each variable are represented by the following letters:

A = 1.00 − Food cost %

B = Item popularity

C = Selling price

D = 1.00 − (Variable cost % + Food cost %)

The mathematical formula for goal value (GV) is calculated as follows: **A × B × C × D = Goal value (GV)**

To illustrate, assume the following information: An Assorted Country Pâté Platter with Loganberry Sauce costs $36 a portion and sells for $144. The food cost percentage is determined by dividing the costs by sales: (FC% = 36/144, or 25%).

Of the five hors d'oeuvre selections listed on the appetizer menu at the Benham Hotel, 60 Pâté Platters are sold of the 400 items sold during the month. During the same month, the variable costs for the garde manger department were 30%.

Using the formula listed above, the following computations can be made to determine the pâté's goal value:

$$(1.00 − 0.25) \text{ A} × 60 \text{ B} × 144 \text{ C} × 1.00 − (0.30 + 0.25) \text{ D} = \text{Goal value (GV)}$$
$$0.75 × 60 × 144 × 0.45 = \text{Goal value}$$

Thus, 2,916 equals its goal value.

A goal value analysis for all of the menu items would appear as follows:

MENU ITEM	A	B	C	D	GOAL VALUE	RANK
Shrimp Cocktail	0.65	130	200	0.35	5,915	1
Smoked Whitefish	0.80	34	200	0.50	2,720	5
Smoked Salmon	0.64	76	225	0.34	3,721	2
Country Pâté	0.75	60	144	0.45	2,916	4
Cheese Platter	0.70	100	125	0.40	3,500	3

However, a comparison by food cost percentage and contribution margin would reveal the following rankings:

MENU ITEM	SP	FC	FC%	CM	FC% RANK	CM RANK
Shrimp Cocktail	200	70	0.35	130	4	3
Smoked Whitefish	200	40	0.20	160	1	1
Smoked Salmon	225	80	0.36	145	5	2
Country Pâté	144	36	0.25	108	2	4
Cheese Platter	125	38	0.30	87	3	5

Perhaps a more revealing answer may be discovered by multiplying the contribution margin (CM) by the popularity (number of portions sold) to determine overall profitability (Total CM) to the operation. Notice this method of ranking mirrors the results determined by the goal value method:

MENU ITEM	SP	FC	CM	POPULARITY	TOTAL CM	RANK
Shrimp Cocktail	200	70	130	130	16,900	1
Smoked Whitefish	200	40	160	34	5,440	5
Smoked Salmon	225	80	145	76	11,020	2
Country Pâté	144	36	108	60	6,480	4
Cheese Platter	125	38	87	100	8,700	3

As can be observed by reviewing these methods of menu analysis, it is up to the chef and other members of management to determine by what method a menu shall be analyzed, and by what criteria shall its menu items be judged.

Pricing the Event

Although it is not primarily the responsibility of the garde manger to price out catered events, the conceptual value in seeing the total picture is always important and beneficial for any department head. In addition to the food that is sold to a client, there are many other products and services that may be contracted out (Figure 2–7).

Sometimes food is prepared by the garde manger department to be sold to a client on a pick-up basis, wherein the client literally picks up the food from the club, caterer, or hotel property. On other occasions, the food is delivered, but with no additional services requested. Most frequently, the food is part of a larger function request for services and becomes only one part of the total cost of the event.

There are four primary methods used in arriving at a price for the sale of food and centerpiece decorations produced by the garde manger department. These methods include (1) cost-based pricing, (2) multiplier-based pricing, (3) competitive-based pricing, and (4) client-based pricing.

COST-BASED PRICING

Cost-based pricing is the most detailed and time consuming, but also the most accurate way to determine the correct price for the work performed. The chef calculates all of the actual costs to produce the function and then adds a predetermined markup to arrive at the sales price. In this case, each applicable projected cost is itemized, including raw food cost, labor cost to prepare the food, transportation costs, consultation fees, peripherals such as rentals and floral arrangements, a pro rata of overhead costs and operating expenses, and the desired profit margin. All of these pro-

RAW INGREDIENTS	UNIT COST ($)	TOTAL DOLLAR COST	REMARKS
2 oz. sliced tenderloin	11.60/lb.	$1.46	Attendant to slice and roll
1 oz. fresh salmon	9.60/lb.	.60	
3 ea. 16/20 ct. green shrimp	10.50/lb.	1.98	
3 oz. pasta 1 oz. cheeses 1 oz. cream	.50/lb. 1.12/lb. 1.60/qt.	.09 – estimate .07 – estimate .05 – estimate	
2 oz. chicken breast, sliced	2.85/lb.	.36	
1 oz. caviar	9.80/lb.	.61	Present one tray at a time
2 oz. chocolate for fondue 2 ea. cubes strawberry, melon, pineapple		.20 – chocolate .08 – estimated fruit	
½ oz. mushrooms 1 oz. herbed cheese	.98/lb. .80/lb.	.03 – mushrooms .05 – cheese	
1 ea. buerreck	4.80/doz.	.40	
1 ea. rumaki	7.80/doz.	.65	
		Projected Total Dollar Cost per Guest: $6.63	

FIGURE 2–7 Sample event costing worksheet

jected costs are detailed in an outline, which can be useful later in evaluating their accuracy (a function of planning and control).

MULTIPLIER-BASED PRICING

One of the more common practices in the food industry is to use a multiplier to arrive at a selling price for an operation's food, known as **multiplier-based pricing**. The operation believes that meeting a specific food cost percentage (FC%) will result in a sufficient profit, after all costs have been satisfied. Once the food cost percentage has been identified, the multiplier can be easily determined. The method is as follows: Divide 100% by the desired food cost % to reveal the multiplier. For example: $100\% \div 25\% = 4$.

Multiply all raw costs by the multiplier 4 to arrive at the selling cost. For example: The Country Pâté Platter costs the hotel $36. Multiplying the raw cost of 36 by the established multiplier of 4 reveals the selling price of $144.

The thinking is that the difference between the raw food cost and the selling price should be, in the mind of the operator, sufficient to cover labor costs, overhead costs, and profit.

COMPETITIVE-BASED PRICING

The third method of establishing the selling price for an item is rather dangerous if used alone for all products sold. **Competitive-based pricing** determines the sales price based on what the competition charges for similar goods and services. The price is not determined by direct costs or profit margin goal setting, but rather by what the market will bear. This method does not consider the obvious differences in overhead costs from one business to the next. However, it is important to be mindful of what the competition is charging for similar services and what the clientele can afford. This method is best used in combination with another method for determining pricing.

CLIENT-BASED PRICING

Probably the hardest method of pricing to administer is known as **client-based pricing**. Essentially, the seller must determine the buyer's perception of the function's value as a foundation for establishing the price. The chef's costs may not directly influence the final price. Rather than considering the food and labor costs, the chef considers the "star quality" of the food and services provided. This system has merit when business is strong for the chef, but it may lead to ruin if the chef is unrealistic about his or her product's value. Enough people are willing to pay top dollar when they perceive the value is evident. However, after a few years of notoriety, chefs may "rest on their laurels" and their product value can decline (along with the chef's reputation). Unfortunately, many Michelin star properties have gone up and down this path, from rise to ruin, before.

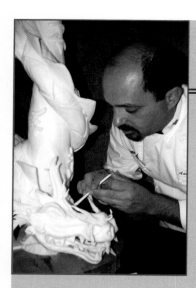

Professional Profile

BIOGRAPHICAL INFORMATION

Name: Andrew Farrugia

Place of Birth: Mellieha, Malta

Recipe Provided: Maltese Octopus and Snail Tian

CULINARY EDUCATION AND TRAINING HIGHLIGHTS

Chef Farrugia received his hands-on work experiences in several superior hotels within his native country of Malta. His apprenticeship began with one of the country's finest pastry chefs, where he developed a love and passion for the culinary arts. He attended school at the Colin Martin School of Sugar Craft in the United Kingdom and at the McCall's Cake Decorating School in Toronto, Canada. Recently, Chef Farrugia finished a 2-year course at the University of Malta, obtaining a Certificate in Tourism Studies. While working in various Maltese hotels, he held positions ranging from commis to executive pastry chef. These experiences prepared Chef Farrugia to later open his own confectionery business. More recently, he has held culinary educator positions at Glion Institute for Hotel and Tourism Studies in Egypt and at the Institute of Tourism Studies in Malta. His creativity and skills in fat carvings landed him a position with the Maltese National Culinary Team and allowed him to successfully compete in international competitions in Malta, England, and Scotland, earning many gold medals and Best of Show awards.

ADVICE TO A JOURNEYMAN CHEF

I've always compared food to fashion. Culinary trends seem to change every year. Most of the times during culinary competitions, chefs experiment with new ideas and concepts so as to catch the eye of the judges, trying to win the gold medals that every competitor dreams of. Such ideas become trendy and we start to see them in books and magazines. As a chef, I advise that we need to keep ourselves aware of these trends. We must change the way we prepare food according to the current trends. We, too, can experiment with new ideas. Although this has an element of risk, as to whether it might be liked or not, when we create something new and innovative, if successful, it will give much satisfaction and pride to the chef.

Andrew Farrugia

RECIPE 2-1

Chef Andrew Farrugia's Maltese Octopus and Snail Tian Served with Warm Dorado Fillet and Red Prickly Pear Dressing

Recipe Yield: 6 portions

MEASUREMENTS		INGREDIENTS
U.S.	**METRIC**	
1½ to 2 pounds	800 g	Octopus
10 ounces	300 g	Snails
1½ to 2 pounds	800 g	Dorado (Lampuki)
10 ounces	300 g	Potato
2 tablespoons	30 mL	Sour cream
1 each	1 each	Red prickly pear
1 each	1 each	Garlic
1 ounce	28 g	Basil
1 ounce	28 g	Mint
½ ounce	14 g	Marjoram
5 ounces	150 g	Tomato
3 tablespoons	45 mL	Balsamic vinegar
6¾ tablespoons	100 mL	Extra virgin olive oil
7½ ounce	200 g	Salad leaves
1⅔ ounce	50 g	Green olives
1¾ ounce	50 g	Capers
⅓ ounce	10 g	Pepper
⅓ ounce	10 g	Salt
2 each	2 each	Lemons

PREPARATION STEPS:

1. Cook octopus in lightly salted boiling water with 1 lemon for approximately 1 hour. When ready, drain and allow to cool.

2. Prepare the snails in the same manner, but cook in a separate pot for 30 minutes.

3. Cut potatoes into small cubes and boil. When barely cooked, drain and cool. Then add the sour cream, salt and pepper, and marjoram.

4. Prepare the Dorado by removing the head and tail, and cut into 6 portions of fillet. Season and squeeze on some lemon juice. Set aside.

5. Cut the octopus into small pieces. Add ½ chopped garlic clove, basil leaves, capers and olives, mint, 30 mL olive oil, and seasoning. Mix all ingredients together and set aside.

6. Take out the snail meat from shell, add chopped tomato, ½ crushed garlic, 20 mL oil, and seasoning. Finally, squeeze a few drops of lemon.

7. For the dressing, blend the prickly pear, 50 mL olive oil, salt, and pepper. When ready, add chopped fresh herbs and lemon juice.

REVIEW QUESTIONS

1. Describe metric measurement.

2. How is temperature converted from Fahrenheit to Celsius?

3. What is a butcher's yield test?

4. Why are standardized recipes important to the chef?

5. What are the three Cs of production reports?

6. List the methods of pricing an event.

7. Explain labor cost percentage.

8. Why is it necessary to estimate hors d'oeurves quantities for a banquet?

9. How do you calculate food cost percentage?

10. List 10 reasons why calculations are important to the modern chef.

A. **Group Discussion**

In groups discuss the advantages of being proficient in both the metric and the U.S. customary measurement systems.

B. **Research Project**

Research how much the metric system is used in the U.S. kitchens and cookbooks. From your findings, report on whether it should be taught.

C. **Group Activity**

Using a raw leg of lamb, complete a butcher's yield test to establish the weight loss due to boning, and the new cost of the raw meat per pound.

D. **Individual Activity**

Select a menu from a local establishment and calculate what you believe to be the highest and the lowest food cost items on the entrée section.

CHAPTER 3

Banquet Organization

After reading this chapter, you should be able to

- discuss the overriding responsibility of the chef.

- explain the importance of ergonomics in kitchen layout and design.

- describe the elements of work triangles in garde manger stations.

- define the concept of ergonomics and activity zones.

- understand the proper methods of dry and refrigerated storage.

- explain the importance of ergonomics in banquet room layout and design.

- create a positive guest experience.

- describe the "now" and "wow!" of form and function.

- use food zones and scatter stations in buffets.

- understand the considerations to make while arranging a buffet.

buffet concept   food scatter stations   guest experience
ergonomics   food zones

*The table, ornamented with incrustations, is round, covered with a fine table-
cloth. The architricline announces that dinner is served. The guests wash their
hands, which they will do again after the first course. They put on special
robes and sandals, unfold their napkins if they have brought them, for the host
does not provide these. Then they take their places around the table.*

EMILE DE LA BÉDOLIÈRE

The Chef's Responsibility

Inherent in the position of garde manger is the moral, ethical, and professional duty of providing
nutritious food in a safe and clean environment. No other responsibility of the chef is greater
than the safety and welfare of employees and customers. The garde manger chef must, at all
times, demand that the staff abide by a strict code of safety and sanitation rules in the preparation, holding, and presentation of food products.

To achieve this objective, all staff should undergo proper and adequate instruction by quali-
fied trainers. These instructors may include selected members of staff, local health department
officials and inspectors, or sanitation educators from a local college or university. There are many
audiovisual aids to assist in this training, including pamphlets, videos, and self-paced programs
available on the Internet.

The local or state department of environmental health often requires the garde manger chef,
as well as other culinary department heads, to complete a specified course in sanitation that covers state requirements for commercial food service operations. Occasionally, the national, state, or
county codes are revised, and all food service operations are required to be in compliance. Therefore, it is vital to the chef and the food service operation that periodic sanitation training sessions
are provided to the staff, because all food service operators are responsible for knowing the health
department regulations for their city and state.

Kitchen Layout and Design

Similar to a domestic kitchen, but on a much larger and more sophisticated level, a commercial kitchen is comprised of various work stations that must flow in concert with the other stations within the operation. The commercial kitchen features both common areas that are used
jointly by several stations, and specialty areas unique to one station of the operation. The
garde manger station is generally associated with the pantry and cold kitchen areas of a food
service operation.

Kitchen Ergonomics

Ergonomics is the science of adapting working conditions to the needs of the worker. The concept suggests that worker productivity and satisfaction is increased when functioning in optimum conditions. Each station of a commercial kitchen or bakery should be designed to support
the principal activities that occur within that work area.

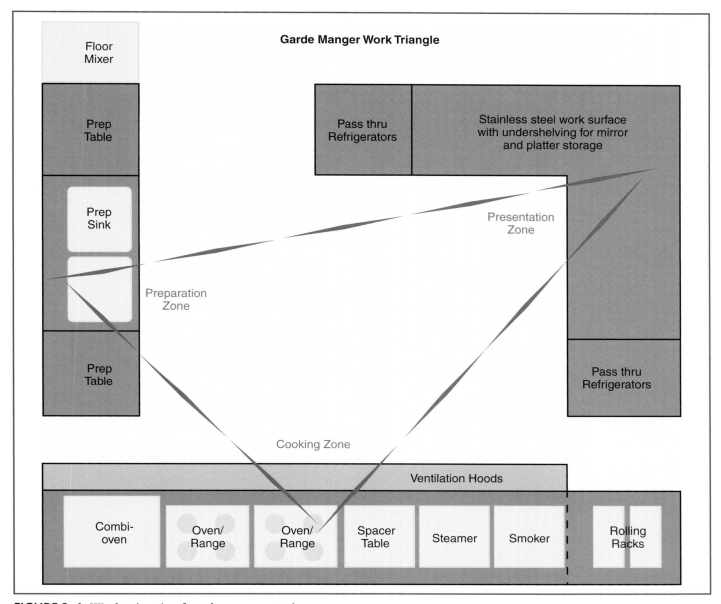

FIGURE 3–1 Work triangle of garde manger station

The principal activity zones in the garde manger station are generally recognized as being food preparation, cooking, and platter presentation. Secondary activities include ware washing and refrigerated storage. The secondary activities are often performed in common areas of the kitchen. Each activity zone needs to include the work surfaces, large equipment, and appliances required for that activity. Additionally, each zone requires the storage space for the utensils, small wares, tabletop platters and mirrors, and ingredients required for the purpose.

In support of the three principal activity zones, it is important to arrange the appliances and worktops in the correct order or sequence. A work triangle is formed between the three zones, as shown in Figure 3–1. For the greatest efficiency, the work triangle should be uninterrupted by through traffic.

PREPARATION ZONE

The garde manger preparation area is where foods are cleaned, trimmed, blended, and possibly ground for further use. Large equipment common to this area includes the following:

- Two-bay vegetable prep sink with drain board and garbage disposal
- Metal-top prep table with utensil drawer
- Meat grinder and sausage horn
- Floor or table-top mixer (usually with optional grinder and shredder attachments)
- Side-by-side reach-in coolers
- Vacuum processing machine

COOKING ZONE

The garde manger cooking zone is where foods are steamed, boiled, sautéed, smoked, baked, or roasted. Large equipment common to this area includes the following:

- Range-top oven
- Combi-oven
- Vegetable steamer
- Smoker
- Microwave oven
- Ice cream freezer
- Pot rack

PRESENTATION ZONE

The garde manger presentation zone is where foods are plated or placed on service platters, mirrors, or trays. Equipment common to this area includes the following:

- Plate racks
- Rolling racks
- Metal-top table with undershelving for mirror and platter storage
- Glass door reach-in coolers, roll-in coolers, or walk-in cooler for distribution
- Conveyor system (for large hotel or banquet operations)

Storage

Since the beginning of structured food service operations, more than several hundred years ago, one of the fundamental responsibilities of the chef garde manger has been the proper care and storage of foodstuffs. As discussed in Chapter 1, the position of garde manger is charged with the broad task of storing many dry and refrigerated products. Storage does not improve the quality of any food. The quality of a food will also not decrease significantly during storage, as long as the food is stored properly and used within the recommended time frame and shelf life (Figure 3–2).

Quality is not the same as safety. A poor-quality food may be safe, such as overripe fruit or mealy sausage. An unsafe food may have good quality in terms of appearance and taste, but have a highly dangerous bacterial count. A product that has been poisoned by cross-contamination does not generally appear tainted to the naked eye. Maintaining a food's quality depends on several factors:

- Quality of the raw product
- Procedures used during processing
- Procedures and method of storage
- Length of storage
- Exposure to light and heat

FIGURE 3–2 Formal storeroom operation

DRY STORAGE

Dry storage is generally less complicated than refrigerated storage. However, there are proper procedures that should be followed to protect the integrity of the food products. The garde manger chef should consider the following procedures:

- Always rotate stock to prevent spoilage. Generally, the policy of "first in/first out" applies.
- Mark "received dates" on containers, and use products before their expiration date.
- Store products in cool, dry areas with adequate ventilation and lighting.
- Keep products off the floors with proper use of washable shelving and storage racks.
- Watch for signs of rodents or bug infestations.
- Keep an inventory of dry storage contents.

REFRIGERATED STORAGE

Refrigerated storage includes both freezer and refrigerator; any storage maintained below 45°F (7.2°C). The following procedures are recommended:

Freezer

- Keep freezer temperature at or below 0°F. A good indication of proper temperature is that ice cream will be frozen solid.
- Use moisture-proof freezer-weight wrap. Examples are foil, freezer bags, and freezer paper. Label and date all packages.
- Food stored beyond the recommended time will be safe to eat, but eating quality (flavor and texture) and nutritive value will be less.
- Keep an inventory of freezer contents.

Refrigerator

- Use a thermometer to check that temperatures remain between 34° and 40°F (1° and 4°C) at all times. Avoid opening the refrigerator door frequently, especially in hot kitchens. The optimum temperatures vary for different foods. When possible, use separate refrigeration to hold the different types of products:

- Meat and poultry: 32° to 36°F (0° to 2°C)
- Fish and shellfish: 30° to 34°F (−1° to 1°C)
- Dairy products: 36° to 40°F (2° to 4°C)
- Eggs: 38° to 40°F (3° to 4°C)
- Produce: 40° to 45°F (4° to 7°C)

- Raw meat, fish, poultry, shucked shellfish, and shrimp should be wrapped securely so they do not leak and contaminate other foods. Place the packages in separate leak-proof washable containers. Clean up leaks with warm soapy water and sanitize with a solution of 1 teaspoon of chlorine bleach to 1 quart of water.
- Cooked meats, platter presentations, and leftovers should be tightly wrapped to prevent drying and loss of flavor.
- Avoid cross-contamination of other foods in the refrigerator.
- Avoid dehydration of products, such as produce and meats, by covering them.
- Avoid transference of odors and flavors. If possible, store dairy products, produce, and meats and seafood in separate coolers.

Banquet Room Layout and Design

In addition to considering the flow of work being performed "behind the scenes" in the kitchen, it is important that the function rooms be designed to meet the needs of the event. When planning the layout and design of a banquet facility, the needs of both the guest and the kitchen and service staffs should be carefully considered.

Banquet Room Ergonomics

Banquet rooms are often the space where the work of the garde manger is introduced to the customer. A successful event begins with a well-conceived space. Many considerations should be made when designing a space for banquet use, regardless of the food. Facility architects and designers generally try to determine the following:

- Access to banquet kitchen and service staff stations
- Relational activities, such as break-out rooms or ice-breaker receptions
- Support needs, such as table, chair, audiovisual equipment, and prop storage
- Expansion needs; the ability to adapt to varying customer counts
- Light and air; the ability to provide comfort
- Audiovisual requirements; the ability to support communication
- Need and placement for access and egress

    Usually, professional banquet managers use a site inspection list to review the condition, appointments, room configuration, and support spaces associated with a banquet room (Figure 3–3).

CUSTOMER AND STAFF SAFETY

When organizing a banquet, among the many important considerations is that of customer and staff safety. The garde manger chef and banquet planner must create an environment that is conducive to service, but unimpeded for safety. Often, operators are tempted to block exits or fill egress hallways with carts, chairs, and banquet tables. Such practices are ill advised and contrary to building safety codes.

```
┌─────────────────────────────────────────────────────────────────┐
│                    SITE INSPECTION CHECKLIST                      │
│                                                                   │
│                                                                   │
│  Today's date _____ By _____ Client _____   │
│  Location _____ Event time _____   │
│  Address _____ Function time ___ am/pm│
│  City _____ Fee/client?_____ Fee/caterer? _____  │
│  Room(s) _____ Cocktails? _____ Dinner _____ Other ____ │
│  Contact _____ Service entrance? ____ │
│  Day phone _____ Elevator? _____  │
│  Emergency # _____ Loading dock? _____  │
│  Access/deliveries _____ am/pm Where? _____ By? _____   │
│  Access/kitchen _____ am/pm Function rooms _____ am/pm  │
│  Burners? _____ Fryers? _____ Floor plan? _____  │
│  Ovens?_____ Broiler? _____ Bar? _____ License? _____  │
│  Gas/power? _____ Fuse box? _____ Tables? _____ Chairs? ___  │
│  Water/sinks? _____ D'wshr? _____ Serving equip.? _____   │
│  Reefers? _____ Walk-in? _____ Other _____    │
│  Urns? _____ Racks? _____ Dance floor? ____ Stage? ___  │
│  Icemaker? _____ Micro? _____ Podium? _____ Lectern? __ │
│  Work space? _____ Tables? _____ A/V? _____ Lighting? ___  │
│  Other? _____ Guest parking? _____ Valet? __ │
│  Cleanup? _____ Directions?_____ Map? _____  │
│  Trash?_____ Bags? _____ Checkrooms? _____ Attendant? _ │
│  Decor/floral/special setup? _____ Restroom? _____ Attendant? _│
│  Music/band/other? _____ In-house staff? _____    │
│  Lockers/restroom for staff? _____  │
│  Rental pick-up arrangements _____   │
│  Notes _____   │
│  _____   │
│  _____   │
│  Follow-up meeting notes to contact _____ Copy attached ____  │
│  Call to reconfirm by _____ (date, 1 wk prior to event)│
└─────────────────────────────────────────────────────────────────┘
```

FIGURE 3–3 Site inspection checklist

Potentially dangerous items such as fuel for chafing dishes or ice sculpture displays are familiar items on buffets. Common sense and professional skill must be employed to ensure a safe event for all.

Buffet Planning and Presentation

According to Margaret Visser (1991):

> At formal Medieval, Renaissance, and Baroque dinners, an edifice of shelves known as a "buffet" was erected to one side of the dining hall; upon it the family silver—which was often far too valuable to be subjected to the hazards of use—was proudly displayed. Later the food was displayed there as well, so that guests could have a preview of what they would be eating, rather as modern restaurants often exhibit dishes of food to tempt their customers. Later still, yet another little room led from the dining room, where guests could visit the buffet. These shelves for display, like the tables, had often been boards set up (*dressées* in French) for special occasions; they are the origin of our "dresser" and "cup boards". . . . Beginning apparently in the nineteenth century, a buffet meal used to be laid out, not on the dining-room table but on the dresser or sideboard. (p. 149) (Figure 3–4)

In modern times, garde manger chefs employ a myriad of platters, mirrors, risers, and table shapes to present their food displays. Whether on a cruise ship, in a five-star hotel, or served in a banquet hall, buffets are intended to intrigue and satisfy their guests (Figure 3–5).

FIGURE 3–4 Antique buffet

FIGURE 3–5 Formal
banquet event. Courtesy of
Getty Images, Inc.

THE GUEST EXPERIENCE

All service organizations, whether they are theme parks, resorts, or hotels, strive to provide the best value and service to their guests. The **guest experience** includes the guest's impressions on the sanitation, safety, facility conditions, staff training, attention to detail, menu quality, service, and all of the other tangible and intangibles associated with the visit. Operators understand that the guest experience means everything to their viability as a company. Satisfied guests translate into repeat business and positive word-of-mouth advertisement. Without positive guest experiences, businesses cannot thrive.

In food service, the garde manger chef realizes that personal skill and artistry can have a direct impact on the guest experience. Well-planned events, skillfully prepared and presented, can make a lasting impression on the dining customer.

FIRST IMPRESSIONS: ENGAGING THE SENSES

As the adage goes, "You only have one chance to make a first impression." The more sensations that can be immediately transmitted to the dining patron upon arrival, the stronger the first impression that is made. Knowledgeable chefs understand that the skillful use of aroma, temperature, and texture can enhance the dining experience far beyond the limiting use of sight and taste.

Experiences that engage multiple senses are richer and more long lasting. Consider the first visit a person has to the seashore and the variety of sensations that are stimulated while standing on the beach watching the waves crash along the shoreline. The smell of the sea and fish washed ashore, the sounds of screeching seagulls, the rhythmic pounding of surf against the beach, the stinging feeling of sand blown against the skin, and the taste of sea salt in the air. The layering of senses causes strong impressions and lasting memories.

A buffet should be planned to employ as many senses as possible. Adding music or the sound of running water, heating aromatics, creating beams of light and shadows, building height and depth in the visual presentations—all of these add to the holistic experience for the guest.

THEMES AND SPECIAL EVENTS

The **buffet concept** is the predominant theme surrounding the event. The theme is communicated repeatedly to the guests by use of music, props, menu selections, and style of service. Each theme brings a unique opportunity for the garde manger chef to express creativity. Creativity in a buffet can be expressed through a variety of methods including the following:

- Buffet concept, such as seasonal, theme, occasion, ethnic, or personal
- Menu selections, arrangements, and zones
- Food aromas, shapes, colors, sizes, and textures
- Front of the house activity, such as tableside cookery or carving stations
- Showmanship with ice, tallow, and vegetable sculptures
- Flourishes and touches, including decorative linen, pottery, statues, paintings, candles

Guests of all ages and socioeconomic backgrounds enjoy theme parties. Structuring entertaining around specific themes or buffet concepts is always popular. The following themes are fashionable and allow creative venues for the garde manger chef:

- Victorian tea parties
- Beach parties and clambakes
- Yacht parties
- Historic mansions
- Nutcracker Christmas
- Safari adventures

■ Masked balls
■ Food art museums, featuring ice, vegetable, tallow, and sugar sculptures
■ Space odysseys
■ Mystery treasure hunts

THE TOTAL EXPERIENCE

The objective when creating a buffet event is to make it exciting and fun for the guest, whether the occasion is a casual affair or an elegant evening. As indicated before, the event must be a total experience, where the senses of the guest are stimulated by a variety of media and they are drawn into the event as a participant, rather than as an observer. The following are methods for creating participative dining:

■ Offering live cooking stations where guests select the ingredients for their custom-made menu item
■ Providing cooking classes or demonstrations on ice or vegetable carvings during the event
■ Holding the event in conjunction with an art show, fashion show, or auction
■ Allowing guests to make their own ice cream sundaes at a "sundae bar"

Building Food Displays: Between Form and Function

Building exciting and functional displays for the food is principal to success in buffet work. Again, the artistic skills of the garde manger are used to communicate and support the concept of the buffet. This is an opportunity for the chef to incorporate new foods and ideas, and to expand more internationally with the menu offerings.

However, it is important during the planning and execution of these buffets to keep functionality in mind. The ability to adequately meet the needs of the customer, by providing properly prepared and presented food in a timely manner, is vital to the event's success. Sometimes the desire to be theatrical with events overshadows the service function. Event planners must work in concert with garde manger chefs and service staff to ensure the proper marriage of function and form.

THE "WOW!" FACTOR

The best compliment that an event planner can receive is a jaw-dropping "Wow!" when guests arrive at a banquet. As Judy Lieberman (1991) writes, "Themes are only limited by the imagination, and the execution of the most outstanding events is restricted only by budget, space, and capabilities of the caterer and staff." The goal of the event planner and chef is to provide a "gift to the senses" when guests first arrive. People generally look forward to attending events, and food service operators need to build on the guest's positive attitude and deliver a quality experience right from the beginning.

THE "NOW" FACTOR

The best thing a garde manger chef can receive is an excited customer. When people are excited about an event, they are receptive to what the chef has prepared for them. According to Ronald Yudd (1990), "Showmanship is what separates à la carte dining from the buffet style of service. The drama, action, colors, props and lighting all enhance the guests' dining experience. The chef's canvas is no longer a single plate, tureen, or side dish but becomes the entire dining area."

Building on the guests' outlook, the chef must deliver a quality product "now," immediately upon their arrival. With heightened senses and anticipation, the guests are searching for a worthy culinary experience to meet their expectations. The garde manger can either satisfy their craving or disappoint them. It is up to the professional chef to meet, and exceed, the customer's expectations.

Traditional Service Patterns

Studying the dining customs of historic eras, ethnic groups, and countries reveals much about the evolution of buffet-style eating. Ancient and medieval civilizations frequently banqueted. And although the French invented the buffet, several cultures have traditions for serving large numbers of people from a common table laden with festive food. In The Netherlands, it's called *rijstaffel*; in Sweden, it's *smörgasbord*; Spain has *tapas*; Russia has *zakuski*; Denmark offers *smorrebrod*.

Modern buffets of Western civilizations tend to arrange their food in traditional dining patterns and for ease of service to the guests. Food is generally presented in zones. Often, the lesser (cheaper) side dishes are offered at the beginning, with more expensive menu items displayed later in the buffet.

COMMUNICATING THE DINING PROCESS

It is important to consider how the meal is to be served and ordered when arranging banquets. Occasionally guests are confused when they attend events that are not designed to clearly communicate the dining process. Often times a member of the wait staff announces the procedure, or in the case of assigned seating, directs specific tables to attend to the buffet in an orderly fashion. The pattern of food zones and scatter stations has a significant impact on the ability of the guest to dine comfortably and on the overall success of the dining process.

FOOD ZONES

Food zones are separate groupings of similar food items. They are generally arranged as mini-sections on a much larger buffet, but can be stations unto themselves. Zones are designed to provide choices to customers from which they may sample some or all of the products. Typical zones include the following:

- Fruit displays
- Cut vegetables and salads
- Relishes and dips
- Cheese displays
- Charcuterie displays
- Smoked meats and seafood
- Hot appetizers
- Cold appetizers
- Petit fours

FOOD SCATTER STATIONS

Food scatter stations are similar to food zones, but they are arranged as a series of separate tables, or islands, within a banquet. Each scatter station can consist of a singular food station, or it can be made up of several stations. The advantage of using a scatter system is the elimination of the long lines that can form along singular buffet tables. Scatter stations allow guests to go directly to the food zones of their interest, and bypass those foods for which they have no interest. Typical stations include the following:

- Pasta bars
- Carving stations
- Hot foods
- Cold foods
- Beverage stations
- Dessert stations
- Wedding cake station
- Bread stations

FIGURE 3–6 Examples of table and service line arrangements

SERVICE LINES AND TRAFFIC FLOW

The size and shape of the service line and food tables can add to the mood of the meal. The use of round, rectangular, square, oval, and serpentine shapes affects the appearance of the buffet and changes the flow of customer traffic (Figure 3–6). It is important to design service lines to adequately accommodate the number of guests in a reasonable period of time.

Chefs also need to design their platters, chafers, and service bowls to provide sufficient portions that require only periodic refreshment; it is often difficult to maneuver around busy buffet lines to replace tired display mirrors and empty hotel pans. A series of backup platters and hotel pans should be readily available for quick replenishment.

ARRANGING THE BUFFET

How the buffet tables are set and how the tables are placed in the banquet facility either enhances or detracts from the guest experience. Table settings and configurations must satisfy two operational concerns: the ease of replenishment by the buffet attendant and the ease of self-service by the guest (Figures 3-7, 3-8, 3-9, and 3-10). After the flow of product and the flow of guests have been decided, then the buffet may be arranged.

Space and time are factors that must be considered when arranging the buffet. The size of the banquet area, the type and style of the function, the allocation of time, and the size and nature of the menu items to be displayed must all be considered. Additionally, the need for fast or leisurely service must be addressed.

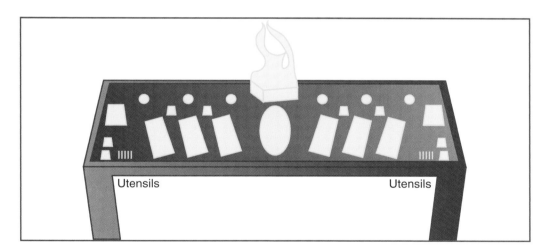

FIGURE 3–7 Single-sided buffet (1 zone)

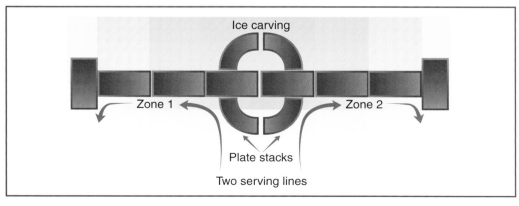

FIGURE 3–8 Single-sided buffet (2 zone)

FIGURE 3–9 Four-sided buffet

When determining the buffet table shape and traffic flow, the chef and event planner must consider the visual impact of the arrangement. They must estimate the space needs for each of the food zones, and how to best use color, height, shape, light, and texture to arrange the display. They should also appraise the following factors:

- *Traffic Flow* The food should be arranged in an order that is logical and progressive. Essentially, the guest should be led through the buffet and given a chance to assemble a meal in the traditional sequence, such as soup through dessert.
- *Space Allocation* In determining how much tabletop is required to build a buffet, the chef should estimate two linear feet of table length for every vessel of food and centerpiece. Given that food containers may vary from small bowls to large mirrors, this number may need to be revised depending on the actual dimensions of the serving dishes. Sufficient room should be

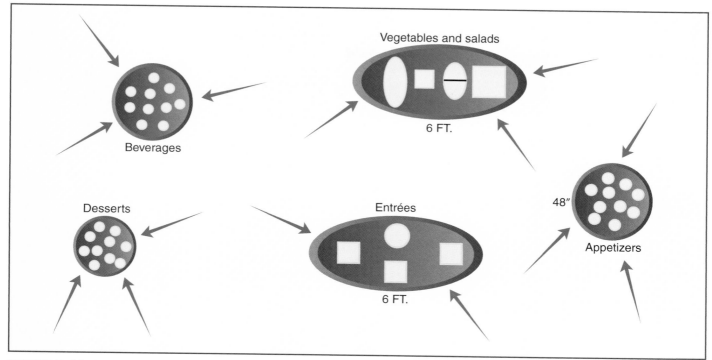

FIGURE 3–10 Scatter zone buffet

provided between each chafing dish, bowl, plate, and platter to prevent a crowded appearance to the overall buffet display.

- *Accessibility* Guests need to be to reach all, or most, of the items on a buffet with ease. Some containers may be laid directly on the table covering, whereas others may be displayed from stands and risers. It is often attractive to tilt display mirrors and trays toward the customer, giving them a better view and access to the food. The skillful use of risers creates interest and drama in a display, while improving the accessibility to items displayed toward the rear of the table.

- *Food Arrangements* The garde manger determines how food is arranged on serving platters and in their serving vessels. Most hot foods will be offered in chafing dishes, or prepared to order at live-cooking scatter stations. Cold foods will be arranged on large display mirrors, silver platters, wooden display boards, marble slabs, or in ceramic dishes. The proper use of height, color, texture, and pattern layout should be addressed. (See Chapter 17 for techniques in platter presentation.)

- *Accompaniments* It is important to place underliners for the serving utensils in front of the food containers with which they will be used. All accompaniments, such as sauces, relishes, and croutons should be placed in close proximity to their dishes.

- *Centerpieces* Centerpieces add great interest and drama to a buffet, and can be memorable for the guests. Functioning centerpieces, such as clamshells of ice filled with chilled shellfish, need to be carefully placed within reach of the customers. Others, such as large floral displays from carved vegetables or ice sculptures of swans, should be displayed with great concern for the safety of the guest. (See Chapter 16 on ice sculpting.) They are often placed out of reach of the guest.

- *Other Decorations* Often, other nonedible centerpieces, props, and decorations bolster a theme display. These items may range from balloon bouquets, to framed artwork on easels, to fish in tanks. Whatever is used must be well cleaned and kept a safe distance from food that is to be eaten. Cross-contamination can occur right under the eyes of the consumer.

■ *Signage* Informational signage is an often overlooked part of a buffet. Done well, in the spirit and style of the event's concept, signage can help guests decide where to go and what to eat. It is particularly helpful for those with food allergies or when unusual menu selections are part of the buffet.

Professional Profile

BIOGRAPHICAL INFORMATION

Name: Jaume Brichs
Place of Birth: Barcelona, Spain

Recipe Provided: Artichoke and Marinated Anchovy Salad

CULINARY EDUCATION AND TRAINING HIGHLIGHTS

Chef Brichs began his training and culinary studies at Sagrado Corazon School in Barcelona, Spain. He then received his 3-year Diploma of Technic in Touristic Companies and Activities at Girona University, St. Pol de Mar, Spain. His education and training led to garde manger positions in several leading Spanish restaurants and hotels, including Restaurant Café de Colombia and Arts Hotel Ritz Carlton. In addition to working as a garde manger, Chef Brichs spent several years as a chef lecturer at Ben Activa and EMFO, Mollet del Vallés. His entrepreneurial spirit and tireless energy enables him to command several diverse positions. He presently owns the Cuiners a Domicili Catering Service, works as a chef at the Restaurant AGUA, consults for the Spanish-based Rafaels Hotel chain, and is a presenter/chef instructor with Culinary Adventures in Barcelona.

ADVICE TO A JOURNEYMAN CHEF

When traveling around the world, it is very important to be mobile. Take the bare essentials with you and be prepared to move on from one place to another. It is so exciting to meet other chefs with new ideas and to be able to share your own experiences, returning home from an enriching journey. It is also important to make a daily chronicle of each trip. A small, handy-sized booklet is useful to write down recipes, meals eaten, ideas, plus anything interesting that occurs each day. And of course, it is useful for collecting the names and addresses of people one meets. A good sense of humor, an open mind, a desire for adventure, plus compassion and patience make an excellent journeyman.

Jaume Brichs

RECIPE 3–1

Chef Jaume Brichs' Artichoke and Marinated Anchovy Salad with Lemon and Vanilla Jelly

Recipe Yield: 8 portions

MEASUREMENTS		INGREDIENTS
U.S.	**METRIC**	
8 each	8 each	Artichokes
8 each	8 each	Fresh anchovies
4 each	4 each	Lemons
1⅓ ounces	40 g	Sugar
1 cup	129 g	Plum tomatoes, ripe
1½ cups	355 mL	Olive oil
½ each	½ each	Vanilla bean, opened
2 cups	472 mL	Bean sprouts
2 cups	472 mL	Red chard, raw
16 each	16 each	Chive sticks
1½ cups	355 mL	White wine vinegar
As needed	As needed	Coarse salt
To taste	To taste	Black pepper (optional)

PREPARATION STEPS:

1. Clean the artichokes. Cut out the choke and outer leaves, and rub the hearts with lemon juice.

2. Either (1) place the artichokes in a plastic vacuum bag with oil, salt, and pepper, then vacuum seal and steam for 25 minutes. Refresh the artichoke in ice water, while remaining in the unopened bag, and set aside. Or (2) cover the artichokes with olive oil, salt, and black pepper and cook gently over low flame until they are soft. Cool in the same oil.

3. Clean the anchovies by taking off their heads and intestines. Separate the two fillets, leaving no bones. Place the fillets in cold salted water for 10 minutes; this will stiffen the fillets and cleanse them.

4. Remove the anchovies from the water and soak them in the vinegar for 5 to 6 hours, depending on the strength of the vinegar and desired taste.

5. Remove the anchovies from the vinegar and place them in olive oil. Set aside.

6. Cut the lemons in half and grill them under the salamander until the skin is browned.

7. In a clean saucepan, add the juice of the lemons. Then add the sugar and heat with the open vanilla bean and its seeds. Cool 30 minutes. Add the olive oil. Leave to macerate 1 day.

8. Blanch the tomatoes and refresh them. Peel and remove seeds, cut into small dice.

9. Line a ring mold with four anchovy fillets seasoned with coarse salt.

10. Place the artichokes and tomato inside.

11. Top with the bean sprouts and red chard leaves.

12. Drizzle with lemon and vanilla dressing.

13. Garnish with long chive sticks.

1. Explain the term *ergonomics*.

2. Why is a site inspection checklist important?

3. How does one engage the senses of the guests at a catering event?

4. How does one ensure that the guests are aware of the theme being portrayed?

5. What is the "Wow!" factor?

6. Explain a food zone.

7. Why is traffic flow important when planning a buffet?

8. Describe the term *space allocation*.

9. What is the role of a centerpiece on a buffet?

10. What are food scatter zones?

A. **Group Discussion**

Discuss the importance of the guest experience when planning an event.

B. **Research Project**

Research an ethnic cuisine and create a menu designing a display to build your food around.

C. **Group Activity**

Plan and design an event around a chosen theme, creating a dramatic guest experience. Give the event a title and plan each element of the guests evening, describing in detail what they would see, eat, and drink.

D. **Individual Activity**

Prepare a diagram of a scatter-zoned buffet to scale for an event catering to 500 people. Illustrate how the guest can easily access the food zones and where they will eat in relation to the food zones.

PART II

Preparation Skills of the Garde Manger

CHAPTER 4

Amuse-Bouches, Appetizers, and Hors d'Oeuvres

After reading this chapter, you should be able to

- describe the distinction between amuse-bouches, appetizers, and hors d'oeuvres.

- provide an overview of various international appetizers and their service customs.

- understand the nature and distinction between hot and cold hors d'oeuvres.

- explain the difference between true caviar and fish roe.

- demonstrate the proper method for caviar service.

- describe the distinction between sushi and sashimi, and the several varieties commonly served.

- discuss a world view of olives and the various methods by which they are cured.

- cite common nuts used in food service.

- explain the difference between dips and spreads.

- describe how canapés are assembled.

amuse-bouches	dim sum	shari
antipasti	dips	spreads
antojitos	do nhau	sulfuring
appetizers	hors d'oeuvres	sushi
canapés	kanto	sushi-zu
caviar	mazza/meze	tapas
chat	mezze	zakuskis
compound butter	rijsttafel	
curing	sashimi	

Hospitality: a little fire, a little food, and an immense quiet.

RALPH WALDO EMERSON

A Taste by a Different Name

Culinary terminology is a vernacular unto its own. Certain words have become part of the broad lexicon of gastronomy, used to describe some element relative to the culinary arts and the practice of dining. Every now and then, words are used interchangeably, making their connotations misleading. Such is the case with amuse-bouches, appetizers and hors d'oeuvres. Each term is associated with small morsels of food, so the public has mistakenly assumed that they are the same. There are slight differences to the terms, and they should be used in the following situations:

■ **Amuse-bouches** (also called *amuse-gueules*). These are the little savory nibbles, to be eaten with one or two bites, offered just for fun as "mouth amusements" before the starter course is served. This practice is generally observed in only the more expensive dining establishments. Chef Thomas Keller's Cornets of Salmon Tartare with Sweet Red Onion Crème Fraîche offered to guests upon their arrival at the celebrated French Laundry in Yountville, California, are meant to lift the diner's spirits. Stan Frankenthaler, executive chef and co-owner of Salamander, an Asian-inspired restaurant in Boston, offers customers amuse-bouches such as Glazed Soy Beans and Chestnuts with Slices of Charred Pasture-Raised Veal as "a welcome, a little tiny starter, a gift of appreciation for their coming to dine with us." Masa's in San Francisco also welcomes customers with amuse-bouches and often presents them on simple white plates, in egg cups, or on silver spoons, depending on the dish.

■ **Appetizers.** Appetizers are the smaller portions of food that appear as the first course served at the table. They are part of the planned meal compared to the hors d'oeuvres, which are never served at the table. "The Romans served many different appetizers to begin their banquets. The most popular items were seasoned eggs and egg-based dishes, vegetables, salad, mushrooms and truffles, assorted shellfish, cheese with herbs, olives, sausages, and even more filling dishes, such as complicated fricassées and casseroles, which today would be considered complete meals in themselves," according to *A Taste of Ancient Rome,* by Ilaria Gozzini Giacosa (p. 49).

■ **Hors d'oeuvres.** The name *hors d'oeuvres* comes from the French and is literally translated as "outside the works." Quoting William and Mary Morris' *Dictionary of Word and Phrase*

Origins, originally it was an architectural term referring to an outbuilding not incorporated into the architect's main design. France's culinary experts borrowed the phrase to distinguish between delicate morsels of food being served apart from the meal versus those being served with the meal. Thus, hors d'oeuvres are quite literally outside the main design of the meal. "Hors d'oeuvres hours" may precede a large meal, or they may be used as a substitution for a meal. They are commonly used for social gatherings to allow guests to mingle while dining.

An International Beginning

All countries and their cuisines share a few culinary commonalities. From dried fruits to chocolate truffles, each has its own variety of sweets. Virtually all offer some form of carbohydrates with the meal, whether noodles, rice, dumplings, bread, or potatoes. And all have common courses that can range from the beginning to the end of the meal.

No matter where one travels, there exists locally some form of starter course or small sampling of food to whet the appetite. And in many countries, the ritual of offering small foods is commonly paired with certain beverages. These culinary customs are part of the social fabric of each country, and they are important for the maintenance of relationships between family and friends. Although each country has its own unique foods, the similarities of ingredients, customs, and cooking methods is noticeable between many cultures.

According to Eve Zibart, "Many of the great cuisines of the world—Chinese, Japanese, Middle Eastern, Spanish, French, and Italian, just for starters—have long recognized that dawdling over small servings of many different dishes, sharing tidbits and discoveries, not only stretches out a pleasant social evening but bonds friends together in a very emotional way. In fact, the very word *companion* comes from the Latin *com panis,* or 'with bread,' meaning the person you share meals with—friendship defined by dining" (p. 26).

The following are examples of various countries' approaches to offering small servings of food at the beginning of a meal, or as the meal itself:

- **Antipasti** (Italy). According to Anne Wilson, "Salads, cooked vegetables, fungi, and some light egg or fish dishes supplied the 'gustus' or hors d'oeuvre at a Roman meal" (p. 326). Today, no one agrees about what should be on an *antipasto* plate. Like everything in Italy, it all depends on where you come from and what the family host or restaurateur wants to do. Often, on one large platter, there will be three varieties of cured meats (a couple of salamis and a mortadella), hard boiled eggs with mayonnaise-based fillings, buffalo mozzarella, prosciutto crudo with melon or figs, pepperoni, olives, celery, and beautifully decorated canapés.
- **Antojitos** (Mexico). In North America, stereotypical Mexican cooking consists almost exclusively of tortilla-based fare such as tacos, quesadillas, and enchiladas. Although hardly a fair representation of the depth and breadth of true Mexican cuisine, these foods are commonly served as appetizers and snacks in Mexico.
- **Chat** (India). *Chat* is the name given to small portions of food eaten at all times of the day and under different situations. *Chai,* India's version of latte, is commonly consumed with chat. The chai is made from strong tea boiled with spices, sugar, and milk.
- **Dim Sum** (China). Chinese dim sum, which means "touch the heart," is served from a cart that is pushed about the dining room with a wide assortment of small sweet and savory dishes to be selected by waiting customers. Originally a Cantonese custom, dim sum is strongly linked to the Chinese tradition of *yum cha,* or drinking tea. Travelers journeying

along the famous Silk Road needed a place to rest, so teahouses began springing up along the roadside. Rural farmers, exhausted after long hours working in the fields, would also head to the local teahouse for an afternoon of tea and relaxing conversation. Today, in the Canton provinces of China, many people gather at teahouses during the morning and early afternoon to socialize or conduct business over small meals. This is known as going to *yum cha* (going to tea) because the drinking of tea is so traditionally associated with the snack foods served. In a single day, some establishments serve more than 100 different varieties of dim sum.

- **Do nhau (Vietnam).** In Vietnam, such snack or drinking dishes are called *do nhau,* which literally translates to "little bites." *Mon an choi* is Vietnamese for hors d'oeuvre or appetizer, and is commonly used as the heading on a menu.

- **Kanto (Thailand).** The Thai, who relish outdoor dining, call their small foods *kanto.* In the traditional Thai household, hors d'oeuvres are not served as part of a cocktail party or before the meal. Instead, Thais enjoy snacks all day and night, whenever their hunger strikes. Street vendors (food hawkers) are common, and the waterways and canals are filled with merchants selling items such as noodles, grilled meats with cucumber pickles, grilled bananas, nuts, and sweets.

- **Mazza/Meze (Arabic countries).** *Meze* is the general category of dishes that are brought in small quantities to start off a meal. These are eaten, along with wine or Raki, the anise-flavored national drink of Turks, for a few hours until the main course is served. *Mazza* tables are served throughout all of the Arabic countries.

- **Mezze (Greece).** At one time in Greece, mezze was served free with the customer's ouzo, like the tapas of Spain. Today, as people sit together, they munch on any one of hundreds of these small dishes that mark the Greek and Middle Eastern dining experience. Mezze can be as simple as a bowl of olives or pickles, or as involved as fried boureks, stuffed tender grape leaves seasoned with sweet onion, dill, and mint. Often as many as a dozen mezze accompany drinks, precede a meal, or are provided for an afternoon coffee with friends.

- **Rijsttafel (Indonesia).** The Dutch word for "rice table" came to mean a sumptuous, multi-course profusion of small, highly seasoned side dishes of vegetables, meat, seafood, and poultry that accompanied hot rice.

- **Tapas (Spain).** The origin of the word *tapa* (from *tapar,* which literally means "cover") appears to go back to the middle of the nineteenth century. It was the name given to the slice of ham, cheese, or bread used to cover the sherry glass served to weary travelers as they arrived at the roadside inn. The tapa protected the wineglass from dust, rain, and flies. In fact, the tapa was free, and the patron only paid for the wine. The bartenders soon discovered the saltiness of the ham spurred beverage sales, and the tradition of tapas was established. Today, every region has its own specialty tapas, although they are rarely eaten in lieu of a main meal (Figure 4–1).

- **Zakuskis (Russia).** The pinnacle of Russian cuisine is the zakuski ceremony. By the early nineteenth century, it became fashionable in sophisticated homes to serve zakuski on a separate table in the dining room or in an adjacent room, where they were eaten buffet-style. Tradition dictates that both cold and hot zakuski entrées are to be served family-style before the meal in order to whet the appetite. In restaurants, appetizers are served one at a time or in groups at the dining table. The first course is completely devoted to providing these morsels of food, including items such as dark bread with butter, cold cuts, smoked salmon, pâté, caviar, salads, and a huge selection of pickled vegetables. Zakuski is often the largest portion of the meal.

FIGURE 4–1 Tapas menu outside of Catalonian restaurant

Hot Hors d'Oeuvres

Hors d'oeuvres that are served warm are generally more substantial than their cold counterparts. They are almost always included when hors d'oeuvres are being served in lieu of an entire meal, and they are generally are offered in combination with cold hors d'oeuvres. When serving any hors d'oeuvres, it is important to consider the ease with which the customer may consume the edible morsel. Food that is cumbersome and challenging to eat while a person is standing can annoy, if not embarrass, the diner. The expression "finger food" is often applied to hors d'oeuvres as an apt description of how they are commonly consumed. The professional garde manger must always consider the ease in which the food can be selected, held, and consumed by the guest.

To ease consumption, most hot hors d'oeuvres are designed for eating without the aid of a cutting knife and often without the need for a fork. The difficulty arises when the diner must hold a plate, the morsel of food, and a cutting utensil at the same time. Add balancing eating while holding a beverage, and it becomes completely impossible to dine gracefully. The banquet sales staff, to ensure a successful event, must closely coordinate the table arrangements and menu selection.

There are some hot foods that can be served in their natural state. Items such as French trimmed (loin bones attached, cleaned and trimmed) baby lamb chops, battered frogs legs, or mussels marinara are both natural presentations and relatively easy to devour. However, many meats and seafood must be highly fabricated for ease of their eating. The following categories of hot hors d'oeuvres are among commonly used for their practicality and customer appeal; they include skewered foods, quiches, strudel and phyllo, and dumpling and buns.

Skewered Foods

Skewered foods are among the easiest and cleanest hot hors d'oeuvres for guests to consume. They require average skill and effort to prepare, but are generally favorites among the diners (Figure 4–2). Many of the skewered foods originated in Asia, the Far East, and the Middle East, and they include any bite-sized foods that are lanced to hold them together. The picks may include wooden or metal skewers, toothpicks, rosemary sprigs, or similar objects (Figure 4–3).

FIGURE 4–2 Threading meat on a skewer to make satays

FIGURE 4–3 Assorted skewered hors d'oeuvres

RECIPE 4-1

INDIAN BEEF KABOBS

Recipe Yield: 16 portions

MEASUREMENTS		INGREDIENTS
U.S.	**METRIC**	
1½ pounds	680 kg	Ground beef, lean
1 each	1 each	Onion, minced
2 tablespoons	30 mL	Gingerroot, grated
3 each	3 each	Garlic cloves, crushed
1 teaspoon	5 mL	Chile powder
1 tablespoon	15 mL	Garam masala
1 tablespoon	15 mL	Cilantro, chopped
1 tablespoon	15 mL	Almonds, ground
1 each	1 each	Egg, beaten
¼ cup	59 mL	Garbanzo bean flour
⅓ cup	79 mL	Yogurt, plain
2 teaspoons	10 mL	Vegetable oil

PREPARATION STEPS:

1. In a large bowl, mix beef, onion, gingerroot, garlic, spices, cilantro, almonds, eggs, and flour.

2. Cover and refrigerate 4 hours to allow flavors to blend and mature.

3. Divide mixture into 16 long oval-shaped portions.

4. Squeeze each portion around one end of a skewer.

5. Mix yogurt and oil, and brush over kebobs.

6. Broil or grill over preheated grill for 18 to 20 minutes, or until no longer pink. Baste kebobs occasionally, and turn for even heating.

RECIPE 4–2

CHICKEN SATAY

Recipe Yield: 16 portions

| MEASUREMENTS | | INGREDIENTS |
U.S.	METRIC	
1 pound	454 g	Chicken breast, boned and skinless
½ teaspoon	2 mL	Sambal oelek (hot pepper paste)
1 teaspoon	5 mL	Gingerroot, fresh grated
2 tablespoons	30 mL	Lemon juice, fresh
3 tablespoons	44 mL	Dark soy sauce
2 tablespoons	30 mL	Honey
1 tablespoon	15 mL	Peanut butter, creamy
½ cup	118 mL	Water

PREPARATION STEPS:

1. Cut chicken into ¾-inch cubes or 16 long strips.

2. Thread equal amounts of cut chicken onto 16 skewers, pushing the meat to one end of the skewer. Cover and refrigerate.

3. In a large sauté pan, combine sambal oelek, gingerroot, lemon juice, soy sauce, honey, peanut butter, and water.

4. Bring mixture to a boil, stirring constantly.

5. Reduce heat, and add as many chicken breast skewers to the pan that will fit without crowding.

6. Simmer 8 to 10 minuted, turning and basting occasionally.

7. Remove cooked skewers from pan, and repeat step 6 with uncooked skewers.

8. Serve warmed or cold.

RECIPE 4-3

LAMB TIKKA

Recipe Yield: 16 portions

MEASUREMENTS		INGREDIENTS
U.S.	**METRIC**	
2 pounds	907 g	Lamb leg, boneless
1 teaspoon	5 mL	Cumin, ground
¾ teaspoon	4 mL	Turmeric, ground
½ teaspoon	2.5 mL	Salt, kosher
⅓ cup	79 mL	Yogurt, plain
½ each	½ each	Onion, medium, minced
2 tablespoons	30 mL	Gingerroot, fresh grated
2 each	2 each	Garlic cloves, crushed
3 to 5 drops	3 to 5 drops	Red food coloring (optional)
1 teaspoon	5 mL	Garam masala

PREPARATION STEPS:

1. Trim fat from lamb, and cut into 1-inch cubes.

2. Combine cumin, turmeric, salt, yogurt, onion, gingerroot, and garlic. Add cubed lamb.

3. Blend meat and spice mixture well, and add enough food color to give a pink color to the meat and sauce.

4. Cover and refrigerate 4 to 6 hours.

5. Drain lamb from marinade and thread tightly onto 16 skewers.

6. Cook kebobs on preheated broiler or grill for 12 to 15 minutes, or until cooked to medium. Baste occasionally with remaining marinade while turning kebobs.

7. Sprinkle with garam masala and serve warmed.

RECIPE 4–4

PORK AND MANGO SKEWERS

Recipe Yield: 16 portions

MEASUREMENTS		INGREDIENTS
U.S.	**METRIC**	
1 pound	454 g	Pork, fresh boneless
½ cup	118 mL	Hoisin sauce
1½ tablespoon	22 mL	Dark soy sauce
⅛ cup	27 mL	Rice wine vinegar
⅛ cup	27 mL	Olive oil
1 tablespoon	15 mL	Gingerroot, fresh grated
1 to 2 each	1 to 2 each	Mangoes

PREPARATION STEPS:

1. Cut the pork into ¾-inch cubes.

2. In a large bowl, combine the hoisin sauce, soy sauce, vinegar, olive oil, and ginger.

3. Add the cubed pork, and marinate overnight under refrigeration.

4. Cut the mangoes into ¾-inch cubes.

5. Thread 2 to 3 pieces of pork and 1 mango cube per skewer.

6. Cook over hot charcoal or charbroiler for 7 to 8 minutes, turning once.

7. Serve warm.

QUICHES

Although most countries create some form of baked egg custard, quiches gained considerable popularity in Western Europe. Quiches are often made with a variety of ingredients to flavor the custard, including seasoned meats, seafood, roasted vegetables, and cheeses. They can be prepared in small pastry cups, tart shells, barquettes, or larger pies. Figure 4–4 shows steps for preparing quiche cups and Figure 4–5 presents assorted quiche shells.

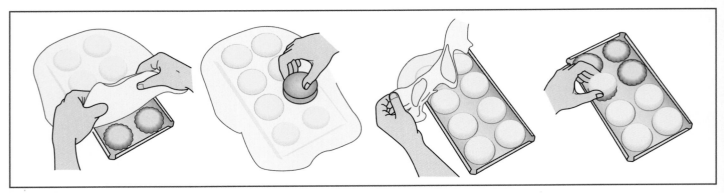

FIGURE 4–4 Making quiche cups

RECIPE 4–5

QUICHE LORRAINE TARTS

MEASUREMENTS		INGREDIENTS
U.S.	**METRIC**	
As needed	As needed	Individual tart shells, prepared
¼ cup	59 mL	Dijon mustard
2 each	2 each	Roma tomato, ripe (optional), thinly sliced
2 cups	473 mL	Half-and-half
3 each	3 each	Eggs, beaten
½ teaspoon	2.5 mL	Salt
½ teaspoon	2.5 mL	Black pepper, fresh ground
1 teaspoon	5 mL	Nutmeg, fresh grated
1 pound	454 g	Bacon lardons, crisp cooked
½ cup	118 mL	Gruyere or French semisoft cheese, shredded

PREPARATION STEPS:

1. Paint the bottom of each tart shell with the mustard. Prick the shell's bottom to prevent it from bubbling up during baking.

2. Add a slice of tomato to each tart shell.

3. Combine the half-and-half, eggs, and spices. Reserve.

4. Sprinkle bacon and cheese into each tart shell.

5. Pour custard mixture into shells.

6. Bake in preheated 375°F (190°C) oven until custard sets, approximately 30 minutes.

7. Serve warm or at room temperature.

FIGURE 4–5 Assorted quiche shells

RECIPE 4–6

SOUTHWESTERN QUICHE CUPS

Note: May substitute Monterey Jack cheese with jalapeños for fresh jalapeños.

MEASUREMENTS		INGREDIENTS
U.S.	**METRIC**	
As needed	As needed	Individual pastry cups, prepared
2 each	2 each	Tomatoes, ripe, small diced
$\frac{2}{3}$ cup	158 mL	Black olives, chopped
10 each	10 each	Scallions, chopped
2 each	2 each	Jalapeños, seeded, minced
1 pound	454 g	Monterey Jack cheese, shredded
8 each	8 each	Eggs, beaten
2 tablespoons	30 mL	Flour
1 cup	237 mL	Half-and-half
$\frac{1}{8}$ teaspoon	2 mL	Nutmeg, grated

PREPARATION STEPS:

1. Combine tomatoes, olives, scallions, and jalapeños.

2. Spoon tomato mixture into each pastry cup.

3. Add shredded cheese.

4. Beat eggs, flour, half-and-half, and nutmeg into custard.

5. Pour custard to nearly fill pastry cups.

6. Bake in preheated 375°F (190°C) oven until custard sets.

7. Serve warm or at room temperature.

RECIPE 4–7

Smoked Salmon with Mushrooms Barquettes

Recipe Yield: 36 portions

MEASUREMENTS		INGREDIENTS
U.S.	**METRIC**	
4 ounces	113 g	Button mushrooms, small, sliced
4 ounces	113 g	Porcini mushrooms, sliced
2 ounces	57 g	Butter
3 each	3 each	Eggs
1 cup	237 mL	Cottage cheese, small curd, cream-style
2 teaspoons	10 mL	Dijon mustard
¾ teaspoon	3 mL	Salt
⅔ cup	158 mL	Half-and-half
12 ounces	340 g	Smoked salmon, flaked
½ cup	118 mL	Carrot, shredded
½ cup	59 mL	Scallion, minced
36 each	36 each	Barquettes, approximately 2 inches

PREPARATION STEPS:

1. Over medium heat, sweat mushrooms and butter in covered pan, reserving liquid. Turn off heat when mushrooms have reduced by one third.

2. Meanwhile, in a large bowl, beat eggs, cottage cheese, mustard, and salt. Reserve.

3. Drain mushrooms and add mushroom liquid and half-and-half to egg mixture. Reserve.

4. Combine mushrooms, salmon flakes, carrot, and scallions, and spoon into barquettes.

5. Add egg custard.

6. Bake in preheated 375°F (190°C) oven until custard sets.

7. Serve warm or at room temperature.

STRUDEL AND PHYLLO

Many countries, particularly those around the Mediterranean and the Middle East, use strudel and phyllo dough to prepare multilayered sweet and savory pastries. The word *phyllo* means "leaf" in the Greek language, describing how the dough is layered like thin leaves. Lamb buerrecks, spanakopita, cabbage and caraway strudel, and phyllo-wrapped prawns with prosciutto are examples of the broad uses for flaky pastry dough. When working with flaky pastry dough, a few rules should be observed:

- When using frozen dough, allow it to thaw overnight under refrigeration.
- Work in an area free of drafts and excess air movement.
- Unroll sheets onto smooth, dry surfaces.
- Work with a few sheets at a time, leaving the remaining pile of sheets covered until needed. Uncovered dough will dry out in a few minutes.
- Brush each layer with melted, but not hot, butter.
- Fillings should be cool and free of excessive moisture.

■ Always brush the outside layer of dough after the filling has been added.

■ Bake on parchment covered sheet pans at 350°F (177°C) until golden brown.

Figures 4–6 through 4–10 show steps for making a variety of phyllo, products, and Figure 4–11 shows assorted hors d'oeuvres in phyllo.

FIGURE 4–6 Folding phyllo dough triangle appetizers

DIRECTIONS FOR PHYLLO TRIANGLES

1. Layer 4 phyllo sheets, brushing each with melted butter, margarine, oil or vegetable spray.

2. A. For small triangles, cut layered phyllo into 8 strips. About 1″ from bottom of each strip, place ½ to 2 teaspoons of cooled filling.
   B. For medium triangles, cut layered phyllo into 4 strips. About 1″ from bottom of each strip, place 1 to 2 tablespoons of cooled filling.
   C. For large triangles, cut layered phyllo in half lengthwise. About 1″ from bottom of each strip, place ¼ cup of cooled filling.

3. Fold one corner of phyllo diagonally across to opposite edge to form a triangle.

4. Continue to fold triangle onto itself. Brush outside with butter.

5. Lay triangles seam side down, at least 1″ apart, on ungreased cookie sheet or baking pan.

6. A. Bake small and medium triangles in preheated 350°F (177°C) oven for about 15-20 minutes or until golden brown.
   B. Bake large triangles at 350°F (177°C) for 25 to 30 minutes or until golden brown.

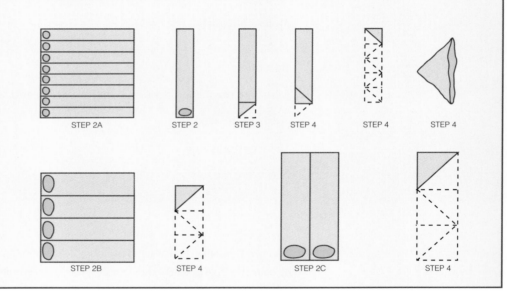

STEP 2A   STEP 2   STEP 3   STEP 4   STEP 4   STEP 4

STEP 2B   STEP 4   STEP 2C   STEP 4

DIRECTIONS FOR PHYLLO POUCHES

1. Layer 4 phyllo sheets, brushing each with softened butter, margarine, oil or vegetable spray. Use 2 tablespoons butter or oil.

2. A. For small pouches, cut layered phyllo into 2 ½ " squares and spoon 1 ½ teaspoons of filling onto center of square.
   B. For large pouches, cut layered phyllo into 6" to 8" squares and spoon ¼ to ⅓ cup of filling into center of square.

3. Brush phyllo from edge of filling to each point of square lightly with water.

4. Gather points of square and pinch together just above filling.

5. A. Brush small pouch with melted butter. Place at least 1" apart on ungreased cookie sheet or baking pan and bake in preheated 350°F (177°C) oven about 12 to 15 minutes or until golden brown.
   B. Bake large pouches in preheated 350°F (177°C) oven about 25 to 30 minutes or until golden brown.

Tips: If you put too much filling into square, it will be very difficult to enclose it with phyllo.

Freeze filled phyllo for 10 minutes before baking to set the shape of the pouch.

To make a phyllo purse, cut circles instead of squares, gather ¾ way to top, spread top edge and tie with blanched scallion strips or other edible purse string.

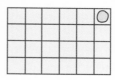
STEP 2A

STEP 2B

STEP 4

POUCH - STEP 4

PURSE - SEE TIP

FIGURE 4–7 Making phyllo dough pouches

FIGURE 4–8 Making round phyllo wraps

DIRECTIONS FOR ROUND OR WRAPPED PHYLLO SHAPES

1. Use 9 sheets of phyllo dough and ¼ cup butter.

2. Brush ½ sheet, lengthwise, with butter and fold in half.

3. Place folded sheet over plate or outline approximate size of food to be wrapped, allowing excess phyllo to overlap and extend beyond outline.

4. Repeat steps 2 and 3 with 6 additional sheets.

5. Prepare 2 more sheets as described in step 2 and place them crosswise directly over center of outline.

6. Place food in center of stack and carefully draw phyllo up and around food.

7. Brush with butter and place at least 1″ apart on cookie sheet or baking pan and bake in preheated 350°F (177°C) oven about 15 to 25 minutes or until golden brown.

Tips: Use with camembert, boursin, brie or any small round cheese.
   You can also wrap pears, apples and other fruit.

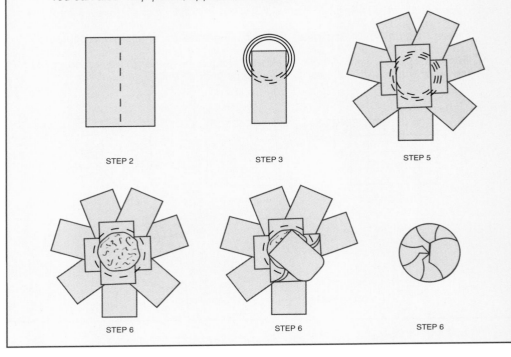

STEP 2   STEP 3   STEP 5

STEP 6   STEP 6   STEP 6

DIRECTIONS FOR PHYLLO CUPS, PIE & TART SHELLS

1. Lightly grease pie, tart, tartlet, or muffin.

2. A. For small tartlet or muffin pan layer 6 phyllo sheets, brushing each with melted butter, oil or vegetable spray.
   B. For large pie or tart pans layer 10 phyllo sheets, brushing each with melted butter, oil or vegetable spray.

3. Cut circles from phyllo using cutters $1^1/_2$ times larger than pan or muffin diameter in which you are baking, so that phyllo circle is large enough to go completely up the sides.

4. Carefully push phyllo into greased pan and press carefully and firmly against bottom and sides. Lightly brush inside of shell with melted butter.

5. A. Bake empty shells in preheated 350°F (177°C) oven about 8 to 10 minutes or until golden brown. Cool 5 minutes, remove and fill.
   B. Bake filled shells in preheated 350°F (177°C) oven 12 to 15 minutes for small cups or shells and 20 to 30 minutes for large cups or shells.

Tips: If shell is for dessert item, lightly sprinkle each layer with sugar after buttering.

For an attractive variation, cut squares instead of circles. When you push them into the pan, leave points sticking up.

Finely chopped or ground nuts between buttered and sugared phyllo layers add wonderful flavor.

Oven-proof bowls can be used to bake phyllo for various sized tart shells. For example, if you need a 5" diameter shell, cut circle $5^1/_2$" diameter and press into 5" bowl. Bake as instructed above.

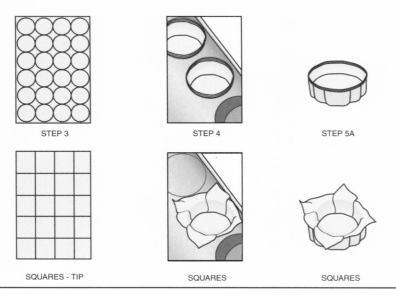

| STEP 3 | STEP 4 | STEP 5A |
| SQUARES - TIP | SQUARES | SQUARES |

FIGURE 4–9 Making phyllo cups

FIGURE 4–10 Making
phyllo rolls

DIRECTIONS FOR PHYLLO ROLLS OR STRUDELS

1. Layer 5 phyllo sheets, brushing each with melted butter, margarine, oil or vegetable spray.

2. A. For small rolls or strudels, cut width of layered phyllo into 4 strips.
   B. For medium rolls or strudels, cut width of layered phyllo in half.
   C. For large roll or strudel, layer 7 phyllo sheets and roll full sheet.

3. A. Place cooled filling at one end of phyllo strip, leaving 1″ from end and ½″ from each side free of any filling for small and medium rolls.
   B. For large roll, leave 1″ from each side free of any filling.

4. Start rolling from edge containing filling.

5. Once filling is enclosed, fold over exposed edges. Do not fold over edges if using a high moisture filling.

6. Continue rolling to end of phyllo strip.

7. Brush outside with butter

8. A. Bake small and medium rolls with seam side down, at least 1″ apart, on ungreased cookie sheet or baking pan in preheated 350˚F (177˚C) oven for 15 to 20 minutes or until golden brown.
   B. Bake large roll or strudel in a preheated 350˚F (177˚C) oven for 25 to 30 minutes or until golden brown.

Tips: *If you are planning to slice roll after baking, score phyllo before putting into oven. This will allow you to slice through cleanly after baking.*

Electric knives work wonderfully when slicing through phyllo products, even when they aren't scored before baking.

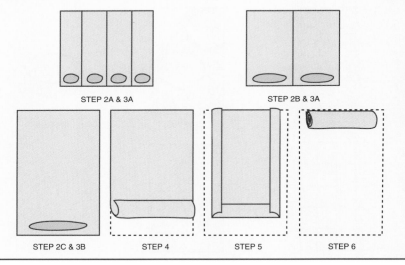

STEP 2A & 3A   STEP 2B & 3A

STEP 2C & 3B   STEP 4   STEP 5   STEP 6

FIGURE 4–11 Assorted phyllo appetizers

RECIPE 4-8

SPANAKOPITA

Recipe Yield: 40 portions

MEASUREMENTS		INGREDIENTS
U.S.	**METRIC**	
10 ounces	283 g	Frozen spinach, thawed and chopped
$\frac{1}{2}$ cup	118 mL	Scallions, chopped
$\frac{1}{2}$ cup	118 mL	Fresh parsley, chopped
$\frac{1}{2}$ cup	118 mL	Fresh dill, chopped
$\frac{1}{2}$ cup	118 mL	Feta cheese, crumbled
$\frac{1}{2}$ cup	118 mL	Farmer cheese, grated
2 tablespoons	30 mL	Kefalotiri, grated (or Parmesan cheese)
2 each	2 each	Eggs
$\frac{1}{3}$ teaspoon	2 mL	Black pepper, fresh ground
10 each	10 each	Phyllo dough sheets, barely thawed
$\frac{3}{4}$ cup	177 mL	Butter, melted

PREPARATION STEPS:

1. Press all excess moisture from the spinach.

2. In a food processor, combine all ingredients except the dough sheets and butter.

3. Process until mixture is coarse, or until desired consistency is reached.

4. Layer 2 sheets of dough on a cutting board, brushing each sheet with melted butter. Cover the remaining dough with a moist towel until needed.

5. Cutting widthwise, divide the dough stack into 8 even-sized rows of dough.

6. About 1 inch from the bottom of each row of dough, place $\frac{1}{2}$ teaspoon (2.5 mL) of cooled spinach filling.

7. Fold dough, like a flag, into a triangle. Brush outside of the stuffed triangle with melted butter.

8. Repeat steps until all portions are prepared.

9. Place triangles seam side down on a parchment-covered sheetpan, and bake at 350°F (177°C) for approximately 15 minutes or until golden brown. Serve warm.

TIP Spinach and cheese triangles may be frozen for future use. Be sure to brush the triangles thoroughly with melted butter before freezing.

RECIPE 4–9

SAUSAGE AND CABBAGE STRUDEL

Recipe Yield: 32 portions

MEASUREMENTS		INGREDIENTS
U.S.	**METRIC**	
12 ounces	340 g	Chorizo sausage, chopped
¼ cup	59 mL	Olive oil
2 cups	473 mL	Onions, chopped
6 each	6 each	Garlic cloves, minced
2 pounds	907 g	Savoy cabbage, thinly sliced
1 teaspoon	5 mL	Caraway seeds
2 tablespoons	30 mL	Fresh savory, chopped (optional)
1 teaspoon	5 mL	Paprika
To taste	To taste	Salt
To taste	To taste	Black pepper, fresh ground
1 each	1 each	Egg
1 cup	473 mL	Ricotta cheese, puréed
¼ cup	59 mL	Parmesan cheese, fresh grated
16 each	16 each	Phyllo dough sheets
½ cup	118 mL	Butter, melted
2 tablespoons	30 g	Poppy seeds

PREPARATION STEPS:

1. In a large sauté pan, brown chorizo over medium heat. Remove meat and place into a large bowl.

2. Add oil to the sauté pan and heat. Add onions and garlic, and sauté for approximately 3 minutes.

3. Add cabbage, caraway seeds, savory, paprika, salt, and pepper. Cook approximately 5 minutes, or until the cabbage is tender. Remove and combine with meat. Drain and cool.

4. Fold egg and cheeses into cooled cabbage and meat mixture.

5. Layer 4 sheets of dough on a cutting board, brushing each sheet with melted butter. Cover the remaining dough with a moist towel until needed.

6. Cutting widthwise, divide the dough stack into 4 even-sized rows of dough.

7. Approximately 1 inch from the edge of each row of dough, place one sixteenth of the filling. Keep ½ inch from each side free of filling.

8. Start rolling each strudel, starting with the edge containing the filling.

9. Continue rolling to end of dough, until strudel resembles an open-ended egg roll.

10. Brush outside of strudel with melted butter. Place seam side down on a parchment covered sheetpan.

11. Sprinkle buttered dough with poppy seeds.

12. Repeat steps until all strudels are formed.

13. If smaller portions are desired, lightly score top of the strudel.

14. Bake in 350°F (177°C) oven for 20 minutes or until golden brown. Serve warm.

RECIPE 4-10

SWEETWATER PRAWNS WITH PROSCIUTTO

Recipe Yield: 48

MEASUREMENTS   INGREDIENTS

U.S.	METRIC	
½ cup	118 mL	Olive oil
3 tablespoons	44 mL	Fresh gingerroot, minced
3 each	3 each	Garlic cloves, minced
1 tablespoon	15 mL	Fresh rosemary, minced
1 tablespoon	15 mL	Fresh lemon thyme, minced
½ cup	118 mL	Vodka
⅔ teaspoon	3 mL	Black pepper, fresh ground
48 each	48 each	Sweetwater prawns, peeled and deveined (approximately 1 ounce each) with tail on
24 each	24 each	Prosciutto slices, cut in half lengthwise
24 each	24 each	Phyllo dough sheets
⅓ cup	79 mL	Butter, melted

PREPARATION STEPS:

1. In a large mixing bowl, combine oil, ginger, garlic, rosemary, thyme, vodka, and pepper. Blend well.

2. Add prawns. Toss to thoroughly coat each prawn. Cover and refrigerate for 3 hours, tossing once while marinating.

3. Remove prawns from marinade, and wrap each with a slice of prosciutto.

4. Layer 4 sheets of phyllo, brushing each sheet with melted butter. Repeat as needed.

5. Cut dough widthwise into 1-inch wide strips.

6. Wrap phyllo around prawns, leaving tail exposed and conforming to the natural shape of the prawn.

7. Brush each wrapped prawn with melted butter, place seam side down on parchment covered sheetpan and refrigerate for 1 hour.

8. Bake in preheated 400°F (204°C) oven for 8 to 12 minutes or until golden brown.

9. Serve warm.

RECIPE 4–11

BOUREK

Recipe Yield: 8 portions

Portion Size: 2 ounces (57 g)

MEASUREMENTS		INGREDIENTS
U.S.	**METRIC**	
¼ cup	59 mL	Vegetable oil
¼ cup	59 mL	Onion, fine dice
½ pound	227 g	Ground beef
2 tablespoons	30 ml	Flat leaf parsley
1 each	1 each	Egg, beaten
⅓ teaspoon	1.5 mL	Salt
⅓ teaspoon	1.5 mL	Black pepper
8 each	8 each	Phyllo pastry sheets
8 each	8 each	Lemon wedges
As needed	As needed	Vegetable oil, for frying Bourek

PREPARATION STEPS:

1. Fry the onion in the oil with the beef, and cook for 20 minutes.

2. Season well with salt and pepper.

3. Add the egg and cook well.

4. Mix in the parsley and cool.

5. Take a sheet of phyllo pastry and place a heaping spoon of the filling onto one end.

6. Fold into a 4-inch long by 1-inch wide tube, tucking in the edges to ensure a good seal.

7. Pan fry the bourek and serve with the lemon wedges.

STEAMED AND FRIED DUMPLINGS AND WRAPS

There are limitless varieties of stuffed pastas and pastries prepared around the world. Asia is famous for its contribution of numerous roll-ups, wraps, and buns made from simple skins of rice flour and water or wheat flour and water. According to Nina Simonds (2000), "In Asia, the wrapper may take the form of a wheat or rice flour skin, some type of leafy green, a square of seaweed, or a steamed or baked bread." A patient chef may make the dough and wrappers, but quality products are generally available at local Asian and other ethnic food markets. To shape, they may be pleated, straight-edged, or open-topped. Then they are boiled, steamed, or deep-fried. No matter the shape, dumplings and wraps always envelope sumptuous fillings made from savory meat, seafood, tofu, tempe, and vegetables. See Figures 4–12 and 4–13 for directions for making spring rolls and pot stickers.

FIGURE 4–12 Directions for preparing spring rolls

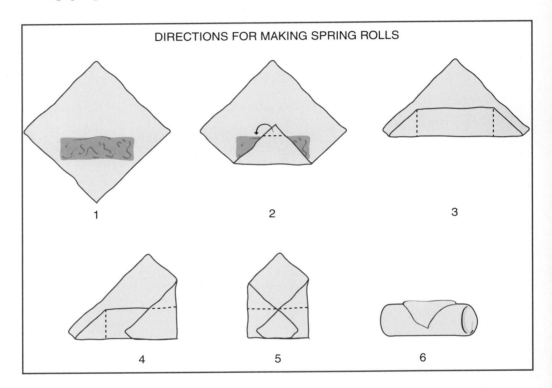

FIGURE 4–13 Directions for preparing pot stickers

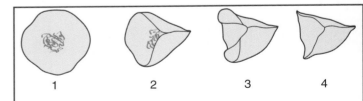

1. Fill dough.
2. Fold in half and seal ¹/₃ of the edges on one side.
3. Bring middle of open edge upward toward center.
4. Seal all edges.

RECIPE 4–12

MARYLAND CRAB CAKE SPOON DUMPLINGS

Recipe Yield: 40 spoon dumplings

MEASUREMENTS		INGREDIENTS
U.S.	**METRIC**	
4 tablespoons	60 mL	Butter, unsalted
½ cup	118 mL	Onion, minced
½ cup	118 mL	Celery, minced
1 pound	500 g	Jumbo lump crab meat, picked over
⅓ cup	79 mL	Bread crumbs, unseasoned, fine
½ cup	118 mL	Mayonnaise
1 teaspoon	5 mL	Seafood seasoning
½ teaspoon	2.5 mL	Worcestershire sauce
⅓ teaspoon	2 mL	Tabasco (or other bottled hot sauce)
2 tablespoons	30 mL	Flat-leafed parsley, minced
½ teaspoon	2.5 mL	Salt
½ teaspoon	2.5 mL	Black pepper, fresh ground

PREPARATION STEPS:

1. In a sauté pan, cook onion and celery in 4 tablespoons of butter over moderately low heat, stirring until tender. Transfer to a bowl. Stir in crab and breadcrumbs.

2. In a small bowl, whisk together mayonnaise, seafood seasoning, Worcestershire sauce, Tabasco, parsley, and salt and pepper and stir into crab mixture until well combined.

3. Line a sheet pan with parchment paper. Form crab mixture into small, flattened rounds, about ¾ ounce (375 g) each, and transfer to baking sheet. Chill crab cakes, covered with plastic wrap, at least 1 hour (and up to 4).

4. In a heavy-bottomed sauté pan, heat 1 tablespoon of butter over moderate heat until foam subsides. Cook half of crab cakes until golden brown, about 1 to 2 minutes on each side, transferring them to a heated platter. Cook remaining dumplings in remaining butter in same manner.

5. Serve room temperature crab cake dumplings on teaspoons or bouillon spoons (Figure 4–14). Consider serving with Sauce Rémoulade (recipe 5–11) or American Seafood Cocktail Sauce (recipe 5–31).

FIGURE 4–14 Maryland Crab Cake Spoon Dumplings

RECIPE 4-13

SPRING ROLL OR WONTON WRAPPER DOUGH

Recipe Yield: 8 portions, 16 wrappers

| MEASUREMENTS | | INGREDIENTS |
U.S.	METRIC	
1 each	1 each	Egg
½ cup	118 mL	Water
2 cups	473 mL	All-purpose flour
½ teaspoon	2 mL	Salt
¼ cup	59 mL	Cornstarch for dusting

PREPARATION STEPS:

1. Mix three fourths of the water with the egg and add it to the flour and salt.

2. Knead into a smooth pliable dough using the rest of the water as needed.

3. Cover and let rest for 1 hour.

4. Divide into two even parts. Using a pasta roller, roll out into very thin sheets, using cornstarch to dust.

5. Cut into desired shapes and stack. Dust between layers with cornstarch.

RECIPE 4-14

DIM SUM DOUGH

Recipe Yield: 8 portions
Portion Size: 1 ounce (28 g)

| MEASUREMENTS | | INGREDIENTS |
U.S.	METRIC	
⅛ cup	29 mL	Sugar
1⅜ cups	325 mL	Warm water 110°F (43.3°C)
½ tablespoon	7 mL	Active dry yeast
½ tablespoon	7 mL	Baking powder
3 cups	710 mL	All-purpose flour
1 tablespoon	15 mL	Shortening

PREPARATION STEPS:

1. Dissolve the sugar in the water and add the yeast. Allow to stand for 10 minutes.

2. Add the baking powder to the flour and pass through a sieve.

3. Rub in the shortening to the flour.

4. Add the water, yeast, and sugar mixture, and knead to a smooth dough.

5. Use the dough after it has rested for 30 minutes.

RECIPE 4-15

Dim Sum Filling

Recipe Yield: 8 portions

MEASUREMENTS		INGREDIENTS
U.S.	**METRIC**	
8 ounces	227 g	Minced pork
5 ounces	142 g	White cabbage, finely chopped
2 ounces	50 g	Chinese black mushrooms, soaked and finely chopped
½ teaspoon	2 mL	Diced ginger
2 teaspoons	10 mL	White sugar
2 teaspoons	10 mL	Salt
Pinch	Pinch	White pepper
2 teaspoons	10 mL	Sesame oil
1 tablespoon	15 mL	Mirin

PREPARATION STEPS:

1. Combine all ingredients well and reserve for 30 minutes to develop flavor.

2. Stuff into the dough and create a nice shape.

3. Steam and serve with soy sauce.

RECIPE 4-16

Steamed Pleated Dumplings

Recipe Yield: 36 portions

MEASUREMENTS		INGREDIENTS
U.S.	**METRIC**	
½ cup	118 mL	Pork, ground
½ cup	118 mL	Chicken, ground
½ cup	118 mL	Shrimp, ground
4 tablespoons	59 mL	Cilantro, minced
2 each	2 each	Scallions, minced
2 each	2 each	Garlic cloves, minced
½ teaspoon	2 mL	Black pepper, fresh ground
2 teaspoons	10 mL	Sugar
2 tablespoons	30 mL	Soy sauce
1 tablespoon	30 mL	Gingerroot, fresh grated
2 tablespoons	30 mL	Coconut cream (optional)
1-pound pkg	454 g	Round wonton skins
As needed	As needed	Cold water
As needed	As needed	Lettuce leaves or silicone paper
		Hot Sesame Soy Sauce
3 tablespoons	44 mL	Soy sauce
3 tablespoons	44 mL	Rice wine vinegar
½ teaspoon	2 mL	Hot sesame oil
1 each	1 each	Scallion, minced

(continued)

PREPARATION STEPS:

1. In a medium bowl, combine all ingredients for the filling, except the wonton skins, water, and lettuce leaves. Blend evenly.

2. Moisten one side of a wonton wrapper with cold water.

3. Place 1 teaspoon (5 mL) of filling onto the center of the moistened wonton skin.

4. Fold the wonton in half and crimp the edges closed. Repeat until all wonton filling has been used.

5. Place the filled wontons in a steamer that has been lined with lettuce leaves or silicone paper to prevent them from sticking to the steamer (Figure 4–15).

6. Cover and steam 15 to 20 minutes or until soft and completely cooked.

7. Serve warmed with Hot Sesame Soy Sauce as a dipping sauce (Figure 4–16).

FIGURE 4–15 Steaming pleated dumplings (note lettuce leaves lining steamer basket)

FIGURE 4–16 Steamed dumplings served with dipping sauce

RECIPE 4–17

Shiu Mai

Recipe Yield: 36 portions

MEASUREMENTS		INGREDIENTS
U.S.	**METRIC**	
¾ pound	340 g	Pork, ground
¾ pound	340 g	Shrimp, ground
9 each	9 each	Chinese mushrooms, soaked to soften and minced
¼ cup	59 mL	Water chestnuts, minced
¼ cup	59 mL	Bamboo shoots, minced
3 tablespoons	44 mL	Cornstarch
¾ teaspoon	4 mL	Salt
½ tablespoon	7 mL	Gingerroot, fresh grated
¼ teaspoon	1 mL	Black pepper, fresh ground
1½ tablespoons	22 mL	Oyster sauce
1½ tablespoons	22 mL	Soy sauce
¾ tablespoon	11 mL	Sesame oil
¾ tablespoon	11 mL	Sherry
½ cup	118 mL	Scallions, minced
1 package	1 package	Wonton skins, round
1 cup	237 mL	Green peas

PREPARATION STEPS:

1. Combine all ingredients for the filling and blend evenly.

2. Place one wonton skin on the cutting board, and portion 1 tablespoon of filling in its center.

3. Using a pastry brush, moisten all around the filling with dabs of cold water.

4. Gather edges upward around the filling to form a pleated pouch with an open top.

5. Gently squeeze the shiu mai with the arc between thumb and index finger, while turning it so the wrapper will cling to the filling. At the same time, press down and level the filling with a small knife or the tablespoon.

6. Garnish the open top of the filling with 1 pea.

7. Repeat steps with remaining filling.

8. Arrange shiu mai on greased steaming tray or one loosely covered with lettuce leaves.

9. Cover and steam 18 to 20 minutes.

10. Serve warm with Hot Sesame Soy Sauce dip if desired.

RECIPE 4–18

Mo-Shu Spring Rolls

Recipe Yield: 24 portions

MEASUREMENTS		INGREDIENTS
U.S.	**METRIC**	
24 each	24 each	Asian pancakes (or spring roll skins or flour tortillas)
As needed	As needed	Toasted sesame oil
		Vegetable Filling
2 tablespoons	30 mL	Corn oil
4 each	4 each	Garlic cloves, minced
3 tablespoons	45 mL	Fresh ginger, minced
10 each	10 each	Chinese black mushrooms, softened in hot water, cut julienne
3 cups	710 mL	Leeks, white part only, cut julienne
2 cups	473 mL	Carrots, julienne grated
5 cups	1.2 L	Napa (Chinese) cabbage, shredded
1½ tablespoons	22 mL	Rice wine or sake
4 cups	946 mL	Bean sprouts, rinsed and drained
		Sauce Mixture
¼ cup	60 mL	Soy sauce
2 teaspoons	10 mL	Toasted sesame oil
½ teaspoon	2.5 mL	Black pepper, fresh ground
¾ teaspoon	3.5 mL	Cornstarch
		Hoisin Sauce
½ cup	118 mL	Hoisin sauce
3 tablespoons	44 mL	Water, warmed

PREPARATION STEPS:

1. Separate the spring roll skins and lightly brush with sesame oil.

2. Fold in half, with oil inside, and steam for 5 minutes.

3. Hold in woven wicker wrapper basket or covered with warmed damp towel.

4. Meanwhile, warm wok or large sauté pan over high heat until very hot. Add oil. Heat for 30 seconds.

5. Add garlic, ginger, and mushrooms. Stir-fry for 10 seconds.

6. Add leeks and stir-fry for 1 minute. Add carrots, cabbage, and rice wine. Continue to stir-fry until vegetables are crisp but tender.

7. Add bean sprouts and Sauce Mixture. Stir until thickened.

8. Transfer to a warmed serving platter.

9. For service, brush inside of pancakes with diluted hoisin sauce, then add filling and fold pancake.

RECIPE 4–19

SHRIMP SAMOSA

Recipe Yield: 36 portions

MEASUREMENTS		INGREDIENTS
U.S.	**METRIC**	
		Samosa Dough
3 cups	710 g	All-purpose flour
¾ teaspoon	4 mL	Salt
½ cup	118 mL	Butter, unsalted
1 cup	237 mL	Yogurt, plain
		Shrimp Filling
¾ cup	177 mL	Onion, minced
2 each	2 each	Garlic cloves, minced
3 tablespoons	44 mL	Vegetable oil
¾ pound	340 g	Shrimp, chopped
½ cup	118 mL	Tomato, chopped
¾ teaspoon	4 mL	Salt
As needed	As needed	Vegetable oil, for frying

PREPARATION STEPS:

1. For the dough, mix the flour and salt, then cut in butter to consistency of coarse meal.

2. Blend in yogurt until dough forms a ball.

3. Wrap tightly and refrigerate for at least 1 hour.

4. While dough is retarding (relaxing and slowing the growth of the dough) in the refrigerator, make the filling.

5. In the oil, stir-fry onion and garlic for 5 to 6 minutes.

6. Add remaining ingredients and continue to stir-fry an additional 5 minutes.

7. Cool and store in refrigerator until ready to form samosas.

8. To make samosas, roll part of the chilled dough onto a floured surface to $\frac{1}{16}$-inch thickness.

9. Cut 4-inch circles into the dough, and then cut each circle in half.

10. Moisten edges of dough, and place 1 teaspoon (5 mL) of filling in center.

11. Fold dough edges to form triangle. Press firmly to seal edges.

12. Repeat until all filling or dough is used. Samosasas may be refrigerated up to 1 day before frying.

13. Deep fry in oil at 375°F (190°C) for about 4 minutes to brown both sides and cook filling.

14. Serve warm.

RECIPE 4–20

CORNISH PASTY DOUGH

Recipe Yield: 8 portions

MEASUREMENTS		INGREDIENTS
U.S.	METRIC	
18 ounces	510 g	All-purpose flour
Pinch	Pinch	Salt
10 ounces	283 g	Softened butter
1 each	1 each	Egg

PREPARATION STEPS:

1. Sieve the flour and salt. Add the butter and egg.

2. Mix well into a smooth dough.

3. Chill for 20 minutes until firm and rested.

RECIPE 4–21

CORNISH PASTY FILLING

Recipe Yield: 8 portions

MEASUREMENTS		INGREDIENTS
U.S.	METRIC	
10 ounces	283 g	Diced lamb or beef
10 ounces	283 g	Potato, finely diced
1/4 cup	59 mL	Onion, finely diced
1 teaspoon	5 mL	Chopped thyme
Pinch	Pinch	Salt
Pinch	Pinch	White pepper
2 tablespoons	30 mL	Lamb or beef stock
1 each	1 each	Egg, separated

PREPARATION STEPS:

1. Mix all ingredients except egg yolk and chill.

2. Bring the pasty dough (see Recipe 4-20) to room temperature and roll it out to 1/8 of an inch thick.

3. Cut into 6-inch circles and fill with the meat filling.

4. Brush one side with some beaten egg yolk and fold over into a semicircle, crimping the edges well to seal.

5. Bake for 40 minutes at 350°F (177°C) until cooked.

RECIPE 4–22

EMPANADA DOUGH

Recipe Yield: 8 portions

MEASUREMENTS		INGREDIENTS
U.S.	**METRIC**	
2½ cups	591 mL	All-purpose flour
1 teaspoon	5 mL	Baking powder
½ teaspoon	2 mL	Salt
2 ounces	57 g	Lard, melted
½ cup	118 mL	Milk or chicken stock

PREPARATION STEPS:

1. Sieve the flour, baking powder, and salt.

2. Add the melted lard and milk, and knead into a firm, smooth dough.

3. Allow to rest for 30 minutes.

4. Roll on a pasta roller to a thickness of ⅛ of an inch.

5. Cut into 3-inch circles.

6. When using the dough, stuff by folding over and sealing.

RECIPE 4–23

EMPANADA FILLING

Recipe Yield: 8 portions

MEASUREMENTS		INGREDIENTS
U.S.	**METRIC**	
1 tablespoon	15 mL	Vegetable oil
1 pound	454 g	Ground beef
2 each	2 each	Onions, finely diced
1 tablespoon	15 mL	Paprika
2 each	2 each	Cloves garlic, finely chopped
1 tsp	5 mL	Oregano, finely chopped
1 cup	237 mL	Beef broth
1 cup	237 mL	Chopped tomatoes
½ cup	118 mL	Chopped green olives
½ cup	11 mL	Diced raisins
Pinch	Pinch	Salt and pepper

PREPARATION STEPS:

1. Heat the oil in a thick-bottomed pan, and brown the beef and onion together.

2. Add the garlic, oregano, and paprika, and cook for another 5 minutes.

3. Add the broth and the tomatoes. Cook for a further 30 minutes, or until almost dry.

4. Add the olives and the raisins, and season well.

5. Cool the mixture completely before stuffing into the pastry dough.

BASIC PASTA DOUGH

Recipe Yield: 8 portions

MEASUREMENTS		INGREDIENTS
U.S.	**METRIC**	
4⅔ cups	1100 mL	All purpose flour
1 teaspoon	5 mL	Fine salt
6 each	6 each	Eggs
2 tablespoons	30 mL	Olive oil
1 tablespoon	15 mL	Water

PREPARATION STEPS:

1. Bring all the ingredients to room temperature.

2. Make a well with the flour and salt on a wooden board.

3. Add the eggs, oil, and water to the well, and start to incorporate the flour into the wet ingredients until a smooth ball of paste is achieved.

4. Knead the dough to a slightly smooth and elastic consistency.

5. Tightly cover the dough with plastic wrap and let it rest 1 hour.

RECIPE 4–25

Pasta Dough, Suitable for Stuffing

Recipe Yield: 8 portions

MEASUREMENTS		INGREDIENTS
U.S.	**METRIC**	
2 cups	473 mL	Semolina flour
2 cups	473 mL	All-purpose flour
1 teaspoon	5 mL	Fine salt
4 each	4 each	Eggs
8 each	8 each	Egg yolks

PREPARATION STEPS:

1. Bring all the ingredients to room temperature.

2. Make a well with the flours and salt on a wooden board.

3. Add the eggs and the egg yolks to the well, and start to incorporate the flour into the wet ingredients until a smooth ball of paste is achieved.

4. Knead the dough to a slightly smooth and elastic consistency.

5. Tightly cover the dough with plastic wrap and let it rest 1 hour.

TIP

The flavor and color of pasta dough can be altered by the addition of richly flavored ingredients that will also impart their color into the dough. The first option the chef has is to change the flour that is used. The use of buckwheat, chestnut, and whole wheat make very good colored pasta. There are many other ingredients that can be used to color and flavor pasta, giving the chef a limitless choice. Adding natural ingredients that have a bright color and rich flavor will give the pasta a new and unmistakable value and versatility.

- Beet juice or beet powder makes a light purple color.
- Saffron creates a very bright yellow color.
- Tomato paste gives a light red color.
- Pureéd blanched spinach or basil makes a bright green color.
- Squid ink turns the pasta a deep black color.
- Turmeric makes a light yellow pasta.
- Herbs that have been blanched and patted dry can be rolled between sheets of dough. This traps them and provides an unusual effect of their being suspended within the pasta.
- Two different colored pastas can be rolled together to form one two-colored dough (either side-by-side or front and back).

Cold Hors d'Oeuvres

Cold hors d'oeuvres tend to be on the lighter side compared to hors d'oeuvres that are prepared and served warm. The ingredients are often less dense and substantial, lending themselves to lighter nibbling. They are often served alone, but they are also commonly served in conjunction with warm hors d'oeuvres.

The variety of hors d'oeuvres that fall under this category is limitless, as are the ingredients from which they are made. They can be as spartan as the olive or dried fruit, or as complex as caviar served with a full complement of garnishes. The following groupings of cold hors d'oeuvres merely scratch the surface, but they are a required part of the modern garde manger's repertoire and often less understood. They include caviar; sushi and sashimi; cold soups; olives, nuts, and dried fruits; dips and spreads; and canapés, toasts, and crackers.

CAVIAR

The tale of **caviar** is perhaps older than civilization. The use of fish roe, as food, exists in the most primitive of societies and the cured form, as we know it, is thought to have originated thousands of years ago in China. Early Persians enjoyed caviar and thought it had "medicinal qualities." Ancient Greek and Roman literature contains many references to the special presentations of the beautiful black caviar beads served at banquets. Today, as then, it is among the most exclusive and expensive of all preserved foods. However, caviar and lesser quality roe are available in specialty food emporiums and neighborhood grocery stores for the masses to enjoy.

The term *caviar* is believed to have its origins in the Turkish word *havyar,* which translates to mean "salted fish roe." True caviar is the salted roe (egg) of the sturgeon. It is literally a salt-cured fish egg. Other fish eggs are erroneously referred to as caviar. These products should be referred to as *fish roe* (e.g., salmon roe, lumpfish roe, paddlefish roe). The term *caviar* should be used to refer exclusively to the eggs derived from the sturgeon.

Types of Caviar

There are basically three different kinds of sturgeon and then variations of each of those. Twenty-five species of sturgeon are known to exist globally, but only in the Northern Hemisphere. To provide less expensive alternatives to satiate the caviar customer, other fish roe are being used. There are several sources for these products, chiefly from North America and the region surrounding the Caspian Sea.

The *beluga* is the largest sturgeon, followed by the *osetra* and then the *sevruga.* Regardless of the size of the fish, the roe is extracted by harvesting the roe sac from the fish. It is best to perform this task while the fish is still alive. Using just the right amount of salt is the key. The skeins of roe are very gently rubbed through a wide-meshed sieve to separate the grains. Bits of the skein are picked out by hand. Then a measured amount of salt is slowly mixed into the roe, by hand, with a gentle churning motion. Only enough salt to draw out some moisture from the eggs, to keep them from clumping, is added. The amount of salt usually ranges between 2.8 and 3.5 percent of the overall weight of the eggs.

Eggs prepared in this manner are referred to as *malossol,* which translates to mean *little salt.* The term is printed on the tin or jar lid and is only used in reference to sturgeon roe. A slightly greater amount of salt is used for lesser quality eggs. The salt is used to preserve the quality and enhance the flavor of the fish eggs. It also performs a third and very important role. It prevents the eggs from freezing when they are stored at their usual holding temperature of 28° to 32°F (−2.2° to 0°C). Once salted, the eggs are packed in tins and allowed to cure for a minimum of 2 weeks before being presented to a consumer (Figure 4-17).

AMERICAN CAVIAR

European settlers of the "New World" discovered sturgeon to be the most prolific fish of the North American continent and by the nineteenth century created the world's greatest

FIGURE 4–17A Female brook trout heavy with roe

FIGURE 4–17B Separating the skin from the roe

FIGURE 4–17C Draining the salted roe

FIGURE 4–17D Canning the cured roe

commercial caviar industry. Until the beginning of the twentieth century, the United States produced 90 percent of the world's caviar, with 60,000 pounds annually coming just from Lake Michigan. At one time, caviar was so common in America that it was served in saloons to encourage thirsty drinkers. Hudson River sturgeon was so plentiful that the flesh was referred to as "Albany beef." A nickel could buy a serving of the best caviar available in New York, and many of the most lavish establishments, including the Waldorf Astoria, offered free-flowing caviar as an amuse-bouche opening to an elegant meal. Caviar was also a common food in California during the gold rush days. But by 1910, the little-understood lake sturgeons were nearly extinct and American production was stopped.

Recently, the United States has made a strong comeback in caviar production. The legendary lake sturgeon caviar is available only from Canada, but the unique paddlefish from the United States produces an amazingly similar egg. Another excellent caviar comes from the hackleback sturgeon, and the roe of many other indigenous species have become refined and popular. In the Pacific Northwest, Native Americans once preserved salmon roe smoked in bags made of deer stomachs; the same area today supplies a huge world market for salmon roe caviar.

U.S. Government law allows only the roe of sturgeon may be called "caviar," whereas the roe of other fish may only be called "caviar" when the name of the fish comes first. The following is a descriptive list of several caviars made from American freshwater fish (See Figures 4–18 and 4–19 for several types of caviar and roe and assorted caviar spoons.)

- *American Sturgeon.* Sturgeon resemble a prehistoric creature, but they are actually the modern relics of an ancient group of fish with fossil records dating back 100 million years. These fish can grow to 10 feet in length and more than 300 pounds in weight. Many fish of 800 to 1000 pounds or more were caught around the turn of the twentieth century, but by the 1920s, the biggest sturgeon were gone.
- *Lake Sturgeon.* Maturing sexually in 15 to 20 years, lake sturgeon can weigh approximately 100 pounds, spawn once every 5 to 7 years, and yield about 25 percent of their body weight in roe. The caviar is comparable in size, color, and flavor to Russian beluga.
- *Hackleback Sturgeon.* This variety is native to the Mississippi/Missouri River system and is faster growing and smaller than most sturgeon, growing to about 38 inches at maturity. Sometimes called the "shovelnose sturgeon" or the "sand sturgeon," it is the most abundant sturgeon in the American wild. The eggs are small, almost always black, or near black, and can have a sweet, buttery flavor reminiscent of beluga caviar.
- *White Sturgeon.* This variety is indigenous to the waters and rivers of North America's Pacific Coast, from southern Alaska to Ensenada, Mexico. It is a massive sturgeon, sometimes measuring 20 feet long and weighing 1500 pounds; it can live for more than 100 years. It is the largest freshwater fish in North America.
- *Paddlefish.* Commonly called "spoonbill," these fish are a cartilaginous cousin to sturgeons. They yield roe ranging in color from pale through dark steel-gray and golden "osetra brown." The paddlefish is found in the rivers of Tennessee, Alabama, and Missouri. The roe are processed in exactly the same manner as caviar from the Caspian Sea. The caviar is smooth and silky with a rich, complex flavor.
- *Salmon.* Salmon roe is sometimes referred to as *red caviar.* Most of the salmon eggs on the market come from chinook or coho salmon caught or raised in the West (including Alaska and Canada). These prized roe are large—sometimes the size of a pearl—and are a glistening orange-red color.
- *Whitefish.* "American golden caviar" comes from small freshwater whitefish found in all Northern countries, including the United States (the Great Lakes region) and Canada. The roe has a fine, firm texture and is a pale-orange, golden color that is almost iridescent. The tiny eggs pop in your mouth, as do sevruga caviar. Whitefish roe has an uncommon subtle flavor and a fine crispy texture.
- *Trout.* The small eggs come from the sac of the rainbow trout and burst with concentrated flavor. Trout roe boasts wonderfully subtle notes of earth, seaweed, iodine, and egg yolk.

FIGURE 4–18 Samples of assorted caviar and roe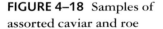

- *Bowfin.* Better known by its Cajun name *choupique,* bowfin is not related to, but is even more ancient than, the sturgeons. The bowfin is the only remaining living specimen of an ancient group of fish that lived more than 180 million years ago. The choupique's name comes from the word *shupik,* a Choctaw Indian word that translates as "mudfish." It is also known as bowfin, swampfish, and cypress trout. The roe is called "Cajun caviar" in Louisiana. This bony fish yields a black roe that makes a good, less expensive substitute for sturgeon caviar. (Unlike sturgeon roe, bowfin roe turns red when heated.) Bowfin caviar is firm and shiny with natural black eggs resembling sturgeon caviar. Clean and nicely separated berries have a distinctive, lively flavor.
- *Lobster.* In contrast to all of the other caviar and roe, lobster roe is cooked before service, creating a totally different texture. It is most commonly used as a topping on other cooked foods, such as she-crab soup, rather than as a tasting unto itself.

IMPORTED CAVIAR

The history of manufacturing black caviar goes back many centuries. Sturgeon is one of the most ancient representatives of the fauna still in existence on the earth. Russian fishermen learned to make caviar in the twelfth century. In the beginning of the eighteenth century, the fish craft in Russia was announced as a state monopoly. By the decree from January 6, 1704, Peter the First ordered "all fish catching undertaken by the Great Tsar." Soon after the occurrence of Peter's decree for the management of the fish craft, a special office was created in Astrakhan.

The Caspian Sea is home to the most popular members of the European and Asiatic sturgeons. In 1925, the Caspian Sea fisheries began commercial production, as we know it today. Two nations, the U.S.S.R and Iran, bordered the Caspian Sea and had successfully managed the fisheries until the breakup of the Soviet Union. Today, five independent nations (Russia, Kazakhstan, Turkmenistan, Iran, and Azerbaijan) border the Caspian Sea and are engaging in a great deal of unregulated production. Seeing caviar as a great source of revenue, newly independent states such as Azerbaijan and Kazakhstan started poaching the depleted Caspian sturgeon. The primary caviar sturgeons from the Caspian Sea are beluga, sevruga, and osetra.

- *Beluga Sturgeon.* The largest and most rare sturgeon, the beluga yields large, translucent, golden-gray berries. Beluga caviar has the largest grains and is the most sought after caviar in the world.
- *Sevruga Sturgeon.* Sevruga caviar is the smallest and most abundant of all the sturgeon caviars. The eggs are predominantly steel-gray.
- *Osetra Sturgeon.* Osetra/asetra caviar is unique in that it varies in size, color, and flavor. The eggs are medium sized and golden brown.
  Other notable imported caviar are tobiko and kaluga:
- *Tobiko Sushi Caviar.* This caviar is commonly used on many sushi dishes for flavor enhancement.
- *Kaluga Caviar.* The color of this roe, from the Amur River system in Manchuria, China, varies from black to golden brown. It has a slightly spicy flavor and an intense salty nature similar to caviar from the Caspian Sea.

Caviar Handling and Service

Caviar roe is tender and fragile, and requires a nimble hand when working with it. The following suggestions are commonly held rules for handling caviar:

- Lift while spooning out.
- If spreading, ease it softly with a teaspoon.
- Caviar should be served from a nonmetal spoon. Caviar spoons are widely available in bone, tortoise shell, and mother-of-pearl (Figure 4–19). Any metal, including silver, imparts a metallic flavor to the granules.
- Salmon caviar has a large "grain" (egg size). When used to decorate canapés, single grains may be set in place with the tip of a paring knife.

FIGURE 4–19 Assorted caviar and roe on mother-of-pearl spoons. Courtesy of Foodpix/ Picture Arts

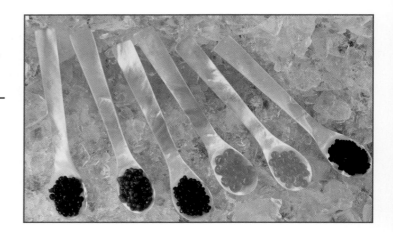

- It is best to rinse lumpfish, whitefish, and salmon roe caviars to prevent any color from running. Turn out caviar into a fine-mesh strainer. Rinse gently with cold tap water. Shake and then turn onto several layers of paper towels to absorb moisture. Then use as directed.
- It is important when serving caviar that the jars be removed from the refrigerator 10 to 15 minutes before serving and then opened immediately before consumption.

CAVIAR SERVICE

The best caviar should be served with simple accompaniments, so as not to compete with or confuse the flavor and texture of the eggs (Figure 4–20). The caviar should be kept in its original tin or placed in glass bowls and set on crushed ice; the proper temperature for service is 28°F (−2.2°C). The lid can be set off to the side where consumers can view the label. Spoons made from mother of pearl, bone, horn, wood, or glass should be used for service. Metal should never be used because the salt will tarnish soft metals and will negatively impact the flavor of the caviar.

FIGURE 4–20 Elegant caviar service. Courtesy of Getty Images/ Foodpix

Accompaniments for the best caviar consist of lightly toasted white bread, dark Russian rye, brioche, or wheat blini, often topped with a small amount of whipped unsalted butter or crème fraîche.

Less expensive roes usually call for the use of fresh lemon, minced onions, finely chopped egg whites and egg yolks, red skin potatoes, sliced cucumbers, and chives and sour cream to accompany the toast points. These accompaniments embellish the flavor and texture of the lesser quality roe.

CALCULATING QUANTITIES FOR SERVICE

Calculating the need for caviar is often tricky and can be particularly challenging to the garde manger chef who is careful about balancing customer expectations with food cost requirements. The following are suggestions for calculating quantities for service:

- For serving caviar straight out of the jar or tin, and true caviar enthusiasts are the customers, estimate at least $\frac{1}{2}$ to 1 ounce (14 to 28 g) per person.
- When serving appetizers with caviar on top, and it is desired to taste the caviar over the other ingredients on the appetizer, use a dollop of a heaping $\frac{1}{2}$-teaspoon (2.5 mL) of caviar. There are between eight and ten $\frac{1}{2}$-teaspoon (2.5-mL) servings per ounce (28 g) of caviar.
- For a touch of caviar, used more as a garnish, use $\frac{1}{4}$ teaspoon (1 mL) or less. There are about twenty $\frac{1}{4}$-teaspoon (1-mL) servings per ounce (28 g) of caviar.
- If caviar is being served by itself, or with crackers or toast points, a 2-ounce (56-g) jar serves approximately 4 people. For party canapés, estimate one 2-ounce (56-g) jar serving approximately 8 people.

Care and Storage

It is best to design the catering sales menu so that caviar is sold by the container size, so no storage of partial leftovers is needed. However, there are many times when caviar needs to be stored for a period of time. When that is the case, the following suggestions are recommended:

- Do not open caviar until needed.
- To avoid having the berries burst, caviar must be refrigerated at 28° to 32°F ($-2.2°$ to 0°C). This can be done either by putting the tin in the coldest part of the refrigerator or by placing the tin in a bowl and surrounding it with crushed ice.
- If caviar is left in the tin, the surface should be smoothed and a sheet of plastic wrap should be pressed directly onto the surface before placing it back in the refrigerator.
- Cover and refrigerate any leftovers promptly and use within a day or two.
- Turn the tin over each day so the oil reaches all of the eggs.
- Fresh caviar can be stored under refrigeration for 15 to 20 days (unopened). Consume caviar within 2 to 3 days once it has been opened. Unopened pasteurized caviar can be kept on the shelf for 6 months.
- Refrigerate but refrain from freezing the caviar. Freezing of caviar can be done, but it is not recommended. Freezing can toughen the caviar roe membrane and alter the flavor. If you will not be able to serve or consume the caviar within the 3 weeks recommended, or if you are not able to refrigerate it properly, freezing is an option. If the caviar is frozen, it must thaw slowly under refrigeration over most of a day before serving. It would be best to serve previously frozen caviar with accompaniments, as listed under Caviar Service.

SUSHI AND SASHIMI

Sushi is commonly described as cold cooked rice dressed with vinegar that is shaped into bite-sized pieces and topped with raw or cooked fish, or formed into a roll with fish, egg, or vegetables and wrapped in seaweed. It is not surprising that some people mistakenly associate the word *sushi* with raw fish, possibly because many sushi varieties are prepared using some type of fish or seafood. Actually, **sashimi** means "raw fish" and is made from slices, slabs, or squares of raw fish served with typical Japanese condiments.

Sushi originated as a way of preserving funa, a type of fish. The fish was salted and allowed to mature on a bed of vinegar rice, after which the rice was discarded. Before long, vinegar rice came to be eaten together with the fish and then other ingredients. Thus, the word *sushi* was derived: the marriage of vinegar rice with other ingredients. Many different combinations of sushi and ways of serving them have evolved.

Types of Sushi

The following list, although not comprehensive, details some of the most common styles of sushi served.

- *Nigiri Sushi.* The word *sushi* alone commonly refers to nigiri sushi, a "hand-shaped" sushi commonly served at sushi restaurants. Nigiri sushi is representative of Tokyo cuisine in that often some type of seafood or fish is incorporated.
- *Oshi Sushi.* The rice merchants of Osaka—the financial capital of Japan—developed oshi sushi, or "pressed sushi." For pressed sushi, vinegar rice is packed into a mold and covered with marinated fish or other ingredients. When unmolded, the resulting loaf of sushi is cut into bite-sized pieces.
- *Maki Sushi.* Maki sushi is a "rolled sushi" with narrow strips of different ingredients (seafood, crisp vegetables, or pickles) layered on a bed of vinegar rice and spread on a sheet of nori, or seaweed, thus calling it "nori-maki sushi." Nori-maki is the most well known sushi in the United States and has the most variety. Just about any ingredient can be rolled into the center, from crisp vegetables to strips of omelet.
- *Bara Sushi.* This is a salad-type sushi.
- *California roll.* A variety of maki sushi, California roll is usually stuffed with avocado.
- *Chirashi Sushi.* This sushi is the easiest type of sushi to make and is made in all Japanese kitchens. It is also known as "scattered sushi" or *chirashi-zushi.* It is simply sushi rice with other ingredients mixed in or placed on the rice. Chirashi-zushi without any seafood often makes its appearance in lunch boxes. It is taken on picnics and is often sold on railway station platforms.
- *Maze Sushi.* Many types of sushi fall into the lunch or snack food category of maze sushi, or "mixed sushi." For example, inari sushi consists of deep-fried bean curd pouches stuffed with mixed vinegar sushi rice. Fukusa sushi or "silk-square sushi" uses square, paper-thin crepes to wrap the vinegar sushi rice. The word *fukusa* means "silk squares" (silk fabric), which are often used to wrap presents or precious articles in Japan.

Sushi Rice

Whether it is hand-packed, mold-pressed, or rolled in seaweed, the one constant ingredient in sushi is vinegar rice. Vinegar rice is slightly harder than plain boiled rice because the rice is cooked with a little less water. Many restaurants cook rice in hot water rather than cold to achieve this effect. Quick cooling while tossing is the key to making good sushi rice. This helps produce the desired consistency and gives the sushi rice a shiny gloss.

Tossing rice is traditionally done in a large wooden tub called a *hangiri.* The size of the tub allows spreading out the rice in a thin layer. The wood absorbs the moisture from the hot rice and makes the rice cool faster, thus preventing the grains from being mashed during tossing. As a substitute for wood, a large plastic tub can be used while stirring with a wooden rice paddle. Avoid using metal because the acidic vinegar reacts with it, creating an unpleasant taste.

COOKING RICE FOR SUSHI

Electric and gas rice cookers are available universally, and are recommended as the most reliable method for achieving consistent results. When not using a rice cooker, follow standard procedures for the preparation of rice, only reduce the quantity of water by about 10 percent.

1. Measure the rice. About 4 ounces (100 g) of rice per person is plenty when making sushi.
2. Wash the rice. The Japanese word for rice washing is *togu*. Its meaning is closer to "polishing" than "washing." First, cover the rice with an ample amount of cold water, which will immediately turn cloudy. Drain the rice right away. Next, with an open palm, rub the rice while it is down against the bottom of the bowl. Agitate and force the grains of rice against each other. Add more water and discard again. Repeat this process several times until the water is almost clear.
3. Soak the rice. Cover the rice with cool water and let the rice soak for 10 to 15 minutes.
4. Drain the rice. Drain the washed rice into a sieve or colander fine enough to catch all of the grains. Let the rice stand for 5 minutes, or until completely dry.
5. Add water. Place the rice in the electric cooker and add water. The standard ratio for Japanese white rice is 1:1. For sushi rice, reduce the amount of water by 10 percent.
6. Cook the rice. Replace the lid and press the ON button, the rice cooker will cook the rice.
7. Wait. This step is critical. When the electric cooker clicks off, remove the lid, stir once, and replace the lid for at least 10 minutes. Patience is important.
8. Add the sushi-zu. See the Shari method (Recipe 4–27).

RECIPE 4–26

SUSHI-ZU

Recipe Yield: 2 liters

MEASUREMENTS		INGREDIENTS
U.S.	**METRIC**	
60.8 ounces	1.8 liters	Rice vinegar (use only real Japanese rice vinegar)
3.1 pounds	1.4 kg	Granulated sugar
14.1 ounces	400 g	Salt
0.7 ounces	20 g	Kombu (kelp)

PREPARATION STEPS:

1. Combine all ingredients in a large, nonreactive pot.

2. Heat over medium flame, stirring occasionally to dissolve sugar and salt.

3. Just before the liquid reaches a boil, remove from heat and cool, removing and discarding the kombu.

4. Store in a covered container at room temperature. Sushi-zu keeps indefinitely.

RECIPE 4–27

SHARI, SUSHI MESHI, SUSHI RICE

Recipe Yield: 2.2 pounds (1 kg)

Note: At restaurants and hotels, unused shari is not served to customers the following day, but is often made into rolled sushi and eaten by kitchen and restaurant staff.

Note: Shari (sushi rice) should be made ahead of time and allowed to cool before use. Only at some very fine sushi counters is warm shari used. In this case, the sushi is to be consumed immediately. Even after only a few minutes, warm rice will damage the quality of most fish used in sushi.

Note: The standard industry recipe is based on 1 kg of cooked rice. Of course, you can prepare a smaller amount. Note that 100 mL of raw rice will become a little more than 100 g of cooked rice. So to make half the standard recipe, start with 400 mL of raw rice.

MEASUREMENTS		INGREDIENTS
U.S.	**METRIC**	
2.2 pounds	1 kg	Cooked rice
4.6 ounces	130 g	Sushi-zu

PREPARATION STEPS:

1. In a wide shallow container (traditional wooden *hangiri* is best), place the hot rice and pour all of the measured sushi-zu over the rice.

2. With a flat paddle (traditional *shamoji* or *miyajima*) or a wooden spoon, toss the rice with a cutting motion, taking care not to overmix. While cutting the rice, cool it with a handheld fan. Cooling helps the rice absorb the vinegar and give the finished shari (sushi rice) a nice sheen.

3. Leftover shari should be covered with a damp cloth; do not refrigerate.

FIGURE 4–21 Typical sushi tools and ingredients

Preparing Sushi

Many people think of raw fish when they hear the word *sushi,* but it is important to understand that sushi is based on the combination of several ingredients (see Figure 4–21 for typical sushi tools and ingredients).

- Fish (raw or cooked)
- Rice
- Seaweed
- Vegetables
- Condiments (including soy sauce, wasabi or rice vinegar)

   (Figures 4–22 and 4–23 illustrate the preparation of a common form of sushi known as maki sushi which is being sliced in Figure 4–24.)

   Sashimi, however, is simply a Japanese dish of raw fish that is often served with the same condiments as sushi; they could include any of the following:

- Soy Sauce
- Wasabi
- Vinegar
- Asian pepper sauce

FIGURE 4–22 Preparing maki sushi

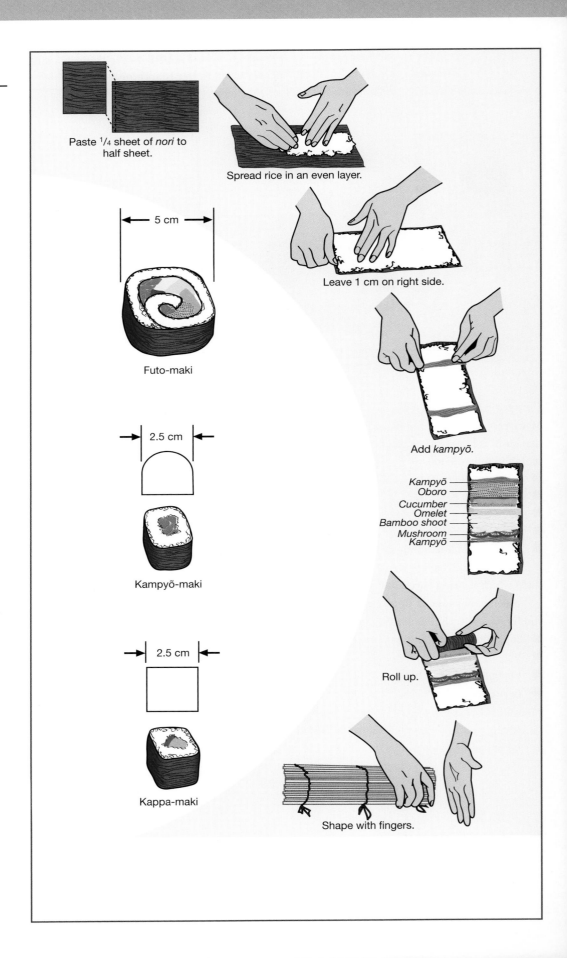

Paste ¹/₄ sheet of *nori* to half sheet.

Spread rice in an even layer.

Leave 1 cm on right side.

Add *kampyō*.

5 cm

Futo-maki

2.5 cm

Kampyō-maki

2.5 cm

Kappa-maki

Kampyō
Oboro
Cucumber
Omelet
Bamboo shoot
Mushroom
Kampyō

Roll up.

Shape with fingers.

FIGURE 4–23A Laying out the ingredients

FIGURE 4–23B Tucking the roll

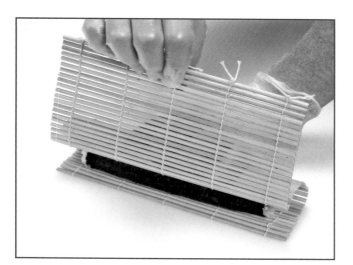

FIGURE 4–23C Using the mat to tighten the roll

FIGURE 4–24 Slicing the maki sushi

Assorted finished sushi (clockwise from left: two styles of nigiri-sushi, chirashi-sushi, maki sushi)

Elegant hors d'oeuvre buffet featuring sushi

COLD SOUPS

Cold soups are often used as a starter course, in lieu of salad or appetizers. Because of their refreshingly cool nature, cold soups are typically served during warmer seasons or in hotter climates. Most cultures have some form of cold soup, seasoned in a way that is typical of their other fare.

Like their warm counterparts, cold soups generally consist of a liquid base, a garnishment, and seasoning. The liquid can be chilled meat stock, fruit purée, vegetable broth, or dairy products. Cold soups may contain meat, poultry, or seafood garnishes, such as with melon soup with proscuitto shavings, or gazpacho with tequila flavored shrimp. However, cold soups are more commonly made with fruits, vegetables, or dairy products.

RECIPE 4–28

COLD CREAM OF LEEK AND TOMATO SOUP

Recipe Yield: 2 quarts

MEASUREMENTS | INGREDIENTS

U.S.	METRIC	
$3/4$ cup	177 mL	Butter
7 each	7 each	Leeks, white, part diced, $1/2$ greens reserved
$12\frac{1}{2}$ ounces	350 g	Onions, large, diced
$1/2$ cup	118 mL	Flour
3 pounds	1.4 kg	Tomatoes, fresh, very ripe, chopped
2 teaspoons	10 mL	Salt
1 teaspoon	5 mL	Pepper
$1/8$ teaspoon	0.5 mL	Granulated sugar
$2\frac{1}{4}$ cups	532 mL	Light cream

PREPARATION STEPS:

1. Julienne cut 1 cup of leek greens. Lightly sweat in $1/4$ cup of butter. Reserve for garnish.

2. Sauté leeks and onions in remaining melted butter until onions are wilted.

3. Sprinkle with flour, stir, and cook for a few minutes without browning to form a roux.

4. Add tomatoes and simmer gently for 2 hours, stirring occasionally.

5. Add salt, pepper, and sugar.

6. Purée mixture.

7. Strain soup, cool, and add cream.

8. Serve cold, garnished with julienned leek greens.

GINGER PEACH SOUP

Recipe Yield: 2 quarts

Note: If peaches are ripe, the skin can be peeled off using a sharp paring knife. If they are less ripe, cut a small cross in the bottom of each peach, blanch them in boiling water for 20 seconds, remove with a slotted spoon and plunge into a bowl of ice water.

MEASUREMENTS		INGREDIENTS
U.S.	**METRIC**	
3 pounds	1.4 kg	Peaches, fresh ripe
¼ cup	59 mL	Lemon juice, fresh
3 cups	710 mL	Buttermilk
1⅓ cups	315 mL	Apple juice
1 teaspoon	5 mL	Fresh ginger, grated
2 teaspoons	10 mL	Honey
1½ teaspoons	7 mL	Kosher salt
8 each	8 each	Aromatic geranium leaves
OR		
24 to 36 each	24 to 36 each	Peach slices, rubbed with lemon

PREPARATION STEPS:

1. Peel and pit the peaches, rubbing them with lemon juice as needed to prevent discoloration.

2. In a food processor, purée the peaches. Scrape the peach puree into a large bowl.

3. Stir in buttermilk, apple juice, ginger, honey, and salt.

4. Refrigerate until cold.

5. Pour into chilled bowls and top each serving with a geranium leaf or several peach slices.

RECIPE 4–30

CANTALOUPE-RUM SOUP

Recipe Yield: 2½ quarts

Note: May substitute ¼ cup orange juice concentrate for orange flavoring.

MEASUREMENTS		INGREDIENTS
U.S.	**METRIC**	
1 each	1 each	Cantaloupe, medium sized, ripe
¼ pound	113 g	Butter, softened
⅛ teaspoon	0.5 mL	Cloves, ground
⅛ teaspoon	0.5 mL	Nutmeg, ground
¼ teaspoon	1 mL	Coriander, ground
⅛ teaspoon	0.5 mL	Salt
1 cup	237 mL	Brown sugar
8 cups	1.9 L	Heavy whipping cream
¼ teaspoon	1 mL	Orange flavoring
½ each	½ each	Lime, juice of
¼ cup	60 mL	Rum, dark
As needed	As needed	Mint leaves

PREPARATION STEPS:

1. Cut, dice, and peel cantaloupe. Over medium heat, cook in saucepot until very soft. Turn off heat. Mash until pulverized.

2. Add butter and spices, and part of the heavy cream to cool the mixture.

3. Blend with mixer until finely puréed.

4. Add orange flavoring, lime juice, and remaining cream.

5. Refrigerate for 1 hour.

6. Before service, drizzle in rum to taste, and garnish with mint leaves.

RECIPE 4–31

AVOCADO-CRABMEAT SOUP

Recipe Yield: 2 quarts

MEASUREMENTS		INGREDIENTS
U.S.	**METRIC**	
¼ cup	59 mL	Butter
¼ cup	59 mL	Flour
1 teaspoon	5 mL	Salt
⅛ teaspoon	1 mL	Black pepper, fresh ground
1 teaspoon	5 mL	Dried mustard
4 cups	946 mL	Milk (or light cream), warmed
2 tablespoons	30 mL	Onion, minced
½ cup	118 mL	Celery, minced
1 cup	237 mL	Avocado pulp, coarsely chopped
2 cups	473 mL	Crabmeat, cooked
As needed	As needed	Chives, fresh
8 each	8 each	Crab claws, cooked, chilled

PREPARATION STEPS:

1. In a saucepot, melt the butter and blend in the flour, salt, pepper, and mustard. Cook roux over low heat until it begins to brown.

2. Gradually stir in warm milk and cook until mixture is smooth and thick, stirring constantly.

3. Blend in remaining ingredients, except chives and crab claws, and transfer to a clean container for refrigeration. Chill for 1 hour.

4. Serve chilled, garnished with fresh chives and crab claws.

RECIPE 4–32

CHILLED GAZPACHO SIPS

Recipe Yield: 1½ quarts

MEASUREMENTS		INGREDIENTS
U.S.	**METRIC**	
2 each	2 each	English cucumber
1 each	1 each	Red onion, finely diced
2 each	2 each	Tomato, ripe, finely diced
1 each	1 each	Green bell pepper, seeded and finely diced
1 each	1 each	Red bell pepper, seeded and finely diced
3½ pints	1.66 L	Tomato juice, canned
2 teaspoons	10 mL	Caster sugar (superfine)
6 tablespoons	90 mL	White wine, dry
6 each	6 each	Garlic cloves, minced

PREPARATION STEPS:

1. Split the cucumbers in half, lengthwise. Scoop out the seeds, using a spoon.

2. Finely dice one cucumber for the soup. Cut the remaining cucumber into thin slivers, as garnishes. Reserve.

3. Blend the diced cucumber, onion, tomato, and bell peppers in a bowl. Reserve.

4. Combine the tomato juice, sugar, white wine, and garlic in a large pitcher. Chill for at least 3 hours, or until ready to serve.

5. To serve, pour juice into 32 shot glasses, then spoon in vegetable mixture.

6. Garnish each glass with a thin sliver of cucumber.

OLIVES, NUTS, AND DRIED FRUITS

The serving of small foods to both family and guests is an ancient tradition. Before complex foods and presentation styles became prevalent, many appetizers consisted of those foodstuffs that were gathered from the nearby orchards, fields, and streams, and set upon the table with little fanfare. Their preparation was basic, if not nonexistent, and their presentation was equally spartan. In the modern world, the simplicity and unadulterated flavors of these traditional foods are still appreciated as they continue to be proudly offered at home, in restaurants, and in banquet settings.

Olives

No other food can rival the olive for its role in the gastronomy of various cultures throughout history. The olive tree is revered as sacred and immortal; its branches symbolize peace; its fruit is regarded as an indispensable food; its oil signifies prosperity and purity, and has been an essential element in religious ritual and culinary tradition across numerous cultures for many centuries.

It is generally believed that the first olive trees came from countries surrounding the Mediterranean Sea. They are among the oldest known cultivated trees in the world, being grown even before the written language was invented. Missionaries brought the olive tree to North America in the 1700s. Franciscan monks planted the first olive trees in California at the San Diego Mission in 1769. The trees were planted throughout California for the purpose of producing oil, but the less expensive traditional European olive oils prevailed in the marketplace. In the late 1800s, Freda Ehmann and her son Edwin began experimenting with new ways to market the olive; they found success with the California-style ripe black olives they were producing.

Today, more than 1200 growers on 35,000 acres grow the olive in the warm inland valleys of California, continuing the long and venerable tradition of the olive. One of the newest regions for producing olives is California, which has the largest olive packers in the world and its trees have the best yields.

The olive has always played a key role in the development of commerce in the Mediterranean cultures. Today, it is a crop of major economic importance in many countries throughout the world. More than 15 million acres of olives are planted worldwide; 90 percent of those border the Mediterranean. The annual olive harvest is nearly 10 million tons. More than 1 million tons are processed as table olives, and the balance is pressed for olive oil.

QUALITY POINTS WITH OLIVES

In contrast to the popular misconception, green and black olives do not come from two different trees. Although there are many varieties of olives, their color indicates the degree of ripeness at which they were picked. Olive packers select their olives based on desired characteristics, and degree of ripeness contributes greatly to an olive's texture and flavor. As a general rule, the darker the color, the riper the olive when it was picked.

In the Northern Hemisphere (and opposite in the Southern), green olives are picked in early autumn, around September and early October. They usually have a firm, crisp texture and nutty flavor. Red or brown-shaded olives are picked in late October and early November. Their texture begins to soften, and their flavor is slightly sweeter than that of green olives. Black olives are allowed to remain on their branches until they have fully ripened (deadened), anytime from late November through January, sometimes later. Like most fruits, olives become sweeter, softer, and richer in flavor as they ripen. In many ways, olives parallel bananas in their stages of ripening—in color, flavor, and texture. They start out firm and green, mature into yellow, and eventually turn soft, sweet, and black.

The overall quality of an olive depends very much on the method of harvesting. In these modern times, commercial growers often use tree shakers to vibrate the fruit from their branches onto tarps lying on the ground below. Less mechanical methods, sometimes used in smaller groves or poorer countries, include beating the branches with sticks to hasten the olive's fall onto

FIGURE 4–25 Picking olives by hand

the tarps. Either method generally results in bruised olives that have been picked at varying degrees of ripeness.

Olives that are handpicked can be carefully plucked or raked at the proper stage of ripeness, and with little bruising to the fruit (Figure 4–25). The resulting benefits include a consistent texture and even flavor, which is crucial to a quality product.

There are several important considerations to make when selecting table olives. It is generally best to decide on olives that are:

- Handpicked (not roughly harvested in batches)
- Naturally cured (never in lye)
- Fresh (not pasteurized)
- Varietals (as compared to generic labels such as "medium black-ripened")

COMMON OLIVES

There are many olives grown throughout the world. As noted, almost 90 percent are grown for the production of oil, while the remaining 1 million tons are produced as table olives. "Walk the markets of any medium-sized town anywhere along the Mediterranean coast and you're sure to see mounds of marvelous local olives glistening like precious stones in the summer sun. Jade green, onyx, amethyst—a wealth of colors, sizes, shapes, and flavors. This fruit helped make the Mediterranean diet famous," notes Ari Weinzweig (2003) of Ann Arbor's celebrated Zingerman's Delicatessen (p. 45).

The following types of olives represent the many table varieties available and include many of the most popular olives on the world market.

Argentina

- The Alfonso olive is large, meaty, and purple, with a uniquely fruity flavor. They are common to the north and west coasts of South America.

France

- Picholines are the crisp, uncracked green olives from southern France. These have a nutty flavor and anise undertones. Besides their use as table olives, they are pleasant additions to stews and fennel dishes.
- Niçoise olives are some of the most coveted of all. Real niçoise olives are tiny and black, and possess a uniquely delicate flavor. They are essential to the preparation of many authentic regional Mediterranean dishes.

- Nyons olives are often expensive, and imitated. They are first dry-cured, then aged in brine to plump them. Authentic Nyons are slightly wrinkled and a bit duller in appearance than the North African imitations that lack the rich flavor.

Greece

- Greece grows more than 1,500,000 acres of olives. Marketplaces generally offer more than 20 varieties from which to choose.
- Greek olives include amfissa, Elitses (tiny), Hondroelia (gigantic), Konservolia (or natural black olives), naflion, and kalamata, which are allowed to ripen on the tree until their skin turns purple-black.

Italy

- Naturally cured Cerignola olives, from the Province of Puglia, are medium-ripe and light brown in color.
- Purple-brown Gaeta olives from central Italy are frequently used on pasta dishes or in pizzas.

Morocco

- The olive is the most predominant fruit tree in Morocco.
- Morocco grows picholine, marocaine, and Zitoun olives, and they are used in both table olives and olive oil.

Spain

- Spain is the world's largest exporter of table olives. The majority of its exports are the Spanish-style pimento-stuffed green olives.
- The varieties grown in Spain are Manzanilla, Hojiblanca, and Gordal. The large, green Gordals (gorda is Spanish for "fat") are known as Queen olives, and have a firm, meaty texture. The cracked brownish green Manzanillas, which are smaller, crisper, and nuttier than Gordal, produce the best tasting table olive, while Hojiblancas are used primarily for olive oil production and make marginal quality table olives.

United States

- California produces nearly all the olives in the United States, growing on more than 34,000 acres.
- Manzanilla olives represent more than 70% of the olives grown in the United States. Other varieties are Sevillano, Mission, Ascolano, and Barouni. Although most olives become the California black ripe olives, the Sevillano are large, brine-cured, green, and meaty, similar to the Spanish-style Gordal.

Assorted olive varieties

CURING OLIVES

Uncommon among fruits and virtually unique to olives, olives are inedible when in their raw state. Their bitterness is caused by a naturally high content of glucosides, necessitating the olives be cured in some fashion before being consumed. There are basically five methods for **curing** olives: water curing, brine curing, lye curing, dry curing, and sun curing.

- Water curing. This method is generally used with the large green olives, right before they turn reddish/brown. The olives are picked, craked with a rolling pin or by machine (to allow the water to penetrate more quickly), and then completely immersed in cold water for 25 days. The water should be changed every day, and stirred occasionally if possible. Store the container in a cool room or area. After 25 days, testing may occur to check for bitterness. Continue water curing until edible.

- Brine curing. This method was the most widely used and effective for curing olives. In past centuries, Greek and Italian families would gather their green, red, or black olives, then place them into a vat with water and salt. The olives would remain immersed in the brine for a minimum of 2 months, and often up to 10 months after the harvest, to kill any active enzymes in the fruit and leach the glucosides out of the olives. Today, a garde manger chef can make his or her own brine-cured olives by first deeply slitting each fresh olive to allow the brine to fully penetrate and intensely flavor the olive. The cut olives are then submersed in a brine solution made of 1 cup canning salt per gallon of cold water. (The olives will have a tendency to float, so a lid and weight are necessary to hold them down.) Stir the vat of brine-covered olives periodically. After 1 week, rinse the olives in cold water and submerge them again in fresh brine. Repeat this weekly process for a minimum of 3 weeks. After testing for bitterness, the olives can continue to be brined until they are acceptable to the chef. This process may take up to 8 weeks.

- Lye curing. This method is widely used among commercial olive producers, because it is the most time- and cost-effective method for curing olives. Invented in Spain in the 1920s, this method involves submerging raw green olives in vats filled with lye solution. In 5 to 15 hours, the bitter glucosides are leached out, as is a lot of the olive's natural flavor. If air is bubbled through the solution, then the green olives will turn black, creating a product akin to the Californian black olive. To make lye-cured olives, the chef should only use larger fresh green olives. Clean the olives and submerge them in a lye solution made of 8 tablespoons of flake lye per gallon of cold water for 12 hours. (Note: the vat should be stainless steel or glass). Test the doneness by selecting a larger olive and cutting it to the pit. The flesh should be yellowish green and easy to cut to the pit. Drain and soak the olives in fresh water for 3 days, changing the water 3 or 4 times per day. Check for flavor on the fourth day; it should taste sweet and fatty with no bitterness. For additional flavor and to improve keeping quality, immerse the olives for 1 week in light brine made of 6 tablespoons of canning salt per gallon of cold water.

- Dry curing. This method is still used widely around the world, although most olives are not cured in this manner. Freshly picked olives, typically older and darker, are rubbed with coarse salt and left to cure for several weeks or months. The salt is later removed by rinsing, although the olives may retain their salt taste. Some producers package them dry; others coat them in olive oil to keep them moist. Dry-cured olives have a concentrated flavor and wrinkled appearance, similar to other dried fruits. A few varieties are dry-cured first and then aged in brine, such as the French-made Nyons olive.

- Sun curing. In a few warm and sun-drenched locations, such as in Greece and Italy, olives may be left to dry on the tree. Not unlike how raisins are made from withered grapes, these dry and wrinkled olives are intensely flavored and ready to eat from the tree without other methods of curing.

Note: Lye is dangerous to the skin and eyes. Wear goggles and rubber gloves when working with lye.

RECIPE 4–33

OLIVES AND ARTICHOKE HEARTS WITH CHILES, GARLIC, AND BALSAMIC VINEGAR

Recipe Yield: 8 portions

MEASUREMENTS		INGREDIENTS
U.S.	METRIC	
13 ounces	406 g	Marinated artichoke hearts, drained
6 ounces	187 g	Olives, imported
1½ teaspoons	7.5 mL	Fresh rosemary, finely minced
3 each	3 each	Garlic cloves, fresh minced
¾ teaspoon	3.5 mL	Red pepper flakes
3 tablespoons	45 mL	Olive oil, extra virgin
3 tablespoons	45 mL	Balsamic vinegar
Optional	Optional	Salt and pepper

PREPARATION STEPS:

1. Slice the artichoke hearts in half.

2. Combine the artichokes and olives in a medium-sized bowl.

3. Add the other ingredients; stir well to blend and coat.

4. Refrigerate to hold, and then warm to room temperature before service.

RECIPE 4–34

GOAT CHEESE STUFFED OLIVES

Recipe Yield: 8 portions

MEASUREMENTS		INGREDIENTS
U.S.	METRIC	
4 ounces	113 g	Goat cheese, softened
1 teaspoon	5 mL	Red pepper flakes, crushed
2 tablespoons	30 mL	Chives, minced
24 each	24 each	Imported olives, large pitted (such as Gordals)

PREPARATION STEPS:

1. Combine the goat cheese, red pepper flakes, and chives in a small bowl. Blend until smooth.

2. Fit a pastry bag with a medium-sized straight pastry tip. Fill the bag with the goat cheese mixture.

3. Pipe just less than 1 teaspoon (5 mL) of filling into each pitted olive.

4. Cover and refrigerate to hold. Bring to room temperature before service.

RECIPE 4–35

NORTH AFRICAN OLIVE AND CITRUS SALAD

Recipe Yield: 8 portions

MEASUREMENTS		INGREDIENTS
U.S.	**METRIC**	
2 each	2 each	Meyer lemons, large
2 each	2 each	Tangerines
3 cups	354 mL	Water
¼ cup	57 g	Sea salt
1 teaspoon	5 mL	Pimentón (ground smoke-dried peppers or smoked paprika)
1 teaspoon	5 mL	Cumin, ground
2 tablespoons	30 mL	Vidallia onion, minced
¼ teaspoon	1 ml	Sugar
4 tablespoons	60 mL	Flat leaf parsley, chopped
4 tablespoons	60 mL	Olive oil, extra virgin
⅓ cup	79 mL	Moroccan violet olives, pitted and minced (about 30)
2 sprigs	2 sprigs	Parsley

PREPARATION STEPS:

1. Peel the lemons and tangerines. Leave whole.

2. Combine the water and sea salt, and add the peeled citrus. Soak for 1½ hours.

3. Remove the fruit from the water and drain well.

4. Remove the membrane. Mince the flesh and reserve in a medium-sized bowl.

5. Add the pimentón, cumin, onion, sugar, and chopped parsley. Splash in the olive oil.

6. Gently blend until well mixed. Cover and refrigerate 1 hour.

7. To serve, place seasoned citrus in serving dish and top with minced olives.

8. Garnish with parsley.

9. Serve as an appetizer.

Nuts

Nuts are defined as "any hard shelled fruit or seed of which the kernel is eaten by mankind." A botanical definition of a nut is a fruit with a hard, dry shell that needs to be cracked open to release the kernel. Technically, all nuts are seeds of a plant, and sunflower, sesame, and pumpkin seeds are simply seeds in the truest sense because they originate at the center of the plant. Because nuts grow on trees, peanuts are considered a legume, because they grow underground.

QUALITY POINTS WITH NUTS

Nuts in the shell have longer storage potential than shelled nuts. Broken pieces are more perishable than halves or whole kernels. A rancid nut has a flat, metallic taste and a lingering aftertaste. Because of their high lipid content, nuts can easily absorb odors from external sources. Thus, they should not be stored around other foods that have strong odors, unless they are tightly stored in sealed containers.

COMMON NUTS

Although there are hundreds of nut varieties consumed around the world, the following nuts are the most common to industrialized societies:

- Almond trees seem to have originated in the Mediterranean region and are now cultivated in the countries surrounding the Mediterranean Sea as well as in California (the largest producer of almonds in the world), Argentina, and Chile. Almonds are used in many desserts but are also used as a garnish for salads, fish, and meat.
- Brazil nuts grow on a tall tree native to the Amazonian rain forest. The name is taken from the fact that most of the forest is part of Brazil. The tree produces large fruit, with each fruit containing up to 2 dozen nuts. Brazil nuts are used in pastries and as snacks.
- Cashews are native to South America and are now cultivated throughout the tropics: South America, Africa, Vietnam, and Eastern India. Cashews are among the most widely cultivated nut. Cashews are a popular snack and are also used in confectionery and cooking.
- Hazelnuts (or filberts) are native to Europe. Hazelnuts grow on small, bushy trees. The United States is one of the largest producers of hazelnuts. Hazelnuts are used in snacks, baking, and confectionery. They are also popular in Spain in the famous dish called *picadas*.
- Macadamia nuts are native to Australia, even though Hawaii is now the leading grower. Macadamia nuts are used as snacks and in exotic dishes.
- Pecans are native to North America where they are cultivated today. Pecans are used in ice cream, baking, and as a topping or filling for some entrées.
- Pine nuts are native to the Mediterranean region and grow on trees, between the leaves of pine cones. Harvesting is laborious and expensive. Pine nuts are a staple in Mediterranean and Middle Eastern cuisine.
- Pistachios are native to Eastern Europe and are now also produced in California (second largest growing region). Pistachio trees produce a green shell with purple skin. Pistachios not only are a popular snack nut but also are used as an ingredient in many desserts. Pistachios are used as a source of green food coloring.
- Walnuts are grown in California, France, Italy, China, and Chile. They are used in baking and snacking, and are an important source for salad oil. Walnuts, both raw and toasted, are popular in salads.
- Sunflower seeds are native to South America and are also grown in Europe. Sunflower seeds/kernels are a popular snack and an important source of cooking oil. Kernels are commonly added to salads for texture.

TOASTING AND ROASTING NUTS

Raw nuts may be heated to warm their centers and release the flavored oils trapped within, and to enhance their textures. The two most common methods for heating raw nuts is to toast them or to roast them. To toast, the nuts are placed in a dry sauté pan over medium heat for 3 to 5

Assorted nut varieties.
Courtesy of Getty Images,
Inc.

minutes and frequently tossed to prevent scorching. Once their centers are warmed, the fragrance of their oils is detectable in the air. Alternatively, they can be heated on a sheet pan in a 350°F (177°C) oven for 10 to 15 minutes until their color is light brown and their oily fragrance is detected. The nuts should be roasted uncovered and stirred occasionally to prevent scorching. Walnuts, pinenuts, almonds, raw peanuts, and raw pistachios are commonly heated to bring out their flavors before being used in pestos, dips, and spreads.

RECIPE 4-36

SPICY CANDIED PECANS

Recipe Yield: 1 pound

Note: The crushed red pepper flakes may be eliminated from the recipe to produce a sweeter candied pecan, often used in salads.

| MEASUREMENTS | | INGREDIENTS |
U.S.	METRIC	
2 quarts	1.9 L	Water
1 pound	454 g	Pecan halves, shelled
⅓ cup	79 mL	Water
¼ cup	57 mL	Sugar
¼ cup	57 mL	Honey
1¼ teaspoons	6 mL	Red pepper flakes, crushed
½ teaspoon	2.5 mL	Salt

PREPARATION STEPS:

1. In a medium saucepan, add the 2 quarts of water and bring to a boil.

2. Add the nuts and boil for 2 minutes. Drain.

3. Meanwhile, in another medium saucepan, combine all of the remaining ingredients. Bring to low boil.

4. Add the drained, warm nuts to the sugar mixture. Increase the heat and stir the nuts, evenly coating them with the mixture, until the liquid disappears.

5. Spread the nuts on a parchment or silpat-covered sheetpan, and bake in a 325°F (163°C) preheated oven. Bake for approximately 20 minutes, or until they become a dark, mahogany color. Turn the nuts over halfway through the baking process.

6. Allow the nuts to cool on the pan. Separate them and place the sheetpan in a freezer. Cover and reserve until needed.

7. To serve, place the frozen nuts in a suitable service bowl.

RECIPE 4–37

SALTY SPICED ALMONDS

Recipe Yield: 1 pound

MEASUREMENTS		INGREDIENTS
U.S.	**METRIC**	
6 tablespoons	90 ml	Vegetable oil
1 pound	454 g	Blanched almonds, whole
2 tablespoons	30 mL	Sea salt, coarse
1 teaspoon	5 mL	Cayenne pepper, ground

PREPARATION STEPS:

1. Add the oil to a medium-sized sauté pan and warm over moderate heat.

2. Add the almonds and cook for 1 to 2 minutes, until the nuts just become golden brown. Do not burn.

3. In a medium-sized bowl, blend the sea salt with the cayenne pepper. Add the hot nuts, and toss to evenly coat.

4. Place coated nuts on a parchment-lined sheetpan to cool.

5. Once cool, store in an airtight container.

RECIPE 4–38

OAXCAN PEANUTS

Recipe Yield: 1½ pounds

MEASUREMENTS		INGREDIENTS
U.S.	**METRIC**	
⅓ cup	79 mL	Peanut oil
⅓ cup	79 mL	Dried hot chiles
7 each	7 each	Garlic cloves, fresh sliced
1½ pounds	680 g	Raw peanuts, shelled (shell, then measure)
1½ teaspoons	7.5 mL	Sea salt, coarse

PREPARATION STEPS:

1. Preheat a large, heavy-bottomed rondeau or wok over medium heat. When hot, add oil.

2. When oil begins to shimmer, add the dried chiles and garlic. Cook for 10 seconds, stirring constantly to flavor the oil.

3. Add the peanuts, and increase the heat to a medium-high temperature. Stir and toss the peanuts frequently until they are evenly golden brown, approximately 6 minutes.

4. Drain the peanuts in a metal colander. Allow peanuts to drain a few minutes.

5. Spread oil-coated peanuts onto paper towels or side towels to dry.

6. Sprinkle with salt and toss the peanuts again to evenly coat.

7. Serve warm, or at room temperature.

8. Store in an airtight container.

Dried Fruits

As long as there have been fruit trees and bushes, there have been dried fruits available for consumption. Hunters and gatherers based much of their diet from the bounty of fruit trees. In North America, the first of these that come to mind are probably tree-hard or top fruits. They consist of two main groups: the *pome fruits,* including the apple and pear-like member, and the *stone fruits,* which include plums, peaches, cherries, and apricots. The pomes have small seeds in a core around which the stalk forms and the flesh encases. The stone fruits have a single seed in a hard shell around which the flesh forms.

Most of these fruits have been cultivated since ancient times, when the Romans spread them throughout their empire. Unfortunately, most of the knowledge of their cultivation was lost during the Dark Ages. A few noblemen, along with monks in their monasteries, maintained fruit gardens and orchards, while the common people reverted to cropping from the wild. For generations, fruits were often seen as lesser fare to meat and more suited as animal feed. If it were not for the ease in which many of these fruits could be fermented for wines or dried for easier carriage and storage, they probably would have been more neglected.

QUALITY POINTS WITH DRIED FRUITS

Because these fruits have no added sugar, the sweetness of the natural fructose in dehydrated fruits make a great appetizer for people who must watch their refined sugar intake. Some dried fruits (e.g., raisins, figs, prunes, dates, and persimmons) are subject to *sugaring* on the surface or within their flesh. Incidence and severity of sugaring increase with storage temperature and time. *Sugar spotting* is a crystallization of sugars under the skin and in the flesh; gentle heating may reverse it.

The lower the moisture content, the longer the post-harvest life. Many dehydrated fruits contain sulfur to prevent natural browning. Because of this, they have a slightly higher sodium content than fresh fruit.

SULFURING AND DRYING FRUITS

Many commercially dried fruits are sulfured with sulfur dioxide (SO_2) or meta-bisulfate to keep them from oxidizing during and after the drying process. This **sulfuring** process preserves their original color. When fruits are dried with no sulfur, they oxidize and change to a brown-to-black color as they dry. Typically, all white, yellow, and orange fruits are sulfured. To sulfur the fruits, they are put into a room in which the mineral rock sulfur is burned to produce sulfur dioxide gas that permeates the fruits. Commonly sulfured fruits include orange apricots, light brown Calimyrna figs, cantaloupe, crystallized ginger, golden raisins, mango, papaya, peaches, pears, and pineapple.

For many thousands of years, people have dried fruits to keep them for long periods of time. The process of drying is intended to reduce the moisture content to below 25 percent of its normal levels, thereby delaying the time for the fruit to spoil. The concentration of sugars (fructose) remaining in the dried fruit acts as a natural stabilizer. There are many ways to dry fruits; some of the most common ways follow.

- Sun dried. Fruits are laid out in baskets or on woven mats to dry in the sunshine. Larger fruits are typically cut in half to hasten their drying. The fruits typically dried this way include apricots, currants, figs, peaches, pears, and raisins.
- Air or tunnel dried. The fruit has warm air blown over it to dry. The fruits commonly dried in this fashion include apples, coconut, raisins, and tomatoes. Fruits dried in this way usually do not oxidize as much as fruit that is sun dried.
- Naturally dried. Fruits naturally dry on the trees before they are harvested. For example, dates dry on the trees before harvesting; technically dates are considered a fresh fruit rather than a dried fruit.

Note: Chefs place sliced fruits on perforated sheet pans in 158°F (70°C) ovens for 45 minutes to simulate sun drying.

Note: Chefs often use kitchen dehydrators to air dry fruits, vegetables, and meats such as jerky.

- Infused with sugar. The fruit is cut and peeled, then placed in a large container. Water with a high concentration of sugar is heated. The containers of fruit are then covered with the 30- to 60-percent (Bricks) sugar syrup and allowed to set for approximately 7 days. During these 7 days, the sugar–water exchanges with the lower viscosity water in the fruit, thereby extracting the water from the fruit. Once the fruits are partially dried this way, they are air dried to complete the drying process. Fruits commonly dried this way include blueberries, cantaloupe, cherries, cranberries, ginger, mango, papaya, pineapple, and strawberries.

- Dried by frying. The fruits are fried in oil to raise the temperature of the water in the fruit to rapidly boil away. This is done with starchy foods, typically vegetables (potato chips). Bananas and plantains are dried this way before the fruits have ripened and their starches are turned into sugar.

Dried fruits with their whole fruit counterparts

TURKISH DRIED APRICOTS WITH GOAT CHEESE AND PISTACHIOS

Recipe Yield: 150 portions

MEASUREMENTS		INGREDIENTS
U.S.	METRIC	
2 pounds	906 g	Dried apricots, pitted, whole or halved
1½ cups	356 mL	Orange juice, fresh
1½ cups	356 mL	Pistachios, shelled
As needed	As needed	Kosher salt
1 pound	454 g	Goat cheese, mild

PREPARATION STEPS:

1. If using whole apricots, they must be split horizontally.

2. Soak the dried apricot halves in the juice for 20 minutes, turning occasionally. Reserve.

3. Chop the pistachios into coarse pieces, and lightly toast them in a dry sauté pan over medium heat, until their oil begins to release and can be smelled. While still warm, lightly season the nuts with kosher salt. Reserve.

4. Drain the apricots on dry cloth or paper, with the cut side up.

5. Top each apricot with a small chunk of cheese.

6. Sprinkle with toasted nuts.

DIPS AND SPREADS

Garde manger departments, for both hand-carried and buffet-style hors d'oeuvres, have long used dips and spreads. The basic ingredients used to prepare them are common in all commercial kitchens, and their ease of preparation and storage coupled with a broad popularity among customers makes them attractive to garde manger chefs.

The chief difference between **dips** and **spreads** is their viscosity. The consistency of a dip must be loose and soft enough to be scooped by brittle chips, but stiff enough to adhere to what is being used as the dipper. The softer dip is usually in textural contrast to the harder item being used as the dipper, and its flavor is often dominant. Dips may be served warm or cold and generally become thicker when refrigerated. They are normally made from blended or chopped vegetables or fruits, and are often bound with yogurt, mayonnaise, or cream cheese.

Spreads are, by and large, stiffer than dips, and savory styles are often made from ground meats, seafood, or cheeses that are blended with flavoring fats such as cream cheese. Sweet spreads are commonly made from puréed fruits, chocolate, or nuts that are sometimes bound with cream cheese or sweetening agents. They are normally applied to canapés, croutons, toasts, or crackers with the aid of a small spatula or knife.

RECIPE 4-40

ENDIVE LEAVES WITH HERBED CHEESE SPREAD

Recipe Yield: 8 portions

MEASUREMENTS		INGREDIENTS
U.S.	**METRIC**	
4 heads	4 heads	Belgian endive, leaves separated
8 ounces	250 g	Ricotta cheese
1/4 cup	10 g	Fresh parsley, chopped
1/4 cup	10 g	Fresh dill, chopped
1 each	1 each	Garlic clove, fresh minced
To taste	To taste	Black pepper, fresh ground
As needed	As needed	Walnut halves

PREPARATION STEPS:

1. Wash and dry the separated endive leaves. Neatly trim the bottom of each leaf.

2. Using a wooden spoon, blend the cheese, herbs, garlic, and pepper in a small bowl.

3. Spoon a small amount onto the base-half (opposite the tip) of each leaf.

4. Garnish with a walnut half.

RECIPE 4-41

SARDINE SPREAD

Recipe Yield: 1½ pints

MEASUREMENTS		INGREDIENTS
U.S.	**METRIC**	
8 ounces	227 g	Sardines, canned
1 pound	454 g	Cream cheese, softened
8 each	8 each	Scallions, chopped
2 bunches	2 bunches	Parsley, chopped
2 each	2 each	Fresh lemons, zest of (juice reserved)
To taste	To taste	Salt
To taste	To taste	White pepper

PREPARATION STEPS:

1. In a food processor, combine all ingredients.

2. Process until desired consistency is achieved.

3. Adjust seasoning.

RECIPE 4–42

CURRIED EGG SPREAD

Recipe Yield: 1½ pints

| MEASUREMENTS | | INGREDIENTS |
U.S.	METRIC	
8 each	8 each	Eggs, hard boiled
2 each	2 each	Pears, cored and chopped
⅓ cup	79 mL	Cilantro, fresh chopped
⅓ cup	79 mL	Parsley, fresh chopped
6 each	6 each	Scallions, chopped
1 cup	237 mL	Mayonnaise
⅓ cup	79 mL	Sour cream
2 tablespoons	30 mL	Dijon mustard
3 tablespoons	45 mL	Mango chutney
1½ teaspoons	7.5 mL	Curry powder
2 teaspoons	10 mL	Salt
2 teaspoons	10 mL	White pepper
8 ounces	250 g	Water chestnuts, chopped
½ cup	118 mL	Peanuts, chopped

PREPARATION STEPS:

1. In a food processor, combine all ingredients except the water chestnuts and peanuts. Pulse the blades to coarsely chop the mixture.

2. Add the water chestnuts and peanuts, and pulse the blades until the mixture is blended, but retains its texture.

3. Adjust seasonings.

4. Refrigerate for at least 30 minutes before use.

Applying a spread to a canapé

RECIPE 4–43

CILANTRO PISTACHIO PESTO

Recipe Yield: 1½ cups

MEASUREMENTS		INGREDIENTS
U.S.	**METRIC**	
1½ cups	355 mL	Packed fresh cilantro leaves blanched, refreshed and squeezed dry
¾ cup	177 mL	Broken pistachio nuts
1 each	1 each	Garlic clove
¼ cup	59 mL	Olive oil
¼ cup	59 mL	Feta cheese

PREPARATION STEPS:

1. Purée the cilantro, garlic, and pistachio nuts with the oil in a food processor.

2. Remove into a bowl and beat in the feta cheese

RECIPE 4–44

ROASTED EGGPLANT-HUMMUS DIP

Recipe Yield: 1 quart

Note: May substitute canned chickpeas.

Note: May also substitute fresh dill or cilantro for parsley, or combine several herbs.

TIP	To increase charred flavor, eggplant may be grilled over charbroiler.

MEASUREMENTS		INGREDIENTS
U.S.	**METRIC**	
1 each	1 each	Eggplant, large
4 cups	946 mL	Chickpeas, cooked and drained
4 each	4 each	Garlic cloves, crushed
½ cup	118 mL	Tahini (sesame paste) (optional)
⅓ cup	79 mL	Lemon juice, fresh (optional)
½ cup	118 mL	Olive oil (optional)
⅓ cup	79 mL	Parsley, fresh chopped
½ teaspoon	2 mL	Cumin, ground
½ teaspoon	2 mL	Coriander, ground
⅛ teaspoon	1 mL	Cayenne
½ teaspoon	2 mL	Black pepper, fresh ground
To taste	To taste	Salt

PREPARATION STEPS:

1. Soak the chickpeas for at least an hour and then cook them until tender.

2. Puncture the eggplant with a fork a few times, turning it as it is being stabbed. Roast the eggplant at 375°F (190°C) for 45 minutes.

3. Remove the eggplant and let cool.

4. Remove the peel from the eggplant.

5. Add the eggplant, along with the other ingredients as desired, into a food processor to purée into a fairly smooth consistency.

6. Adjust seasoning.

7. Use immediately.

8. Store refrigerated in a covered container.

POTTED BLACK BEAN DIP

Recipe Yield: 1 quart

MEASUREMENTS | INGREDIENTS

U.S.	METRIC	
1 pound	454 g	Black turtle beans
2 tablespoons	30 mL	Olive oil
1 each	1 each	Onion, large
2 each	2 each	Bay leaf
6 cups	1.4 L	Chicken stock
1 cup	237 mL	Sherry
4 each	4 each	Garlic cloves, minced
6 each	4 each	Green chiles, seeded and chopped
1 cup	237 mL	Cilantro, chopped
2 cups	473 mL	Walnuts, chopped
⅔ cups	158 mL	Scallions, chopped
2 tablespoons	30 mL	Chile powder
2 tablespoons	30 mL	Cumin
To Taste	To taste	Salt

PREPARATION STEPS:

1. Wash the beans and soak overnight. Drain.

2. In a large saucepan, add the garlic and oil and sauté until soft.

3. Add the bay leaf, stock, sherry, and beans. Bring to a boil, and then reduce to a simmer for 45 minutes or until tender.

4. Drain the beans and add to a food processor.

5. Add the chiles, cilantro, walnuts, scallions, and seasonings.

6. Process until the mixture reaches the desired consistency.

7. Adjust seasoning.

8. Using a casserole or large ramekin, heat in oven at 350°F (177°C) until hot.

9. Serve warm.

RECIPE 4-46

GUACAMOLE

Recipe Yield: 1 quart

MEASUREMENTS		INGREDIENTS
U.S.	**METRIC**	
8 ounces	227 mL	Fresh tomato, peeled and seeded (or diced tomatillos)
4 ounces	113 mL	Onion, diced
2 teaspoons	10 mL	Jalapeño, seeded and minced
2 each	2 each	Garlic cloves, minced
4 ounces	113 g	Lime juice, fresh
3 tablespoons	45 mL	Cilantro, fresh chopped
1½ pounds	680 g	Avocado pulp, fresh
To taste	To taste	Salt

PREPARATION STEPS:

1. In a food processor, combine the tomato, onion, jalapeño, garlic, lime juice, and cilantro. Purée to a coarse consistency.

2. Place the avocado pulp in a large bowl, and fold in the tomato salsa mixture.

3. Add salt and lime juice to taste.

4. Serve fresh with tortilla chips.

RECIPE 4-47

CUCUMBER RAITA

Recipe Yield: 1½ pints

MEASUREMENTS		INGREDIENTS
U.S.	**METRIC**	
1 each	1 each	Cucumber, large
2 cups	454 mL	Plain yogurt
2 Tablespoons	30 mL	Cilantro leaves, chopped
2 Tablespoons	30 mL	Mint leaves, chopped
2 each	2 each	Green chile, seeded, minced
To taste	To taste	Salt
To taste	To taste	Pepper
2 teaspoon	10 mL	Cumin seeds
2 teaspoon	10 mL	Mustard seeds
As needed	As needed	Cilantro or mint, to garnish

PREPARATION STEPS:

1. Cut cucumber into matchstick-sized pieces and place in a bowl.

2. Add yogurt, cilantro, mint, chile, and salt. Blend gently.

3. Add pepper. Refrigerate at least 30 minutes before serving.

4. Garnish with herb leaves.

RECIPE 4–48

TAPENADE

Recipe Yield: 1½ pints

MEASUREMENTS		INGREDIENTS
U.S.	**METRIC**	
2 cups	474 g	Black olives, oil-cured (pitted, then measured)
16 each	16 each	Anchovy fillets, drained and chopped
½ cup	118 g	Capers, drained
¼ cup	59 mL	Dijon mustard
2⅔ tablespoons	40 mL	Lemon juice, fresh
2 teaspoons	10 mL	Dry mustard
1 teaspoon	5 mL	Black pepper, fresh ground
1 cup	237 mL	Olive oil, extra virgin

PREPARATION STEPS:

1. Combine the pitted olives, anchovies, and capers in a food processor. Briefly process until mixture is coarse.

2. Add the mustard, lemon juice, dry mustard, and pepper. Blend briefly.

3. Slowly add the oil, and process to form a paste.

4. Store in covered container in the refrigerator for up to 2 weeks.

RECIPE 4–49

CRISP VEGETABLE CHIPS

Recipe Yield: 8 portions

MEASUREMENTS		INGREDIENTS
U.S.	**METRIC**	
3 pints	1.5 L	Vegetable oil
3 each	3 each	Sweet potatoes, peeled
4 each	4 each	Potatoes, russet, peeled
3 each	3 each	Jicama, peeled
4 each	4 each	Parsnips, peeled
3 each	3 each	Beetroots, peeled
2 tablespoons	30 mL	Sea salt, coarse

PREPARATION STEPS:

1. In a large heavy-bottomed sautoir or deep fryer, heat the oil until it reaches 375°F (190°C).

2. Working with one vegetable at a time, saving the beets until last, thinly slice the peeled vegetables lengthwise using a mandoline or electric slicer.

3. Deep fry small batches of vegetables for about 1 minute, or until they are golden.

4. Using a spider-strainer or slotted spoon, remove the fried vegetables and drain on paper towels.

5. Sprinkle the chips liberally with sea salt.

6. Serve alone or with dips.

Note: Choose small, medium, or large sizes of vegetables.

Colorful vegetable dips

Care and Storage

Generally, dips and even spreads benefit from being prepared several hours in advance of their use. As with most blended food items, time is needed to allow flavors to fully mesh. For example, a Roquefort, garlic, Worcestershire, lemon juice, and hot red pepper dip bound with mayonnaise will need a few hours, if not overnight, to fully mature. When held overnight under refrigeration, items bound with mayonnaise or dressings tend to loosen, while mixtures bound with cream cheese or butter always stiffen.

It is imperative that the mixtures be stored in airtight containers under refrigeration with appropriate name and date labeling. Besides the evaporation that occurs while foods are stored in coolers, the transfer of odors occurs. Dairy products are magnets for odors, and any product made with cream cheese, milk, mayonnaise, or butter is going to absorb undesirable "cooler smells" when not properly stored.

After removal from the cooler, the dips and spreads benefit from being briefly stirred or whipped. Whipping aerates the product and ensures an even and desirable consistency, allowing for an effective use as a dip or spread.

CANAPÉS TOASTS, AND CRACKERS

Canapés are savory hors d'oeuvres made with a bread, cracker, or pastry base, such that they can be picked up with the fingers and eaten in one or two bites. The word *canapé* originally meant a canopy of mosquito netting over a couch or bed. In time, it came to mean the bed or davenport itself. Eventually, the word became synonymous with its present meaning of a morsel of bread or cracker with a tasty mixture of meat, cheese, or fish spread on it.

Elements of the Canapé

The basic components to the canapé include the *base, adhesive spread, body,* and *garnish* (Figure 4–26). Although canapés have evolved from their classic origins with greater varieties of bases and adhesive spreads in use, their assembly remains very traditional.

FIGURE 4–26 Four parts of a canapé

garnish

body

adhesive

base

BASE

The base originally was prepared from sliced white, rye, brioche, or pumpernickel bread that was slowly toasted until dry and crunchy. Contemporary creations have expanded to include bases made of crackers, toasted polenta, bagel chips, wonton skins, phyllo dough, tortillas, fried vegetables, potatoes, and other crunchy edibles.

ADHESIVE SPREAD

The purpose of the adhesive spread is three-fold. First, it is the covering for the base that allows the body and garnish to remain attached. It must be firm enough to hold its shape when spread or piped, but soft enough to adhere to the base and body. Second, the adhesive seals the base and keeps it from absorbing excess liquid from the body. Adhesive spreads are often made with a high degree of fat such as butter, compound butters, or pastes to form a moisture barrier. Third, adhesive spreads provide a flavor profile that either complements the body or provides a distinguishing characteristic of its own.

BODY

The body of the canapé is made from premium ingredients including shellfish, meats, sausages, vegetables, eggs, smoked seafood, or poultry. The body is generally the primary focus of the canapé, around which the other ingredients are designed.

GARNISH

The garnish adds the final appearance and taste to the canapé. It often creates the focal point to the finished piece through the use of color and shape. It may also serve as a contrasting texture or flavor to the other elements. Most herbs quickly wilt and are not practical for use. Typical garnishes include piped sour cream, pâté a choux ornaments, vegetables, capers, sliced pickles and cornichons, caviar, olives, nuts, and chopped eggs.

RECIPE 4–50

MAÎTRE D' HÔTEL BUTTER

Recipe Yield: 2 cups

MEASUREMENTS		INGREDIENTS
U.S.	METRIC	
1 pound	454 g	Unsalted butter, softened
3 tablespoons	45 mL	Lemon juice, fresh
3 tablespoons	45 mL	Fresh parsley, minced
2 teaspoons	10 mL	Fresh thyme, minced (optional)
1 teaspoon	5 mL	Kosher salt
½ teaspoon	2.5 mL	White pepper, fresh ground

PREPARATION STEPS:

1. Using a handheld mixer or food processor, blend the butter until smooth.

2. Add the remaining ingredients and blend for about 1 minute longer until fully incorporated.

3. Lay a piece of plastic wrap on the countertop.

4. Scrape the butter onto the plastic and, using the plastic wrap as a guide, form the butter into an even log.

5. Fold the ends closed and refrigerate until ready to use.

RECIPE 4–51

GORGONZOLA-SCALLION COMPOUND BUTTER

Recipe Yield: 2 cups

MEASUREMENTS		INGREDIENTS
U.S.	METRIC	
8 ounces	227 g	Unsalted butter, softened
8 ounces	227 g	Gorgonzola cheese, crumbled, at room temperature
1 teaspoon	5 mL	Balsamic vinegar
½ cup	118 mL	Green onions, chopped
2 teaspoons	10 mL	Black pepper, coarsely fresh ground

PREPARATION STEPS:

1. Using a handheld mixer or food processor, blend the butter and Gorgonzola until smooth.

2. Add the scallions, pepper, and vinegar and blend for about 1 minute longer until fully incorporated.

3. Lay a piece of plastic wrap on the countertop.

4. Scrape the butter onto the plastic and, using the plastic wrap as a guide, form the butter into an even log.

5. Fold the ends closed and refrigerate until ready to use.

RECIPE 4–52

GARLIC-HORSERADISH COMPOUND BUTTER

Recipe Yield: 2 cups

MEASUREMENTS		INGREDIENTS
U.S.	**METRIC**	
1 pound	454 g	Unsalted butter, softened
3 tablespoons	45 mL	Roasted garlic, puréed
1/4 cup	59 mL	Prepared horseradish
1 tablespoon	15 mL	Lemon juice, fresh
1/4 teaspoon	1 mL	White pepper
1 teaspoon	5 mL	Kosher salt

PREPARATION STEPS:

1. Using a handheld mixer or food processor, blend the butter until smooth.

2. Add the remaining ingredients and blend for about 1 minute longer until fully incorporated.

3. Lay a piece of plastic wrap on the countertop.

4. Scrape the butter onto the plastic and, using the plastic wrap as a guide, form the butter into an even log.

5. Fold the ends closed and refrigerate until ready to use.

RECIPE 4–53

GARLIC-MUSTARD COMPOUND BUTTER

Recipe Yield: 2 cups

MEASUREMENTS		INGREDIENTS
U.S.	**METRIC**	
1 pound	454 g	Unsalted butter, softened
8 each	227 g	Garlic cloves, peeled and chopped
1/2 cup	118 mL	Tarragon leaves, stems removed and chopped
1/4 cup	59 mL	Whole-grain mustard
1 teaspoon	5 mL	Kosher salt
1 teaspoon	5 mL	Black pepper, fresh ground

PREPARATION STEPS:

1. Using a handheld mixer or food processor, blend the butter until smooth.

2. Add the remaining ingredients and blend for about 1 minute longer until fully incorporated.

3. Lay a piece of plastic wrap on the countertop.

4. Scrape the butter onto the plastic and, using the plastic wrap as a guide, form the butter into an even log.

5. Fold the ends closed and refrigerate until ready to use.

RECIPE 4–54

LEMON DILL BUTTER

Recipe Yield: 2 cups

Note: **Compound butter** is made from blending unsalted butter with any number of ingredients. The butter may contain only one ingredient, as with anchovy butter, or a myriad of herbs, spices, and flavors. The butter is softened, the ingredients are finely chopped or ground, and they are blended together. Afterward, the butter is stored in a tightly closed container or rolled, in a tubular shape, in plastic wrap or parchment paper, then stored in the refrigerator or frozen for later use. Compound butters may be used as spreads for canapés and crackers, as an additive to a sauce, or as a flavor agent over a meat or vegetable.

MEASUREMENTS		INGREDIENTS
U.S.	**METRIC**	
1 pound	454 g	Unsalted butter, softened
¼ cup	59 mL	Fresh dill weed, minced
¼ cup	59 mL	Fresh lemon zest, minced
To taste	To taste	Kosher salt
To taste	To taste	Black pepper, fresh ground

PREPARATION STEPS:

1. Using a handheld mixer or food processor, blend the butter until smooth.

2. Add the remaining ingredients and blend for about 1 minute longer until fully incorporated.

3. Lay a piece of plastic wrap on the countertop.

4. Scrape the butter onto the plastic and, using the plastic wrap as a guide, form the butter into an even log.

Note: The adhesive spread should be applied before cutting the pullman slices to their desired canapé size and shape.

Note: Canapés tend to be hand-carried or placed on buffets, often for extended periods of time. It is important for the health and well-being of the customer that they be refrigerated, or held in a safe manner, just before service.

Guidelines for Assembling Canapés

As typical when preparing any involved recipe or mixture, the chef must have the mise en place organized before beginning canapé assembly. The following steps are recommended for canapé production:

1. Prepare a clean, clear work area. Mass production of canapés generally requires a large work surface and a rolling rack with sheet pans.

2. All of the four components for the canapés should be prepared and readily available. It is best to work on one component at a time, starting with the base. If the same base is to be used for several different canapés, with different adhesives or bodies, then space will be the determining factor for how many varieties can be prepared at the same time.

3. If toasted bread is to be used as the base for the canapé, it is best to work with pullman loaves that have been sliced lengthwise ⅙-inch to ¼-inch thick, then slowly toasted in the oven. They must be dried to the point of crispness, but not brittle. Pullman loaves provide a larger work surface that is more productive for the chef.

4. The adhesive spread is applied to the base in an even fashion, covering the entire surface of the base unit. If there is limited work area, or if the canapés are to be stored before being assembled on display trays or mirrors, then they should be placed on sheet pans with parchment paper.

5. The body is placed on the canapé in the same location for each unit.

6. The garnish is then applied with the same attention to uniformity.

7. If the canapés are to be stored for any length of time, the sheet pans are covered tightly with film wrap and placed on the rolling rack, while the chef prepares the next set of canapés.

Pullman loaf being sliced

Assorted canapés

RECIPE 4–55

PEARS AND PROSCIUTTO CANAPÉ

Recipe Yield: 40 pieces

MEASUREMENTS | | INGREDIENTS

U.S.	METRIC	
1 each	1 each	Baguette
6 tablespoons	90 mL	Olive oil
4 each	4 each	Pears, firm but ripe
20 slices	20 slices	Prosciutto, thinly sliced, cut in half
8 ounces	250 g	Gorgonzola, crumbled
40 each	40 each	Arugula leaves

PREPARATION STEPS:

1. Preheat broiler or chargrill.

2. Thinly slice the baguette to make 40 circles.

3. Lightly brush both sides of the bread with oil and toast on both sides for 1 to 2 minutes.

4. Slice pears lengthwise and trim to fit the crouton. Lightly brush with oil and lightly grill for 1 minute on each side.

5. Top each toasted bread crouton with a half slice of prosciutto, pear slice, and a little crumbled cheese.

6. Garnish with an arugula leaf.

RECIPE 4–56

GRENOBLE CANAPÉ

Recipe Yield: 48 pieces

MEASUREMENTS | | INGREDIENTS

U.S.	METRIC	
¾ cup	79 mL	Walnut pieces
½ pound	227 g	Butter, unsalted, softened
To taste	To taste	Salt
To taste	To taste	White pepper, fresh ground
48 each	48 each	Square croutons
48 each	48 each	Gruyère cheese, thinly sliced, 1½ × 1½ inch
48 each	48 each	Walnut, whole toasted

PREPARATION STEPS:

1. Toast the walnut pieces on a sheet pan or sauté pan in a preheated moderate oven.

2. When fragrant (about 10 to 15 minutes), pulverize with a mortar.

3. Blend with butter and spices until smooth.

4. Neatly spread walnut butter on croutons.

5. Place one slice of cheese on the covered crouton.

6. Garnish with the whole toasted walnut.

RECIPE 4–57

BLUE CHEESE AND DUCK BREAST CANAPÉS

Recipe Yield: 48 pieces

MEASUREMENTS		INGREDIENTS
U.S.	**METRIC**	
½ pound	227 g	Butter, unsalted
½ cup	118 mL	Blue cheese (Maytag, Stilton, Bel Paese, Roquefort)
To taste	To taste	White pepper, fresh ground
To taste	To taste	Salt
48 each	48 each	Croutons, round, dried
3 each	3 each	Duck breast, sliced 1½ inches
96 each	96 each	Mandarin orange segments
1 cup	237 mL	Pistachios, chopped toasted

PREPARATION STEPS:

1. Combine butter, blue cheese, salt, and pepper. Blend until smooth.

2. Spread cheese paste evenly onto croutons.

3. Neatly place a piece of duck breast onto the covered crouton.

4. Place two orange segments onto the duck.

5. Sprinkle oranges with chopped pistachios.

6. Serve immediately.

Cajun Shrimp Canapés

Recipe Yield: 26 to 30 portions

MEASUREMENTS		INGREDIENTS
U.S.	**METRIC**	
6 each	6 each	Jalapeño peppers
2 tablespoons	30 mL	Olive oil
1 cup	237 mL	Cream cheese
To taste	To taste	Salt
To taste	To taste	Black pepper, fresh ground
As needed	As needed	Round cayenne pepper cracker
1 pound	454 g	Shrimp, peeled and deveined, tail removed, $^{26}/_{30}$ count
2 tablespoons	30 mL	Cajun spice blend
½ cup	118 mL	Cream cheese
30 each	30 each	Cilantro leaves

PREPARATION STEPS:

1. Rub the peppers with olive oil and roast in preheated 400°F (204°C) oven for about 30 minutes, or until they start to turn dark brown. Place in an airtight heavy plastic food bag and allow to condensate for 15 minutes. Rub skin to clean pepper, and remove burnt skin and seeds, reserving flesh. Mince or mash the flesh.

2. Blend the roasted pepper flesh with the cream cheese and seasonings.

3. Spread evenly on a cracker. Reserve.

4. Sprinkle cleaned shrimp with Cajun spices, and quickly sauté over high heat. Cool shrimp quickly.

5. Neatly place one shrimp on each coated cracker.

6. Garnish with cream cheese and cilantro.

RECIPE 4–59

ROMANOFF TARTLETS

Recipe Yield: 24 tartlets

MEASUREMENTS		INGREDIENTS
U.S.	**METRIC**	
½ cup	118 mL	Butter, softened
½ cup	118 mL	Cream cheese, softened
⅓ cup	79 mL	Anchovy paste
4 tablespoons	59 mL	Chives, minced
24 each	24 each	Tartlet shells, pre-baked
1 cup	237 mL	Bowfin or sturgeon roe (black roe)
1 cup	237 mL	Salmon or trout roe (red roe)
8 ounces	227 g	Smoked salmon, thinly sliced
½ cup	118 mL	Sour cream
24 each	24 each	Chive spears

PREPARATION STEPS:

1. Combine butter, cream cheese, anchovy paste, and minced chives.

2. Spread paste on the inside bottoms of the tartlet shells.

3. Line the black caviar over one edge of the tartlet, covering one third of the paste.

4. Line the red caviar over the opposite side, covering one third.

5. Place the smoked salmon down the center, covering the remaining third.

6. Garnish with a dollop of sour cream and chive spear.

Specialty Presentations

Note: See Chapter 17 for reference on platter presentations.

Assorted smoked seafood. Courtesy of Foodpix/ Picture Arts

Seafood raw bar. Courtesy of Foodpix/Picture Arts

Professional Profile

BIOGRAPHICAL INFORMATION

Name: K.F. Seetoh   **Recipe Provided:** Singapore Chili Crab
Place of Birth: Singapore

CULINARY EDUCATION AND TRAINING HIGHLIGHTS

K.F. Seetoh did not realize that his training in media and a deep reverence for the food of his country would later lead him to a successful career in food journalism. Born in Singapore in 1962 to parents of Cantonese decent, Seetoh (as his friends call him) earned a diploma in design and photography from Singapore's Baharuoldin Institute. While serving 2 years of National Military Service, he became a staff photographer and performed a variety of photography related jobs for the military. After leaving the service, Seetoh worked from 1983 to 1990 as a photojournalist for *The Straits Times,* Singapore's largest English daily paper. This led to opening his own studio, K.F. Seetoh Photography in 1990, a business he still maintains today. In 1995, Seetoh established an Asian stock photo library, ImaginAsia, selling photos depicting the scenery, food, and culture of Asia via the Internet. While traveling extensively to acquire his photos, Seetoh gained a deep appreciation for Asia's varied cuisine. In 1996, he founded his company Makansutra to promote and celebrate Asian cuisine, primarily its street food. In 1998, he published the *Makansutra Food Guide to Singapore,* highlighting who he considered to be the best street food vendors in Singapore. (Singapore has 121 public market areas, each featuring between 100 and 150 "food hawkers.") An immediate success, it has become Singapore's Bible of food with revised editions being printed every 18 months. In 2003, Seetoh expanded his company and created a similar guide, *Makansutra Indonesia,* highlighting the street food and culinary heritage of Jakarta. In 2005, *Makansutra Malaysia* was released, with future food guides planned for Vietnam, various cities in China, and other Asian countries. In addition to his publications, in 2001 Seetoh joined forces with Channel "i" as the colorful and energetic host of the television show *Makansutra,* which features Singapore's street food vendors and their exciting culinary creations. In 2004, he opened Soul Food by Makansutra, Singapore's first hawker bistro. It is Seetoh's mission to promote the food culture as a lifestyle and to recognize the "culinary soldiers" that defend Singapore's food heritage.

ADVICE TO JOURNEYMAN CHEF

As my own heritage is Cantonese, by default I am a "foodie." I believe Canton is the gastronomical center of China. They cook anything that slithers, crawls, swims, or flies. To the Chinese, everything is about taste. Appearance is important, but taste is everything. Money should never be a reward or reason to cook. The smile of the customers when they eat your food is the reward. People will know when you are true to that philosophy, and they will come back to your restaurant. When you are a student, don't just learn the recipe; learn the soul and reason behind the dish you are making. Be possessed by the spirit of the recipe. Don't cook it! Feel it!

K.F. Seetoh

RECIPE 4-60

CHEF K.F. SEETOH'S SINGAPORE CHILI CRAB

MEASUREMENTS		INGREDIENTS
U.S.	METRIC	
16 ounces	453 g	Soft shelled crabs or mud crabs
4 tablespoons	60 mL	All-purpose flour (for soft shell crabs, only)
3 tablespoons	45 mL	Vegetable oil
8 each	8 each	Garlic cloves, minced
8 each	8 each	Red Chilies, fresh, minced
1 each	1 each	Egg, whole
2 each	2 each	Spring onions (scallions), 3-inch pieces
1 teaspoon	5 mL	Lime juice, fresh
1 bunch	1 bunch	Cilantro
		Sauce
8 ounces	227 g	Water
5 tablespoons	75 mL	Tomato ketchup
2 tablespoons	30 mL	Sugar
1½ teaspoons	7 mL	Corn flour
1 teaspoon	5 mL	Dark miso (or substitute with pounded brown preserved soy beans)
¼ teaspoon	1 mL	Salt

PREPARATION STEPS:

1. Prepare sauce first by mixing all sauce ingredients in a bowl. Set aside to allow flavors to blend.

2. Heat oil in shallow saucepan or wok until very hot. Add the garlic and stir-fry for 1 minute. Add the chilies and continue to stir-fry until they become fragrant, 15 to 30 seconds.

3. If using mud crabs, add them to the oil at this time. Fry well, turning as needed, until shells start to turn red. Add sauce, stir well, reduce heat then cover and simmer until shells are totally red. About 1–3 minutes.

4. Add cracked whole egg to mixture and streak the egg with a fork; continue to simmer until it is cooked.

5. Add fresh squeezed lime juice and scallions.

Note: If using soft-shelled crabs, cut them into quarters. Then dredge the pieces through the flour and deep-fry until crispy golden brown. Add the sauce over the fried soft-shelled crabs just prior to service.

1. What are amuse-bouches?

2. Explain the term *hors d'oeuvres.*

3. Describe antipasti.

4. In which country are chat served?

5. What does *dim sum* mean?

6. Where are kanto normally served?

7. Describe what is meant by the term *mezze.*

8. How did tapas originate?

9. What is caviar?

10. What is the difference between sushi and sashimi?

A. **Group Discussion**

Discuss how the beginning courses of various countries have found their way into mainstream restaurants.

B. **Research Project**

Research a country that is not mentioned in the chapter and report on what they use for small courses to begin a meal.

C. **Group Activity**

Plan a canapé platter containing five original items describing the uses, color, texture, and presentation.

D. **Individual Activity**

Create recipes for four original amuse-bouches suitable for a fine dining restaurant.

Salads with Vegetables, Fruits, and Grains

After reading this chapter, you should be able to

- discuss the functional origins of salad in modern gastronomy.

- list the classification of salads.

- explain how salads are composed.

- evaluate the appropriate selection and use of salads.

- prepare a variety of international cold sauces and dressings.

- demonstrate techniques in the preparation of salad ingredients.

KEY TERMS

as-purchased (AP)	fruit	salad base
bulbs	grains	salad body
combination salad	legumes	salad dressing
complex salad	microgreens	salad garnish
edible portion (EP)	roots	simple salad
emulsion	salad	tubers

It was heavy, it was tall, it sprouted,

it eared...it nodded, it hung...

Indeed the lucky grains were sent down to us, the black millet,

the double-kernelled, millet, pink-sprouted and white.

—HOU CHI

A Bountiful Pantry

We broadly define salads as any food served with a dressing. This wide-ranging and encompassing description provides the chef with a bountiful pantry from which to work. The world has become much smaller from improved communication, travel, and product distribution, and the chef's pantry has benefited greatly. Many more products are either locally grown and harvested, or available with certainty from distant shores. Garde manger departments have limitless varieties of ingredients from which to prepare their many salad creations.

Classical Function of Salads

The Anglican word **salad** originated from *sal,* the Latin term for salt. The Romans believed salting greens would counteract their bitterness. *Herba salata* is the Latin equivalent of "salted greens." It is believed that the earliest salads were mixtures of greens, sometimes pickled, that were seasoned with salt. Later, in the seventeenth and eighteenth centuries, lettuces became the foundation on which other vegetables and meats were placed. Simple dressings made from oil and vinegar were added to the salad ingredients for moisture and flavor.

When Georges-Auguste Escoffier wrote his venerable work, *Le Guide Culinaire,* in the early twentieth century, he devoted relatively little space to the subject of salads. In his words, "Salads are of two kinds: simple or compound. Simple, or raw salads always accompany hot roasts; compound salads, which generally consist of cooked vegetables, accompany cold roasts" (p. 639). Even though Escoffier created the *salads des fines gueules* (salad for those with fine palates), made from celery, partridge breast, and truffles dressed with Provençal virgin olive oil and Dijon mustard, he generally limited his use of salads. Proteins were the primary source of sustenance at the finer establishments and homes in continental Europe, and the primary role of salad was to support and complement the main courses.

Modern Types of Salads

Nearly a century has passed since Escoffier wrote those words, and modern food service operations have embraced the cuisines of many distant cultures and lands where the role of salads has been more prominent. Indeed, our modern sense of gastronomy considers salads to be much broader in scope and content. Salads have evolved from being served only as an accompaniment to being served as a course unto their own. Whether as a starter course, accompaniment, main course, separate course, dessert, or entire meal—salads are served at any meal of the day.

Categorizing salads is as imperfect as classifying cheeses (Chapter 15). As **salads** have evolved beyond the limitations of classical cuisine, they no longer fit the definitions advanced by Escoffier and his contemporaries. Opening four different modern texts on culinary fundamentals will reveal as many different categories for salads. Some garde manger chefs believe that organizing them by their ingredients is most useful, whereas others would argue that organizing them by their purpose makes the most sense. We choose to categorize them broadly into three primary types according to their content and structure: *simple, complex,* and *combination.*

SIMPLE

Simple salads are those that are basic in nature and composition. They include light salads made from a variety of one or more greens, fruits, pastas, or grains, but not in combination with each other. They are generally dressed with mild seasonings, such as vinaigrette, and are used as a course served around the entrée. Examples include coleslaw, macaroni salad, spinach salad, gelatins, and fruit cup.

COMPLEX OR MIXED

Complex salads, also known as mixed salads, are heartier in character than simple salads, and are composed of raw or cooked vegetables, fruits, meats, seafood, game, or poultry. **Complex salads,** are seasoned with flavorful dressings or marinades and usually contain multiple ingredients from the same categories of foods. They are served as the salad course or as appetizers, accompaniments, or desserts. Examples of complex salads include seafood salad, grilled marinated vegetables, and fruit plate with citrus yogurt dressing.

COMBINATION

In **combination salads**, all of the ingredients are not blended together as one homogenous flavor as they are with simple or complex salads. They are generally comprised of several different categories of ingredients that are seasoned separately but presented together on the same plate. Combination salads are the most substantial of salads and are generally featured as the main course. Examples of combination salads include the chef's salad, pasta primavera, warmed grilled duck breast over greens with walnut vinaigrette and Gorgonzola, and boiled chicken and macadamia nut salad in pineapple wedges (Figure 5–1).

Composition and Evaluation of the Salad

Since the early days of salad preparation, the name for the salad has either been given in recognition of its main ingredients or in recognition of an individual for whom or by whom the salad was created. Although the name has some significance, the importance is found in its composition and evaluation of quality points. Salads may be analyzed according to their structure and content, along with their wholesome value. The garde manger chef should consider the following variables when assessing the creation and presentation of salads.

COMPOSITION

Salads traditionally have been structured with the following elements for hundreds of years. These parts are fundamental to most salads, but may not exist in all. Their uses, to a large degree, are subject to the needs of the meal event and the opinion of the garde manger chef. The four basic parts include **base, body, dressing** and **garnish** (Figure 5–2).

■ Base. The base serves as both the underline upon which the salad is built and the canvas with which to frame the remainder of the ingredients. From a functional standpoint, the base gives definition to the placement of the salad, often cups the body of the salad, and collects and retains excessive dressing. Common bases include lettuce cups, chiffonade of lettuce, rice noodles, pineapple wedges, and avocado halves.

FIGURE 5–1 Three categories of salads (from left: simple, complex, and combination)

■ Body. The body is the most important part and focal point of the salad, and is placed on the base. It generally represents the most substantial part of the salad and often lends its name to the salad itself. Examples of the salad body include tuna salad, chilled shellfish, fruit salad, and smoked duck salad.

■ Dressing. The dressing is used to moisten and flavor the other ingredients, thereby enhancing the body of the salad. Dressings can add substantial flavor to mild salads, serve to contrast with existing flavors, or delicately enhance the dominant tastes found in the salad. Dressings may be served on the side (often a request of the diet-conscious customer), drizzled over the body, or blended with the body of the salad. Common examples include mayonnaise, vinaigrette, blue cheese dressing, poppy seed dressing, and boiled dressing.

FIGURE 5–2 Four components of the salad (base, body, dressing, and garnish)

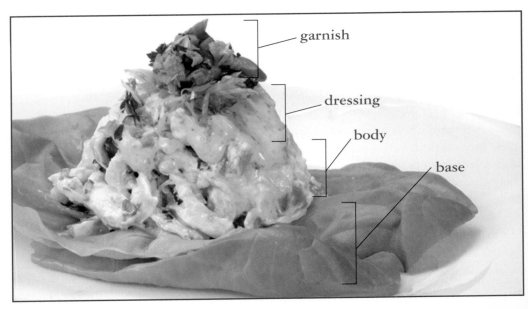

■ **Garnish.** The purpose of the garnish is to add contrast in taste, texture, color, height, and aroma to the body of a salad. Early green salads were garnished with spiced apple rings, parsley sprigs, black olives, tomato wedges, and cubed croutons. Although functional, they lacked in imagination and intensity. In modern times, the body of the salad is flavored sufficiently to stand on its own merits. However, garde manger chefs still use garnishes such as carmelized walnuts, crostini spread with fine herbs and Boursin cheese, toasted sesame seeds, Parmesan crisps, and lavash triangles to add texture, height, and drama to their salads, as well as contrasting flavors.

EVALUATION OF QUALITY POINTS

When analyzing the merits of a salad, the chef must think about several variables that contribute to the salad's potential success both as a salad and as an item within a larger menu. The variety of ingredients and styles available to the chef allows for endless options. Careful consideration should be given to the following quality points:

■ **Appropriateness.** This consideration is broad and encompassing. For example, it may not be appropriate to serve certain foods to people of specific religions, medical conditions, age, or ethnic backgrounds. Or it may be redundant to repeat the use of certain ingredients in a multicourse meal, including items such as seafood, mushrooms, cheese, or peppers.

■ **Taste, Flavor, and Aroma.** If a food item lacks in taste and flavor, it should not be considered; there is little reason to suffer with anything less than palatable food. Likewise, as salads are frequently used as palate cleansers and the bridge between other foods or courses, they should possess an appealing aroma to aid in stimulating the appetite.

■ **Appearance.** Appearance is key to food appreciation. As the familiar expression reminds us, "people eat with their eyes." The appearance should not only be pleasing but also be in keeping with the theme and style of the meal.

■ **Texture and Sound.** Although the senses of hearing and touch are often overlooked when meal planning, they are important attributes to the dining experience. Variety and contrasts in texture and sound while chewing the food add interest and value to most dishes.

■ **Nutritional Value.** Customers are increasingly concerned over their personal health and nutrition. It is both responsible and good business to address their interests by evaluating recipes with these concerns in mind and by making reasonable ingredient or cooking method substitutions when appropriate.

■ **Portion Size.** Portion size is relative to the entire meal and its purpose. The human body has both limited capacity and need for food in its digestive tract. Chefs must consider what is an appropriate quantity to serve. Additionally, the size of the ingredients should be evaluated for ease of consumption and appearance. Salad ingredients should be neither too large to gracefully manipulate while dining nor too small for a pleasant appearance.

■ **Cost.** Financial constraints are a reality of business. It is prudent and responsible to evaluate all salads and other foods according to their cost contribution and margin for profit. Chefs must maintain a conceptual awareness relative to food costs and profitability.

■ **Practicality.** When presenting quality food that is both flavorful and attractive, chefs should also consider the practical nature of the food. The salad must be designed to meet the needs of the customer and the abilities of the food operation and its staff.

Salad Ingredients

Salad ingredients are not limited to any classification. There are few absolute rules in cooking, but it is safe to say that *anything edible* can be used in salads. From the body of the salad, to the sauce or dressing used to moisten and flavor the mixture, salads contain an endless variety of seasonings and ingredients. In addition to cold sauces and dressings, this chapter discusses those vegetarian items not addressed in other chapters, namely (1) vegetables, (2) fruits, (3) grains, and (4) starches.

FIGURE 5–3 Pesto being spread on bruschetta

The remainder of this chapter is devoted to the skills used in preparing these ingredients. The products and skills selected for description are representative of the ingredients and techniques commonly used in garde manger departments.

COLD SAUCES AND DRESSINGS

The garde manger department, in the preparation of appetizers, salads, and sandwiches, uses three primary cold sauces and dressings: basic French, mayonnaise, and boiled dressing. They are primarily used to enhance the flavor of other foods to which they are matched. Cold sauces and dressings are almost always applied to cooked foods or as an accompaniment to those foods such as vegetables that are consumed raw. In some instances, a dressing could serve as a marinade, but that is not its primary function. Each of these sauces and dressings is used either at room temperature or chilled (Figures 5–3 and 5–4).

TECHNIQUES IN PREPARING COLD SAUCES AND DRESSINGS

The principal difference between the sauces is that the *mayonnaise* and *basic French* dressing are **emulsion** dressings, whereas the *boiled dressing* is not.

BASIC FRENCH DRESSING

Of the three primary cold dressings, basic French is the least complicated. Its ingredients consist of oil and vinegar in a ratio of three parts oil to one part vinegar. The seasonings are also uncomplicated, using only salt and pepper. The emulsion created by rapid blending of the oil and vinegar is only temporary and must be quickly reworked before use.

FIGURE 5–4 Assorted salad dressings with green salad

RECIPE 5-1

BASIC FRENCH DRESSING

Recipe Yield: 1 pint

MEASUREMENTS		INGREDIENTS
U.S.	**METRIC**	
½ cup	118 mL	Wine vinegar
⅔ teaspoon	3 mL	White pepper, fresh ground
1½ teaspoon	7 mL	Salt
1½ cup	355 mL	Salad oil

PREPARATION STEPS:

1. Add vinegar, pepper, and salt in a medium-sized stainless steel mixing bowl, whisking to dissolve (Figure 5–5A).

2. Drizzle in the oil, whisking rapidly to form an emulsion (Figure 5–5B).

3. Store in covered container. Whisk again before use.

FIGURE 5–5A Measuring ingredients for an emulsified dressing

FIGURE 5–5B Whisking oil into the French dressing

OTHER DRESSING RECIPES

RECIPE 5–2

ORIENTAL VINAIGRETTE

Recipe Yield: 1 pint

MEASUREMENTS		INGREDIENTS
U.S.	**METRIC**	
¾ cup	177 mL	Rice wine vinegar
¼ cup	59 mL	Soy sauce, light
1¾ cup	412 mL	Salad oil
⅛ cup	30 mL	Sesame oil
¼ teaspoon	1 mL	Garlic, fresh minced
1 tablespoon	15 mL	Ginger, fresh grated
2 teaspoons	10 mL	Black pepper, fresh ground
½ teaspoon	2.5 mL	Tabasco (or other hot pepper sauce)
To taste	To taste	Salt

PREPARATION STEPS:

1. Combine all ingredients, except salt and oil, in a large bowl.

2. Blend thoroughly, using a whisk.

3. Add oil to ingredients.

4. Season with salt.

5. Allow dressing to stand for flavors to blend.

6. Whisk again immediately before serving.

RECIPE 5–3

BASIL VINAIGRETTE

Recipe Yield: 1¾ cups

| MEASUREMENTS | | INGREDIENTS |
U.S.	METRIC	
5 tablespoons	74 mL	Red wine vinegar
1½ teaspoons	7 mL	Dijon mustard
1 each	1 each	Garlic clove, minced
2 teaspoons	10 mL	Fresh basil, minced
1 teaspoon	5 mL	Salt
½ teaspoon	2.5 mL	Black pepper, fresh ground
1 cup	237 mL	Olive oil, extra virgin

PREPARATION STEPS:

1. Combine all ingredients, except oil, into a jar or food processor and quickly blend.

2. Add oil and blend together.

3. Hold until ready to serve.

4. Whisk again before use.

RECIPE 5–4

CAPER AND HERB VINAIGRETTE

Recipe Yield: 1 quart

| MEASUREMENTS | | INGREDIENTS |
U.S.	METRIC	
2 ounces	57 g	Watercress
2 ounces	57 g	Italian parsley
2 ounces	57 g	Dill sprigs
2 ounces	57 g	Chives
2 ounces	57 g	Pommery mustard
4 ounces	118 mL	Red wine vinegar
1 teaspoon	5 mL	Salt
½ teaspoon	2.5 mL	Black pepper, fresh ground
3 cups	711 mL	Olive oil
3 ounces	84 g	Capers

PREPARATION STEPS:

1. Wash and clean the herbs. Shake or press dry.

2. Trim most of the stems.

3. Pulverize in a food processor.

4. Add the mustard and pulse until mixture becomes like a paste.

5. Add the vinegar and seasonings. Pulse briefly to combine.

6. Slowly add the oil while the processor is running, building the vinaigrette.

7. Add the capers.

8. Cover and refrigerate until serving.

RECIPE 5–5

FRENCH ROUILLE

Recipe Yield: 3 cups

MEASUREMENTS		INGREDIENTS
U.S.	**METRIC**	
2 each	2 each	Red pepper, roasted and peeled
4 each	4 each	Garlic cloves
2 each	2 each	White bread slice, torn
2 each	2 each	Pasteurized egg yolks, warmed
2 tablespoons	30 mL	Dijon mustard
2 each	2 each	Lemons, juice of
1 cup	237 mL	Olive oil
To taste	To taste	Salt
To taste	To taste	Black pepper

PREPARATION STEPS:

1. In a food processor, combine all the ingredients, except the olive oil, salt, and pepper.

2. Purée until smooth.

3. With the machine running, slowly add the olive oil.

4. Season the emulsion with salt and pepper.

5. Refrigerate in a tightly covered container.

RECIPE 5–6

LEMON AIOLI

Recipe Yield: 2½ cups

TIP	The aioli may be kept in the refrigerator for up to a day in a covered container.

MEASUREMENTS		INGREDIENTS
U.S.	**METRIC**	
4 each	4 each	Large egg yolks, warmed
½ cup	118 mL	Lemon juice, fresh
8 each	8 each	Garlic cloves, minced
1 teaspoon	5 mL	Kosher salt
2 tablespoons	30 mL	Lemon zest, fresh grated
⅛ teaspoon	½ mL	White pepper
1½ cups	355 mL	Olive oil, extra virgin

PREPARATION STEPS:

1. Combine egg yolks, lemon juice, garlic, salt, zest, and pepper in a blender.

2. With blender running, slowly drizzle in olive oil and blend until an emulsion forms.

3. Adjust seasoning, as needed.

4. Refrigerate in a tightly covered container.

RECIPE 5–7

CAESAR DRESSING

Recipe Yield: 1½ cups

MEASUREMENTS		INGREDIENTS
U.S.	**METRIC**	
3 each	3 each	Garlic cloves, crushed
1 cup	237 mL	Olive oil
1 tablespoon	15 mL	Lemon juice, fresh
3 tablespoons	44 mL	Worcestershire sauce
2 tablespoons	30 mL	White wine vinegar
To taste	To taste	Salt
To taste	To taste	Black pepper, fresh ground
2 each	2 each	Eggs, medium
¼ cup	59 mL	Fresh Parmesan cheese
2 each	2 each	Anchovies, mashed

PREPARATION STEPS:

1. Combine garlic and olive oil; allow to infuse for 2 hours. Alternatively, warm the oil slightly and add the garlic; allow to infuse 10 minutes.

2. Remove the garlic and rapidly blend the olive oil, lemon juice, Worcestershire sauce, vinegar, and salt to create an emulsion. Add pepper to taste.

3. Boil the eggs for 1 minute, allowing the outside to set and leaving part of the white and yolk loose.

4. Cut off the top of the eggs and scoop out the mixture. Rapidly blend with the vinaigrette.

5. Add mashed anchovies and Parmesan cheese. Blend rapidly.

MAYONNAISE

Mayonnaise is often used by itself as a flavoring agent or as a base for other ingredients. It is probably the most common of cold sauces and dressings because its flavor and function are so versatile. Care must be taken to adequately cook the egg yolk without scrambling the egg. Sometimes yogurt is used as a substitute for mayonnaise-based recipes.

RECIPE 5–8

MAYONNAISE

Recipe Yield: 1 pint

MEASUREMENTS		INGREDIENTS
U.S.	**METRIC**	
2 each	2 each	Egg yolks, large
1½ tablespoons	22 mL	Lemon juice, fresh
¼ teaspoon	1 mL	Salt
1½ teaspoons	7 mL	Dry mustard
Pinch	Pinch	Black pepper, fresh ground
1 pint	474 mL	Salad oil

PREPARATION STEPS:

1. Over a warm water bain marie, whisk together egg yolks, lemon juice, salt, mustard, and pepper in a medium bowl until smooth and light.

2. Whisk oil in, by drops, until the mixture starts to thicken and stiffen.

3. As the sauce begins to thicken, when about one third of the oil has been added, whisk in the oil more steadily, making sure each addition is thoroughly blended before adding the next.

4. If the oil stops being absorbed, whisk vigorously before adding more. Add drops of cold water if dressing is too thick.

5. Serve immediately or refrigerate in a covered jar for 1 to 2 days.

RECIPE 5–9

CHANTILLY DRESSING

Recipe Yield: 2½ pints

MEASUREMENTS		INGREDIENTS
U.S.	**METRIC**	
1 cup	237 mL	Heavy cream
1 quart	0.95 L	Mayonnaise

PREPARATION STEPS:

1. Just before use, whip the heavy cream until it reaches stiff peaks.

2. Fold the whipped cream into the mayonnaise carefully to retain volume.

3. Use immediately.

RECIPE 5-10

TARTAR SAUCE

Recipe Yield: 1½ pints

MEASUREMENTS		INGREDIENTS
U.S.	**METRIC**	
2 cups	473 mL	Mayonnaise
½ cup	118 mL	Sweet pickle relish, drained
2 each	2 each	Hard boiled eggs, minced through coarse sieve
4 tablespoons	59 mL	Shallots, minced
4 tablespoons	59 mL	Bottled capers, drained
1 teaspoon	5 mL	Tarragon leaves, dried and crushed (optional)
4 tablespoons	59 mL	Parsley leaves, fresh minced (optional)
4 tablespoons	59 mL	Dijon or Creole mustard
2 teaspoons	10 mL	Lemon juice, fresh
To taste	To taste	Salt
To taste	To taste	White pepper, ground

PREPARATION STEPS:

1. Combine all ingredients and blend well.

2. Adjust seasoning.

3. Refrigerate in tightly covered container.

RECIPE 5-11

SAUCE RÉMOULADE

Recipe Yield: 1 quart

MEASUREMENTS		INGREDIENTS
U.S.	**METRIC**	
3 cups	711 mL	Mayonnaise
2 ounces	57 g	Cornichons, chopped
2 ounces	57 g	Capers, drained and chopped
3 tablespoons	44 mL	Tarragon, fresh minced
3 tablespoons	44 mL	Chervil, fresh minced
3 tablespoons	44 mL	Parsley, fresh minced
3 tablespoons	44 mL	Chives, fresh minced
1 teaspoon	5 mL	Anchovy paste
1 tablespoon	15 mL	Dijon mustard
⅛ teaspoon	1 mL	Tabasco sauce
½ teaspoon	2.5 mL	Worcestershire sauce
To taste	To taste	Salt

PREPARATION STEPS:

1. In a large bowl, blend all ingredients, except salt.

2. Taste and then add salt as needed. Adjust Tabasco sauce and Worcestershire sauce as needed.

3. Cover and refrigerate until serving.

RECIPE 5-12

BLUE CHEESE DRESSING

Recipe Yield: 1 quart

MEASUREMENTS		INGREDIENTS
U.S.	**METRIC**	
2½ cups	572 mL	Mayonnaise
½ cup	118 mL	Buttermilk, divided
1 teaspoon	5 mL	Onion, minced
1 teaspoon	5 mL	Garlic, minced
8 ounces	227 g	Blue cheese, crumbled
1 each	1 each	Lemon, juice of
1 tablespoon	15 mL	Tabasco sauce
1 tablespoon	15 mL	Worcestershire sauce
½ teaspoon	2.5 mL	White pepper
To taste	To taste	Salt

PREPARATION STEPS:

1. In a large bowl, blend mayonnaise, part of the buttermilk, onion, and garlic.

2. Fold crumbled blue cheese into mixture.

3. Season with lemon juice, Tabasco sauce, Worcestershire sauce, pepper, and salt.

4. Thin with remaining buttermilk to reach desired consistency.

5. Cover and refrigerate at least 4 hours before using, allowing flavors to blend.

RECIPE 5-13

RUSSIAN DRESSING

Recipe Yield: 2½ cups

MEASUREMENTS		INGREDIENTS
U.S.	**METRIC**	
4 tablespoons	59 mL	Red bell pepper, minced
4 tablespoons	59 mL	Green bell pepper, minced
2 each	2 ea	Celery stalk, finely chopped
2 cups	473 mL	Mayonnaise
2 tablespoons	30 mL	Tomato paste
2 tablespoons	30 mL	Worcestershire sauce
⅙ teaspoon	1 mL	Hot pepper sauce

PREPARATION STEPS:

1. Blend minced/chopped vegetables with the mayonnaise

2. Add the tomato paste, Worcestershire sauce, and hot pepper sauce.

3. Adjust seasoning.

RECIPE 5–14

PARMESAN DRESSING

Recipe Yield: 1 pint

MEASUREMENTS		INGREDIENTS
U.S.	**METRIC**	
¾ cup	177 mL	Olive oil
4 tablespoons	60 mL	White wine vinegar
1 cup	237 mL	Mayonnaise
1 each	1 each	Garlic clove, crushed
1 cup	237 mL	Parmesan cheese, fresh, finely grated
To taste	To taste	Salt
To taste	To taste	Black pepper, fresh ground

PREPARATION STEPS:

1. In a medium-sized mixing bowl, alternately blend the olive oil and vinegar with the mayonnaise.

2. Stir in the garlic.

3. Add the cheese.

4. Season to taste.

RECIPE 5–15

RANCH DRESSING

Recipe Yield: 1 quart

MEASUREMENTS		INGREDIENTS
U.S.	**METRIC**	
1½ pints	0.75 L	Sour cream
2½ cups	572 mL	Buttermilk
½ cup	118 mL	Wine vinegar, white
3 ounces	88 mL	Lemon juice
3 ounces	88 mL	Worcestershire sauce
3 tablespoons	45 mL	Parsley, minced
2 tablespoons	30 mL	Chives, minced
3 each	3 each	Garlic, fresh minced
2 each	2 each	Scallions, minced
1 ounce	28 g	Prepared mustard
1½ teaspoons	7.5 mL	Celery seed

PREPARATION STEPS:

1. Combine all ingredients in a large bowl.

2. Blend thoroughly.

3. Cover and refrigerate at least 4 hours to allow flavors to blend.

HONEY LIME YOGURT DRESSING

Recipe Yield: 1 quart

MEASUREMENTS		INGREDIENTS
U.S.	METRIC	
3 cups	711 mL	Yogurt, nonfat, plain
¾ cup	170 mL	Honey
1 tablespoon	15 mL	Lime juice
½ teaspoon	2.5 mL	Salt
2 tablespoons	30 mL	Poppy seeds

PREPARATION STEPS:

1. Combine all ingredients in a mixing bowl.

2. Blend thoroughly.

3. Cover and refrigerate until serving.

LOWFAT CREAMY DRESSING

Recipe Yield: 2 quarts

MEASUREMENTS		INGREDIENTS
U.S.	METRIC	
16 ounces	512 g	Ricotta cheese, low fat (part skim)
20 ounces	566 g	Nonfat yogurt
5 ounces	141 g	Red wine vinegar
3 tablespoons	45 mL	Chives, fresh chopped
3 tablespoons	45 mL	Parsley, fresh chopped
3 tablespoons	45 mL	Tarragon, fresh chopped
2 each	2 each	Garlic cloves, fresh minced (optional)
2 each	2 each	Scallions, minced

PREPARATION STEPS:

1. Purée ricotta in blender until almost smooth.

2. In a medium-sized bowl, blend ricotta mixture and remaining ingredients.

3. Cover and refrigerate until serving.

BOILED OR COOKED DRESSING

The boiled dressing was developed in early America, when vegetable oil was a scarce commodity. The other dressings were created in the kitchens of Europe, where olive and other oils were more readily available. Rather than being held by an emulsion of oil, the boiled dressing uses milk thickened by cooked flour and eggs.

RECIPE 5–18

BOILED DRESSING

Recipe Yield: 1 pint

MEASUREMENTS		INGREDIENTS
U.S.	**METRIC**	
$\frac{2}{3}$ ounce	19 g	All-purpose flour
$\frac{1}{8}$ ounce	3 g	Sugar
$1\frac{1}{2}$ teaspoons	7.5 g	Dry mustard
$1\frac{1}{2}$ teaspoons	7.5 g	Salt
Pinch	Pinch	White pepper, fresh ground
$\frac{1}{4}$ teaspoon	1 mL	Paprika
3 each	3 each	Large egg yolks, beaten
2 cups	473 mL	Milk
1 ounce	28 g	Lemon juice, fresh
2 ounces	57 g	White vinegar
$1\frac{1}{2}$ ounces	43 g	Butter

PREPARATION STEPS:

1. Blend together the dry ingredients.

2. Add beaten egg yolks to a clean stainless mixing bowl that will later fit over a double boiler.

3. Whisk in the blended dry ingredients until mixture is smooth.

4. Add the milk and mix until smooth.

5. While whisking the mixture, drizzle in the lemon juice and vinegar.

6. Place the mixing bowl over the simmering water of a double boiler.

7. Cook over low heat, stirring constantly until mixture is thickened.

8. Remove bowl from heat and whisk in butter.

9. Adjust seasoning, as needed.

10. Transfer to a clean stainless steel or glass container for storage. The dressing will continue to thicken as it cools.

RECIPE 5–19

COLESLAW DRESSING

Recipe Yield: 1¾ cups

MEASUREMENTS		INGREDIENTS
U.S.	**METRIC**	
1 tablespoon	15 mL	All-purpose flour
1 tablespoon	15 mL	Granulated sugar
1 teaspoon	5 mL	Mustard powder
1 teaspoon	5 mL	Salt
Pinch	Pinch	Cayenne pepper
3 tablespoons	44 mL	White vinegar (or cider vinegar)
1 tablespoon	15 mL	Butter
1 each	1 each	Egg yolk, medium
1 cup	237 mL	Heavy cream, evaporated milk, or sour cream

PREPARATION STEPS:

1. In a small saucepan, combine the flour, sugar, mustard powder, salt, and cayenne pepper.

2. Add the vinegar and cook until the mixture boils, stirring constantly.

3. Remove pan from heat and rapidly stir in the butter and egg yolk.

4. Whisk in the cream and beat the mixture until light and fluffy.

5. Thin the mixture with vinegar if needed.

RECIPE 5-20

CITRUS SALAD DRESSING

Recipe Yield: 1 quart

MEASUREMENTS		INGREDIENTS
U.S.	**METRIC**	
1 ounce	28 g	Cornstarch
6 ounces	168 g	Sugar, granulated
4 each	4 each	Eggs, large
1 cup	237 mL	Orange juice
1 cup	237 mL	Pineapple juice
½ cup	118 mL	Lemon juice
1 cup	237 mL	Sour cream

PREPARATION STEPS:

1. Combine the cornstarch and sugar in a large bowl. Mix well.

2. Add the cracked eggs and beat until the mixture is creamy.

3. Combine the fruit juices in a stainless steel saucepan and bring the liquid to a boil.

4. Gradually whisk the hot liquids into the egg mixture.

5. Return the whole mixture to the saucepan and gradually bring it to a boil, stirring constantly.

6. When the mixture has thickened, pour it into a clean stainless bowl and chill over an ice bath. Allow to thoroughly cool.

7. Whisk the sour cream into the chilled citrus mixture.

8. Cover well and refrigerate until ready to use.

OTHER SAUCES

The garde manger department uses other important cold sauces and dressings. Many of these other sauces were developed in countries outside of Western Europe, where the three classical cold dressings were first created. They are made by methods other than those previously listed, and several are grouped collectively here as *other sauces*.

RECIPE 5–21

GREEK TZATZIKI

Recipe Yield: 1½ pints

MEASUREMENTS		INGREDIENTS
U.S.	**METRIC**	
3 each	3 each	Cucumbers, medium-sized
To taste	To taste	Salt
6 to 8 each	6 to 8 each	Garlic cloves, crushed
2 tablespoons	30 mL	Vinegar
2¼ cups	532 mL	Plain yogurt, strained to thicken
1 tablespoon	15 mL	Olive oil (Greek preferred)

TIP As an option, add ¼ teaspoon of finely chopped mint.

PREPARATION STEPS:

1. Skin the cucumbers and cut into small dice using food processor.

2. Drain the cucumbers and squeeze well to remove excess moisture. Add salt.

3. Add garlic, vinegar, and yogurt. Mix well. Add olive oil, if desired.

4. Refrigerate in a tightly covered container.

RECIPE 5–22

GREEK SKORDALIA

Recipe Yield: 2½ cups

MEASUREMENTS		INGREDIENTS
U.S.	**METRIC**	
12 each	12 each	Garlic cloves, peeled
½ cup	118 mL	Vinegar, malt or white
2 cups	473 mL	Bread (crusts removed), diced, and soaked in water
To taste	To taste	Salt
1 cup	237 mL	Olive oil

PREPARATION STEPS:

1. Marinate peeled garlic in vinegar for 5 hours.

2. Soak bread in water. Strain and squeeze, removing excess water from bread.

3. Strain the garlic and reserve the vinegar in a separate bowl. Crush the garlic and add the soaked bread.

4. Add 2 tablespoons of the garlic-flavored vinegar and some salt to the garlic and bread mixture. Blend all ingredients in a food processor.

5. Drizzle in oil slowly, while blending all ingredients. Thin with water if needed.

RECIPE 5–23

THAI PEANUT SAUCE

Recipe Yield: 1 quart

| MEASUREMENTS | | INGREDIENTS |
U.S.	METRIC	
4 each	4 each	Garlic cloves, minced
1 each	1 each	Spanish onion, chopped
4 tablespoons	59 mL	Lime juice, fresh
1 tablespoon	15 mL	Brown sugar
1 teaspoon	5 mL	Curry powder or coriander
1 teaspoon	5 mL	Crushed hot red pepper flakes
1½ cups	355 mL	Peanut butter, creamy
1½ cups	355 mL	Canned coconut milk

PREPARATION STEPS:

1. In a food processor or blender, combine garlic, onion, lime juice, brown sugar, curry powder, and hot pepper flakes, processing until garlic and onion are finely chopped.

2. Add peanut butter and coconut milk, processing until smooth.

3. Refrigerate in a tightly covered container (Figure 5–6).

FIGURE 5–6 Peanut sauce over pork satays

RECIPE 5–24

CHINESE DUCK SAUCE

Recipe Yield: 1 pint

| MEASUREMENTS | | INGREDIENTS |
U.S.	METRIC	
2 cups	474 mL	Purple plums, peeled, pitted, mashed
1½ cups	355 mL	Peaches, peeled, pitted, mashed
½ cup	118 mL	Strawberries, mashed
½ cup	118 mL	Red wine vinegar
1 cup	237 mL	Sugar, granulated

PREPARATION STEPS:

1. In a large saucepan, combine all ingredients. Stir to blend.

2. Gently simmer for 1½ hours.

3. Add additional vinegar, sugar, or water to taste when serving.

RECIPE 5–25

SWEET AND SOUR SAUCE

Recipe Yield: 1 quart

MEASUREMENTS		INGREDIENTS
U.S.	**METRIC**	
1 tablespoon	15 mL	Salad oil
2 each	2 each	Garlic cloves, fresh minced
1 cup	237 mL	Scallions, finely chopped
3 cups	711 mL	Pineapple, small diced
¼ cup	59 mL	Ketchup
1 tablespoon	15 mL	Rice vinegar
To taste	To taste	Salt
To taste	To taste	Black pepper, fresh ground

PREPARATION STEPS:

1. Heat oil in a large sauté pan. Add the garlic and scallions.

2. Stir in the pineapple and ketchup.

3. Add the vinegar.

4. Season with salt and pepper.

5. Simmer for 10 minutes.

6. Let cool.

RECIPE 5–26

NAM PLA SAUCE

Recipe Yield: 1 cup

MEASUREMENTS		INGREDIENTS
U.S.	**METRIC**	
2 each	2 each	Kaffir limes, juice of
½ cup	118 mL	Fish sauce
1 teaspoon	5 mL	Palm sugar (or brown sugar)
6 tablespoons	90 mL	Water
5 each	5 each	Scallions, cut into fine rings
4 each	4 each	Red chiles, cut into fine rings, seeds removed
2 each	2 each	Kaffir lime leaves, julienne cut
4 teaspoons	20 mL	Ginger, minced
2 each	2 each	Garlic cloves, minced
2 tablespoons	30 mL	Cilantro, chopped

PREPARATION STEPS:

1. In a medium bowl, combine the lime juice, fish sauce, sugar, and water until the sugar dissolves.

2. Add the scallions, chiles, lime leaves, ginger, garlic, and cilantro.

3. Mix well. Reserve.

RECIPE 5–27

SOY CHILE DIPPING SAUCE

Recipe Yield: 3 cups

TIP Can be used as a condiment for any Asian-Pacific dishes. May be topped with chopped peanuts, garlic, fish sauce, or shredded carrot.

MEASUREMENTS		INGREDIENTS
U.S.	**METRIC**	
1 cup	237 mL	Rice vinegar
5 tablespoons	75 mL	Soy sauce
1 cup	237 mL	Sugar
6 tablespoons	90 mL	Thai chili sauce, sambal or roasted, puréed
6 tablespoons	90 mL	Cilantro, finely chopped

PREPARATION STEPS:

1. Blend vinegar and soy sauce.

2. Add sugar and dissolve.

3. Blend in chili and cilantro.

4. Mix well. Reserve.

RECIPE 5–28

CHINESE BLACK BEAN SAUCE

Recipe Yield: 2½ pints

MEASUREMENTS		INGREDIENTS
U.S.	**METRIC**	
¼ cup	59 mL	Peanut oil
8 each	8 each	Garlic cloves, peeled and minced
½ inch	1.25 cm	Fresh ginger, peeled and minced
½ cup	59 mL	Fermented black beans, salted
4 teaspoons	20 mL	Sugar, granulated
½ cup	118 mL	Sherry
4 teaspoons	20 mL	Soy sauce, light
1 quart	946 mL	Chicken stock
4 teaspoons	20 mL	Soy sauce, dark
¼ cup	59 mL	Cornstarch
8 each	8 each	Scallions, sliced

PREPARATION STEPS:

1. Add oil to wok or large saucepan. Bring to high heat.

2. Add garlic, ginger, and black beans. Stir-fry for 15 seconds.

3. Add sugar, sherry, and light soy sauce. Continue to stir-fry for 15 seconds.

4. Combine and dissolve cornstarch in chicken stock and dark soy sauce. Slowly stir into heated mixture, bringing to a boil while continuously stirring. Reduce heat and simmer 3 minutes.

5. Remove from heat, adding garnish of sliced scallions.

RECIPE 5-29

CHILE-LIME SAUCE

Recipe Yield: 3¼ cups

MEASUREMENTS		INGREDIENTS
U.S.	**METRIC**	
1 cup	237 mL	Water
1 cup	237 mL	Sugar
5 each	5 each	Limes, juice of
1 cup	237 mL	Chile paste
¼ cup	59 mL	Fish sauce

PREPARATION STEPS:

1. In a large heavy-bottomed saucepan, combine water and sugar. Heat at medium-high until boiling. Boil for 5 minutes.

2. Remove from heat. Pour syrup into a medium-sized mixing bowl to completely cool.

3. Add the lime juice, chile paste, and fish sauce. Whisk to combine. Reserve.

RECIPE 5-30

ITALIAN SALSA VERDE

Recipe Yield: 2 cups

MEASUREMENTS		INGREDIENTS
U.S.	**METRIC**	
1 cup	237 mL	Fresh parsley, chopped
9 each	9 each	Anchovies in olive oil, drained and chopped
2 tablespoons	30 mL	Capers, rinsed, drained, and chopped
2 tablespoons	30 mL	Cornichons, chopped
2 each	2 each	Garlic cloves, chopped
1 each	1 each	Lemon, juice of
1 cup	237 mL	Olive oil
To taste	To taste	Salt
To taste	To taste	Black pepper, fresh ground

PREPARATION STEPS:

1. In a food processor, combine all ingredients, except olive oil.

2. Purée until smooth.

3. With the machine running, slowly add the olive oil.

4. Season the emulsion with salt and pepper.

5. Refrigerate in a tightly covered container.

RECIPE 5–31

AMERICAN SEAFOOD COCKTAIL SAUCE

Recipe Yield: 2½ cups

MEASUREMENTS		INGREDIENTS
U.S.	**METRIC**	
1 cup	237 mL	Ketchup
1 cup	237 mL	Chili sauce
½ cup	118 mL	Prepared horseradish
½ to 1 each	½ to 1 each	Lemon, juice of
Dash	Dash	Hot pepper sauce
1 teaspoon	5 mL	Worcestershire sauce

PREPARATION STEPS:

1. Combine all ingredients and blend well.

2. Adjust seasoning.

3. Refrigerate in a tightly covered container.

RECIPE 5–32

ITALIAN PESTO

Recipe Yield: 1 quart

MEASUREMENTS		INGREDIENTS
U.S.	**METRIC**	
6 cups	1.4 L	Basil leaves, fresh with stems removed
⅓ cup	79 mL	Pine nuts (or walnuts), lightly toasted
½ cup	118 mL	Parmesan cheese, fresh grated
¼ cup	59 mL	Romano cheese, fresh grated
8 each	8 each	Garlic cloves
2 cups	473 mL	Olive oil, extra virgin
To taste	To taste	Kosher salt
To taste	To taste	Black pepper, fresh ground

PREPARATION STEPS:

1. Pulse basil leaves, pine nuts, cheese, and garlic cloves in a food processor until finely chopped.

2. With the machine running, pour in olive oil in a thin, steady stream, blending until the mixture is well combined and emulsified.

3. Add salt and pepper to taste.

4. Refrigerate overnight.

RECIPE 5–33

ASIAN
BARBECUE
SAUCE

Recipe Yield: 1 quart

MEASUREMENTS		INGREDIENTS
U.S.	**METRIC**	
2 tablespoons	30 mL	Soy sauce, sweet
1 cup	237 mL	Hoisin sauce
½ cup	118 mL	Oyster sauce
½ cup	118 mL	Apple cider vinegar
¼ cup	59 mL	Honey
¼ cup	59 mL	Sweet Sherry
2 tablespoons	30 mL	Asian chili sauce
½ cup	118 mL	Plum sauce
¼ cup	59 mL	Ginger, minced
¼ cup	59 mL	Garlic, minced
¼ cup	59 mL	Green onion, minced
¼ cup	59 mL	Saké
2 tablespoons	30 mL	Basil, minced
1 tablespoon	30 mL	Sesame oil

PREPARATION STEPS:

1. Combine all the ingredients and store under refrigeration for 24 hours before use.

2. Brush chicken, pork, or beef with the sauce, marinating for 4 hours.

3. Barbecue, as desired, until very tender.

4. Baste with the sauce throughout the cooking period.

RECIPE 5–34

VINEGAR-BASED
BARBECUE
SAUCE

Recipe Yield: 1½ pints

MEASUREMENTS		INGREDIENTS
U.S.	**METRIC**	
2 cups	473 mL	Cider vinegar
1 cup	250 mL	Butter
8 tablespoons	118 mL	Brown sugar
4 tablespoons	59 mL	Crushed red pepper flakes

PREPARATION STEPS:

1. Combine all the ingredients and boil together for 5 minutes.

2. Baste the meat with the sauce while cooking and serve it as a sauce for the meat when it is cooked

3. This is suitable for beef and pork ribs and chicken.

RECIPE 5-35

SOUTHERN BARBECUE SAUCE

Recipe Yield: 1½ pints

MEASUREMENTS		INGREDIENTS
U.S.	METRIC	
½ cup	118 mL	Onion, finely diced
2 tablespoons	30 mL	Garlic, finely diced
2 tablespoons	30 mL	Butter
1 tablespoon	15 mL	Vegetable oil
1 cup	237 mL	Brown sugar
1 cup	237 mL	Water
2 cups	473 mL	Ketchup
2 teaspoons	10 mL	Prepared mustard
2 teaspoons	10 mL	Worcestershire sauce
2 teaspoons	10 mL	White vinegar
½ teaspoons	2 mL	Salt
2 teaspoons	10 mL	A.1. steak sauce

PREPARATION STEPS:

1. Sauté the onion and garlic in the butter and oil until soft.

2. Add all the other ingredients and simmer for 15 to 20 minutes.

RECIPE 5-36

SPICY TEXAS PIT BARBECUE SAUCE

Recipe Yield: 1½ pints

TIP This is an excellent sauce for basting meats, as a dipping sauce, or even with 2 cups of water added as a marinade.

MEASUREMENTS		INGREDIENTS
U.S.	METRIC	
2 cups	473 mL	Ketchup
¾ cup	177 mL	Worcestershire sauce
1 cup	237 mL	Apple cider vinegar
½ cup	118 mL	Dark brown sugar
2 tablespoons	30 mL	Prepared mustard
2 tablespoons	30 mL	Grated horseradish
3 tablespoons	44 mL	Tabasco sauce
5 ounces	142 g	Butter
1 tablespoon	15 mL	Ground black pepper
1 cup	237 mL	Dark beer
¼ cup	59 mL	Lemon juice

PREPARATION STEPS:

1. Combine all the ingredients and simmer for 40 minutes.

2. Chill and reserve for service.

RECIPE 5-37

CITRUS BARBECUE SAUCE

Recipe Yield: 3½ cups

MEASUREMENTS		INGREDIENTS
U.S.	METRIC	
⅔ cup	158 mL	Lemon juice, fresh squeezed
1½ cups	355 mL	Orange juice, fresh squeezed
⅔ cup	158 mL	Ketchup
½ teaspoon	2 mL	Chilli powder
4 teaspoons	20 mL	Prepared horseradish
2 teaspoons	10 mL	Kosher salt
2 tablespoons	30 mL	Worcestershire sauce
½ teaspoon	2 mL	Tabasco sauce
2 tablespoons	30 mL	Garlic clove, minced
2 tablespoons	30 mL	Dry mustard
½ cup	118 mL	Honey

PREPARATION STEPS:

1. Combine all ingredients and blend well.

2. Refrigerate in a tightly covered container.

FIGURE 5–7 Shopping for produce at the Mercado Boqueria in Barcelona

VEGETABLES

The importance of vegetables as guardians of good health has been legendary; they have been recognized by folklore and used in primitive medicines. Early on, doctors advised patients that vegetables would help to keep their "humours of the body" in balance long before the science of nutrition existed. Modern biochemists have found that, in addition to supplying energy-giving carbohydrates, vegetables provide almost all of the specialized vitamins and minerals necessary for good health.

Of the more than 300,000 plant species on the earth, we have discovered almost 6,000 to be edible. Of these, only about 150 vegetables are consumed with any regularity on the world market (Figure 5–7). Vegetables can best be understood when classified into categories: greens; vine; bulbs, roots, and tubers; and legumes. These are discussed relative to their preparation.

TECHNIQUES IN PREPARING VEGETABLES

Although there are many similarities in the plant and vegetable world, the techniques used to select, wash, and prepare them differ to some degree. The color, texture, and flavor of each vegetable must be preserved for optimum guest appreciation. Harmony, in both flavor and appearance, is paramount to a salad's success.

GREENS

Table 5–1 lists commonly available greens and divides them by categories of variety and taste. Create blends of greens from different columns in the table to create salads with various flavor profiles and appearances.

Table 5–1   Common Salad Greens

LETTUCES	SPINACH	SPICY GREENS	BITTER GREENS	CABBAGES	HERBS AS GREENS
Iceberg	Spinach	Watercress (mildest variety)	Watercress (peppery variety)	Napa	Parsley
Red Leaf	Baby Chard	Dandelion (mildest)	Dandelion (peppery variety)	Savoy	Basil
Green Leaf	Baby Beet	Arugula (aka Rocket)	Escarole	Green	Mint
Oak Leaf		Mizuna	Chicory (aka Curly Endive)	Red	Cilantro
Bibb (aka Limestone)		Baby Kale	Komatsu		
Boston		Baby Collards	Tatsoi		
Mâche (aka Lamb's Lettuce)		Belgian Endive (aka Witloof Chicory)	Radicchio		
Romaine (aka Cos Lettuce)			Frisée		
			Baby Mustards		
			Baby Turnip Greens		
			Treviso		

PREPARING SALAD GREENS   A wide range of greens is used to prepare salads. The rules by which a chef may select particular greens, accept them from a supplier, and prepare them for use are generally common knowledge. The following points are worth considering in these decisions:

- Freshness is the basis for good eating; buy only greens that look the freshest. The leaves and stems should be free of rust spots and should be crisp when bent (Figure 5–8).
- Greens begin to lose their moisture immediately upon being plucked or cut from their roots. When possible, buy greens with their roots still attached to delay the process of oxidation.
- The best heads of lettuce are tightly closed but relatively firm to the squeeze. An immature head of lettuce feels like a puff of air when squeezed, because it mostly is.
- The best broccoli or cauliflower has tightly packed flower heads, firm stems, and no sign of yellow or brown.
- Cabbage heads start to open up when they are left on the plant too long. The same is true for Brussels sprouts.

FIGURE 5–8 Lettuce leaf showing rust and bruising

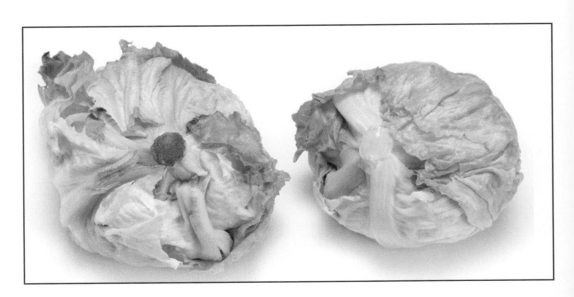

FIGURE 5–9 Greens being washed and soaked in cold water

FIGURE 5–10 Drying greens under towels

- Remove any bruised or wilted greens to prevent decay in others.
- Remove any metal or rubber bands binding the greens, because they cause bruising and rust.
- Soaking and floating them in cold water can revive greens that are slightly limp by filling their cell walls again. It is also the best and often, only, way to remove sand, dirt, and insects (Figure 5–9).
- After swirling the water around the greens to aid in cleansing, gently lift the greens from the water bath to allow the dirt to sink to the bottom, away from the greens.
- Greens must be dried thoroughly after being washed. Any water remaining on the greens will dilute the dressing and prevent the dressing from adhering. Using a salad spinner, or rolling the whole leaves in dry towels, works well to dry them (Figure 5–10).
- Leaves should be stored whole, and cut only when ready to be used. They should be stored in a humid environment, away from air movement like the fan of a walk-in cooler. They should be used within four days of washing.

Chefs are always interested in determining yields of trimmed ingredients and prepared foods. This information is vital to the garde manger chef for costing, purchasing, and production. The Comparison of Salad Green Yield (Table 5–2) depicts the average **as-purchased (AP)** quantity of various greens and the **edible Portion (EP)** quantity after the greens have been washed and trimmed. The table lists the greens as they are commonly purchased, often as heads or bunches.

BULBS, ROOTS, AND TUBERS

Plant foods commonly known as root vegetables are not always actually roots, but rather swollen stems and nutrient reservoirs that grow below ground. **Bulbs** are subterranean buds consisting of both stems and leaves. They have flat, round, or elongated fleshy bodies with roots growing downward from their undersides and shoots and leaves sprouting from their tops. Chives, garlic, onions, and leeks are familiar examples of bulbs. **Roots** are typically the parts of the plant that grow downward away from the sun into the soil to absorb nutrients and moisture. Carrots, turnips, radishes, wasabi, and beets are common examples of roots. **Tubers** are the swollen tips of underground stems that store energy in the form of starch. Tubers are the more exotic of the underground vegetables and include manioc, jicama, yam, kohlrabi, breadroot, and taro.

Table 5–2   Comparison of Salad Green Yields

INGREDIENT NAME	AS PURCHASED WEIGHT (U.S.)	AS PURCHASED WEIGHT (METRIC)	EDIBLE PORTION BY WEIGHT (U.S.)	EDIBLE PORTION BY WEIGHT (METRIC)	EDIBLE PORTION BY VOLUME (U.S.)	EDIBLE PORTION BY VOLUME (METRIC)
Arugula	6 ounces	168 g	3.5 ounces	98 g	2.5 cup	591 mL
Boston	5.25 ounces	147 g	4 ounces	112 g	2.5 cup	591 mL
Green Cabbage	2 lb	896 g	1 lb, 14 ounces	840 g	8 cup	1.89 L
Napa Cabbage	2 lb	896 g	1 lb, 12 ounces	784 g	7 cup	1.66 L
Dandelion Greens	6.5 ounces	182 g	3.25 ounces	91 g	3 cup	709 mL
Escarole	12.5 ounces	350 g	9 ounces	252 g	6 cup	1.42 L
Frisée	12 ounces	336 g	7.5 ounces	210 g	6 cup	1.42 L
Green Leaf Lettuce	11 ounces	308 g	7 ounces	196 g	4 cup	946 mL
Radicchio	9.5 ounces	266 g	4.5 ounces	126 g	2 cup	473 mL
Romaine	14 ounces	392 g	9.25 ounces	259 g	4.5 cup	1.06 L
Spinach, flat leaf	11 ounces	308 g	7.5 ounces	210 g	7 cup	
Watercress	5 ounces	140 g	3 ounces	84 g	2.5 cup	

Of the relative handful of farm-raised produce, about one third grows underground. Only about 25 of these find their way into Western food markets. This area of underused vegetables is an opportunity for the garde manger chef to add interest and variety to the menu.

TRADITIONAL VEGETABLE CUTS (FIGURE 5–11)

- brunoise—A cube-shaped item with dimensions of ⅛ inch (0.3 cm) square.
- small dice—A cube-shaped item with dimensions of ¼ inch (0.62 cm).
- medium dice—A cube-shaped item with dimensions of ⅜ inch (0.93 cm) square.
- large dice—A cube-shaped item with dimensions of ⅝ inch (1.56 cm) square.
- paysanne—A very thin, square-shaped slice with dimensions of ½ inch (1.25 cm) square.

FIGURE 5–11 Assorted vegetable cuts (clockwise from top left: paysane, julienne, batonette, oblique, large dice, medium dice, small dice, brunoise)

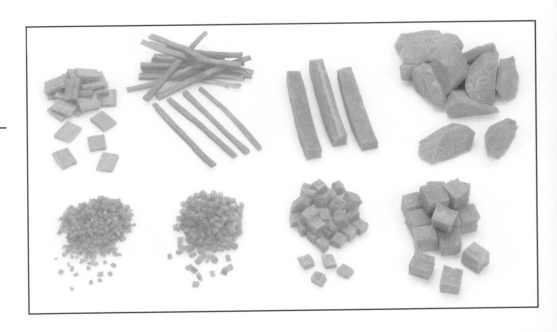

FIGURE 5–12 Nine styles of tourné vegetables. (From left: classical, elongated, château, pointed, bullet, angular bullet, fondant, olivette, Parisienne)

- **julienne**—A thin strip measuring ⅛ inch (0.31 cm) × ⅛ inch (0.31 cm) × 1 to 2 inches (3 to 5 cm). When used with potatoes, this cut is sometimes referred to as an *allumette* (French for matchstick).
- **batonnet**—A stick-shaped item with dimensions of ¼ inch (0.62 cm) × ¼ inch (0.62 cm) × 2 inches (5 cm).
- **oblique cut**—Normally used in the preparation of carrots and parsnips, this cut is a slice through the length of the vegetable at an angle then the cut side is cut at the same angle until the vegetable is completely sliced.

The term *tourné* is a cutting technique that results in a football-shaped finished product with seven equal sides and blunt ends (Figure 5–12). The size of the finished product may vary, the most common being 2 inches (5 cm) long.

There are nine basic styles of tourné:

- classical
- elongated
- château
- pointed
- bullet
- angular bullet
- fondant
- olivette
- Parisienne

PROCEDURE FOR GRATING GINGER ROOT Fresh ginger root can be easily grated by the following method (Figure 5–13):

1. Wrap a cheese grater with two layers of cellophane clear film.
2. Rub a cut end of the fresh ginger root across the film-covered grater (the film will not cut or rip).
3. Collect the grated ginger by wiping it clean from the film wrap.
4. Remove film wrap from grater. Cleaning of the grater is seldom required.

VINE

Although the tomato is commonly associated with Italian food, the plant is native to the western coast of South America and was not introduced in Europe until the early 1500s when returning Spanish colonists brought it from the New World. Originally cultivated by the Aztecs and Incas as early as 700 AD, the tomato is native to the Americas, specifically the Andes region of Chile, Colombia, Bolivia, and Peru. It is believed that the tomato was first domesticated in Mexico.

Tomatoes quickly became popular in the Mediterranean countries, but were not so readily received farther north. The British, in particular, considered the fruit to be beautiful but poisonous. This fear was shared in the American colonies and it was years before the tomato gained widespread acceptance. By the middle of the nineteenth century, tomatoes were in use across America. Like the pumpkin, botanically speaking, the tomato is a fruit, not a vegetable; however, it is served and prepared as a vegetable (Figure 5–14).

FIGURE 5–13 Grating fresh ginger on plastic wrap coated grater

FIGURE 5–14 Tomatoes on vines

Cucumbers, squash, gourds, and pumpkins are all members of the squash family, and although they are very different vegetables, their culture is very similar. The plants are generally grown in mounds and send out vines that run over the ground. The following information applies to vine-grown vegetables:

- Generally speaking, when the vegetable provides little fragrance, it also provides little taste.
- Baseball-like hardness generally indicates immaturity.
- Vine vegetables like peas, green beans, and cucumbers or squash are best when very young.
- The ideal zucchini squash or cucumber is of moderate, uniform diameter, like a sausage.
- Pumpkins and winter squash that are left on the vine until the plant is dead and the shell is hard will keep all winter long.
- The peas in snow peas and the beans in green beans should be barely perceptible or not visible at all.
- Pear-shaped tomatoes are better for cooking than for eating.
- The walls of chiles and peppers thicken and get sweeter—or hotter—as they get older.
- Green peppers are immature; nearly all peppers turn some other color—anything from white to chocolate brown—when fully ripe. Reds, oranges, and yellows seem to be the sweetest.
- Potatoes should be firm but not green. Cut off any green portion of a potato.

PROCEDURE FOR ROASTING PEPPERS Use the following method for preparing roasted bell peppers (Figure 5–15):

1. Inspect peppers for bruises or rot; remove damaged area with a paring knife.
2. Using kitchen tongs, place peppers over open flame, allowing peppers to rest on the range grate.
3. Turn peppers periodically to allow peppers to char evenly on all sides.
4. Place charred peppers in strong plastic bag (such as a food storage bag) and close bag, causing peppers to sweat for a few minutes.
5. Remove peppers from bag and wipe charred skin with clean cloth, or rinse away skin under running water.

PROCEDURE FOR PEELING AND SEEDING TOMATOES Use the following procedure for preparing tomato concassée (Figure 5–16):

1. Fill a medium sauce pan two-thirds full of water.
2. Fill a medium bowl with ice water.

FIGURE 5–15 Roasting bell peppers to enhance their flavor

FIGURE 5–16 Making tomato concassée

3. Using a paring knife, remove the core of the tomato and cut a shallow "X" through the skin on the bottom of the tomato.
4. Using a spider or slotted spoon, blanch the tomato in the boiling water for approximately 15 to 20 seconds. (Ripe tomatoes require less time.)
5. Immediately shock the tomato in the ice water.
6. Using a paring knife, peel the skin to remove.
7. Cut the tomato in half horizontally and squeeze to remove the seeds.
8. If concassée is desired, dice the tomato to required size.

DRIED LEGUMES

Salads that use dried **legumes**, in particular those that combine them with grains, achieve a wonderful nutritional balance. The combination of legumes and grains contains essential nutrients, including numerous complex carbohydrates, dietary fiber, minerals, vitamins, and complete proteins.

Dried beans, and to a lesser extent peas, are notorious for creating intestinal gas. Dried legumes contain complex sugars called *oligosaccharides,* which are not easily broken down by the stomach's digestive enzymes. As undigested oligosaccharides travel through the gut, they eventually reach the lower intestinal tract where resident bacteria eagerly devour these complex sugars. Gas is the waste product of this process. Rinsing beans well and cooking them thoroughly can easily remedy the problem.

PROCEDURE FOR SOAKING DRIED BEANS   Most dried beans and peas (except for lentils and split peas) need to be soaked in water before they can be cooked. Soaking dried beans before cooking rehydrates and softens them, and helps break down those indigestible oligosaccharides. Presoaking reduces the cooking time by at least half.

To soak beans, use four parts water to one part beans. Most beans need to soak for 4 hours, although very old beans may need longer. Soybeans and broad beans, such as fava beans, have very tough seed coats that require 12 to 24 hours of soaking. Do not add salt to the water; salt reacts with the seed coat, forming a tough barrier that inhibits the absorption of liquid. The following methods of soaking are general; the time may need adjustment depending on the age and variety of legume.

Overnight Soaking Method
1. Wash, sort, and place dried beans in a large saucepan. Cover generously with cold water.
2. Let stand 8 hours or overnight.

Quick-Soak Method
1. Place washed and sorted beans in a large saucepan. Cover with 2 inches of fresh, unsalted water or 3 times their volume.
2. Bring to a boil. Reduce heat to medium and boil for 10 minutes.
3. Drain the beans and cover with 2 inches (or 3 times their volume) of fresh, cool water.
4. Allow soaking for 30 minutes. Discard soaking water, rinse, and the beans are reconstituted and ready for further cooking.

FRUITS

Fruit is a seed-bearing structure of a flowering plant. Fruits develop from the female part of the plant, the ovary of the flower. Depending on the type of plant, the fruit may be juicy and fleshy, like a peach, plum, apple, or blueberry. Or the fruit may be a dry fruit, such as an acorn, chestnut, wheat, corn, or rice. Certain foods thought of as vegetables, including tomatoes, squash, peppers, and eggplant are really fruits because they develop from the ovary of the flower.

Fruits are composed of three layers. The outermost layer is called the *exocarp,* the middle layer is the *mesocarp,* and the inner layer, the *endocarp.* The layers can be seen when cutting a peach in half. The soft fuzzy skin, the juicy flesh, and the pit represent the three layers.

FIGURE 5–17 Selecting fresh fruit from an open-air market

Traditionally, chefs have selected fruits when they are in season, because they are at their best then (Figure 5–17). For some fruits, this is true. Locally grown peaches, berries, and apricots fare best when they are not shipped. However, some melons, bananas, apples, and citrus fruits can travel well with careful packing and transportation. With a few exceptions, like bananas, fruit should be refrigerated. Unless otherwise specified, fruit (and other produce) should be kept between 40° and 45°F (4° and 7°C) with a relative humidity of 80 to 90 percent.

TECHNIQUES IN PREPARING FRUITS

- **Apples.** If possible, try to use wax-free apples. Wax increases shelf life but toughens the skin and tastes unpleasant. Apples should be crisp and firm. Soft, mushy apples do not juice well. Always store apples loosely in the refrigerator, which increases shelf life six-fold. After they are cut or peeled, apples (and other light-colored fruit such as pears and peaches) will begin to oxidize and turn dark. To prevent browning, drop apple rings into a bowl of cold water (about 2 quarts) containing $\frac{1}{2}$ teaspoon ascorbic acid powder (1,500 mg), or use equivalent in finely crushed vitamin C tablets or $\frac{1}{2}$ cup lemon juice. Vitamin tablets contain filler, which may turn the water cloudy but is not harmful.
- **Apricots.** Cut in half and remove the pit. Apricots may be eaten skin and all. To peel, blanch for 10 to 20 seconds in boiling water and plunge immediately into cold water. The peel will slip off.
- **Cantaloupe.** To check for ripeness, gently press against the vine end of the melon (larger indentation). The melon should give gently and should smell sweet. Wash the cantaloupe be-

fore cutting to prevent food poisoning from cross-contamination. Gently scoop out seeds with a spoon and cut the fruit into wedges, or use a Parisian scoop.

- **Cherries.** Using a cherry pitter, pit the cherries over a bowl to be sure that each cherry is successfully pitted. Look for heavy, firm cherries with a tight skin.
- **Grapes.** Wash the grapes thoroughly before using. Look for bunches with the grapes held tightly to the stems. Make sure to buy seedless varieties and store covered in the refrigerator.
- **Honeydew melon.** The melons should be heavy for their size and give slightly when pressed with the fingers. They should also smell sweet, like honey. Wash them well before slicing and gently scoop out the seeds.
- **Kiwi.** The black seeds are edible. Peel the skin and slice to serve.
- **Lemons.** Lemons should feel heavy for their size. Skin should be smooth with no green spots; green spots signal high acid content. Store loosely in refrigerator. To get the most juice, gently roll the lemons on the countertop to break down the cell structure. Or, prick the lemons with a knife and microwave them on high for 30 seconds, then juice.
- **Mangoes.** Mangoes should smell sweet and be soft when pressed with the fingers, but not wrinkled. Color is not a reliable indicator of ripeness. To prepare mangoes, hold them upright and cut down along one of the fat sides, curving the knife to avoid the large oval pit. Repeat on the other side. Score the flesh with a knife in a crisscross pattern. Gently press the scored halves to turn inside out and cut off the chunks of fruit from the peel. Trim the rest of the fruit off the pit and peel.
- **Papaya.** The large black papaya seeds are edible, although they have a strong peppery taste. Test the fruit to make sure it gives slightly with pressure from the hand's palm, then wash, peel, scoop out the seeds, and slice.
- **Peaches.** These fruit should be firm, appear yellow with a red or pink blush, smell sweet, and give slightly to palm pressure. Be sure to select freestone peaches, not cling, because the pit is very difficult to remove from the cling peach. Peaches can be blanched for 30 seconds in boiling water and then plunged into cold water. The skins will slip off.
- **Pears.** When pears are perfectly ripe—not too soft and not too firm—they are at their most delicious. Look for slightly soft flesh around the stem area. The juice from a pear is thick and sweet and can be diluted with apple juice.
- **Pineapple.** The ripe fruit should smell ripe and give very slightly when pressed at the bottom. Also, a leaf should pull freely from the crown when the pineapple is ripe. To prepare, wash, then firmly grasp the leaves and twist off. Cut the pineapple into four sections. Using a curved knife, cut the flesh away from the prickly peel. Remove the hard core and slice the fruit or cut it into chunks.
- **Raspberries.** Raspberries need no preparation—just wash gently and quickly, then serve. Carefully check packages before buying to ensure there are no damaged or moldy fruits in the bottom of the container.
- **Starfruit.** Starfruit are ripe when yellow and give slightly to pressure. The entire lemony fruit is edible. Wash and slice to create the fruit's star-shaped pattern.
- **Strawberries.** Fresh strawberries should be red, firm, and plump. The best are chosen from pick-your-own farms or from farmer's markets. To prepare, wash thoroughly, then cut out the leaves (hull) and any white part at the top, or shoulder, of the strawberry. Then slice or chop.
- **Watermelon.** Seedless varieties of watermelon are not really seedless, but have very small, tender, edible seeds. Watermelon is best cut into large wedges, then sliced. When ripe, the watermelon gives a hollow thud when thumped and smells sweet. Watermelon rind is a popular ingredient for use in pickled fruits.

> **TIP** Never wash berries until just before use.

CITRUS FRUITS

Citrus fruits are one of the more important families of fruits. Their good flavor (sweet, sour, or bitter) and their large capacity in juice make them unique for salads, cooking, confectionery, and distillery. They are a good source of vitamin C, so they are nutritious as well as delicious. Citrus fruits originated in Asia 20 million years ago, but now they are cultivated all over the planet and are the world's largest crop. Citrus was first mentioned in literature in 2400 BC. The most common citrus fruits are oranges, tangerines, grapefruits, lemons, and limes.

PROCEDURE FOR SEGMENTING CITRUS FRUITS Removing the bitter membrane from citrus fruit is an essential step in many salads and tarts. The task is made easier when using a newly sharpened medium-sized paring or vegetable knife (Figure 5–18).

1. To cut citrus fruits such as oranges into segments, first trim the top and bottom flat to expose the inner fruit.
2. Set the fruit so it is standing on the work surface and use a paring knife to remove the skin and bitter white pith.
3. Follow the natural round shape of the orange and take care not to cut off too much of the flesh of the fruit. Turn the citrus as each cut is made.
4. Trim off any white areas that may remain before segmenting the fruit.
5. Insert the blade of the paring knife between the membrane and the pulp of each segment and cut toward the center.
6. Flip the cutting edge away so that the blade is parallel to the next membrane.
7. Push out to remove segment.

BERRIES

One of America's most well-liked fruits—berries—has origins in both Europe and the United States. The people of Troy (modern day Turkey) were the first to note an appreciation for raspberries, and strawberries were cultivated in ancient Rome. Blueberries are native to North America, and the black raspberry was created in New York state by crossing the blackberry and raspberry. Types of berries range from smooth-skinned varieties like blueberries, to berries that have fleshy segments like raspberries and blackberries. Strawberries are not considered a true berry, in that they do not grow from a flower ovary but rather from the base of the plant.

PROCEDURE FOR HOLDING BERRIES

1. After purchasing berries, check the fruit and toss out any moldy or deformed berries.
2. Immediately use the overripe berries within 24 hours.

FIGURE 5–18
Segmenting an orange

3. Arrange unwashed berries on sheet pans lined with paper towels. (Only wash berries just before their use.) The berries may be topped with a paper towel to absorb any additional moisture.

4. Plastic wrap the entire container to ensure the fruit retains its freshness, but generally berries should be eaten within 1 week.

5. When using commercially frozen berries (which are already prewashed and prepared) it is best to use them while still frozen. For salads, fresh berries are much better unless the berries are coated with a dressing.

6. Because berries have a short shelf life, an alternative to enjoy them all year round is to buy them fresh and freeze them. The secret to successful freezing is to use unwashed and completely dry berries before placing them in a single layer on a sheet pan in the freezer. Once the berries are frozen, transfer them to plastic bags or freezer containers. Frozen berries should last approximately 10 months to 1 year.

7. If berries are to be served alone, thaw until they are pliable and serve partially frozen.

GRAPES

Grapes are an edible fruit in the buckthorn family and of the vines that produce the fruit. The European grape has been used as food since prehistoric times. Fossil traces show that the grape existed in Africa during the Old Stone Age 10,000 years BC. The Egyptians cultivated the grape 5,000 years BC, and also drank grape wine. Ancient Greeks and Romans were particularly fond of grape wine. Grape seeds have been found in remains of lake dwellings of the Bronze Age in Switzerland and Italy and in tombs of ancient Egypt.

Grapes are one of the world's most important fruits. The European grape is commercially cultivated in warmer regions all over the world, particularly in Western Europe, the Balkans, California, Australia, New Zealand, South Africa, and parts of South America. Either with or without seeds, grapes are juicy fruits that grow in berry clusters on vines. Although a high percentage are devoted to wine grapes, the table grape is also commonly consumed. Table grapes differ from wine grapes, being fuller fleshed, crunchy, and often seedless, like the Sultana grape, early red Cardinal, or green Thompson Seedless.

Although there are some classic foods that use grapes as an integral part of the dish, they are more commonly consumed as an accompaniment or garnish, such as in salads or cheese plates.

PROCEDURE FOR SUGAR COATING GRAPES Crystallized grapes work well as garnishes for desserts and in fruit and cheese platters. The following procedure works well with any grape (or citrus fruit) (Figure 5–19).

1. Wash and dry small clusters of grapes.

2. Whip raw egg white until it is slightly frothy. (*Note:* Lemon juice may be safely used as a substitute for egg whites.)

3. Dip the small clusters into the egg whites or lemon juice, a few at a time, or use a small pastry brush to lightly coat them. Gently knock off excess liquid.

4. Gently dip the wet grapes into fine granulated sugar, gently bouncing them in a sieve to remove excess sugar.

5. Place the coated grape clusters on parchment-covered sheet pans. Place pans in cool areas, allowing clusters to harden for an hour.

FIGURE 5–19 Mise en place for sugar coating grapes

GRAINS

No food is more widely consumed than **grains**. Grains of one kind or another are a staple part of cuisines from every corner of the earth. For thousands of years they have been valued for their simplicity of preparation and nutritional value. When early humans began planting seeds, and turned from societies of hunters and gatherers to agrarians, they settled together to protect their crops. The grasses they planted have been known to every culture. Every culture has its own uses for this irreplaceable staple, such as Asian rice dishes (rice), Italian pastas (wheat), and Central American tortillas (corn).

Strictly speaking, grains are the fruit (kernels and seeds) and plant material from cereal grasses. They are relatively inexpensive and readily available. The most common and significant are rice, wheat, and corn. Milling removes the most nutritious parts of the grain—the bran and germ. Over time, these polished grains were perceived as status symbols, much to the demise of the common people's health. The result was widespread vitamin deficiency diseases such as beriberi.

Fortunately, *whole grains* are making a comeback as an important part of a healthy diet.

Because most of the nutrients in grains are contained in the outer layers, whole-grain products are the most nutritious. The best grains are those whose kernels are intact—not broken, scratched, or damaged. Milled grains that are broken into pieces and particles are called *cracked.* Further milling creates *meal,* and the finest grind is called *flour.* In all cases, grains should be stored in a cool, dry area in a sealed glass or plastic container, away from air and moisture.

Table 5–3 Methods of Cooking Grains

GRAIN	HOW TO COOK (1 CUP/.24 LITERS GRAIN)	CHARACTERISTICS
Amaranth	2½ cups water for 20 minutes	Prized crop of the Aztecs; adds a crunchy texture to breads, cookies, and casseroles.
Whole Barley	3 cups water for 60 minutes	Has a pleasant, chewy texture; can be substituted for brown rice.
Buckwheat	2 cups water for 10 minutes	Toasted buckwheat (kasha) has a robust, hearty flavor; good cold-weather fare.
Bulgur	2 cups water for 10 minutes	Made by soaking and cooking the whole wheat kernel, then removing 5% of the bran and cracking the remaining kernel into small pieces. It can be used in salads, soups, breads, and desserts.
Corn	Cook 1 cup coarse yellow or white cornmeal in 4 cups water for 30 minutes to make polenta	Top polenta with yogurt and/or maple syrup for breakfast, layer with sauce and vegetables for lasagna or pizza.
Couscous	2½ cups water for 10 minutes	Made from durum wheat, it tastes like pasta. Often served as a pilaf, couscous is a good source of protein.
Millet	3 cups water for 30 minutes	A welcoming change to rice; light toasting gives it a pleasing aroma and almost nutty flavor. It adds texture and flavor to breads or can be ground and used like cornmeal.
Oats	Whole oats in 4 cups water for 60 minutes; rolled oats in 2 cups water for 10 minutes; steel cut oats in 3 cups water for 30 minutes	Oats are rich in protein and minerals. Although they have appeared on the breakfast table for many years, oat flour is a tasty addition to breads and baked goods.
Quinoa	2 cups water for 15 minutes	Although not a true grain, this prize of the Incas is a superior source of protein, as well as calcium, iron, vitamins, and potassium. Tasty and quick cooking, it's a welcoming addition to almost any dish, from salads to desserts.
Rice	2 cups water, bring to a boil for 8 minutes, then reduce to low for 35 minutes	Many different varieties abound, each with unique flavor and texture characteristics.
Rye	2 cups water for about 10 minutes	With a hearty flavor, it can be eaten like rolled oats or added to bread for chewiness.
Spelt Berries		Use in place of wheat in baked goods, cereals, and other recipes. Related to wheat but frequently tolerated by those with wheat allergies.
Teff	in 3 cups water for 15 minutes	The smallest grain in the world, its size prohibits it from being hulled, thus retaining all the whole grain nutrients. It's delicious in combination with other grains.

TECHNIQUES IN PREPARING GRAINS

Grains are often prepared in different ways for salads; some differences are more subtle than others. The recommended methods for cooking the various types of grains are explained in Table 5–3.

PROCEDURE FOR PREPARING COUSCOUS

1. Pour out the couscous (semolina) into a bowl.
2. Cover it with warm salted water. Soak for 15 minutes.
3. Mix couscous around with hands to make sure all the grains are evenly wet and to begin softening them. Set it aside.
4. Fill the bottom half of a couscousier (couscous pot) with water and bring to a boil.
5. Drain the couscous, toss with 1½ tablespoons of olive oil, and place the couscous into the top half of the couscousier. Cover tightly and allow the pot to steam for about 15 to 20 minutes, until the couscous is tender.

FIGURE 5–20 Garnishing a
grain salad

6. When the steam comes through the couscous, pour the couscous out onto the platter.
   Sprinkle with ½ to 1 cup cold water to moisten, and fluff with a fork.
7. Add the couscous to the top of the steamer, cover, and steam again for 15 minutes.
8. Fluff finished couscous with salted olive oil to flavor and break up the clumps of cooked
   semolina (Figure 5–20).

RECIPE 5-38

TABBOULEH

Recipe Yield: 8 portions

| MEASUREMENTS | | INGREDIENTS |
U.S.	METRIC	
2 cups	350 g	Fine bulgur wheat, soaked for 20 minutes
16 each	16 each	Scallions, green tops removed
6 each	6 each	Tomatoes, peeled
2 bunches	2 bunches	Fresh parsley, stalks removed
2 bunches	2 bunches	Fresh mint, stalks removed
5 each	5 each	Lemons, juice of
1½ cups	350 mL	Olive oil
To taste	To taste	Salt
To taste	To taste	Black pepper, fresh ground

PREPARATION STEPS:

1. Drain the bulgur wheat and press dry.

2. Finely chop the scallions, tomatoes, and herbs. Blend with the bulgur wheat.

3. Whip the lemon juice and drizzle in the oil until emulsified. Season with salt and pepper.

4. Pour dressing over salad ingredients. Adjust seasoning, adding more lemon juice if needed.

5. Allow salad to stand 30 minutes, until grains are tender and flavor develops.

RECIPE 5-39

CAPONATA

Recipe Yield: 8 portions

U.S.	METRIC	INGREDIENTS
1½ pounds	680 g	Eggplant, peeled and cut into ½-inch cubes
As needed	As needed	Salt
½ each	½ each	Spanish onion, diced
2 ounces	59 mL	Olive oil
1 pound	454 g	Roma tomato concassée
1½ tablespoons	22 mL	Raisins, yellow
2 tablespoons	30 mL	Kalamata olives, pitted and chopped
2 tablespoons	30 mL	Green Sicilian olives, pitted and chopped
1½ tablespoons	22 mL	Capers
1½ tablespoons	22 mL	Pine nuts
To taste	To taste	Salt
To taste	To taste	Black pepper, fresh ground
2 ounces	57 g	Olive oil
3 each	3 each	Celery stalks, peeled and diced
4 ounces	125 g	Red wine vinegar
1 tablespoon	15 mL	Sugar
To taste	To taste	Salt
To taste	To taste	Black pepper, fresh ground

PREPARATION STEPS:

1. Place eggplant in a colander and lightly coat with salt. Cover and press with a heavy plate. Let stand for 30 minutes.

2. In a sauté pan, combine ¼ cup olive oil and onions, and lightly brown over low heat.

3. Add the tomato concassée, raisins, olives, capers, and pine nuts. Simmer for 30 minutes. Season with salt and pepper.

4. Meanwhile, rinse the eggplant to remove the salt and dry with paper towels.

5. In a separate sauté pan, heat the 2 ounces of olive oil over medium heat and add the eggplant. Sauté the eggplant until deep golden brown.

6. Add cooked eggplant to simmering tomato mixture.

7. In the empty sauté pan, quickly sauté the diced celery, adding oil if needed. Add celery to the tomato-eggplant mixture.

8. Deglaze the empty sauté pan with red wine vinegar over high heat, and reduce wine vinegar to half. Add sugar. Add deglazed wine mixture to tomato-eggplant mixture.

9. Adjust seasoning of sweet and sour flavors; adjust sugar and vinegar.

10. Allow salad to stand overnight to develop flavors.

RECIPE 5–40

VEGETARIAN SALMAGUNDI

Recipe Yield: 8 portions

MEASUREMENTS		INGREDIENTS
U.S.	**METRIC**	
1½ each	1½ each	Cucumber, young, peeled
3 each	3 each	Green bell pepper, blanched, ribs and seeds removed
1 each	1 each	Onion, large, peeled
1 each	1 each	Cabbage, medium-sized head, cleaned
As needed	As needed	Salt
3 each	3 each	Green tomatoes, medium-sized, diced
⅔ cup	158 mL	White vinegar
⅔ cup	158 mL	Water
⅓ teaspoon	2 mL	White pepper
⅓ teaspoon	2 mL	Sugar

PREPARATION STEPS:

1. Remove ribs and seeds from the bell peppers.

2. Wash and dice the tomatoes.

3. Thinly slice or shred the cucumbers, green peppers, onion, and cabbage. Separate each into different bowls. Mix a sprinkle of salt into each vegetable, allowing them to macerate for 1 hour.

4. Press out all excess moisture from the salted vegetables. Blend together and add the diced tomatoes.

5. Mix the vinegar and water with ⅔ teaspoon of salt. Add the white pepper and sugar, and bring to a light boil. Remove from the heat source and allow cooling.

6. After the dressing has cooled, blend with the vegetables. Allow the vegetables to steep in the dressing for a few hours before serving to develop their flavor.

RECIPE 5–41

PICKLED CUCUMBER SLAW

Recipe Yield: 6 portions

MEASUREMENTS		INGREDIENTS
U.S.	**METRIC**	
1½ cups	356 mL	English cucumber, skin on, sliced 1/8-inch thick
1 tablespoon	15 mL	Kosher salt
4 tablespoons	60 mL	Mirin (sweet rice wine)
4 tablespoons	60 mL	Rice vinegar
1 tablespoon	15 mL	Pickled ginger juice

PREPARATION STEPS:

1. Place sliced cucumbers in a bowl. Sprinkle evenly with salt and mix well to coat. Reserve cucumbers at room temperature for 30 minutes.

2. Rinse cucumbers under cold running water, for about 20 minutes, until all salt is washed off. (Taste to test saltiness.) Drain and reserve.

3. Combine the mirin, vinegar, and pickled ginger juice in a separate bowl. Pour over the cucumbers to cover. Marinate a minimum of 12 hours.

4. Serve chilled or at room temperature (Figure 5–21).

FIGURE 5–21 Pickled Cucumber Slaw

RECIPE 5–42

GREEN SALAD WITH BLUE CHEESE, WALNUTS, AND FIGS

Recipe Yield: 8 portions

| MEASUREMENTS | | INGREDIENTS |
U.S.	METRIC	
1 pound	453 g	Salad greens, cleaned, drained and cut to bite size; dried, and chilled
½ cup	118 mL	Walnut oil
1 teaspoon	5 mL	Salt
½ cup	118 mL	Blue cheese, crumbled
8 each	8 each	Fresh figs, sliced (or 10 dried figs)
½ cup	118 mL	Walnuts, chopped and lightly toasted
⅓ teaspoon	2 mL	Black pepper, fresh ground
4 each	4 each	Lemons, cut into 1/4 wedges

PREPARATION STEPS:

1. Place the cleaned and drained greens in a large bowl. Sprinkle on the walnut oil and salt, tossing gently to evenly coat.

2. Add the blue cheese, cut figs, and walnuts, tossing gently to distribute evenly.

3. Grind in the pepper.

4. Serve with 2 lemon wedges each (Figure 5–22).

FIGURE 5–22 Green Salad with Blue Cheese, Walnuts, and Figs

RECIPE 5-43

Caesar Salad

Recipe Yield: 8 portions

MEASUREMENTS		INGREDIENTS
U.S.	**METRIC**	
2 cups	473 mL	Italian bread, ½-inch cubed
2 each	2 each	Garlic cloves, mashed
¼ cup	59 mL	Olive oil (Italian preferred)
2 each	2 each	Romaine lettuce heads, leaves of
2 each	2 each	Garlic cloves, mashed
¼ teaspoon	1 mL	Salt
3 tablespoons	45 mL	Olive oil (Italian preferred)
2 each	2 each	Fresh eggs
4 tablespoons	59 mL	Olive oil (Italian preferred)
¼ teaspoon	1 mL	Salt
1 tablespoon	15 mL	Whole black pepper, fresh ground
2 teaspoons	10 mL	Olive oil (Italian preferred)
1 each	1 each	Fresh lemon, juice of
6 drops	6 drops	Worcestershire sauce
2 each	2 each	Anchovies (optional), mashed
¼ cup	59 mL	Fresh Parmesan cheese, grated

PREPARATION STEPS:

1. Spread cubed bread on sheet pan and dry in medium oven. Baste drying bread with mixture of 2 mashed garlic cloves and ¼ cup olive oil.

2. Strip whole leaves from the heads of Romaine lettuce. Gently wash leaves in cold water and shake thoroughly dry. Reserve and refrigerate in plastic bag for later use.

3. Purée 2 cloves of garlic with ¼ teaspoon of salt and 3 tablespoons of olive oil. Strain mixture. Add strained oil to large sauté pan. Add croutons. Heat briefly while tossing croutons to evenly coat and pour coated croutons into serving bowl.

4. Boil eggs for exactly 1 minute.

5. Place the reserved lettuce leaves in a large wooden serving bowl. Add 4 tablespoons of olive oil, scooping the lettuce to coat the leaves. Sprinkle with ¼ teaspoon of salt, the freshly ground black pepper, and 2 more teaspoons of olive oil. Toss mixture again to coat leaves.

6. Add fresh squeezed lemon juice, the Worcestershire sauce, the mashed anchovies (if desired), and coddled eggs. Toss mixture and add fresh grated Parmesan.

PIG'S KNUCKLES WITH MEAT AND SPICED AUTUMN SQUASH

Recipe Yield: 8 portions

MEASUREMENTS		INGREDIENTS
U.S.	**METRIC**	
6 each	6 each	Pig's knuckles
2 tablespoons	30 mL	Pickling spice
2 cups	473 mL	White wine vinegar
2 tablespoons	20 mL	Salt
2 pounds	907 g	Squash (Hubbard, butternut, or acorn), unpeeled, cut into 1-inch cubes
2 each	2 each	Onion, medium, minced
4 each	4 each	Dill pickles, chopped
2 each	2 each	Green bell pepper, seeded and minced
2 tablespoons	30 mL	Prepared mustard
2 tablespoons	30 mL	White wine vinegar
4 tablespoons	59 mL	Olive oil, extra virgin
1 head	1 head	Red tip leaf lettuce, washed and separated

PREPARATION STEPS:

1. Cook the pig's knuckles in lightly salted boiling water, uncovered, for 1½ to 2 hours until they are tender.

2. Meanwhile, in a medium stockpot, combine 1 gallon of boiling water, the pickling spice, vinegar, salt, and squash. Cook for about 20 minutes, until the squash is tender but firm. Drain and remove the skin of the squash. Cool and reserve in a large bowl.

3. Drain the pig's knuckles when cooked, and remove their outer skin. Trim away the meat from the bones, cutting into small dice. Cool.

4. Combine the squash and cubed meat.

5. Blend the onion, pickle, bell pepper, mustard, vinegar, and olive oil. Thoroughly mix with the squash and pig meat.

6. Place washed lettuce leaves on plates, and spoon over with marinated squash and pig mixture.

RECIPE 5–45

WARM SHRIMP AND ARTICHOKE SALAD WITH ORZO

Recipe Yield: 8 portions

MEASUREMENTS		INGREDIENTS
U.S.	**METRIC**	
		Orzo Salad
1½ cups	355 mL	Orzo
4 each	4 each	Scallions, chopped
2 tablespoons	30 mL	Fresh dill, chopped
2 tablespoons	30 mL	Fresh parsley, chopped
2 tablespoons	30 mL	Capers, drained
⅓ cup	79 mL	Pine nuts, lightly roasted or toasted
1 tablespoon	15 mL	Fresh mint, chopped
1¼ cups	296 mL	Basil Vinaigrette (see Basil Vinaigrette recipe 5–3)
		Shrimp and Artichokes
1½ pounds	680 g	Shrimp (approx. 31/35 count) peeled and deveined, tails removed
¾ pound	340 g	Artichoke hearts, cooked
2 tablespoons	30 mL	Butter, unsalted
To taste	To taste	Salt
To taste	To taste	Black pepper, fresh ground
3 heads	3 heads	Soft lettuce (red tip leaf, Boston, bibb)
1 tablespoon	15 mL	Pernod
1 teaspoon	5 mL	Dijon mustard
2 tablespoons	30 mL	White wine vinegar
2 tablespoons	30 mL	Pine nuts, toasted
¼ cup	59 mL	Mint leaves

(continued)

FIGURE 5–23 Warm Shrimp and Artichoke Salad with Orzo

PREPARATION STEPS:

1. Cook orzo in 2 quarts of boiling water for 7 to 8 minutes, or until just tender. Drain and shock thoroughly with cold water. Drain again, eliminating all water.

2. Meanwhile, make Basil Vinaigrette by combining all vinaigrette ingredients in a covered jar and shaking vigorously or by whisking in a bowl.

3. Toss cooled orzo with Basil Vinaigrette. Reserve.

4. In a sauté pan, toss shrimp, artichokes, salt, and pepper with butter until shrimp is pink and tender.

5. Arrange the salad plates with the greens as a base.

6. Portion the Orzo Salad onto one side of the plate.

7. Carefully spoon the shrimp and artichokes from the pan, reserving the juices, and place them next to the Orzo Salad.

8. Heat the pan juices with the Pernod, mustard, and vinegar, stirring constantly. Reduce a little.

9. Pour pan sauce over salads.

10. Garnish with toasted pine nuts and mint leaves (Figure 5–23).

RECIPE 5-46

JAMAICAN SALSA SALAD

Recipe Yield: 8 to 12 portions

> **TIP** Boiling water will soften the onion and remove some of its acidic "bite."

> **TIP** While being refrigerated before serving, place the avocado pits throughout the salad to help retain the avocados' color. Then remove the pits before serving.

MEASUREMENTS		INGREDIENTS
U.S.	**METRIC**	
1 cup	237 mL	Red onion
3 each	3 each	Serrano chiles, seeded and finely minced
¾ cup	178 mL	Lime juice, fresh squeezed
2 each	2 each	Mangoes, ripe, small dice
3 cups	711 mL	Fresh pineapple, small dice
3 cups	711 mL	Jicama, minced
8 each	8 each	Avocados, ripe, peeled and diced
4 teaspoons	20 mL	Garlic, fresh minced
1¼ teaspoons	6 mL	Salt
1 teaspoon	5 mL	Cumin, ground
⅓ cup	79 mL	Cilantro, fresh minced
⅓ cup	79 mL	Mint, fresh minced

PREPARATION STEPS:

1. Pour boiling water over the onion. Drain onion well and then mince.

2. Combine onion, chiles, and lime juice in a large bowl.

3. Add the mango, pineapple, jicama, and avocados. Toss gently to evenly blend.

4. Add the garlic, salt, cumin, cilantro, and mint. Toss gently to evenly distribute all ingredients.

5. Cover well and chill for at least 2 hours before serving.

Professional Profile

BIOGRAPHICAL INFORMATION

Name: Muaynee (Marnee) Siriyarn   **Recipe Provided:** Green Papaya Salad
Place of Birth: Bangkok, Thailand

CULINARY EDUCATION AND TRAINING HIGHLIGHTS

As the third child in a family of eight children, Muaynee had to help her parents with the family restaurant at an early age. By the age of seven, she had begun doing odd jobs in the kitchen and dining room, and by the time she was 12 years old, she was cooking. Muaynee's father died when she was 13, making it necessary for her to have her own food court in the local market. As a result of working to help her family, Muaynee never finished her high school education. However, with the urging of her family, she completed a one-year course in facial and hair beauty care. Upon graduation, and after four months on the job, Muaynee realized that she could not earn a proper living in that field. So she returned to work at her mother's restaurant in Bangkok until she was 23. In 1977, Muaynee moved to Seattle to help open a restaurant, although it never opened. She soon moved to Florida to work in a Chinese restaurant, and then later worked at the King Arthur Inn for two years. In 1979, Muaynee moved to San Francisco to help her relatives with their hotel and restaurant, the Hotel Union. She worked in different properties until she married Chai in 1982. Chai also grew up in the restaurant business in Bangkok, so in 1986 they opened their own Siamese restaurant, Marnee's Thai, in San Francisco. Working 18-hour days, from kitchen to dining room, Chai and Muaynee poured their lifeblood into their food and service. Their restaurant's acclaim and success led them to open a second Marnee's Thai, in 2003. Today, their restaurant walls are lined with personal honors and glowing restaurant reviews, while customers line the streets waiting for tables. Marnee's Thai has been one of the top Thai restaurants in the San Francisco *Zagat Survey* for the last 13 years. Chai, who has gone on to earn an MBA degree, was voted "International Chef of the Year" at the Awards of the Americas. In 2003, he won the "Thailand Super Chef Award" from the Prime Minister of Thailand. Muaynee continues to run the restaurant and greet the customers with her infectious smile, while their two children work by their sides, learning the trade as they did.

ADVICE TO A JOURNEYMAN CHEF

We all can be successful, if we are willing to work hard. In this business, you cannot be lazy. You must love what you are doing because you must give your business a lot of attention. Running a restaurant is like raising a child. You nurture it and meet its needs, often 18 hours a day.

Muaynee (Marnee) Siriyarn

RECIPE 5-47

CHEF MUAYNEE SIRIYARN'S GREEN PAPAYA SALAD

Recipe Yield: 8 portions

| TIP | Peel the papaya and cut in half lengthwise. Scoop out the seeds and discard. Use a Japanese mandoline or box grater to shred the papaya into thin, long strands (about 1/16 inch [0.16 cm] thick.) |

MEASUREMENTS		INGREDIENTS
U.S.	**METRIC**	
		Salad Dressing
¼ cup	59 mL	Fish sauce (Thai preferred)
½ cup	118 mL	Palm sugar (or light brown)
½ cup	118 mL	Lime juice, fresh squeezed
		Salad Body
4 each	4 each	Garlic cloves
3 to 6 each	3 to 6 each	Thai bird chiles, chopped (depending on desired heat)
¼ cup	59 mL	Dried shrimp
6 cups	1.4 L	Green papaya, peeled and shredded
1 cup	237 mL	Long beans, raw, cut into 1½ inch pieces (or green beans)
16 each	16 each	Cherry tomatoes, halved
		Base
½ cup	118 mL	Green cabbage, cut into thin wedges
		Garnish
1½ cups	356 mL	Roasted peanuts

PREPARATION STEPS:

1. Combine all of the dressing ingredients. Blend well and reserve.

2. Place the garlic and chiles in a large mortar and pound until they are broken down.

3. Add the dried shrimp and pound slightly. Add the shredded green papaya and pound slightly to release some juices. Add the green beans and pound again, then add the tomato halves and pound.

4. Toss in the salad dressing. Keep pounding the mixture while continuously turning with a spoon to bruise the papaya.

5. Stir in the peanuts and mix well. Adjust seasonings.

6. Layer a platter or wide-mouth bowl with the cabbage wedges, then top with the salad (Figure 5–24).

FIGURE 5–24 Green Papaya Salad

1. What was the original function of salads on a menu?

2. What is a simple salad?

3. Explain a combination salad.

4. When composing salads, what are the four basic elements used?

5. List four quality points that should be observed when preparing a salad.

6. Discuss the handling of salad greens.

7. List ten greens available for salad preparation.

8. What is the reason for presoaking dried beans?

9. What is the procedure for holding fresh berries for salads?

10. How do you segment an orange?

A. **Group Discussion**

Discuss the role played by grains in salad preparation; report on new ways of presenting them.

B. **Research Project**

As a group, research the use of **micro greens**, reporting on the quality and how functional they are as salad greens.

C. **Group Activity**

Compose three new combination salads using drawings to illustrate their construction.

D. **Individual Activity**

Create two recipes for salad dressing using fresh fruits oils and vinegars suitable for dressing bitter greens.

CHAPTER 6

Sandwiches

After reading this chapter, you should be able to

- define a sandwich.
- outline the components of a sandwich.
- describe how to build a sandwich.
- list the different types of sandwiches.
- demonstrate how to make various cold sandwiches.
- demonstrate how to make various hot sandwiches.
- describe an Indian taco.
- describe a burger.
- describe a Monte Cristo sandwich.
- describe a club sandwich.
- describe a knuckle duster.
- describe a barbecue sandwich.
- describe a Philly steak sandwich.
- describe a French dip sandwich.
- describe a Rueben sandwich.
- describe a Cajun po'boy sandwich.
- describe a soft shell crab sandwich.
- describe a Cuban sandwich.
- describe a Swedish smörgåsbord.
- describe British tea sandwiches.
- describe a Greek gyros.
- describe Döner kebabs.
- describe Mexican tacos.
- describe an Italian panini.
- describe a croque monsieur.

Never eat more than you can lift.

MISS PIGGY

Sandwich: What's in a Name?

The majority of early European farmers took meats and bread out to the fields with them to eat as a meal during their rigorous day's work. It would be sensible to assume, then, that the **sandwich** was invented without being recorded as a gastronomic event. However, according to culinary history, the fourth Earl of Sandwich, John Montague, was responsible for making popular what we now know as the *sandwich*. The Earl had a celebrated naval career during which he is said to have discovered the Sandwich Islands, as well as many other small islands in the South Pacific.

When he retired, he spent much of his time enjoying his passion for playing cards. He played for very long periods of time and detested leaving the table for any reason. As legend has it, to counter the problem of having to interrupt the game for meals, the Earl taught his servants how to encase salt beef and other ingredients between two slices of toasted bread. They were instructed to bring the new creations to the players so the card games could continue without interruption. Whether the story is true or not, the innovation bears his name; and to this day, sandwiches remain one of the most popular forms of food consumption.

Composition of a Sandwich

Sandwiches generally contain three specific component parts that, when put together, constitute the completed dish. The measure of each component is vital to the overall success of the sandwich. Each element must make a strong impression, and each ingredient must be equally worthy in quality and flavor to the others.

These components can vary considerably depending on the style and origin of the sandwich. Each part can contain one or more ingredients, which can be either hot or cold. Some of the modern ingredients may seem as though they do not belong on a sandwich; however, the accepted ingredients have changed dramatically since the time of the fourth Earl of Sandwich. Sandwiches may have many toppings, garnitures, sides, dips, and accompaniments served with them or aside them. These **component parts of a sandwich** include (1) the structure or base, (2) the moistening agent, and (3) the filling (Figure 6–1).

■ The structure or base is the part on which the ingredients are placed or in which they are wrapped. The base normally consists of some form of bread or dough product that is whole or sliced and that acts as the carrier of the other ingredients. It should be sturdy enough to hold the ingredients without becoming limp or broken.

■ The moistening agent lubricates and binds the sandwich, and the added moisture improves the flavor and texture. In some cases, the moistening agent acts as the protective layer between the filling and the structure, preventing the filling from softening or wetting the

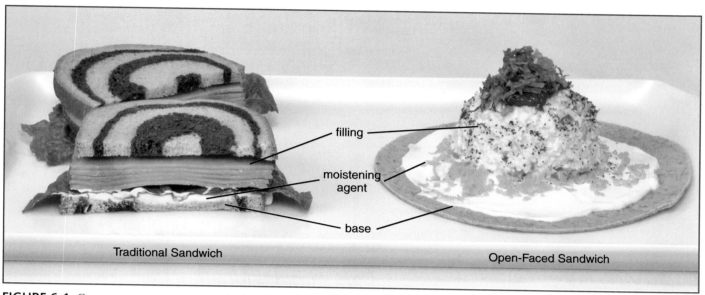

FIGURE 6-1 Component parts to a sandwich (base/structure, moistening agent, and filling)

bread. Agents can be in the form of mayonnaise, softened butter, sour cream, mustard, dips or spreads, or an appropriate cold or hot sauce. It can also be a relish, salsa, chutney, or some form of pickled vegetable or fruit.

- The filling is "the main event" and can consist of one or more ingredients that are stacked, layered, or folded within or on the structure to form the sandwich. The varieties of fillings are endless and should be carefully selected in keeping with the rest of the components. The filling can be any form of cooked or cured meat, fish, shellfish, poultry, game, vegetable, fruit, cheese, salad, or a combination of any of these. The fillings can be hot or cold, and sometimes both.

Note: The quantity and variety of fillings in American sandwiches are often larger in comparison to those in other parts of the world and may appear extreme to the uninitiated.

Building a Sandwich

When building a sandwich, it is important to have a defined area of the kitchen designated specifically to that task. It should either be incorporated into an existing station or have its own station, depending on the volume of business. A sandwich station in a kitchen should have its own refrigerated area, containing all of the prepared items in the correct order of assembly. Often, it is located within the salad station and is operated in conjunction with that station because as that station shares many of the same ingredients involved in sandwich preparation (Figure 6–2).

When building sandwiches, there are some simple rules for assembly, which if taken into consideration, will improve the process of production considerably. This attention to detail will ultimately assist the staff in their consistency, portion control, stock rotation, productivity, and sanitation. Some suggestions for sandwich stations include the following:

- Have a designated area with lots of table space available.
- Design a flow chart for the assembly of the sandwich.
- Make sure that the ingredients have all been cut to the correct size for the assembly.
- Make sure all the fresh ingredients are of the highest quality available.
- Have all the equipment that is required for producing the sandwich at hand and ready for use.
- Do not omit the seasoning during the production.
- The knife used to cut the sandwich should be the correct length and sharpened regularly.

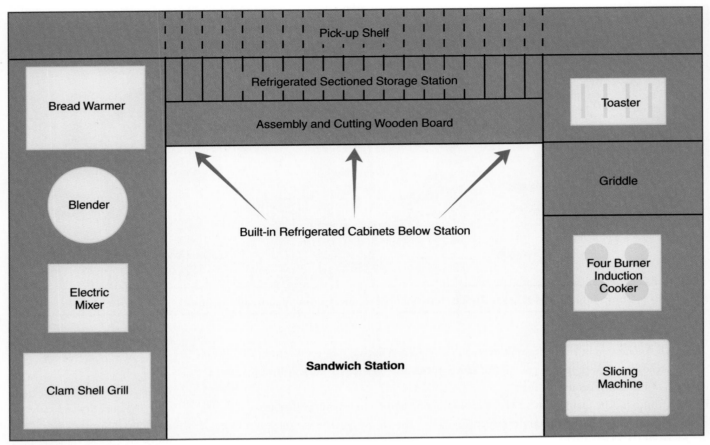

FIGURE 6-2 Mise en place of a sandwich station

- Rotate stock efficiently.
- Ready-to-eat food should always be produced with gloved hands.

Types of Sandwiches

Sandwiches are largely divided into two main categories: cold and hot. Although the garde manger is solely responsible for the production of all cold sandwiches, he or she also prepares the cold assembly for sandwiches that are to be later served hot. This text includes a variety of both cold and hot preparations, in that the garde manger department is involved, to some degree, in all sandwich preparation. Each of these two categories, cold and hot, is further divided into many types of sandwiches.

COLD SANDWICHES

Producing **cold sandwiches** can be an efficient way to feed many people in a short time, making it a popular choice on lunch and dinner menus for busy chefs. The choice of international breads and rolls, as well as the variety of fillings available, makes for a multitude of choices for the customer. The cold sandwich is the ideal food for carriage, and is commonly eaten in areas not otherwise conducive for food consumption.

Filled Roll or Bun Sandwiches

The **filled roll** or **bun sandwich** is one of the most popular of all sandwich varieties and is a globally common method of preparing a quick meal. Rolls are available in a myriad of shapes and sizes, and the sandwiches are served hot or cold for lunch, as a packed lunch, as a lunch

FIGURE 6-3 Lobster
Avocado Lemon Roll

buffet, for a light snack, or as a light dinner. Rolls can be toasted, grilled, or griddled. Bun-filled sandwiches are commonly carried for meals that are not eaten while sitting at a table, such as at picnics, while traveling, or during sporting events. This makes them a versatile and beloved form of food that will only grow in popularity as the pace of life continues to accelerate and the need for "on-the-go" food continues.

Making the filled roll or bun sandwich (Figure 6–3):

1. The bread rolls can include hoagie rolls, burger buns, soft rolls, potato rolls, croissants, baps, cotters rolls, sub buns, or brioche buns.
2. The fillings can be a combination of any protein, cheese, salad, or vegetable. The fillings can be layered or chopped, and bound in a dressing or sauce.

RECIPE 6–1

LOBSTER
AVOCADO
LEMON ROLL

Recipe Yield: One sandwich

MEASUREMENTS		INGREDIENTS
U.S.	**METRIC**	
1 each	1 each	6-inch piece of French stick split open to form a V-hollow
1 tablespoon	15 mL	Butter, softened
1 teaspoon	15 mL	Lemon juice
2 each	2 each	Leaves of romaine lettuce heart, cores removed
2 tablespoon	30 mL	Mayonnaise
4 ounces	113 g	Cooked lobster meat, chunked
4 each	4 each	Slices of ripe avocado
1 tablespoon	15 mL	Chopped chives
As needed	As needed	Salt and black pepper
¼ cup	39.5 mL	Celery, small diced (optional)

PREPARATION STEPS

1. Whip the lemon juice into the butter and spread onto the bread.

2. Fill the V-shaped hollow of bread with the lettuce.

3. Combine the lobster, mayonnaise, chive, and avocado and season with salt and pepper.

4. Fill into the lettuce and close the sandwich.

Cold Wrapped Sandwiches

Cold wrapped sandwiches have become very popular because they are easy to eat and simple to assemble. They use a variety of ingredients that are moistened, then wrapped and rolled to encase the fillings. Wrapped sandwiches are then held together with wooden picks and cut decoratively for service. When the sandwiches are sliced, the colors and layers of the filling are exposed, providing the sandwich with wonderful eye appeal and a multitude of presentation possibilities. They can also be cut into small thin slices that are held by a wooden pick and stood on end for presentation. This style is often referred to as *pinwheel sandwiches*.

Wraps are often made with large leafy greens, using one or several overlapping leaves as the wrapping material. These can be a little unstable, but when secured properly with wooden picks, they are a very effective solution for creating lower carbohydrate sandwiches.

Making the cold wrapped sandwich (Figure 6–4):

1. Wrapped breads are flat and pliable, and can include lawash (also known as lavash), pita, tortilla, chapatti, and naan breads. The leaves of romaine, radicchio, and oak leaf lettuces work well, as do the soft and tender leaves of Napa, Savoy, and red cabbages.
2. The fillings for wraps should be reasonably sturdy and contain dressing or sauce that is slightly thicker in consistency than for normal sandwiches. The viscosity becomes important during the rolling and cutting process, because it is likely that liquid will be squeezed from the sandwich, negatively affecting the presentation.
3. Thinly sliced meats, fish, and shellfish combined with salad items, vegetables, and cheeses are ideal for the fillings.

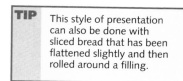

TIP This style of presentation can also be done with sliced bread that has been flattened slightly and then rolled around a filling.

FIGURE 6-4 Grilled Caribbean Shrimp Wrap with Lime Sour Cream and Red Cabbage

RECIPE 6–2

GRILLED CARIBBEAN SHRIMP WRAP WITH LIME SOUR CREAM AND RED CABBAGE

Recipe Yield: One sandwich

MEASUREMENTS		INGREDIENTS
U.S.	METRIC	
6 each	6 each	21 to 25 shrimp, medium
1 tablespoon	15 mL	Garlic, finely minced
1 tablespoon	15 mL	Olive oil
1 teaspoon	5 mL	Minced thyme
1 teaspoon	5 mL	Scotch bonnet pepper, minced
1 teaspoon	5 mL	Salt
¼ cup	59 mL	Sour cream
1 tablespoon	15 mL	Lime juice
1 each	1 each	14-inch tortilla shell
½ cup	118 mL	Red cabbage, finely shredded
1 tablespoon	15 mL	Olive oil
1 tablespoon	15 mL	Champagne vinegar
1 each	1 each	Romaine lettuce leaf
1 each	1 each	Sliced tomato

PREPARATION STEPS:

1. Marinate the shrimp with the garlic, olive oil, thyme, Scotch bonnet peppers, and salt for 1 hour.

2. Grill the shrimp until just done.

3. Slice the shrimp into strips.

4. Beat the lime juice into the sour cream.

5. Spread the tortilla with the sour cream.

6. Toss the cabbage in the oil and vinegar.

7. Add the shrimp, cabbage, lettuce, and tomato to the tortilla and roll tightly, securing with skewers.

8. Trim the ends and cut the roll at an angle one third of the length then cut straight at the next third.

9. Present standing up.

Open-Faced Sandwiches

The **open-faced sandwich** traditionally begins with a base made from traditionally shaped, sliced bread of any flavor. The bread can be strengthened by grilling or roasting, firming it as a base for the sandwich. The base is then spread with a moistening agent, and topped and garnished with an arrangement of complimentary ingredients that form the body of the sandwich. The toppings for open-faced sandwiches are either cold or hot, or both. These sandwiches are regularly served as entrées both at lunch and dinner.

Making the open-faced sandwich (Figure 6–5):

1. The breads that are used for open-faced sandwiches are sliced from large loaves and are often cut at an angle for shape and size. They come from loaves that are sturdy in structure and that can withstand grilling or roasting. These include sourdough loaves, Pullman loaves, baguettes, large Italian loaves, large whole meal loaves, ciabatta loaves, and granary loaves.

2. The toppings can range from the complex, like pesto with grilled sardines and chèvre, to the simple, like egg mayonnaise. Generally the toppings, sauces, spreads, and dips that are used bring color and diverse flavor to this type of sandwich, giving it a very clean and easy-to-eat appeal.

FIGURE 6-5 Smoked Duck with Caramelized Oranges Open-Faced Sandwich

RECIPE 6–3

SMOKED DUCK WITH CARAMELIZED ORANGES OPEN-FACED SANDWICH

Recipe Yield: One sandwich

MEASUREMENTS		INGREDIENTS
U.S.	**METRIC**	
3 each	3 each	Orange slices, peeled
2 Tablespoons	30 mL	Sugar
1 each	1 each	Large oval slice of Italian bread, lightly toasted
1 Tablespoon	15 mL	Cream cheese, softened
$^1/_2$ cup	79 mL	Curly endive lettuce
1 teaspoon	5 mL	Chives, fine dice
4 ounces	113 g	Smoked duck breast, sliced

PREPARATION STEPS:

1. Sprinkle sugar onto the orange slices and torch until caramelized.

2. Spread the cream cheese and chives onto the bread and coat with the lettuce leaves.

3. Place the carved duck breast in a perfect shingle down the middle of the sandwich.

4. Top with the caramelized oranges.

Closed or Filled Sandwiches

Closed sandwiches are the type that the fourth Earl would have probably received from his servants. Typically they are made with sliced breads that have been buttered or spread with some moistening agent, and then filled and topped with more bread. They can be made with two, three or even four slices of bread, with filling in between each layer. It is common to cut the sandwich in half or into small triangular quarters for presentation. In finer establishments, the crusts are removed from the bread to make them daintier to eat and to enhance their presentation.

Making the closed or filled sandwich (Figure 6–6):

1. The breads that are used in this sandwich are normally the softer breads that are sliced thin. They include Vienna loaves, potato bread, bloomer, large Italian loaves, and a variety of brown, wheat, and white loaf breads. Sourdough, baguettes, and ciabatta loaves can also be used, but they must be very fresh.

2. The fillings consist of meats of all kinds, including the cured and smoked specialties of hams and salami. Other common fillings include roasted meats, poultry and game, as well as cheeses of all kinds. Meat and fish spreads are ideal for these sandwiches, as are the inclusion of salad material and raw salad vegetables.

FIGURE 6-6 Egg Salad Sandwich

RECIPE 6–4

EGG SALAD SANDWICH FILLING

Recipe Yield: Fills one sandwich

| MEASUREMENTS | | INGREDIENTS |
U.S.	METRIC	
3 each	3 each	Hard-boiled eggs, grated finely
1 teaspoon	5 mL	Parsley, chopped
1/4 cup	59 mL	Mayonnaise
1 teaspoon	5 mL	Capers, finely diced
Pinch	Pinch	Salt
Pinch	Pinch	Cayenne pepper

PREPARATION STEPS:

1. Combine all the ingredients together and use to fill sandwiches.

2. This can also be stuffed into the egg white, using only the grated egg yolk for the filling.

RECIPE 6–5

FRESH LOX SANDWICH FILLING

Recipe Yield: Fills one sandwich

| MEASUREMENTS | | INGREDIENTS |
U.S.	METRIC	
2 ounces	57 g	Cream cheese, softened
4 ounces	113 g	Smoked salmon (lox), diced
1 tablespoon	15 mL	Red onion, fine dice
1 tablespoon	15 mL	Capers, fine dice
1 tablespoon	15 mL	Chive, fine dice
Pinch	Pinch	Salt
Pinch	Pinch	Black pepper

TIP The salmon will crumble during the mixing process. The filling will stiffen when refrigerated, so it is best to spread it on the bread before chilling.

PREPARATION STEPS:

1. Soften the cream cheese by rapidly mixing in a bowl.

2. Carefully blend all of the ingredients together, and use immediately.

RECIPE 6–6

CHICKEN SALAD SANDWICH FILLING

Recipe Yield: Fills one sandwich

MEASUREMENTS		INGREDIENTS
U.S.	**METRIC**	
½ cup	118.5 mL	Mayonnaise
¼ cup	59 mL	Green onion, finely chopped
1 tablespoon	15 mL	Lemon juice
1 tablespoon	15 mL	Flat leaf parsley, chopped
6 ounces	170 g	Smoked chicken cut into julienne pieces
1 tablespoon	15 mL	Gherkins, very fine dice

PREPARATION STEPS:

1. Combine the onions, mayonnaise, lemon juice, and parsley. Beat together.

2. Fold in the chicken and the gherkins.

3. Use for sandwich filling.

The Sandwich Buffet

The sandwich buffet is an efficient method of feeding many people at the same time, while allowing them the opportunity to choose their own ingredients. It requires the setting up of lines of sandwich materials, all in the correct order of the sandwich's construction. The food is set out on buffet tables, allowing the guests to walk down each side, building their own sandwich as they go. Salad materials, accompanying sauces, relishes and other condiments are served individually packaged at the end of the buffet. This system of service speeds up the process and allows the customer to assemble some parts of the sandwich at their table, rather than slowing down the buffet line. The sandwich buffet is commonly used for cold sandwiches, but it can also be used for hot sandwiches with simple fillings (Figure 6–7).

FIGURE 6-7 Assorted sandwiches served buffet style. Courtesy of Irina Terentjeva/Shutterstock

HOT SANDWICHES

The **hot sandwich** has become very popular as an easy meal at lunch or dinner, giving the guest the option of eating quick and light. There are many dishes throughout the world that can be classified as hot sandwiches, although the name does not always imply their status as a sandwich. Being that the item is covered, wrapped, or surrounded by bread or another dough product, this suggests that it truly is a sandwich of some sort. However, what really classifies an item as a sandwich is the fact that it consists of the three sandwich components and it can be picked up and eaten by hand as a meal.

RECIPE 6–7

WARMED CLUB SANDWICH

Recipe Yield: One sandwich

MEASUREMENTS　INGREDIENTS

U.S.	METRIC	INGREDIENTS
3 each	3 each	White bread slices, freshly toasted
½ cup	118 mL	Mayonnaise
4 each	4 each	Bib lettuce leaves
6 each	6 each	Tomato slices
6 each	6 each	Bacon slices, crisp and hot
4 ounces	113 g	Smoked turkey breast, hot

PREPARATION STEPS:

1. Spread one side of each slice of bread with mayonnaise.

2. Arrange half the lettuce, tomato, turkey breast, and bacon on one slice of toast.

3. Top with another slice and repeat.

4. Close the sandwich with the last slice and cut into triangles, holding them together with skewers.

RECIPE 6–8.1

TORTA DE PAVO

Recipe Yield: 8 portions

MEASUREMENTS		INGREDIENTS
U.S.	**METRIC**	
1 each	1 each	Long Cuban bread or sourdough baguette, 6 inches in length
1 tablespoon	15 mL	Butter, softened
2 tablespoons	30 mL	Chipotle Purée recipe
2 tablespoons	30 mL	Refried Black Beans recipe
2 each	2 each	Tomato slices, thinly sliced
4 ounces	113 g	Freshly roasted turkey, still hot
To taste	To taste	Salt
To taste	To taste	Pepper
3 tablespoons	44 mL	Sour cream

PREPARATION STEPS:

1. Split the roll and coat the inside with the butter, chipotle purée, and refried beans.

2. Roast in preheated oven for 2 minutes at 350°F (177°C).

3. Layer the warm roll with the turkey and tomato. Season to taste.

4. Top with the sour cream.

RECIPE 6–8.2

CHIPOTLE PURÉE

Recipe Yield: 8 portions

MEASUREMENTS		INGREDIENTS
U.S.	**METRIC**	
1 cup	1 cup	Chipotle peppers, roasted
2 tablespoons	30 mL	Vegetable oil
1 teaspoon	5 mL	Lime juice

PREPARATION STEPS:

1. In a food processor or blender, purée the ingredients together to a smooth paste.

RECIPE 6–8.3

REFRIED BLACK BEANS

Recipe Yield: For one sandwich

MEASUREMENTS		INGREDIENTS
U.S.	**METRIC**	
1 tablespoon	15 mL	Vegetable oil
¼ cup	59 mL	Onions, chopped
1 cup	118 mL	Black beans in their cooking liquid
As needed	As needed	Salt and pepper

PREPARATION STEPS:

1. Heat the oil and sauté the onions until golden brown.

2. Add the beans and mash them in the pan, cooking until almost dry.

3. Season with salt and pepper.

Hot Wrapped Sandwiches

There are many international **hot wrapped sandwiches**, and like their cold counterparts, they are simply flat bread containing a hot filling. The fillings can also feature cold accompaniments, including salad materials, relishes, salsas, sauces, and vegetable and cheese toppings. For example, the Mexican taco is a warm tortilla wrapped around skillet roasted meats, fish, or chicken and vegetables, and then topped with or accompanied by salsa and sour cream. This dish is not normally called a "sandwich," although it is picked up and eaten by hand and is enclosed in ethnic bread (Figure 6–8).

FIGURE 6-8 Example of a filled taco

RECIPE 6–9

TANDORI CHICKEN NAAN WRAP

Recipe Yield: One sandwich

MEASUREMENTS		INGREDIENTS
U.S.	**METRIC**	
1 each	1 each	Warm naan bread
6 ounces	170 g	Grilled Tandori chicken, very hot and broken into pieces
½ cup	118 mL	Yogurt, plain
1 tablespoon	115 mL	Mint, chopped
¼ cup	59 mL	Fresh tomato, chopped
1 each	1 each	Large lettuce leaf

PREPARATION STEPS:

1. Lay out the naan bread and layer with the lettuce and the chicken.

2. Combine the yogurt, mint, and tomato.

3. Dress the chicken with the yogurt mixture and wrap in the naan.

4. Serve immediately while still hot.

Grilled

The **grilled sandwich** is made by encasing a filling between two slices of bread or by placing the filling on the top of bread slices (Figure 6–9). The sandwich is then placed onto an open grill or chargrill for cooking, giving it the color and crispness desired. Although the process of using a chargrill is somewhat more difficult than using a flat grill, the phenomenal flavor of a char-grill provides the most authentic taste for grilled sandwiches.

To accelerate this process and to control the cooking, the sandwich can be cooked on a clamshell grill. This piece of equipment traps the sandwich between two hot corrugated plates. The result is a nicely grilled sandwich that has the grill marks on the bread left by the hot ridges of the corrugated plates. This kind of sandwich appears in many cultures, under a host of names, but they are all essentially grilled sandwiches. They are sometimes called *toasted sandwiches* when they are cooked on the clam shell grill.

FIGURE 6-9 Grilled Eggplant Sandwich with Tahini Sauce

RECIPE 6-10

GRILLED BEEF GORGONZOLA SANDWICH

Recipe Yield: One sandwich

MEASUREMENTS		INGREDIENTS
U.S.	**METRIC**	
1 ounce	28 g	Gorgonzola cheese
¼ cup	59 mL	Sour cream
2 each	2 each	6-inch sourdough bread, sliced and lightly toasted
4 ounces	113 g	Rare roast beef, shaved
1 ounce	28 g	Red onion, shaved
1 each	1 each	Tomato slice
1 each	1 each	Romaine leaf

PREPARATION STEPS:

1. Mix the cheese and the sour cream together and spread onto the bread.

2. Layer the beef, onions, tomato, and lettuce between the bread slices.

3. Place the two pieces of bread together.

4. Brush with butter and grill.

5. Cut into quarters.

RECIPE 6-11

GRILLED EGGPLANT SANDWICH WITH TAHINI SAUCE

Recipe Yield: One Sandwich

MEASUREMENTS		INGREDIENTS
U.S.	**METRIC**	
2 each	2 each	Garlic cloves
¼ cup	59 mL	Tahini (ground sesame seeds)
¼ cup	59 mL	Fresh lemon juice
2 tablespoons	30 mL	Water
1 each	1 each	6-inch herb focaccia split and toasted until golden brown
3 each	3 each	Slices of eggplant, brushed with olive oil and grilled
3 each	3 each	Slices of beefsteak tomato, brushed with olive oil and grilled
⅛ cup	30 mL	Sweet onion, shaved

TIP Slices of eggplant may be marinated in basil vinaigrette and then drained and grilled.

PREPARATION STEPS:

1. In a blender, purée the garlic, tahini, lemon juice, and water to make a sauce.

2. Spread the sauce on the focaccia and layer one side with the eggplant, tomato, and onions.

3. Top with the remaining side and cut securing with wooden skewers

Fried

Fried sandwiches are made with any filling placed between two slices of bread and cooked by shallow frying or sautéing. They can be dipped in a coating for added texture and color. The coatings generally consist of a combination of dry ingredients that adhere to the sandwich using an egg-milk batter. Bread crumbs, seasoned flour, and ground nuts, as well as herb and spice coatings are popular (Figure 6–10).

FIGURE 6-10 Grilled Portobello Roasted Pepper Sandwich

RECIPE 6-12

GRILLED PORTABELLO ROASTED PEPPER SANDWICH

Recipe Yield: One sandwich

MEASUREMENTS		INGREDIENTS
U.S.	**METRIC**	
1 each	1 each	Portabello mushroom, gills removed and peeled
¼ cup	59 mL	Olive oil
Pinch	Pinch	Salt
¼ cup	59 g	Mayonnaise
1 tablespoon	15 mL	Roasted garlic, puréed
Pinch	Pinch	Salt
Pinch	Pinch	Black pepper
1 each	1 each	Small round focaccia (approximately 4 inches), halved and toasted
4 each	4 each	Sun dried tomatoes
1 each	1 each	Roasted red pepper
4 each	4 each	Basil leaves

PREPARATION STEPS:

1. Brush the mushroom with olive oil and sprinkle with salt.

2. Grill over a hot grill until cooked through, cool thoroughly.

3. Mix the mayonnaise and the roasted garlic. Season well with salt and black pepper.

4. Spread the garlic mayonnaise on both slices of bread.

5. Spread the sliced mushroom, sun dried tomato, roasted red pepper, and basil leaves on the focaccia slice and top with the other half.

6. Shallow fry the sandwich in hot olive oil until golden brown on both sides.

7. Cut in half and secure with wooden skewers.

Griddled

A **griddled sandwich** is very much the same as a fried sandwich, but it is cooked on a flat heated griddle or cast iron comal (Figure 6–11). These sandwiches can be coated with raw egg, flour, and bread crumbs. They are also commonly dusted with spices or sugar when cooked.

FIGURE 6-11 Grilled Cheese Sandwich

RECIPE 6–13

MONTE CRISTO SANDWICH

Recipe Yield: One sandwich

MEASUREMENTS		INGREDIENTS
U.S.	**METRIC**	
2 slices	2 slices	White bread
1 tablespoon	15 mL	Butter, softened
2 ounces	57 mL	Sliced turkey breast, cooked
2 ounces	57 mL	Shaved ham, cooked
2 each	2 each	Gruyère or Emmentaler cheese slices
1 each	1 each	Egg
¼ cup	59 mL	Milk

PREPARATION STEPS:

1. Spread the butter on both pieces of the bread.

2. Layer the meats and cheeses between the bread slices, keeping the cheese closest to the butter.

3. Beat the milk and egg together.

4. Dip the sandwich into the egg mixture and allow to soak.

5. Deep fry or griddle the sandwich until golden brown.

6. Cut into halves or quarters and serve with fruit jam or cranberry jelly for dipping.

INTERNATIONAL SANDWICHES

It is important to identify from where some of the most popular international sandwiches originated in order to appreciate their history. Sandwiches are commonplace in any international kitchen, but they may have their roots in another continent. A few sandwiches discussed in this chapter have an amazing story behind them, while others are shrouded in mystery. It is difficult to give credit to a specific person for a creation when there are conflicting tales about the origin of the food. The following accounts are presented as accurately as possible.

American Sandwiches

The United States is famous for its "larger-than-life" characters, and the same applies to the size and of its sandwiches. The following are some of the more traditional sandwiches on American menus, representing a cross section of the whole country.

INDIAN TACO

The Indian taco was originally known as *Navajo Taco,* but it has become synonymous with all Native American tribes. This sandwich requires no plates or silverware, and is just rolled up in the bread and eaten as is. The Indian taco is a combination of beans, cooked ground beef, chopped lettuce, sliced tomato, and shredded cheese, all topped with green chile. It sits on a plate-sized circle of crispy Navajo or Indian fry bread (Figure 6–12).

THE BURGER

The burger is one of the most popular sandwiches in the world, but its origin is shrouded behind the curtains of time. There are numerous stories and claims to the source, but no one can be absolutely sure which one is the original. One story relates the adventures of Charlie Nagreen of Seymour, Wisconsin, who in 1885 sold meatballs from his ox-drawn food stand at the Outagamie County Fair. As a very astute 15 year old, he soon realized that the meatballs were not selling because his customers were having difficulty eating them while strolling around the fairground. To counter this problem, he flattened the meatballs and placed them between two

FIGURE 6-12 Example of an Indian taco

slices of bread, selling his new creation as "the hamburger." He was dubbed *Hamburger Charlie,* and he returned to the fairgrounds every year until his death in 1951.

Another tale has the burger's roots in a waterfront restaurant in Hamburg, Germany, where an aspiring cook named Otto Kuasw made a special sandwich for visiting American sailors. The creation was comprised of a thin sausage patty fried in butter, topped with egg, and placed between two slices of bread. As the legend goes, the sailors shared the recipe with New York port restaurateurs in 1894, who in turn started serving the burgers to their clients. It is possible, of course, that both stories are true and that Hamburger Charlie got his name just a little later than recorded.

The burger generally consists of one, two, or multiple thin meat patties that are griddled or chargrilled. The burger is commonly served on a sesame bun with pickle, mayonnaise, mustard, ketchup, lettuce, and tomato. Variations include the burger being topped with melted cheese to form the popular cheeseburger, or other items such as cooked bacon slices, onion, green olives, and barbecue sauce (Figure 6–13).

MONTE CRISTO

The Monte Cristo sandwich varies considerably from restaurant to restaurant, and there is no documentation as to its origin. Its derivation is thought to be a variation of the classic French sandwich croque monsieur and was first served in southern California in the early 1950s. There is no known explanation for its name.

The sandwich itself consists of a combination of ham, turkey or chicken, and Swiss cheese between two slices of white square bread. It is then dipped in beaten egg and deep-fried, pan fried in butter, or griddled until golden brown. The sandwich is normally accompanied with a side of maple syrup, currant jelly, strawberry jam, or cranberry sauce for dipping (Figure 6–14).

THE CLUB SANDWICH

The club sandwich is one of the most common international sandwiches and seems to be available in some form or other anywhere in the world. The origin of its name is a matter of speculation and guesswork, but its strong association with hotels and country clubs from around the world is likely the genesis. The most popular theory contends that this sandwich originated in men's social clubs, most notably the Saratoga Club in Saratoga, NY.

FIGURE 6-13 Example of a double cheeseburger with lettuce and tomato set up

FIGURE 6-14 Monte Cristo Sandwich

The club sandwich definitely existed in the United States by the late nineteenth century and was a favorite of former King Edward VII of England and his wife, Wallis Simpson. The first appearance of the club sandwich in print was in Ray L. McCardell's *Conversations of a Chorus Girl* in 1903, and recipes were printed in Fannie Farmer's *Boston Cooking-School Cookbook* in 1906, indicating the item had been popular for some time.

Some food historians believe that the sandwich was originally only a double-decker and that it originated aboard the double-decker "club cars" that existed then on the trains in America that traveled from New York to Chicago between the 1930s and 1940s. The sandwich itself is made with cooked chicken breast, mayonnaise, bacon, lettuce, and tomato. It is served between toasted bread slices and can be single, double, or triple layered. It is normally cut into triangles and skewered with wooden picks (Figure 6–15).

FIGURE 6-15 Example of a club sandwich

FIGURE 6-16 Example of a lobster knuckle sandwich

KNUCKLE DUSTER

The knuckle duster sandwich is a relatively new creation and is generally found in coastal areas of the United States where lobsters are harvested or farmed. It is a creative means for using the knuckle meat found between the claws and body of the lobster. These parts are odd shaped, and have their own distinct flavor and texture, which many believe is even better than that found in the lobster tail. The sandwich consists of seasoned lobster knuckle meat that is sandwiched between a crusty roll or bun with lemon mayonnaise spread and green salad ingredients (Figure 6–16).

BARBECUE SANDWICH

The barbecue sandwich can be made with pulled, sliced, or chopped barbecued meats or poultry. It consists of selected meats that have been added to sautéed onions and garlic, then moistened with a piquant barbecue sauce and placed on a sturdy bun or roll (Figure 6–17).

FIGURE 6-17 Example of a barbecue sandwich

Only sufficient sauce to bind the meat is required to prevent the filling from falling out. One very old version of this sandwich is the "sloppy Joe," which is made with ground meat cooked down and moistened with barbecue sauce or flavored tomato-based sauce. The term *sloppy* is given to the sandwich to describe the nature of the filling, which is very loose and heavily sauced, making it a particularly interesting challenge to cleanly consume.

PHILLY STEAK

The Philly steak sandwich was created during the 1930s in the Italian immigrant section of South Philadelphia. Pat Olivieri, who was the owner of a hot dog and sandwich stand, created "the Philly." During free time one day, he decided to make lunch for himself but could only find a slab of steak too large for cooking on his hot dog grill. He sliced the steak thinly, placed it on the grill, added sautéed onions for flavor; and stuffed in a roll, making a tasty sandwich.

Just then, he became busy and ended up selling his lunch to a cab driver who had been attracted to the stand by the smell of the cooking meat and onions. The cab driver was delighted with his sandwich and advised him to give up selling hot dogs and start selling the steak sandwich. Olivieri did, and it became known as the Philly steak sandwich. Around 20 years later, Joe Lorenzo (an employee) added cheese to the sandwich, and the modern Philadelphia cheesesteak sandwich was born (Figure 6–18).

FRENCH DIP

The French dip sandwich actually has its place among American sandwiches. Although the sandwich itself is not French, it was created by Philippe Mathieu who was. In 1918, he owned the delicatessen and sandwich shop called "Philippe the Original" in Los Angeles. As legend has it, when preparing a sandwich one day for a policeman, he accidentally dropped the sliced French roll into the drippings of a roasting pan. It was an instant success, and the policeman returned the next day with some friends to order more of the same dipped sandwich. The sandwich is made by placing thinly sliced roast beef on a long crusty French roll, and it is served with a side dipping sauce made from the roasted meat's au jus (Figure 6–19).

THE REUBEN

Arnold Reuben, the founder of Reuben's Restaurant and Delicatessen in New York City, created the Reuben sandwich around 1927. It is said that he created it for a hungry leading lady of actor Charlie Chaplin. She turned up late one night to the restaurant and asked him to make her a combination sandwich, because she was so hungry that she could "eat a brick." After having consumed his creation, she joked that he should name the sandwich after her. However, she was promptly told that he would name the sandwich after himself instead.

FIGURE 6-18 Example of a Philly cheese steak sandwich

FIGURE 6-19 Example of a French dip sandwich

The grilled Reuben sandwich is made with shaved corned beef, Swiss cheese, sauerkraut, and Russian or Thousand Island dressing, and placed between two slices of rye bread (Figure 6–20).

CAJUN PO'BOY

A po'boy sandwich is normally made with long loaves of French bread filled with an assortment of meats with gravy or fried seafood, and dressed with condiments and salads (Figure 6–21). It is thought that the po'boy sandwich was invented in New Orleans in the 1920s. Two poor Cajun boys, Benny and Clovis Martin, created the sandwich to help the 1000 streetcar workers who were on strike in 1914. The workers could not afford anything else to eat but the "poor boys" sandwiches, from which they eventually got their name.

FIGURE 6-20 Example of a reuben sandwich

FIGURE 6-21 Example of a Cajun po'boy sandwich

SOFT-SHELL CRAB

The soft-shell crab sandwich is a relatively new creation, but it has become common all over the United States. The crabs are harvested just after molting, and are very tender and succulent when fried or roasted. They are generally coated in a liquid such as buttermilk, dusted with cracker crumbs or cornmeal, and then fried. They are served with a moistening agent of choice between toasted muffins or hero rolls (Figure 6–22). Soft-shell crab can also be served in open-faced sandwiches in many creative ways.

FIGURE 6-22 Chesapeake Bay Soft-Shell Crab Sandwich

CHESAPEAKE BAY SOFT-SHELL CRAB SANDWICH

Recipe Yield: One sandwich

| MEASUREMENTS | | INGREDIENTS |
U.S.	METRIC	
½ cup	118 mL	Flour
½ teaspoon	2 mL	Salt
½ teaspoon	2 mL	Pepper
1 each	1 each	Soft-shell crab
1 each	1 each	Egg
½ teaspoon	2 mL	Tabasco sauce
1 cup	237 mL	Cracker crumbs
As needed	As needed	Peanut oil for deep frying
½ cup	118 mL	Tartar sauce
1 each	1 each	Kaiser roll
1 each	1 each	Lemon wedges

PREPARATION STEPS:

1. Combine the flour, salt, and pepper and dust well over the crab.

2. Beat together the egg and Tabasco sauce.

3. Dip the crab in the egg mixture and roll in the cracker crumbs.

4. Deep-fry until golden brown and cooked.

5. Drain well on paper towels.

6. Toast the buns and top with the crab.

7. Spoon tartar sauce over the crab, and serve with lemon wedge.

FIGURE 6-23 Example of a Cuban sandwich

CUBAN SANDWICH

The Cuban sandwich, which is also known as the *cubano,* originated in Cuba and is a popular meal in southern Florida where many Cuban immigrants have settled. In Miami, a community known as Little Havana hosts a large Cuban population that fled to the United States after the 1959 Cuban revolution. The community strives to keep alive the rich culinary traditions of Cuba, and it has strongly influenced the gastronomy of South Florida.

The sandwich consists of ham and roast pork, mayonaise, mustard, dill pickles, and swiss cheese on Cuban bread, a light French bread (Figure 6–23). The pork for the sandwich is normally marinated in mojo sauce before being roasted and shaved. The sandwich is then buttered and pressed with a sandwich press called a *plancha,* melting the cheese and toasting the bread. The sandwich can be served hot or cold.

SWEDISH SMÖRGÅSBORD

Smorgasbord simply means a sandwich board or buttered bread table, and comes from the Swedish word *smörgåsbord.* The real Swedish smorgasbord is steeped in tradition and ceremony, and contains dishes that represent the sun, the sea, and the sky (Figure 6–24). It is normally served in several courses and takes the form of a buffet table in a restaurant. It can also be a multicourse meal in the home, and is commonly served during holiday periods. The food is generally served on small Swedish crisp breads, giving them the appearance of being open sandwiches, with lots of colorful treats adorned with garnishes and salads.

BRITISH TEA SANDWICHES

British tea sandwiches are closed sandwiches that are trimmed and neatly cut into nice shapes, with the crusts removed. They are a feature on afternoon tea menus and some high tea menus that are popular in Great Britain. Tea sandwiches are very dainty and little, and great care is taken over their production. They also have their own area on the afternoon tea stand for their service, and they are commonly the first thing eaten by guests.

FIGURE 6-24 Example of a Swedish smörgåsbord. Courtesy of Chad Ehlers/ Alamy

The bread used is basic white and brown bread that comes in a presliced soft loaf, which is common all over the British Isles. The fillings are traditional and include egg and mustard cress, cucumber, smoked salmon and cream cheese, and cheddar and chutney (Figure 6–25).

GREEK GYRO SANDWICH

The gyro is a popular Greek sandwich specialty and is a common street food served as a snack, a lunch, or a dinner entrée. The meat used for the gyro is well-seasoned lamb or a combination of pressed beef and lamb. The meat is layered onto a large cylinder with a central pivot or skewer, which slowly rotates vertically next to a heat source. The heat source is a vertical charcoal firebox or an electric or gas fire that slowly cooks the meat from the outside.

FIGURE 6-25 Selection of tea sandwiches

FIGURE 6-26 Example of a gyro sandwich

The name *gyro* relates to the circular spinning motion of a gyroscope, similar to the meat as it is being cooked. The person making the sandwich slices off long strips of the hot cooked meat to order, and then serves it folded in pita bread, which is heated on a griddle or grill. The meat is topped with a yogurt sauce, lettuce, and tomato, and the sandwich is complete (Figure 6–26).

TURKISH DÖNER KEBAB

Döner kebab is the name given to the same technique as described in the Greek gyro sandwich. It simply means "to turn," which relates to the turning of the meats on the spit, called a shawarma. The meats used change from country to country and can include lamb, beef, turkey, veal, or chicken. They can be in a whole form or ground and spiced. The meat is served in a similar fashion as the gyro, wrapped in a style of flat bead and topped with sauces and salad ingredients.

MEXICAN TACO

The Mexican taco is a combination of succulent charcoal grilled meats, onions, peppers, and smoky chiles, all wrapped in a warm flour or corn tortilla (Figure 6–27). As with many other

FIGURE 6-27 Example of a grilled beef taco

sandwiches, it can be made with a mixture of almost any meat and vegetables. Considered a convenience food to the Latino population of the world, it can be eaten as a snack or as a meal, and is a great food for traveling. Tacos are commonly served with roasted tomato salsa and accompanied with a spicy chile sauce.

ITALIAN PANINI

The word *panini* means "little rolls" or "little bread" and is synonymous with sandwich-making in Italy. The panini is a peasant-style enclosed sandwich, which is normally grilled in a clamshell grill. This sandwich, which has gained widespread acceptance outside of Italy, can be filled with any variety of ingredients and is normally served on large slices of bread (Figure 6–28).

FIGURE 6-28 Chicken Red Pepper Brioche Panini

RECIPE 6–15

CHICKEN RED PEPPER BRIOCHE PANINI

Recipe Yield: Makes one sandwich

MEASUREMENTS		INGREDIENTS
U.S.	**METRIC**	
1 each	1 each	Brioche bun, approximately 4 inches in diameter
¼ cup	59 mL	Mayonnaise
½ cup	118 mL	Roasted chicken, pulled
¼ cup	59 mL	Roasted red peppers
1 each	1 each	Sweet onion slice
2 each	2 each	Sharp cheddar cheese slices
2 tablespoons	340 mL	Olive oil

PREPARATION STEPS:

1. Slice the brioche in half and spread with the mayonnaise.

2. Layer with the chicken, cheese, red peppers, and onion.

3. Close the sandwich and brush with the olive oil.

4. Grill on a panini-styled grill until golden brown.

FRENCH CROQUE MONSIEUR

The croque monsieur could be considered the original grilled cheese sandwich. It consists of Gruyère cheese and lean ham layered between two slices of crustless white bread and fried in clarified butter (Figure 6–29). It is thought that the sandwich originated in 1910 in a Paris café where it quickly became a favorite, its popularity spreading rapidly throughout the city. This sandwich is still served as a popular snack or casual meal throughout France. It is usually made in a special sandwich grilling iron, much like a clamshell grill, which cooks equally from the bottom and the top. In most Paris cafés, the croque monsieur is now prepared as a one-sided sandwich made with a large single slice of bread from a round loaf.

FIGURE 6-29 Example of croque monsieur

Professional Profile

BIOGRAPHICAL INFORMATION

Name: Daniel Wilson

Place of Birth: Auckland, New Zealand

Recipe Provided: Open-Faced Vegetarian Sanger

CULINARY EDUCATION AND TRAINING HIGHLIGHTS

A native of New Zealand, chef Wilson has traveled to many parts of the world, during his young life, in search of experience and knowledge. After graduation from high school in Auckland, he moved to the United States to enroll in the Culinary Arts program at Grand Rapids Community College (GRCC). While a student at GRCC, his skills and culinary acumen earned Daniel positions on both the Central Region (USA) Hot Food Competition Team that went to the ACF National Competition in San Antonio, Texas and the Central Region (USA) Knowledge Bowl Team that competed in the ACF National Competition in Las Vegas, Nevada. He also traveled to Greece, Turkey, Israel and Egypt with the authors (Garlough and Campbell) on their annual culinary study-tour. While a student in Grand Rapids, Chef Wilson worked as a chef de partie at the Egypt Valley Country Club. Upon graduation, Daniel moved to Melbourne, Australia to work as the sous chef at Blake's Restaurant. He then returned to the United States to work as the chef de partie (in garde manger) at the celebrated Restaurant Daniel in New York City. This outstanding learning experience led to his return to Melbourne, Australia to work as the executive chef of Blake's Cafeteria and his current position as the executive chef of Arintji, an upscale Australian restaurant. His stellar work in Melbourne earned Chef Wilson recognition as the "2003 Young Chef of the Year."

ADVICE TO JOURNEYMAN CHEF

I would say that the most important thing is to think of yourself as a sponge and absorb as much as you possibly can. I love to travel and have spent a lot of time in Malaysia and Thailand. The approach to food there is amazing, and the best food comes from the least likely kitchens. The more techniques and methods you learn are simply more arrows in your quiver. Do not get stuck in your ways, but embrace new ideas and take something from everyone and everything—you never know when you might use it. Also, immerse yourself in the culture in the country where you are, and learn from those people who have had the tradition passed down; they will give you insight that you will never find in a book. The food culture from all countries fascinates me; I love to watch masters of their own cuisine and see how it is really done. Lastly, *taste everything!*

Daniel Wilson

RECIPE 6–16

CHEF DANIEL WILSON'S OPEN-FACED VEGETARIAN SANGER

Recipe Yield: 8 portions

MEASUREMENTS		INGREDIENTS
U.S.	**METRIC**	
1 each	1 each	Global aubergine (eggplant), small, peeled
2 each	2 each	Tomatoes, large, very ripe
		Vegetable Spread
2 tablespoons	30 mL	Olive oil
2 each	2 each	Garlic cloves, minced
2 each	2 each	Anchovy fillets, rinsed and chopped
$1/4$ teaspoon	1.2 mL	Black pepper, coarsely fresh ground
$1/4$ teaspoon	1.2 mL	Granulated sugar
$1/4$ teaspoon	1.2 mL	Thyme, dried
$1/4$ cup	59 mL	Water
		Assembly
1 each	1 each	Focaccia loaf, or other dense herb bread
As needed	As needed	Olive oil
8 ounces	250 g	Fresh goat's milk cheese, such as Feta
As needed	As needed	Black pepper, fresh ground

PREPARATION STEPS

1. Chop the aubergine into small pieces and simmer in heavily salted water for 5 minutes. Drain and rinse under cold running water. Thoroughly pat dry with paper towels. Reserve.

2. Peel, seed, and coarsely chop the tomatoes. Reserve.

3. Heat the oil in a medium saucepan and add the garlic, cooking until fragrant and lightly colored.

4. Stir in the chopped aubergine, tomatoes, and anchovy.

5. Add the black pepper, sugar, thyme, and water.

6. Gently cook the vegetable spread, covered, for 20 minutes.

7. Remove cover and cook for an additional 10 minutes to produce a thick, chunky spread.

8. Set spread aside to cool.

9. Slice the focaccia to produce 8 or 16 slices. Gently brush both sides of the sliced bread with the olive oil and toast in a preheated 220°F (104°C) oven.

10. Cover one cut side with vegetable spread and sprinkle with cheese crumbles. Season with black pepper.

11. Serve at room temperature, or reheat in oven for 10 minutes before service.

REVIEW QUESTIONS

1. What is a sandwich?

2. Describe the component parts of a sandwich.

3. What is a wrapped sandwich?

4. What is an open-faced sandwich?

5. List the hot sandwiches available to the chef.

6. Describe an Indian taco.

7. What is a Monte Cristo sandwich?

8. Where did the gyro get its name?

9. What are typical fillings for British tea sandwiches?

10. Where did croque monsieur originate?

ACTIVITIES AND APPLICATIONS

A. **Group Discussion**

Discuss how sandwiches that are often eaten "on-the-run" can be changed for the nutritional benefit of the public.

B. **Research Project**

As a group, research the most popular sandwich in your immediate area of the country. Use a sample of 10 to 12 assorted menus from differing styles of restaurant.

C. **Group Activity**

Create a new form of sandwich that has never been seen before. Consider the component parts of a sandwich and construct your idea from there.

D. **Individual Activity**

Research a country's cuisine and find some more unusual sandwiches recipes. Present your findings to the class.

PART III

Fabrication Skills of the Garde Manger

CHAPTER 7

Poultry and Game Birds

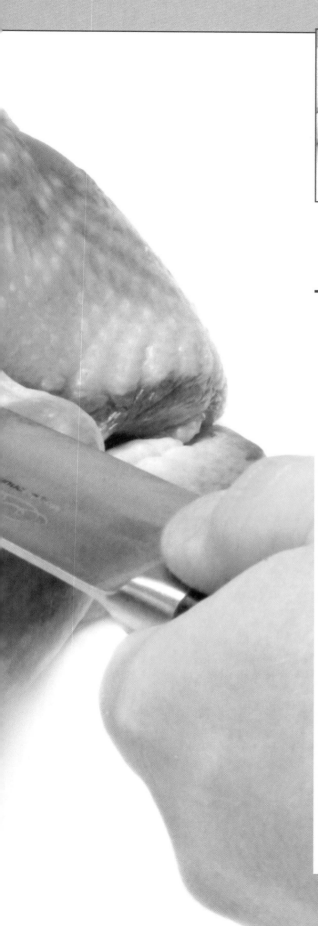

After reading this chapter, you should be able to

- pluck and singe poultry and game birds.

- recognize quality in poultry and game birds.

- fabricate and eviscerate poultry and game birds.

- explain the basic techniques associated with poultry and game birds.

- bard, lard, truss, stud, and truffle poultry and game birds.

- prepare a crapaudine.

- demonstrate how to prepare a poultry chop.

- demonstrate how to prepare a ballontine three ways.

- demonstrate how to prepare a galantine three ways.

- cut and stuff an airline suprême.

- explain how to cut for sauté and roasting.

- demonstrate how to make an ostrich escalope.

KEY TERMS

aging	eviscerating	sauté
ballontine	(to) French	studding
bard	galantine	suprême
chop	game birds	truss
crapaudine	larding	truffling
escalope	poultry	

Cuisine is when things taste like themselves.

CURNONSKY

Chapter Format

To illustrate the techniques associated with **poultry** and **game birds**, this chapter deals with the step-by-step methods involved in the execution of each technique. The techniques highlighted, although demonstrated using only one example of game bird or poultry, can also be applied to any of the birds shown, unless otherwise stated.

Recognition of Quality Points

The quality of domesticated poultry differs from that of game birds because when the wild birds are killed, there is no guarantee of their age. There are, however, some general quality points that are worth learning and that apply to all birds that are going to be consumed:

- Even distribution of fat especially in the cavity
- No obvious abrasions, blood spots, or bruises
- Pliable or bendable breast bone and beak
- Small, underdeveloped flight muscles in the wings
- Hairless legs and feet
- Firm flesh with straight plump breasts
- No scaling on the feet
- Pleasant odor, especially from the cavity

The quality of feathered game is directly related to the age of the animal and the diet available in its environment. This can be why the flavor of duck, raised in a coastal delta environment, has a somewhat fishy taste. These variations in diet have much to do with the cooked flavor of the birds.

Aging the meat imparts tenderness. With all feathered game, the flavor improves by hanging it for a few days in a cool, well-ventilated place. Care should be taken with water birds not to get them too *high* (an expression referring to a certain smell from hanging). After plucking, game birds are prepared very much like poultry, except that, in the case of game, larding and barding is more common.

Like cheese and wine, meat benefits from a certain period of **aging**, or slow chemical change, before it is consumed. Its flavor improves while it gets tender. As lactic acid accumulates in the tissue after slaughter, it begins to break down the walls of the cell bodies that store protein-attacking enzymes. As a result, these enzymes, whose normal function is to digest proteins in a controlled way for use by the cells, are liberated and attack the cell proteins indiscriminately. Flavor changes are probably the result of the degradation of proteins into individual amino acids, which generally have a strong flavor.

Some additional quality points for wild game include the following:

- The comb and wattles should be bright and undamaged.
- The spurs should not be too long.
- The feet should be smooth and slender.
- The bird should have bright full eyes.
- The stomach should be intact.
- All damaged parts should be rubbed well with flour and black pepper.
- The gall bladder in birds must be carefully removed to avoid rupture.

The following quality point is associated with individual game birds and is usually a sign of youth:

- For geese and ducks, the chef should be able to grasp the windpipe and join fingers together, clicking fingernails on the neck side of the pipe.

Evisceration of Poultry and Game Birds

Great care must be taken when **eviscerating** any kind of bird. Removing the entrails can cause sanitation problems with the finished product, so contamination must be foremost in the mind of the kitchen staff and butcher.

1. Pluck the bird by holding it by the feet, spreading out the wings and plucking the breast feathers opposite the way they grow. The legs should then be done and finally the wings should be removed.
2. Scalding or blanching the bird for 10 to 15 seconds aids in the plucking, but take great care not to damage the skin or flesh.
3. Rub the bird down with a dry towel. Singe the tiny hairs of the skin over an open flame, while rubbing them off as you singe.
4. Remove the head and neck from the back of the bird, leaving the breast side of the long piece of skin intact (Figure 7–1A). Remove the gullet, the crop, and the windpipe from the inside of the skin.
5. Remove the vent and the vent fat from around the opening between the legs and loosen the whole interior gut. Remove as whole as you can without puncturing the intestine.
6. Clean the interior of the cavity ensuring the lungs are removed and discarded. They are located in the spinal column.
7. Separate the intestines from the liver, kidney, gizzard, and heart (Figure 7–1B).
8. Wash the organs well and store for other uses. Wash the bird well, taking great care to thoroughly clean the interior.

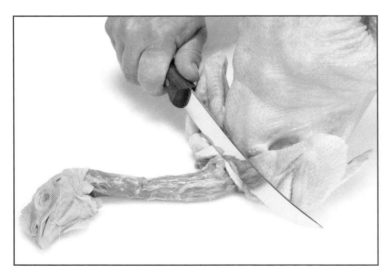

FIGURE 7–1A
Removing the head
and the neck

FIGURE 7–1B Separating the intestine

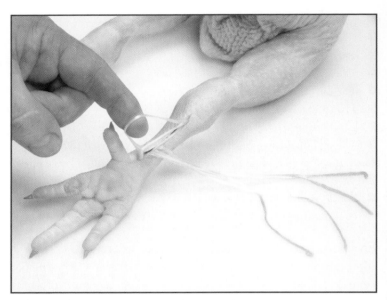

FIGURE 7–1C Pulling sinew from legs

9. Remove the feet at this time; however, if the legs are being served whole, the sinew should be removed using the feet as leverage. A small incision is made on the back of the leg, down by the claws, and the sinew is drawn from there using a towel to grip them (Figure 7–1c).

Dressing of Poultry and Game Birds

The techniques involved in the use of poultry and game birds are shown on individual birds, although they apply to most others. Specific quality points associated with individual game birds are also discussed.

CHICKEN

The term **ballontine** is normally associated with the leg and thigh of raw poultry and game birds, and although not as popular as it has been in the past, it is starting to regain its popularity because the cost of constraints associated with a lunchtime menu. There are three styles or techniques used in the production of ballontines.

Method One

1. Remove the leg from the body of the chicken, leaving as much skin attached as possible during the process. Be sure to remove the oyster that is located in the spinal column of the bird (Figure 7–2A).
2. Remove the skin completely from the leg, starting at the thigh and working toward the knuckle end of the drumstick. Take care not to damage it in any way while doing so.
3. Soak the skin in salted water for 10 minutes, remove any unwanted fat, and dry the skin well.
4. Bone the flesh from the leg and thigh (Figure 7–2B).
5. Beat lightly with a mallet until the piece of meat is evenly flattened, then wrap around a small nugget of chosen forcemeat, stuffing, or other ingredients of your choice.
6. Wrap the meat in the skin and shape into a small ham (Figure 7–2C).
7. Sew the opening with string and chill for cooking.

Method Two

1. Remove the leg from the body of the chicken, leaving as much skin attached as possible during the process. Be sure to remove the oyster that is located in the spinal column of the bird.

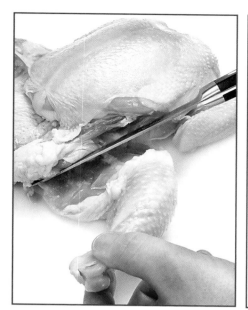

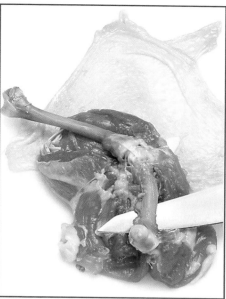

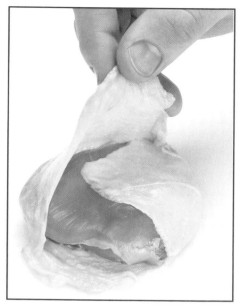

FIGURE 7–2A Removing the leg and thigh

FIGURE 7–2B Boning the flesh from leg and thigh

FIGURE 7–2C Wrapping skin around flesh and forcement

2. Remove the skin completely from the leg and thigh (Figure 7–3A).

3. Locate the thighbone and clean thoroughly, removing the marrow and blanching in boiling water for 3 minutes; trim with a small saw to about 2 inches (5 cm) in size.

4. Make flavored, garnished forcemeat with the flesh and stuff back into the skin, shaping like a small ham; sew the opening as before (Figure 7–3B and C).

5. Stuff the bone with some of the forcemeat and place in the aperture at the end of where the drumstick was (Figure 7–3D). Chill and reserve for cooking.

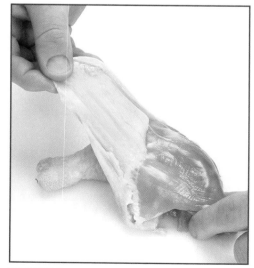

FIGURE 7–3A Removing skin from the leg

FIGURE 7–3B Stuffing the leg skin

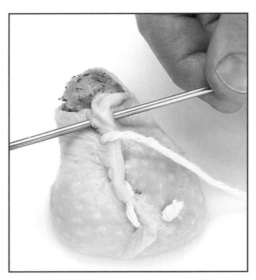

FIGURE 7–3C Sewing the ballontine

FIGURE 7–3D Inserting the bone to simulate original shape

Method Three

1. Remove the leg from the body of the chicken, leaving as much skin attached as possible during the process. Be sure to remove the oyster that is located in the spinal column of the bird.
2. Turn the leg and thigh skin side down and very carefully remove both the thigh bone and the drumstick bone, ensuring that the sinews are removed at the same time (Figure 7–4A).
3. The end of the drumstick bone can be sawed off close to the opening, leaving the knuckle intact. Make sure there are no splinters of bone left during the sawing.
4. Place some forcemeat or stuffing into the place where the bone was and wrap the skin over the flesh, securing by sewing together with butcher's twine (Figure 7–4B and C).
5. Chill for service.

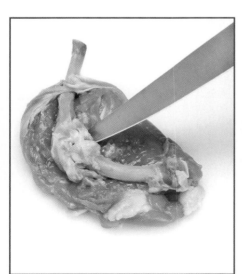

FIGURE 7–4A Deboning the leg and thigh

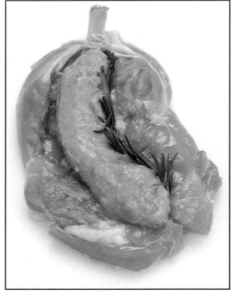

FIGURE 7–4B Placing forcemeat in the bone pocket

FIGURE 7–4C Wrapping the leg for sewing

The **suprême** includes the breast and wing of the bird and can be completed in several ways, although the removal of the breast is always the same.

1. Remove the legs, thighs, and wishbone from the bird.
2. Cut into the bone, close to the breast, at the joint of the wing, circling the bone to loosen the flesh. Scrape the wing clean, breaking at the first joint so as to have the cleaned wing bone attached to the breast (Figure 7–5A).
3. Skinning the breast at this time is optional; leaving it on would create an "airline breast."
4. With a boning knife, make an incision along the breastbone from end to end, either side of the breast ridge.
5. Detach the flesh from the bone, carefully starting at the pointed end and following the breastbone closely until the wing joint is reached.
6. Cut through the joint and remove the breast with the cleaned wing attached (Figure 7–5B).
7. Remove the fillet from the underneath and remove the sinew. This can be used for something else, it can be lightly beaten to act as a patch for stuffing the breast, or it can simply be placed into the breast (Figure 7–5C).

FIGURE 7–5A
Cleaning the wing bone

FIGURE 7–5B Separating the breast from the rib cage

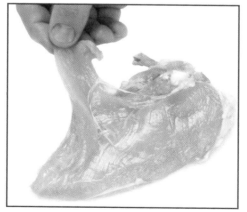

FIGURE 7–5C Trimming the airline suprême

To stuff a chicken breast, great care must be taken to ensure that the stuffing is well contained within the flesh. Perfecting the technique will allow for the addition of various stuffings that will become liquid after they are cooked.

1. Make an incision lengthwise along the thick part of the wing and form a pocket by cutting underneath the incision, into the flesh of the breast (Figure 7–6A). (The fillet alone can then be inserted.)
2. Flatten the filet with a mallet or side of a French knife and place the stuffing into the pocket, and close using the lightly flattened filet as a patch to close the breast and seal the stuffing inside (Figure 7–6B and C).
3. A small dusting of flour between the filet and the supreme helps seal the patch if a stuffing that will become a liquid during cooking is being used.
4. Trim any skin or sinew from the sûpreme.

The **sauté** cut can be used for a host of different cooking styles and many international dishes. In the past it was associated with the classic sautéing process, but today it appears in dishes as diverse as southern American barbecue to Caribbean fried chicken.

1. Remove the wishbone from the bird.
2. Remove the legs and the thighs from the carcass.
3. Separate drumstick and thigh at the natural joint (Figure 7–7A).
4. Remove the wing from the drummet and trim off the tip (Figure 7–7B).
5. Cut off what would be half suprêmes with the drummets attached.
6. Remove the breast from the carcass. Trim and then cut crosswise into two equal parts (Figure 7–7C).
7. Trim the carcass and cut into three parts, one with the blade bones of the bird, one with oysters (two bone cavities at back of bird), and one with the "parson's nose."
8. What remains are two breast pieces with drummet attached, two pieces of breast, two wings, two drumsticks, two thighs, and three sections of carcass.

By **trussing** a bird before cooking, the chef can determine the shape of the bird when cooked. The specific shape is determined by the way the twine is used to tie the bird. An important rea-

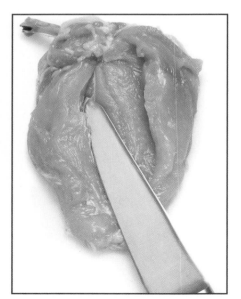

FIGURE 7–6A Making a pocket in the breast meat

FIGURE 7–6B Flattening the fillet with a meat mallet

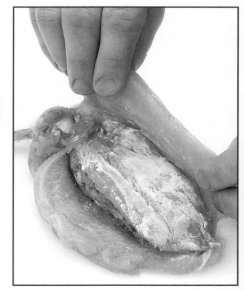

FIGURE 7–6C Stuffing and sealing the suprême

FIGURE 7–7A Separating the drumstick and thigh at their joint

FIGURE 7–7B Trimming half suprême with drummet attached

FIGURE 7–7C Cutting the breast into portions

son to **truss** is to create a specific shape for presentation when serving a bird whole, especially when the bird is stuffed. Accuracy and portion yield through consistent carving may be considered to be even more important because of the financial benefits of good portion control.

Trussing can be done using a trussing needle, but this can be time consuming and inefficient because there are many pieces of twine to be removed after cooking. The following describes how to truss using a needle and without using a needle, as a more efficient, more modern technique.

Trussing with a needle:

1. Remove wishbone: This greatly assists the carving of the bird, particularly if it is stuffed (Figure 7–8A).
2. Scrape the wishbone lightly with the tip of a small knife, then free the bone from the joint of the carcass with the fingertips and twist the bone out whole without damaging the flesh.
3. Break the backbone in the region of the wings, with a smart tap with the back of a knife so that the bird will lie flat on the board or dish.
4. Using an 8-inch (20-cm) trussing needle threaded with string, push the needle immediately behind the legs at the hinge between drumstick and thigh, through the body, and out the other side in the same position (Figure 7–8B).
5. Turn the bird on its side and pass the needle through the middle portion of the wing, draw the skin of the neck over the back to close the aperture, pass the needle through the skin and the middle of the backbone, and then pass the needle through the middle portion of the wing (Figure 7–8C). The ends of string can now be tied firmly together.
6. Pass the needle with a second string through the natural cavity in the thigh bone, through the carcass from one side to the other, then over the folded leg between the skins, through the end of the body without touching the breast, then through the skin of the opposite leg. Tie the ends firmly just below the knuckle.

FIGURE 7–8A Removing the wishbone

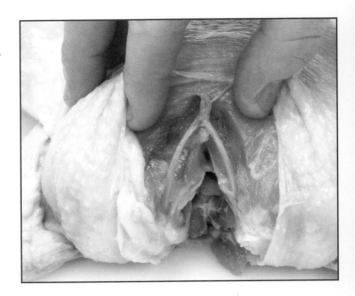

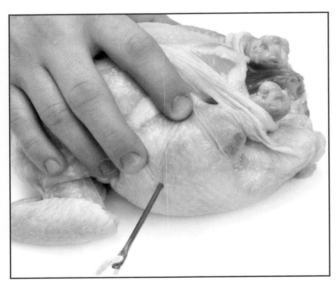

FIGURE 7–8B Pushing the needle through the thighs

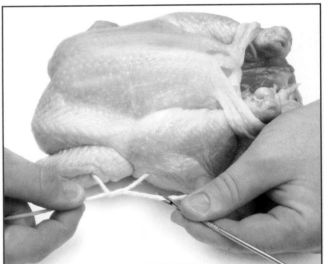

FIGURE 7–8C Pushing the needle through the middle wing and drawing the string

Trussing with no needle can be a much faster and more efficient way of tying the birds into shape. The advantage of this method is that it only uses one piece of twine that is easily removed when the bird is cooked.

1. Cut a length of string about 3 feet (90 cm) or about four times the length of the bird that you are trussing.
2. Hold the center 6 inches (15 cm) of the string with both hands and place the string under the flap of skin, closing the aperture of the neck and tucking under and into the space between the body and the wings (Figure 7–9A).
3. Pull the two ends of the string up toward the breast, crossing them over when they reach the end of the drumstick.
4. Loop them around the end of the drumsticks and pull back down to where they came from (Figure 7–9B).

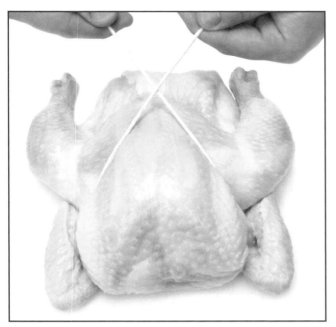

FIGURE 7–9A Tucking the string under the neck

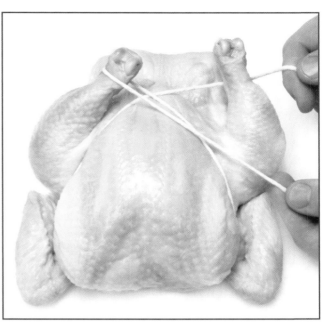

FIGURE 7–9B Crossing the string between the legs

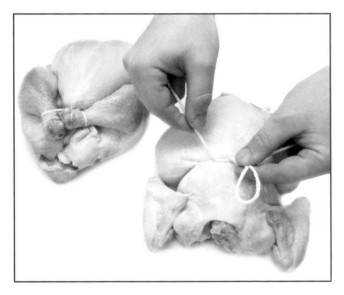

FIGURE 7–9C Looping the string around the bird and tying

5. When you reach the middle of the bird, flip it over while still holding the string ends and tie securely against the backbone (Figure 7–9C).
6. The advantage of this method is that you only have to cut the string once and you have the bird free for carving.

TURKEY

Turkeys are usually delivered with some feathers on the neck and wings. Pluck carefully and remove fine feathers as for chicken. Singe and draw the sinews from the legs. This operation is essential because the sinews will spoil the flesh of the legs and make it impossible to carve them.

Cut the skin all round, just above the foot, break the bone, twist the foot, and place it in a sinew hook. Pull steadily on the leg, pulling all sinews out that are attached to the foot. Trim the tips of the wings, and clean and dress the turkey as you would a chicken.

Note: For larger birds, a hook attached to the wall may be used.

This method greatly enhances the carving of the turkey legs; however, it is not often that birds arrive with their feet. A good substitute for this technique is the boning and removing of the sinew from the legs before cooking. This gives a dual benefit in that the carving of the legs becomes much easier and both the breast and the legs cook more evenly, rendering a moist product.

Whether to stuff a turkey has been the subject of much debate because of the potential for health problems. Turkeys should only be stuffed just before going into the oven. Under no circumstances should poultry be stuffed and then held without immediate cooking.

The following steps should be followed when stuffing a turkey:

1. Remove the wishbone by scraping down with the tip of a small knife until the wishbone frees from the flesh.
2. With the tip of your knife, snap through the sinew where the ends of the wishbone join the carcass.
3. Twist off the whole bone at the tip of the breastbone, doing as little damage as possible to the skin or the flesh.
4. Pack the stuffing into the breast end of the bird, covering it with the flap of skin (Figure 7–10A and B).
5. Pull down the skin tightly over the neck end and secure on the spine with a wooden skewer, or truss using a needle with some twine (Figure 7–10C).
6. Roast immediately.

Cutting a **chop** from a turkey can be tricky and somewhat wasteful; however, if the turkey is being boned for use as an escalope or strip for frying, then it makes sense.

The steps for cutting a chop from a turkey are as follows:

1. Remove the legs and thighs from the turkey.
2. Remove the carcass from underneath the breast, leaving the rib skin and the breastplate intact.
3. Split the breast down the middle, removing the center breast plate with powerful poultry scissors. You should have the two breasts, part of the plate, and all the ribs intact.
4. **French** the ribs and cut the chops using a small saw to cut the plate.

Stuffing and rolling a deboned bird is a very common technique used by garde manger chefs. Poultry, such as turkey, is healthy, cost effective, and popular among consumers; however, it is difficult for chefs to carve roasted birds in such a way as to produce consistent looking servings

FIGURE 7–10A Inserting the stuffing into the cavity

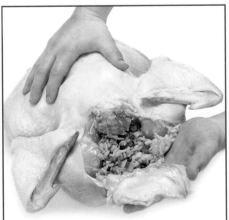

FIGURE 7–10B Wrapping the skin flap under the bird

FIGURE 7–10C Securing the skin flap with a wooden skewer

that are of equal portion size. Also, inevitably, there is wasted meat that is unable to be utilized due to the natural contours of the poultry.

To utilize the full bird and to make consistently sized portions (with the added benefit of stuffing, which reduces the portion cost) chefs often choose to bone and roll whole breasts. Figure 7–11 illustrates the manner in which a turkey may be flattened, stuffed, and rolled for roasting.

DUCK

In young ducks, the breast is pliable, the feet are soft and small, and the underside of the wings are downy. Soft, flexible quills with a decided point are also a good indicator of young age. The older the bird, the more rounded the tips.

Wild duck can taste "fishy" or "muddy" because the duck feeds in coastal areas. Duck flesh tends to retain the flavor of what the duck eats most often. The ducks can be cleaned and soaked in a gallon of water with a cup of vinegar for 24 hours. Or, a sliced onion and a sliced orange may be placed in their cavities until and while they are cooking. Both of these steps help remove the smell and flavors developed by their eating habits.

There are many ways to make **galantine** of duck, and the technique involves completely removing the skeleton of the bird. The flesh is left intact or partially intact, or it may be totally used for the forcemeat that will eventually be rolled up in the skin. Although a galantine can be made without cutting the skin, it is more commonly begun by first cutting the skin. The galan-

FIGURE 7–11A Splitting the breast down the middle

FIGURE 7–11B Laying the stuffing

FIGURE 7–11C Rolling the stuffed breast

tines can be poached in rich broth made from the skeleton that was removed, often with added gelatin, to aid in binding. (For more information, see Chapter 14.)

The following procedures are used to make a galantine:

Method One

1. Cut along the center of the back of the bird, from neck to tail, through the skin (Figure 7–12A).
2. Carefully remove the flesh from the bone, leaving it attached to the skin. Use a boning knife to release the flesh from the bone where necessary (Figure 7–12B).
3. When you reach the wings and thighs, find the joints and cut through the ligaments to detach them from the carcass.
4. Continue toward the breast, carefully trimming away the fillets from the breastbone.
5. Be careful on releasing the skin from the ridge of the breastbone because it is very thin at that point and tends to tear.
6. Bone out the thighs, drumsticks, and wings, channeling the bone from the winglet.
7. Make forcemeat with the desired garnish and place a log shape of it down the center. Roll and tie securely for poaching (Figure 7–12C and D).

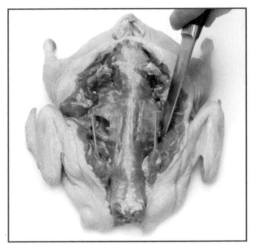

FIGURE 7–12A Boning the bird from back, leaving the skin attached

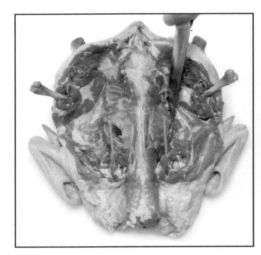

FIGURE 7–12B Removing flesh from bone, leaving it attached to the skin

FIGURE 7–12C Spreading the forcemeat over the flesh and skin

FIGURE 7–12D Rolling the galantine

Method Two

1. Cut along the center of the back of the bird, from neck to tail, through the skin.
2. Carefully remove the flesh from the bone, leaving it attached to the skin. Use a boning knife to release the flesh from the bone where necessary.
3. When you reach the wings and thighs, find the joints and cut through the ligaments to detach them from the carcass.
4. Continue toward the breast, carefully trimming away the fillets from the breastbone.
5. Be careful on releasing the skin from the ridge of the breastbone because it is very thin at that point and tends to tear.
6. Bone out the thighs, drumsticks, and wings, channeling the bone from the winglet.
7. Using a stuffing, forcemeat, or another smaller stuffed bird, reshape the duck to its original shape, sewing up the back and trussing well (Figure–13A, B, and C).
8. This technique can be used on many birds and is a great delight at the table because it is carved all the way through with no bone to stop the process.
9. When done with this technique, turkey can be deep fried or roasted to create *Turducken*.

Note: Turducken is made by thoroughly boning out a turkey, duck, and chicken. The turkey is laid open, with the duck spread open onto the turkey. The chicken is laid onto the duck. The three are rolled up into a poultry roll with sausage and cornbread stuffing, and tied into a roast.

FIGURE 7–13A Placing a ball of forcemeat onto the flesh, with its skin still attached

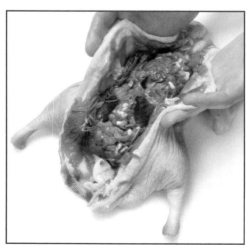

FIGURE 7–13B Shaping the galantine back to the original contour of the bird

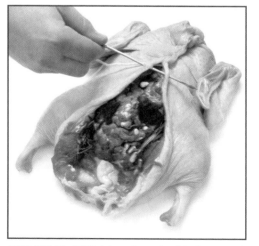

FIGURE 7–13C Trussing the bird to hold its shape

Method Three

1. Completely skin the duck, cleaning the skin thoroughly of all blood spots and excess fat (Figure 7–14A).
2. Cut into as large a rectangle as possible with the amount of skin that you have.
3. Layer duck forcemeat and the reserved duck breast with garnishes of your choice (Figure 7–14B).
4. Roll and tie for poaching (Figure 7–14C).

Note: Galantines can also be roasted or braised for serving as a hot entrée.

PHEASANT

When checking pheasants for quality, there are some important steps to consider:

- The hen bird is prized as the more tender meat.
- Both the hen and the cock respond well to hanging, which is generally done when the birds are still intact, in a cool, well-ventilated room for 6 to 12 days.
- Fat buildup is better toward the end of the season.
- Commercially purchased birds have normally been adequately aged before their sale.

The **larding** process is necessary to introduce much needed fat to the game birds that are going to be roasted or braised. This allows the meat to baste during the cooking process, ensuring that they are moist and flavorful when done. Some game birds if shot closer to the end of the season, may contain more fat, so the chef must decide which birds might need to be larded.

For larding, thin strips of fat pork, salt pork, or salted and smoked bacon are inserted neatly into the tender breast of feathered game. Larding is done using a larding needle, a tool specifically used for this purpose. In modern larding techniques, the ingredients used to lard are not

FIGURE 7–14A Completely removing the skin from the bird

FIGURE 7–14B Layering the flesh with forcemeat

FIGURE 7–14C Wrapping the flesh with skin to form the galantine

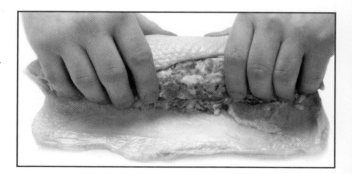

necessarily fat. They could be blanched vegetables or other meats. This trend is more for presentation and flavor rather than for the introduction of fat.

The following procedures are used to lard a pheasant:

1. Cut some very cold pieces of salt pork fat into strips 4 inches by $^1/_4$ inch by $^1/_4$ inch (10 cm by .6 cm by .6 cm) (Figure 7–15A).
2. Place the chilled strips into a larding needle (Figure 7–15B).
3. Remove the skin from the breast without breaking it by loosening it from the vent end to the neck and lifting it gently upward to expose the flesh of the breast.
4. Remove the wishbone from the bird.
5. Lard the breasts by inserting the needle into the flesh at even intervals, pulling the needle completely through and out the other side and leaving the fat neatly in the flesh of the bird (Figure 7–15C).
6. Repeat until the breasts are completed.
7. Replace the skin securely over the flesh and truss the bird for cooking.

GROUSE

The quality points specific to grouse are that grouse should be hung for 3 to 4 days only in a dry, cool, well-ventilated room. Grouse has very dark meat with normally very little fat content, although they do contain more fat toward the end of the season.

To **bard** a game bird, thin slices of fat pork, salt pork, or even bacon are used to cover the tender breast of the game. The coating of fat protects the lean meat during cooking, ensuring that it does not overbrown or dry out. Sometimes herbs, leaves, or rubs are put between pork fat and breast to give a distinctive flavor.

The use of smoked bacon must be done with care, because the smoky flavor can easily overpower the subtle characteristic flavor of game bird.

1. Shave the fat on a slicing machine to get it very even and thin (Figure 7–16A).
2. Cover the breasts with the slices and truss tightly to support the barding during the cooking process (Figure 7–16B and C).

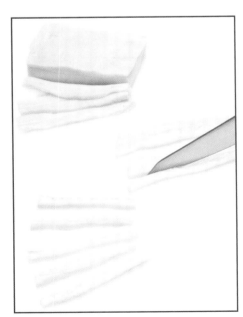

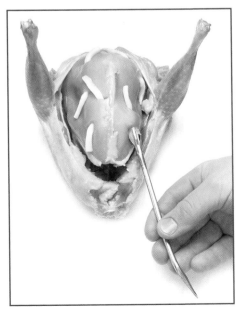

FIGURE 7–15A Slicing salt pork for larding

FIGURE 7–15B Placing pork fat into a larding needle

FIGURE 7–15C Inserting the fat into the flesh

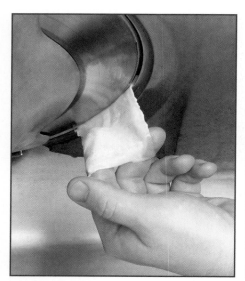

FIGURE 7–16A Shaving pork fat on an electric slicer

FIGURE 7–16B Covering the breast with slices of fat

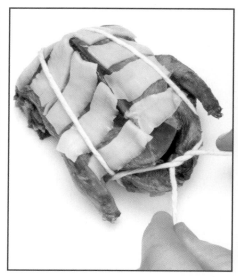

FIGURE 7–16C Trussing the bird to secure the fat

PARTRIDGE

When choosing partridge, their quality is determined by their age, with the younger birds being superior.

Truffling is accomplished by inserting thin strips of truffle into incisions made in the breast, or by pushing thin slices of truffle with the finger under the breast skin of the bird. **Studding** is accomplished in the same manner, but with a host of other ingredients. It may be done with roasted garlic, shallots, or morels. The technique is always the same, but the ingredients differ with seasonal availability.

The following procedures are used for truffling and studding:

1. As with larding, expose the breast from the skin in order to work on the flesh (Figure 7–17A).
2. Make some lateral incisions with the tip of a small knife to create small pockets to receive the special ingredients. Six or seven incisions on the whole breast are sufficient (Figure 7–17B).
3. Place the small nuggets into the pockets and smooth the flesh back over to cover the hole (Figure 7–17C).
4. Replace the skin and truss the bird to hold all fast as it cooks.

WOODCOCK

Woodcock are much better later in the season when they have had a chance to develop a little fat. They should be hung for 4 to 5 days. When they are plucked, always leave the skin on to protect the very delicate meat during cooking and to allow the moisture to stay within the meat.

When grilling birds, especially larger ones, great care must be taken to ensure the full cooking of the legs and thighs. The use of indirect heat when using the grill helps ensure accurate cooking. The technique used for preparing birds for the grill is specifically designed to flatten them to make them easier to cook.

The following procedures are used for grilling birds:

1. Split the bird down the back along the backbone with a large poultry knife or poultry sheers, removing the spinal column from the neck to parson's nose (Figure 7–18A).
2. Flatten the bird well with a mallet and remove the ribs (Figure 7–18B).
3. Tuck the legs into the body and cut small holes in the breast skin to secure the drumstick end.

FIGURE 7–17A Removing the skin to expose the flesh

FIGURE 7–17B Making incisions with the tip of a knife

FIGURE 7–17C Inserting (studding) the truffle pieces into the cuts

4. Trim around where the backbone had been, ensuring that there are no bone fragments (Figure 7–18C).
5. The bird can now be marinated, or just seasoned and grilled.

GUINEA FOWL

When considering the quality points specific to guinea fowl, age should be of the first importance because they do toughen considerably after 8 to 10 weeks of age. Guinea hens are considered by most to be poultry, but like game they are best hung for 2 to 3 days in a cool, well-ventilated area before plucking, dressing, and cooking.

The method **crapaudine** is another means of preparing a bird for the grill. The bird is split in such a way that, when cooked in this manner, it resembles a frog or toad. In some recipes, the unusual presentation is further enhanced by the addition of sliced hard-boiled eggs to represent the eyes.

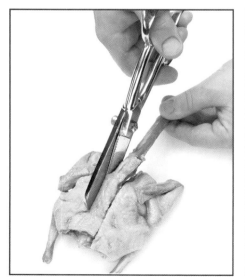

FIGURE 7–18A Splitting the bird down the backbone

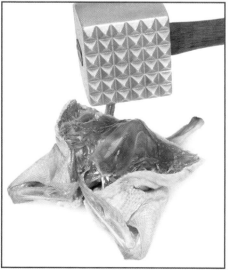

FIGURE 7–18B Flattening the bird with a meat mallet

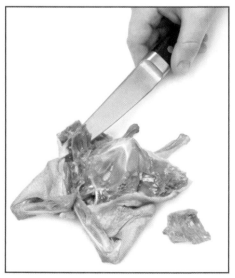

FIGURE 7–18C Trimming the bones from the interior of the bird

The following procedures are used for crapaudine:

1. Spread the legs out as far as they will go without tearing the skin.
2. Place the bird breast side up on a cutting board and with a sharp heavy knife make a cut down from the tip of the breast through the ribs to the where the wing joins the spine, making sure that you avoid cutting the legs (Figure 7–19A).
3. Open the bird up using the joint of the wings and the breast end as a hinge (Figure 7–19B).
4. Using the flat side of the French knife or the palm of the hand, press firmly to flatten the breast (Figure 7–19C).
5. Trim around the cut areas, ensuring there are no bone fragments.
6. Chill and prepare to grill.

SNIPE

Long-beaked birds, such as snipe, should not be completely drawn; only the gizzard, the intestines, and gallbladder are removed from a small opening under the leg.

FIGURE 7–19A Cutting through the ribcage

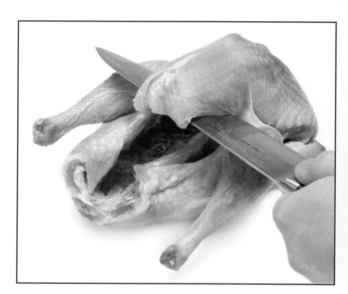

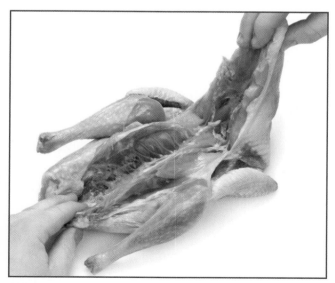

FIGURE 7–19B Opening the bird to form the crapaudine

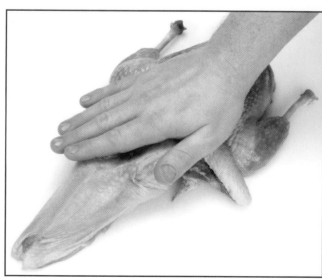

FIGURE 7–19C Pressing lightly to flatten

The following procedures are used for self-trussing a bird:

1. Clean the bird and remove the gizzard, intestines, and gallbladder.
2. Singe off any fine hairs around the neck.
3. For trussing:

    If the bird has a head, press the thighs close to the body and use the long beak to break through the leg and thigh where they join, all the way through the body and out the other side again between the leg and thigh joint. The beak now serves as a trussing implement to hold the bird together during cooking, replacing the normal job of a trussing needle and string (Figure 7–20). *Otherwise,* use a knife to make an incision in one leg (Figure 7–21A) and then push the end joint of the other leg through the incision on the first leg (Figure 7–21B).
4. Garnish with herbs, and shape the bird for roasting (Figure 7–21C).

Ostrich

Using a larger part of the ostrich leg, the meat can be cut into **escalopes.** These thin slices may be stuffed or cooked by the sauté method. Ostrich does not yield a large amount of useful meat cuts, so most of the carcass weight goes for grinding.

The following steps are used to prepare escalopes of ostrich:

1. Cut across the grain of the meat and slice the ostrich into 2- to 3-ounce (57- to 85-g) pieces (Figure 7–22A).

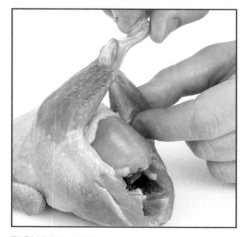

FIGURE 7–20 Self-trussing technique used when head is attached

FIGURE 7–21A Making an incision in one leg

FIGURE 7–21B Inserting second leg through incision in first leg

FIGURE 7–21C The shaped bird, ready for roasting

FIGURE 7–22A Cutting meat into slices

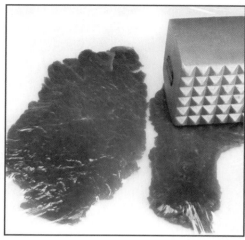

FIGURE 7–22B Flattening the meat slices to form escallops

2. Lightly beat the meat with a mallet until an even escalope is formed (Figure 7–22B).
3. Spread with a chosen stuffing, leaving about 1 inch (2.5 cm) around the edge uncovered.
4. Dip the next escalope into flour and pat off the excess.
5. Place one escalope on top of the other, making sure they fit exactly.
6. They may be dressed in flour, breadcrumbs, or cornmeal and reserved for cooking.

Note: Game birds that are caught in the wild and ungraded birds are generally forbidden from being processed in licensed kitchens (Figure 7–23). In most food service operations, inspected game must be purchased from licensed game purveyors or distributors.

FIGURE 7–23 Wild game birds

Professional Profile

BIOGRAPHICAL INFORMATION

Name: Steve Stallard

Place of Birth: Pontiac, Michigan

Recipe Provided: Roulades of Partridge with Pear-Honey Sauce

CULINARY EDUCATION AND TRAINING HIGHLIGHTS

Born in Pontiac, Michigan, Chef Stallard grew up gathering and harvesting local indigenous ingredients for his parents' table. As a result, he later became passionate about hunting wild game, fishing for anything that was in season, and continually foraging for wild mushrooms during their appearance. Learning and understanding these ingredients, their flavors, their nuances, and how they interact with other ingredients has led him to a remarkable culinary career. After courses of study at the Culinary Institute of America, Chef Stallard traveled to The Taillevant Restaurant in Paris, France, where he was steeped in the culinary traditions that he longed to understand. Upon returning to the United States, he took positions as sous chef at the Greenbrier Hotel in 1989, executive chef at Redbirds Food Service, and later as sous chef at the Amway Grand Plaza Hotel from 1990 to 1995. These positions helped solidify his management and culinary talents, gaining him the "five-star award" in management for 5 consecutive years. In 1997, Chef Stallard became the executive chef of the Racket Club of Memphis, and in 1999, he was hired as the corporate chef of the Dow Chemical Group in Midland, Michigan. During this time, Chef Stallard has had the honor of cooking for four American Heads of State, including President Reagan, President Ford, President Carter, and President Bush (senior). He has judged many international culinary competitions, including those at Hotel Olympia in London, as well as many national and local competitions in the United States. His present position of president and chief executive officer of BLIS, LLC allows him to continue with his core passion of creating high-end products for discerning chefs around the world, including producing some of the finest caviar products available on the world market, all from wild-caught fish.

ADVICE TO A JOURNEYMAN CHEF

Fundamentals are absolute in cooking. The constant focus on what has been done during the evolution of culinary arts cannot be avoided or ignored. No matter how proficient or successful a chef has become, that chef must be able justify why he or she has developed the cuisine style of personal choice. The broader the base of knowledge and skills, the more the individual can expand on his or her craft. This is particularly crucial as we mature as culinary artists.

Steve Stallard

RECIPE 7–1

CHEF STEVE STALLARD'S ROULADES OF PARTRIDGE WITH PEAR-HONEY SAUCE

Recipe Yield: 8 portions

MEASUREMENTS		INGREDIENTS
U.S.	**METRIC**	
		Pear-Honey Sauce
1 cup	237 mL	Honey
4 sprigs	4 sprigs	Thyme
4 cups	948 mL	Pears, fresh
1 cup	237 mL	Chicken stock
2 cups	474 mL	Heavy (whipping) cream
½ cup	118 mL	Butter, unsalted
		Partridge Roulade
2 each	2 each	Garlic cloves, minced
10 ounces	286 g	Chicken breast, chopped
2 each	2 each	Eggs
1 cup	237 mL	Heavy cream
As needed	As needed	Salt
As needed	As needed	Black pepper
5 ounces	143 g	Spinach, fresh
2 tablespoons	30 mL	Butter
1 cup	237 mL	Chanterelles (or other wild mushroom variety)
¼ cup	59 mL	Brandy
4 each	4 each	Partridges, bones removed, skin on (or boneless breast of duck)
As needed	As needed	Caul fat (or butcher's twine)

PREPARATION STEPS

1. For the sauce, combine the honey and thyme in a saucepan. Cook until the honey darkens slightly, stirring constantly. Do not caramelize the honey. Add the pears and stock, and cook until liquid is reduced by half. Add cream and boil until slightly reduced. Add butter and cook gently until sauce is smooth. Reserve.

2. For the roulade, combine the garlic, chicken, egg, heavy cream, and seasonings in a food processor.

3. Cook the spinach in a small amount of water until wilted, about 2 minutes. Drain and then squeeze dry. Chop coarsely and add to the food processor.

4. Purée until almost smooth. Reserve in a mixing bowl.

5. Sauté mushrooms in the butter until tender.

6. Add the brandy and cook over high heat until brandy is absorbed and slightly glazed. Fold into puréed mixture.

7. Lay each boneless bird on plastic wrap with skin side down and pound out evenly like a galantine. Spread mushroom mixture over surface of meat. Roll, jellyroll fashion, in plastic wrap. Remove plastic wrap and encase in caul fat or tie with string.

8. Place roulades on rack in roasting pan and brush with melted butter. Bake at 450°F (232°C) for 15 to 20 minutes.

9. Slice into medallions and serve warm with Pear-Honey sauce, or chill and serve with Cranberry-Ginger Chutney.

1. List six general points of quality of domesticated poultry.

2. Describe the production of one style of ballontine.

3. What is a suprême?

4. How do you cut whole chicken for sauté?

5. Describe the advantages of trussing a bird before cooking.

6. Describe the method crapaudine.

7. What is a galantine?

8. What is larding?

9. How does barding enhance flavor?

10. What is the purpose of splitting a bird, as for spatchcock, when grilling?

A. Group Discussion

In groups discuss the differences between domesticated poultry and wild game, in particular the differences between their flavors when cooked.

B. Research Project

In groups, research the seasons for wild game birds in your state, reporting to the rest of the group when the birds can be killed and how many are hunted per annum.

C. Group Activity

Using a fresh chicken, remove all the edible flesh and calculate the weight loss due to boning.

D. Individual Activity

Create a sausage recipe of your own using proteins other than pork and beef.

CHAPTER 8

Game Meats

After reading this chapter, you should be able to

- explain how fat affects flavor.

- describe how to tenderize meat by aging.

- explain how to care for a carcass after it has been killed.

- describe how the slaughter affects the meat.

- demonstrate how to tenderize game meats.

- understand the use of marinades.

- explain what affects the quality of game meats.

- demonstrate how to add fats.

- demonstrate how to apply a marinade.

- explain the difference between farmed game and wild game.

- demonstrate how to lard and bard.

- demonstrate how to skin a rabbit.

- demonstrate how to remove the rack from rabbits.

- demonstrate how to remove the baron from hare.

- demonstrate how to stuff a frog's leg.

Game, under the command of the knowing chef, undergoes a great number of cunning modifications and transformations, and supplies the main body of those highly flavored dishes which make up truly gastronomical cookery.

JEAN ANTHELME BRILLAT-SAVARIN

Game's Coming of Age

While game has gained a strong resurgence in interest from the dining public, as well as the modern chef, there is really very little that is new with game cookery. Game was the primary meat source on land or in the air, before the domestication of animals and birds, since the Neolithic cook grilled the first steak. American Indians are credited with being the first avid game hunters and cooks, and European settlers to the New World documented their legendary hunting skills. With the migration of these settlers came their own culinary traditions and recipes that merged with the methods devised by the Indians.

Although game cookery is widely practiced in many U.S. food service operations, some of the information contained in this chapter may not be applicable in modern commercial kitchens. American health code laws restrict the presence of field-dressed and ungraded meats in most licensed kitchens. However, game is often dressed and prepared in many American private clubs as well as numerous foreign food service operations. For that reason, the following information on game preparation is included.

Game Meats

The French word *gibier* applies to all animals being hunted and eaten, and is the recognized word in cooking wild game. It is derived from the verb *gibercer,* which means "to hunt." However, in modern times, there is such a shortage of **wild game** that it is unusual to find a restaurant that solely uses actual wild-caught game. The price and the difficulty of maintaining continuous quality are too troublesome.

Internationally, there are a few places, very close to the sources, that claim their game is wild. However, with the stringent restrictions of modern health codes and the fact that most game is controlled in some way, there is a question as to the validity of their claims of "truly wild." According to John Ash (1991), "The resurgence in game cooking in America would not be possible were it not for the recent availability of conscientiously processed, farm-raised game birds and meats from quality-oriented purveyors, both in America and abroad."

In the wild, animals have to work hard to sustain themselves and they tend to develop tough lean muscles, with very little fat. Wild animals also eat a varied diet, so the flavor of their meat has an intensity that most people refer to as "gamey," or strong. With the advent of farming game, the term *game* has become a little distorted in its meaning, in that there are definite distinctions between eating true wild game and farm-raised game.

Wild and farm-raised meats, although they have subtle variations, both taste earthy and complex. This offers a refreshing change on menus that normally deal only with their seemingly bland

FIGURE 8–1 A well-presented game meat entrée

cousins. They have wonderful eating qualities, and when they are cooked with an understanding of their potential challenges, they create superb eating experiences for the guest (Figure 8–1).

How Fat Affects Game Animals

To render the meat as perfect for cooking as possible, great care has to be taken with the handling of the meat, all the way from the slaughtering to the pot. Whether the animal is farmed or wild, the challenge to the chef is the fact that all these animals have a distinct lack of pure body fat, making them one of the most difficult foods to render moist, flavorful and delicious as a meal.

This lack of even fat distribution can be overcome by careful additions of fats during the preparations. These can be in the form of rich marinades, and during the cooking, in the form of larding and barding. These additions compensate for the lack of flavor and moisture—to an extent. However, the flavor must be further developed by the addition of aromatics in the form herbs, spices, wines and spirits, and other fats such as butter and oils, all carried in the perfect sauce rendered from the animal's bones.

Age

When the chef is choosing animals for the menu, it is important that they be as young as possible. Their **age** will contribute to the chef's likelihood for success, ensuring a tender, flavorful piece of meat every time. As mentioned, the one drawback in the case of wild game is their distinctive lack of body fat, especially in the form of **marbling** in the meat fibers.

Fat contributes to the tenderness of meat by acting as a "shortening" agent, much as it does in pastry, allowing the meat to stay moist. When it is melted during cooking, it penetrates the tissue and helps separate the fibers of meat, lubricating the tissue and so making it easier to cut and eat. Without much fat, otherwise tender meat becomes tough, dry, and unappealing to the palette. Surface or deposit fat is much less successful in penetrating the tissue than marbled fat, giving marbled fat a huge advantage over surface fat. This is why so much value is placed on creating good marbling in steaks.

The younger animals have a more finely fibered tissue, hence more connective tissue is found in the meat, making it tougher and difficult to render edible. However, the collagen of young animals is much more easily denatured than older connective tissue. So, the chef must find an acceptable balance between the age and the relative disadvantages of a young animal, especially with this classification of meat.

When using completely wild game, it becomes a bit of a guessing game as to how old the animal is, especially when they are still at large or in the sights of the hunter. Using good judgment and generally following the correct procedures, a chef will end up with a fine product.

With farmed game, the chef is somewhat safe because the animals are slaughtered at the farm when they are at their best age for killing. Farm-raised game is meatier and more tender than wild game, and it has a bit more fat, although noticeably less than beef or pork. The fat in game animals is concentrated on the back or under the spine, rather than being dispersed as marbling throughout the meat. An approximate guide for all these animals is that they should be between 3 and 5 years old for the best results.

THE SLAUGHTER OF THE GAME

Many people who hunt for wild game fail to bring out the best in the meat, which can be ruined by poor field dressing and butchering, and improper preparation on the way to the table.

By far, the most humane methods of **slaughter** are those that result in the highest quality of meat. Any kind of stress on the animal immediately before slaughter, such as fasting, duress in transport, fear, or being chased or shot has an adverse effect on the final product. This is due to the body consuming its own energy stores when muscles are active, which creates a waste product known as *lactic acid* that is carried away by the blood or further oxidized while an animal is still alive.

For a time after an animal is killed, the muscles continue to work at maintaining body temperature, and because blood is no longer flowing, lactic acid accumulates. Stress in animals, however, results in muscular tension depleting the supply of lactic acid. The muscles accumulate less lactic acid after death and the meat has a lower acid content, thus spoiling quicker. Commercial slaughter is carried out as humanely as possible, with the animal being stunned, usually with a blow or electrical discharge to the head. It is then hung up and bled from one of the major blood vessels. This creates less stress on the animal and results in much longer lasting and stable meat.

Rules for Preparing Wild Game Meats

When preparing wild game meats in the field, there are some very important rules that should be followed.

- Wear plastic gloves at all times.
- Clean your knife continually, especially when you are working inside the carcass, to prevent cross-contamination.

- Keep the carcass clean by getting it off the ground as quickly as possible.
- Always use clean equipment during dressing.
- Remove the intestines, lungs, liver, and heart as soon after the kill as possible.
- Carefully remove any musk glands that exude a powerful acid that quickly ruins the game.
- Protect the cavity from insect invasion.
- Ensure that the organs are not pierced and that the animal's hair is kept free of exposed flesh at all times.
- Cool the carcass quickly and keep it cool during processing and transportation.
- At this time, it is advisable to hang the animal to drain and dry out. (The U.S. Department of Agriculture recommends hanging meat by the hind legs.)
- Wipe out any excess blood in the gutted cavity with paper towels and fresh water, making sure to remove any loose hairs.
- Dry the cavity well and prop it open until the cavity is very dry. This is enhanced by good air circulation.
- To prevent severe spoilage, hang the meat at the proper temperature.

AGING LARGE GAME MEAT

Like cheese and wine, meat benefits from a certain period of **aging**, or slow chemical change, before it is cooked and consumed. The flavor improves as it begins to tenderize, and it becomes much easier to cook with. The technique of aging meat is the practice of holding the carcass or the larger cuts of meat at temperatures of 34° to 37°F (1.1° to 2.8°C) for 7 to 14 days and even as long as 30 days to allow the enzymes in the meat to break down some of the complex proteins in the carcass. Aged meat is always more tender and flavorful, and aging is a necessary procedure with game meat. However, there are some basic rules:

- Do not age any game carcass if it was shot during warm weather and was not chilled rapidly enough.
- Do not age animals that were severely stressed before being killed.
- If the wounds are too extensive over the body, aging is not recommended.
- The animal should be at least 1 year of age before aging will make any difference.
- Aging is not recommended for carcasses with little or no fat covering.
- After about 2 weeks, tenderization slows down and bacterial slime begins to develop. This should be avoided at all costs.

One of the results of this aging process is the production of a rather pungent flavor that to some people is desirable in all game, but which puts others off the idea of consuming wild game. Even after aging, the strong flavor or gaminess can be somewhat eliminated by soaking the meat in a mixture of salt, vinegar, and water overnight. Use $\frac{1}{2}$ cup (118 mL) of vinegar to 1 gallon (3.8 L) of water with $\frac{1}{4}$ cup (59 mL) of kosher salt.

The Tenderizing of Game Meat

The most common method of **tenderizing** meat is to damage it physically by cutting, pounding, and grinding. These are all ways of breaking down the structure of the muscle bundles, and rendering them more edible and digestible to the human being. Grinding the meat is the most powerful method, because it breaks down the tissue into tiny shreds and is normally the way 40 to 50 percent of the large animal carcasses are processed.

Tenderizing can also be done with the application of certain plant enzymes or acid marinades. These enzymes have been used for hundreds of years, and are still used today in many ways. In Mexico, meat has been tenderized by wrapping it in papaya leaves before cooking. Today, a derivative of those leaves is the enzyme papain, which is available commercially, diluted in salt and sugar.

Several other plants, including the fig, the pineapple, and some fungi produce protein-digesting enzymes that can break down muscle and connective proteins in meat. There is, however, a problem with most of these tenderizers. They generally tenderize only the surface area of the meat on which they are applied, so they really need to be injected into the meat to help their penetration. These tenderizers do not commonly react until they reach a temperature of 140° to 175°F (60° to 79°C), so they do not help the raw tenderization of meat, but they do help during the cooking of a braise or a fine stew.

The Marinating of Game Meats

After the aging process, the flavor and tenderness of the game meat can be further enhanced by the use of marinating. However, the marinade should in no way mask or dramatically change the integrity of the natural flavor of the meat. There are many types of marinades, but the function of the marinade should be understood before deciding what to marinate and what to use. Consider the following:

- The marinade will tenderize the muscle fibers of certain meats.
- The marinade is intended to improve the flavor of the meat by penetrating the meat fiber.
- The marinade does preserve the meat for a small period of time.
- There are two basic types of marinade:
    - A cooked marinade, is generally used for larger pieces of meat and stays on the meat for a long period of time, up to 4 to 5 days.
    - A uncooked marinade is used for smaller pieces of meat, and is normally on the meat a shorter period of time, between 2 and 8 hours.
- The marinade generally contains some kind of oil that helps protect the meat during marinating.
- The ingredients used should be compatible with the meat that they are intended for.
- The ingredients can include herbs, spices, acid liquids, salts, fruits, alcohol and wines, flavored oils, and pungent vegetables.
- The meat needs to be turned regularly in the marinade.
- Any alcohol, including wine, used in a marinade should first be brought to a boil and chilled before use, because the raw alcohol tends to burn the surface of the meat.
- The marinade can be used as part of the cooking of the dish that it was used to marinade.
- The acids in a marinade that contains vinegar, citric juices, or other acidic liquids break down protein chains in meats, making them more tender. However, food should not sit in a marinade for too long.

RECIPE 8–1

COOKED MARINADE

Recipe Yield: To marinate a 6- to 8-pound (2.7- to 3.6-kg) piece of meat

> **TIP** This marinade can be used for 6 hours up to a few days, depending on the toughness and thickness of the meats.

MEASUREMENTS		INGREDIENTS
U.S.	**METRIC**	
¼ cup	59 mL	Olive oil
2 cups	473 mL	Mixed root vegetables: carrots, onions, shallots, garlic, fennel, celery, and any other of your choice, finely chopped
1 bunch	1 bunch	Mixed herbs of choice, including chives, thyme, and tarragon
¼ cup	59 mL	Mixed toasted whole spices of choice
1 cup	237 mL	Red wine vinegar
3 cups	710 mL	Red wine
1 cup	237 mL	Port

PREPARATION STEPS

1. Heat the oil in a pan and sauté the vegetables to a light golden brown.

2. Add the liquids, herbs, and spices and simmer for 30 minutes.

3. Cool completely before adding to the meat.

RECIPE 8-2

UNCOOKED MARINADE

Recipe Yield: Marinates 2 pounds (1 kg) of meat

> **TIP** This marinade can be used for steaks, chops, or any small cut of game meat that is around 1 inch in thickness for 3 to 4 hours, turning regularly.

MEASUREMENTS		INGREDIENTS
U.S.	**METRIC**	
1 each	1 each	Carrot, small dice
1 each	1 each	Celery stalk, small dice
4 each	4 each	Shallots, fine dice
1 each	1 each	Clove garlic, minced
2 each	2 each	Parsley stalks
1 each	1 each	Rosemary sprig with the stalk removed
1 each	1 each	Thyme sprig with stalk removed
4 each	4 each	Juniper berries, crushed
2 each	2 each	Star anise, roughly crushed
¼ cup	59 mL	Olive oil
¼ cup	59 mL	Red wine vinegar
1 tablespoon	15 mL	Salt

PREPARATION STEPS

1. Combine all the ingredients and let the mixture sit at room temperature for 1 hour before applying to the meat.

Adding Fat to the Game Meat

Because of the distinct lack of fat in most game animals, it becomes necessary for the chef to think about adding fats to the meats before cooking, during cooking, and after cooking in the form of an accompanying sauce. The addition of fat to the meats improves its overall flavor and gives the meat moisture to develop the overall eating qualities.

- Barding and larding are both of great importance when preparing game meat for cooking, because they are one of the only ways of introducing the much-needed fat into recipes for game meat. It is imperative to have a supply of very fine pork fat available, as well as a generous amount of fatty bacon. Thin slices and strips of both of these fats are used to either cover (bard) or insert (lard) into the tender meat of the game. This process is used to protect the meat from overbrowning and from getting dry during cooking. (See Chapter 7 for more on larding and barding.) Sometimes, certain herbs or vine leaves are put between and around the pork fat and the meat to give a distinctive flavor. Wrapping in bacon is suitable for smaller cuts of meat. Steaks, for example, can be wrapped in bacon before cooking, to allow the meat to baste naturally during the cooking process.

- Stuffing with a fatty product involves placing a very nice stuffing, containing a good quantity of fat, down the center of a very lean piece of well-marinated game meat and coating it with bacon or pork fat. This greatly enhances your chances of rendering the meat moist and flavorful.

- Basting while cooking with a good quality fat during all dry methods of cooking helps achieve a nice, moist result.

- Tumbling marinade and fats into game meats can be an excellent alternative method of introducing fats and flavor into game meats. Tumbling is a modern technique and challenges traditional methods; it can provide excellent results. Vacuum tumbling is a method of marinating meat. The tumbling massages the product, creating more relaxed meat, and then the vacuum causes the product to absorb more marinade, which results in a moister, easier to cook, and more tender product. Tumblers, or tumbler-massagers, provide mechanical agitation for the meat and the marinade solution. The meat and a solution or dry mix are placed into a revolving canister and allowed to tumble, ideally at 33° to 35°F (0.6° to 1.7°C). Game meats are ideal for this process because of their stronger connective tissue. Meats are tumbled until they pick up sufficient moisture; the length of time depends on the type of meat, the cut, and the tumbler speed. Putting a vacuum on a tumbler facilitates the uptake of liquid into the meat tissues. It further tenderizes the meat by expanding the meat muscle and breaking some of the muscle fibers.

FURRED GAME

The term **furred game** is generally given to a larger animal that has a fur-covered skin. Almost all large game meats fall into this category. Because the larger and smaller cuts are similar to veal, lamb, and goat, they are covered in Chapter 9.

This category can be split into two smaller classifications: hoofed and pawed. (In some countries, they are called furs and pelts.)

Venison

Venison is the meat of the red, fallow, or roe deer. These are ruminant (cud-chewing), even-toed, hoofed animals with antlers that are shed each year. Fallow deer are the most common, prized for their tenderness and flavored meat. All venison, whether wild or farm-raised, is low in fat and cholesterol, yet still high in protein.

In the wild, the animal has to work hard to sustain itself and, as a result, develops tough muscles. Wild animals also eat a varied diet, so the flavor of their meat has an intensity that most

FIGURE 8–2A Combining ingredients for marinade

FIGURE 8–2B Submerging the venison in the marinade

FIGURE 8–2C Applying weight to ensure that the meat remains covered, top to bottom

people refer to as "gamey." Most farm-raised deer live on natural, planted pastures and eat a stable diet. The meat should be dark and firm, with clear white fat. Because there is only a little fat on venison, the meat tends to be dry, so additional fat or liquid is added for cooking. Venison is a fine, delicately textured, and virtually fat-free meat, ideal for imparting fat and flavor through marinating (Figure 8–2).

Elk

American elk, or *wapiti,* is native to the northern part of the Western Hemisphere from southern Canada to northern Mexico. They have dark brown fur on the head and neck, and creamy gray fur on the back and flanks. A full-grown stag stands up to 4.9 feet (1.5 m) high at the shoulder, and weighs up to 750 pounds (340 kg).

Elk meat tastes like mild (almost sweet) beef, with only a very faint venison flavor. Elk can be substituted equally for venison in most standard venison recipes and not only has a very pleasant taste but is also very low in both cholesterol and fat (Figure 8–3).

Bison

Bison is the largest terrestrial animal in North America, where it is commonly referred to as *buffalo.* The bison is recognized by the hump over the front shoulders and the short, sharply pointed horns curving outward and up from the sides of the enormous head and slimmer hindquarters. Mature bulls of the North American bison are about $6\frac{1}{2}$ feet (2 m) tall at the hump and 9 to 12 feet (2.7 to 3.7 m) long. They weigh 1800 to 2400 pounds (850 to 1100 kg); the female being the smaller. The head, neck, forelegs, and front parts of the body have a thick coat of long, dark hair. The rear part of the body is covered with much shorter hair.

Bison has an unforgettable sweet flavor and a rich texture; it can be cooked similarly to beef (Figure 8–4). Bison is a highly nutrient, dense food because of the proportion of protein, fat, minerals, and fatty acids compared to caloric value. It also contains large amounts of iron, zinc, phosphorus, and several essential fatty acids. It is less fatty than beef, and because of that, has almost a cleaner taste.

FIGURE 8–3A A piece of elk, larded using a large needle

FIGURE 8–3B Stuffing the larded loin of elk

FIGURE 8–3C Barding the rolled meat

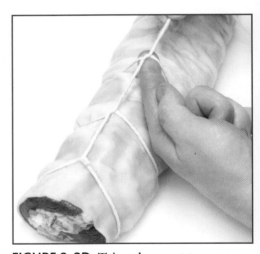

FIGURE 8–3D Tying the meat to secure the fat

FIGURE 8–4 Securing bacon around a steak of bison

PAWED

The term **pawed game** applies to animals that have pawed feet as opposed to hoofed feet. They are the smallest of the game meats and are not used as often as they probably should be; they have great flavor and are very versatile when cooking.

Rabbit

Rabbits were known to be used for meat as far back as 1500 BC. Their meat is white and has its own distinctive flavor, comparing favorably to the flavor of veal. It has many advantages to the chef, including fewer calories, a lower percentage of fat, and less cholesterol than chicken, beef, lamb, or pork. Because of these traits, rabbit is gaining popularity with health-conscious consumers and is becoming a popular menu item. When preparing fresh-caught rabbits, it is important to understand the technique of removing the skin.

1. Slice through the skin, up toward the inside of one of the legs, cutting around the paws and pulling the leg out of the skin (Figure 8–5A). Repeat for each leg.
2. Remove the tail and, using the tip of the knife to loosen the tissue between skin and flesh, draw the skin down toward the head (Figure 8–5B).
3. Cut around the front paws and continue to draw the skin over the head (Figure 8–5C).
4. Cut off the ears and trim the skin off the head. Now the skin should be completely removed and you can start to remove the gut.
5. Make an incision from vent to ribs, along the middle of the belly, and through the rib cage, making sure not to go so deep that you perforate the intestines.
6. Break through the skin of the diaphragm, separating the belly from the organs. Lift out the lungs, liver, heart, and intestines, and separate those you wish to use.
7. Wash the rabbit well, taking particular care with the cavity, and pat dry.

Once skinned, the rabbit can be cut into pieces for cooking as a chicken would be cut for sauté; however, one of the more popular yet difficult pieces to remove is the rack:

1. Remove the hind and fore legs and reserve for further use.
2. Remove the saddle from the carcass, leaving the rib cage intact.
3. Chine the rack with poultry scissors, splitting into two racks (Figure 8–6A).
4. Break the rib bones half way to their ends with a sharp blow of a heavy knife (Figure 8–6B).

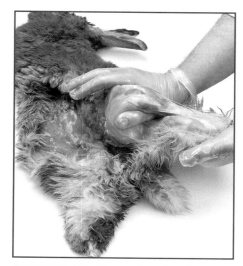

FIGURE 8–5A Removing the skin from rabbit legs

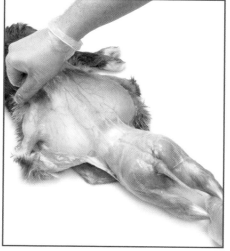

FIGURE 8–5B Pulling the loosened skin off the carcass

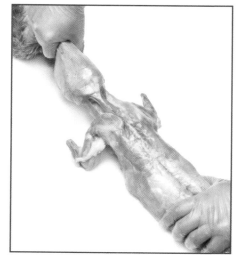

FIGURE 8–5C Removing the ears and the skin over the head

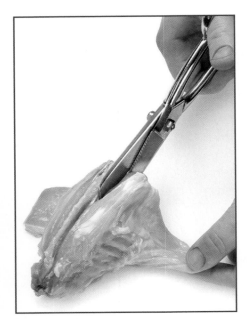

FIGURE 8–6A Removing the chine bone to create the racks

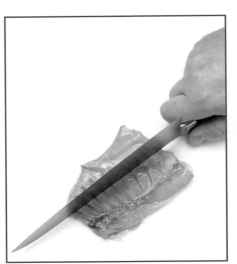

FIGURE 8–6B Breaking the rib bones evenly to assist in the cleaning of the rack

FIGURE 8–6C Pulling the flesh down to free the rack bones

5. Pull the excess ribs over the ones attached to the flesh, removing the connective tissue and flesh from between the bones to create the French look (Figure 8–6C).
6. Cut the excess off at the eye of the meat, leaving the rack.

Hare

The wild hare looks similar to a large rabbit and, although they are from the same family, they are very different. Hares are much larger in size and have a very dark, rich gamey flesh.

Hares are recognized as being good for eating if they have very tender ears when torn and when they have short, easily broken claws. Another sign is their harelip, which in young animals is not as defined as when they are older. They are at their best at the age of 7 to 8 months and weighing then about $2\frac{1}{2}$ to 5 pounds (1.1 to 2.3 kg). Very young hares (leverets) may be roasted whole, but for larger hares, the body alone is used and is known as a *saddle* or *baron* of hare. The young hares do not need to be hung, because they are generally very tender. After gutting, the older animals can be hung by their hind legs for 5 to 7 days to improve the flavor.

Hares should be bled as soon as possible after killing; the blood is then saved for further use in the sauces and stews that are particular only to hares. One of the unique cuts that can be removed from this animal is the baron, which is the name given to a saddle and a rack still affixed and cooked together.

To prepare a baron of hare:

1. Remove the hind legs and loins from the body.
2. Remove the bones from the loins, and the aitchbone from the legs.
3. Remove the sinew and silverskin from the back of the baron.
4. Stuff with a sausage and fruit dressing, and wrap the baron with backfat.
5. Tie securely with butcher's twine.

WEBBED

The only member of the **webbed game** family is the frog, which are caught wild in many countries of the world. There are many types of frog available for cooking, and they are much prized for their flavor. Frogs can be cooked in a host of different ways and appear on the menu as appetizers, salads, and entrées.

Frog

Frog species including the green frog (*Rana clamitans*), the leopard frog (*Rana pipiens*), and the pickerel frog (*Rana palustris*) are farmed or harvested from the wild for the table. Without doubt, the bullfrog (*Rana catesbeiana*) has the greatest yield of meat and potential for culinary development. The common bullfrog, sometimes referred to as the "giant frog" or "jumbo frog," is the largest native North American species, often reaching 8 inches (20 cm) in body length. The frog's legs are removed just below the pelvis and are normally bought skinless. When they are large, the legs can be stuffed and served as an appetizer or several may be served as an entrée.

To prepare stuffed frogs' legs:

1. Remove the meat from the foot end of the leg, peeling it back without breaking it and leaving it still attached to the bone (Figure 8–7A).
2. Beat lightly to form a flattened piece of meat (Figure 8–7B).
3. Fill with stuffing and roll around the thigh.
4. Wrap in caul fat and cook as desired (Figure 8–7C).

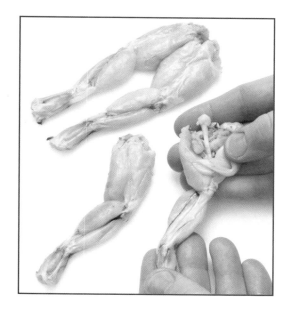

FIGURE 8–7A Loosening the meat around the frog's thighs

FIGURE 8–7B Flattening the meat into an even size before stuffing

FIGURE 8–7C Wrapping the stuffed leg in sheets of caul fat

Note: Table 8–1 merely illustrates recommendations as set forth by the USDA; however, the times and temperatures listed are not necessarily recommended by the authors or others represented in this text.

Cooking Game

The scope of this textbook lies within the realm of the garde manger kitchen. As such, there are several products that are butchered or otherwise preprepared by the garde manger chef and then distributed to other stations of the food service operation to be cooked. Table 8–1 reflects the suggested cooking times for game birds and meats by the U.S. Department of Agriculture (USDA). Chefs may choose to follow these guidelines as they deem appropriate.

Table 8–1   Game Cooking Chart (Times Approximate)

For tenderness and doneness, the USDA recommends cooking whole game birds to 180°F (82°C) as measured in the thigh using a food thermometer. Cook breast meat to 170°F (77°C). Ground meats and other cuts of game meat should reach 160°F (71°C). Approximate cooking times for use in meal planning are given on the chart below.

TYPE OF GAME	ROAST	GRILL/FRY DIRECT HEAT	SMOKE/ INDIRECT HEAT	BRAISE/STEW IN LIQUID; COVERED
Game Birds				
Whole bird, 4 to 6 lb (1.8 to 2.7 kg) (Do not stuff)	30 to 35 min/lb (450 g)	Not preferred	2½ hours	Not preferred
Breast or parts	350°F (177°C) 1 to 1¼ hours	20 to 40 min	2 hours	60 to 75 min
Whole small birds	350°F (177°C) 45 min	30 min	1 to 1½ hours	45 to 60 min
Game Animals				
Rib roast, bone in 4 to 6 lb (1.8 to 2.7 kg)	325°F (163°C) 27 to 30 min/lb (450 g)	Not recommended	Not recommended	Not recommended
Rib roast, boneless, rolled 4 to 6 lb (1.8 to 2.7 kg)	32 to 38 min/lb (450 g)			
Chuck roast, brisket 3 to 4 lb (1.4 to 1.8 kg)	Not recommended	Not recommended	Several hours	325°F (163°C) 2 to 3 hours
Round or rump roast 2½ to 4 lb (1.1 to 1.8 kg)	325°F (163°C) 35 to 40 min/lb (450 g)	18 to 25 min/lb	2½ to 3 hours	325°F (163°C) 2 to 3 hours
Whole leg (boar, deer) 6 to 8 lb	375°F (191°C) 2 hours	Not recommended	3 to 4 hours	Not recommended
Tenderloin Whole, 4 to 6 lb (1.8 to 2.7 kg) Half, 2 to 3 lb (1 to 1.4 kg)	425°F (218°C) 45 to 60 min total	12 to 15 min/side	Not recommended	Not recommended
Tenderloin Half, 2 to 3 lb (1 to 1.4 kg)		10 to 12 min/side		
Steaks, ¾-inch (2-cm) thick	Not recommended	6 to 7 min/side	Not recommended	Not recommended
Ground meat patties	Not recommended	6 to 8 min/side	Not recommended	Not applicable
Meat loaf, 1 to 2 lb (.45 to 2.7 kg)	350°F (177°C) 60 to 90 min	Not recommended	Not recommended	Not applicable
Stew or shank cross cuts 1 to 1½ inches (2.5 to 3.8 cm) thick	Not recommended	Not recommended	Not recommended	Cover with liquid; simmer 2 to 3 hours
Ribs, 4 inches (10 cm)	375°F (191°C) 20 min	8 to 10 min/side	Not recommended	Parboil 1 hour; then grill or roast

Professional Profile

BIOGRAPHICAL INFORMATION

Name: Martin Couttie

Place of Birth: Paisley, Scotland

Recipe Provided: Venison Mincemeat

CULINARY EDUCATION AND TRAINING HIGHLIGHTS

As apprentices have traditionally done over centuries of European occupational instruction, Martin Couttie finished normal school at age 15 to begin his training in the catering and tourism industry. He started working in a local hotel in his hometown of Paisley and attended Reid Kerr College, before moving to Ayrshire and the stately Marine Highland Hotel overlooking the Royal Troon championship golf course and the Firth of Clyde. In conjunction with working full time at the hotel, he also attended classes at the Glasgow College of Food Technology. Working through all of the stations in the kitchen, including larder, sauce, vegetables, and pastry, Chef Couttie rose to the position of executive sous chef overseeing the daily operations of three fine dining restaurants and banquets. In the early 1990s, he left the Marine Hotel to take the position of club manager/executive chef of the exclusive private links-style golf course, Prestwick St. Nicholas Golf Club. He left Prestwick and opened the Southern Gailes Golf Course on the Ayrshire coast. From there, he started his own catering business. Chef Couttie has shared his expertise as an adjunct chef-instructor at several Scottish culinary schools, and coached many of his apprentices to success in culinary competitions. One of Chef Couttie's trainees became the "British Young Chef of the Year."

ADVICE TO A JOURNEYMAN CHEF

Learn everything you can about the preparation of meats, poultry, game, fish, and seafood. Experiment with flavors using herbs, spices, and marinades. Never stop learning—read, listen, discuss and, if possible, travel. Be brave and do not be afraid to try out your own ideas. Cookery, at its best, is an art form and as with any art form requires passion. Conveying your passion to your audience is extremely fulfilling. Working with a team or likeminded craftsmen is also a wonderful experience. Be sure to pass on your passion and knowledge to young people entering the industry as others did for you.

Martin Couttie

RECIPE 8–3

CHEF MARTIN COUTTIE'S VENISON MINCEMEAT

Recipe Yield: 8 portions

TIP	This mixture is used by both the pastry and garde manger stations to make pies or tarts.

MEASUREMENTS		INGREDIENTS
U.S.	**METRIC**	
4 pounds	1.89 kg	Venison, boneless
1 pound	473 g	Beef suet, chopped
½ cup	118 ml	Candied orange peel
½ cup	118 ml	Lemon juice
3 tablespoons	45 ml	Orange rind, grated
½ teaspoon	3 ml	Black pepper, ground
4 teaspoons	20 ml	Ground allspice
4 teaspoons	20 ml	Cinnamon
2 teaspoons	10 ml	Ground cloves
½ cup	118 ml	Candied lemon peel
1 cup	237 ml	Citron, chopped
1 cup	237 ml	Orange juice
4 teaspoons	20 ml	Lemon rind, grated
1 teaspoon	5 ml	Salt
4 cups	948 ml	Sugar
2 teaspoons	10 ml	Mace
2 teaspoons	10 ml	Nutmeg
2 quarts	1.89 L	Apple cider
3 cups	711 ml	Raisins
2 quarts	1.89 L	Sour cherries, pitted
3 pounds	1.42 kg	Apples, peeled, cored, and chopped
1 pint	473 ml	Brandy

PREPARATION STEPS:

1. Place the venison in a large saucepan or small stockpot. Add enough water to cover the meat by 1 inch (2.5 cm).

2. Bring to a boil over high heat, then reduce to simmer, partially covered for 2 hours, or until meat is tender.

3. Drain the venison, patting dry with side towels. Cut into medium cubes.

4. Mince the meat by putting through the medium die of the meat grinder (Chapter 13).

5. Transfer the venison to a large saucepan. Add all of the remaining ingredients, except the brandy. Bring to a boil over high heat, then lower the heat and simmer partially covered for 1½ hours, stirring frequently.

6. Remove from the heat and stir in the brandy; immediately ladle into hot, sterilized jars and process under boiling water for 10 minutes (Chapter 12).

7. Store at room temperature for at least 2 weeks before using.

1. Explain how the fat in meat affects its flavor.

2. How does aging affect meat?

3. How does the correct procedure for slaughtering an animal affect its eating quality?

4. What is a marinade?

5. What does marinating do to meat?

6. What affects the quality of game meats?

7. Describe the term *barding*.

8. Describe the term *larding*.

9. What is vacuum tumbling?

10. Explain how wild game animals differ from domesticated animals.

A. **Group Discussion**

In small groups, discuss the care that needs to be taken when dressing a piece of wild game in the field.

B. **Research Project**

In groups, research how much game is consumed in your local area by reporting on what is the most popular animal and the most common cooking methods.

C. **Group Activity**

Compare the eating qualities of a piece of meat that has had fat and marinade added as compared to one that has not.

D. **Individual Activity**

Write a four-course wild game dinner menu including appetizer, soup, salad, and entrée. Select appropriate wines to accompany the meal.

CHAPTER 9

Meats

After reading this chapter, you should be able to

- describe the process of dry and wet aging.
- identify the location of the primal cuts.
- describe the common cuts.
- identify the grades of beef.
- explain what loins are and where they are from.
- explain what steaks are and where they come from.
- explain what cutlets are and where they come from.
- explain what racks are and where they come from.
- explain what chops are and where they come from.
- explain what tenderloins are and where they come from.
- explain what crowns are and where they come from.
- demonstrate how to clean a beef tenderloin.
- demonstrate how to bone a leg of veal.
- demonstrate how to make an escalope.
- identify the specialty cuts of pork.
- demonstrate how to prepare a rosette.
- demonstrate how to bone and stuff a baron of goat.

chops

common cuts

crown

cutlets

dry aging

loins

loin cuts *or* steaks

primal cuts

racks

small cuts

tenderloins

veal

wet aging

The way you cut your meat reflects the way you live.

CONFUCIUS

The Commonality of Meat

In contrast to Chapter 8, which discusses game meats, this chapter is concerned with those meats that are from domesticated livestock, that is, farm animals raised for use and consumption. Their ease of domestication was a key factor in their development as a chief source of protein. These meats are derived from three species of animals: *bovine, ovine,* and *porcine.* They are all naturally different in size, shape, and taste, yet they share many common characteristics that are useful to know in their butchering and preparation.

Dry and Wet Aging

The **dry aging** of meat is not a new concept. In fact, it is what was done with all meat before the vacuum packing of meats at packing plants. The main reason that dry aging is not done universally any more is because of the costs involved:

- There is a certain amount of weight loss during the aging process caused by the evaporation of water from the carcass, thus driving up the price even more.
- Time, space, labor, and refrigeration are involved.
- The best meat for aging is the meat that contains the most marbling or fat distribution. Because only the highest grades have the necessary marbling, this becomes a hindrance because of the expensive cuts of meat.

The aging of meat is simply the holding of a carcass, or the primal cuts, at refrigerated temperatures to allow the meat to naturally improve in flavor and tenderness. The muscle of meat undergoes progressive changes after slaughter that affect tenderness. It goes into rigor—a shortening and stiffening process—that can last between a few hours and 1 or 2 days. While muscle is undergoing changes associated with tenderness, chemical breakdown of certain muscle and fat constituents occurs, resulting in a more intense flavor and aroma that are desirable to most consumers. However, undesirable flavors and aromas can develop during aging, mainly due to the effects of microbial growth, rancidity of the fat, and absorption of other odors present in a walk-in cooler.

Temperature, relative humidity, air movement, and general sanitation of the aging room are essential considerations in successfully aging beef. The temperature of the aging room should be maintained at approximately 34° to 36°F (1.1° to 2.2°C) with a relative humidity at 85 to 90 percent and airflow of 15 to 20 linear feet per minute at the surface of the product. Aging allows for natural enzymes to break down the hard connective tissue in meats and for water to evaporate away, concentrating the flavor.

The **wet aging** of meat is the less expensive alternative to dry aging and involves the meat being shipped from packing plants to butchers in vacuum packaging. Butchers can set this packed meat aside in their refrigerators and allow it to age. Because the meat is packed in its own juices, the enzymes break down the connective tissues and make it tender. However, because no liquid is lost from the package, the concentration of flavor does not take place, leaving a relatively bland product.

Fabrication Charts

The fabrication charts are useful in locating the **primal cuts** in the carcasses, and showing the similarities in the animal structure (Figure 9–1). Although the obvious differences are the size and weight, the position of the cuts and generally what they are used for are similar at least in the cooking method and classification. As we note the animal's structure, as chefs, we should evaluate it according to how it needs to be cooked. Understanding what to do with the meat should be of paramount importance to the chef. And evaluating the animal's structure, by dividing it between dry and moist methods of cooking, would make sense.

The charts show the primal cuts equated to their most favorable cooking method. There are always going to be exceptions, but if you understand the coloration between the animals and how they are cooked, you are already at an advantage.

The smaller cuts, from the primal cuts, are focused on here, because the correct cutting and preparation of these are most crucial to the cooking method and presentation.

Note: It is not advisable to age meat that has been under vacuum for a long time, because it will result in dangerous spoiling in the meat.

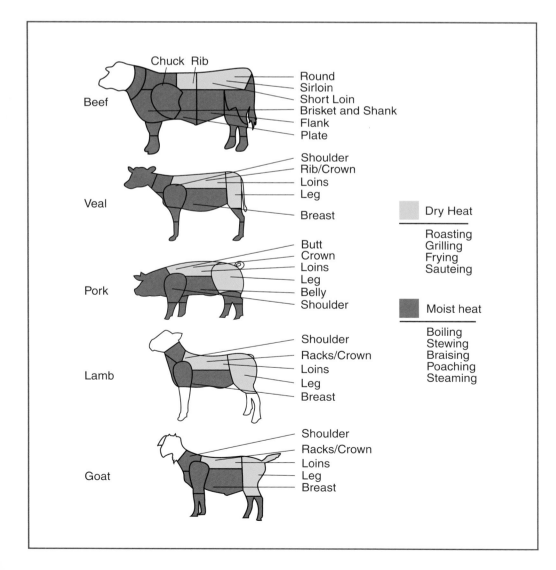

FIGURE 9–1 Comparative chart indicating the recommended method of cooking, by location in the animal's body

FIGURE 9–2 Left to right: loin of beef, loin of pork, loin of veal, loin of lamb, loin of goat

Common Cuts

The **common cuts** are the cuts that are available from all the animals. They are taken from the exact same place in the animals and are cut from the primal cuts in the same way. The only real difference is the size of the cut from one larger animal to one much smaller. They are also very similar in the way they would be cooked; a dry method cut on one animal generally translates to the same cut being cooked in the same way on a different animal.

Remove the **loins** by boning and cleaning the racks and saddles of the animal (Figure 9–2). They are cleaned of all sinew and gristle, then used whole for roasting. They can be replaced back onto the bone and tied in their original position before cooking, often with aromatic herbs placed between the bone and the meat.

The **tenderloins** are removed from the opposite side from where the loins were located, within the saddle (Figure 9–3). They are the only piece of meat that is located within the bone structure of the animal.

The **loin cuts** or **steaks** are achieved by removing the loin from the saddle region of the animals (Figure 9–4). This is then sliced through the eye of the meat, leaving a neat layer of fat still attached.

All **chops** are made by cutting through the loin and the tenderloin (Figure 9–5). The cut is made through the bone separating both pieces, such that each portion has meat on either side of a bone shaped like a T.

The **racks** are cut by chining the spine out of the animal, cutting through the ribs on either side, and leaving the rib bones attached to the eye of the loin. These ribs are then shortened and cleaned to reveal the classic rack appearance of exposed multiple ribs attached to an eye of loin (Figure 9–6).

The **cutlets** are cut from these cleaned racks by slicing into one, two or three pieces. They are slightly beaten to create a better shape and for even cooking (Figure 9–7).

When preparing a **crown** of meat, the racks are tied together back to back and pulled into a crown shape (Figure 9–8). Loosening the sinews in between the cutlets aids this process. Beef tends to be too large for this cut.

Note: In the larger animals, the tenderloins are big enough to be further processed.

FIGURE 9–3 From the top: The common tenderloins of beef, veal, pork, goat, and lamb. Note the similarity and differences in size and shape.

FIGURE 9–4 Clockwise from the top: New York strip, pork medallion, goat noisette, lamb noisette, and veal grenadine. Note: Although the names change, the cuts are removed from the same areas of the animals.

FIGURE 9–5A Clockwise from left: Barnsley lamb chop, lamb chop, pork chop, goat chop

FIGURE 9–5B Clockwise from left: Porterhouse, T-bone, veal chop

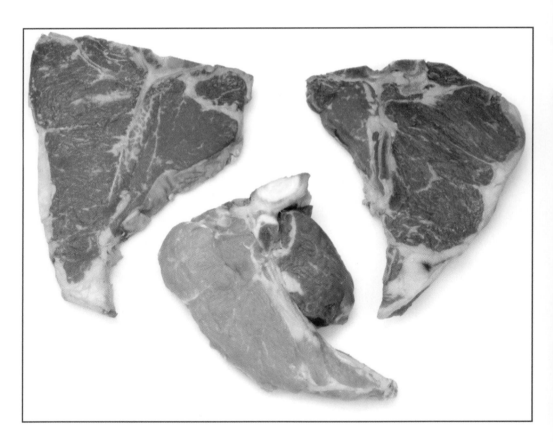

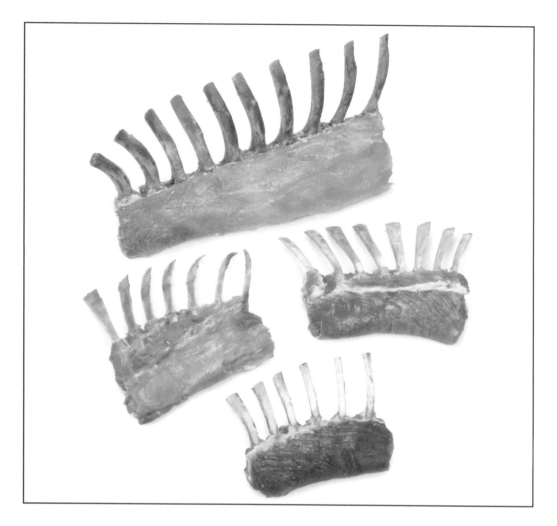

FIGURE 9–6 Clockwise from top: Rack of veal, rack of pork, rack of lamb, and rack of goat

FIGURE 9–7 Clockwise from top left: Veal cutlet (rib steak), beef cutlet (cowboy steak), pork cutlet, goat cutlet, lamb cutlet

FIGURE 9–8 Clockwise from top left: Crown of pork, crown of goat, crown of lamb, crown of veal

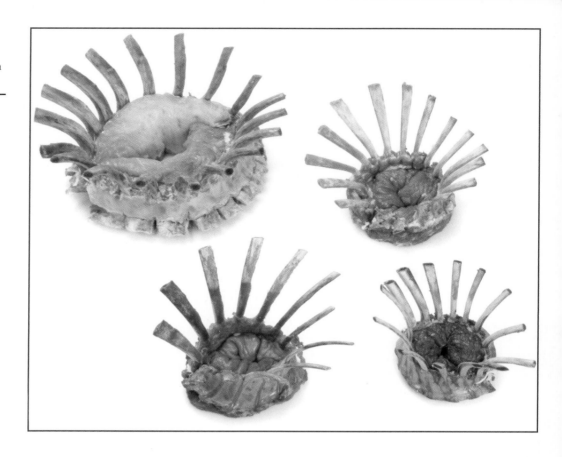

FIGURE 9–8 Clockwise from top left: Crown of pork, crown of goat, crown of lamb, crown of veal

Specialty Small Cuts

The **small cuts** of meat are those that are most often broken down by the modern chef. Rarely does a menu support all the cuts available on a whole lamb or pig unless, they are being bought whole as suckling pigs or baby lambs. It makes sense, therefore, to discuss the smaller cuts that are taken from pieces of the animal rather than from the whole animal.

The carcasses are split into primal cuts, which are the basic division between the muscles in the carcass. The subprimal cuts are a further breakdown, which are then cut into small cuts or fabricated cuts. Specific cuts are described under their different meat headings.

BEEF

Note: Maturity and marbling are the major considerations in beef quality grading.

Beef is the meat of domesticated cattle reared to supply meat to the commercial market. Most beef consumed today comes from steers that are castrated as calves and reared as beef cattle.

The marbling in the meat of beef is the fine flecks of fat that are evenly distributed amongst the lean meat. Marbling is one of the major considerations when grading beef for quality (Figure 9–9). It is evaluated visually in the rib eye muscle between the twelfth and thirteenth ribs, and although it contributes only slightly to meat tenderness, it does contribute to juiciness and flavor. The other consideration is the age or maturity of the beef. The U.S. Department of Agriculture (USDA) grades beef into eight standards (Table 9–1): (1) prime, (2) choice, (3) select, (4) standard, (5) commercial, (6) utility, (7) cutter, and (8) canner.

The tenderloin of beef is the most sought after part of the animal and is the only piece of tender meat located on the inside of the bone structure. Breaking down tenderloin can be beneficial to the chef because it yields some trim that, if cleaned well, can be used.

Table 9-1   Quality and Maturity of Beef

	9 TO 30 MONTHS	30 TO 42 MONTHS	42 TO 72 MONTHS	72 MONTH AND ABOVE
High marbling	Prime	Prime	Commercial	Commercial
Medium high	Choice	Choice	Commercial	Commercial
Slight marbling	Select	Commercial	Utility	Utility
Few traces	Standard	Utility	Cutter	Cutter
No marbling	Standard	Cutter	Canner	Canner

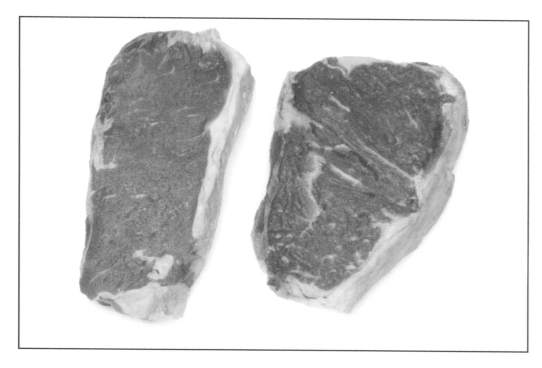

FIGURE 9–9 Comparison of a well-marbled steak (right) versus a leaner cut

Breaking Down Tenderloin

The following steps are used to trim down a tenderloin:

1. Remove the fat surrounding the tenderloin, being careful not to cut into the meat.
2. Locate the chain or fillet string that runs down parallel to the tenderloin and remove it all the way down to the thickest part of the meat.
3. As you remove it from that end make sure you feel around that area and locate exactly how it attaches itself.
4. Remove it without taking large chunks of the meat off.
5. Clean the silverskin of the whole tenderloin and clean away any fat from the other side (Figure 9–10A).
6. The meat can now be cut into three distinct areas: the head, the heart, and the tail (Figure 9–10B).
7. It can now be cut into the smaller cuts.

VEAL

Veal, as we know it today, has its origins in Europe (mostly in The Netherlands, Spain, France, and Italy) where the young animals were fed on skim milk, whey, and fat. To remain efficient milk producers, all dairy cows must give birth once a year. Their female calves, or heifers, are raised as dairy cows, while the male calves (bulls) are marketed to veal farmers to produce the

FIGURE 9–10A
Removing the chain meat and the silverskin from a beef tenderloin

FIGURE 9–10B Separating the head, heart, and tail meats of the tenderloin

veal we use today. Most of the veal is slaughtered between 8 and 16 weeks of age. The meat slaughtered after 5 months is called *calf meat* and does not qualify as veal. The following are the three grades of veal:

■ Special-fed veal consists of calves that are fed a nutritionally complete milk supplement until they reach 18 to 20 weeks of age and typically weigh from 400 to 450 pounds (181 to 204 kg). Their meat is ivory or creamy pink in color with a firm velvety texture.

■ Bob veal calves are solely fed on milk and their meat is a light pink color and has a soft texture.

■ Grain-fed veal calves are initially fed milk then receive a diet of grain, hay, and nutrition formulas. These calves are usually marketed at 5 to 6 months of age and weigh from 450 to 600 pounds (204 to 272 kg). Their meat tends to be darker in color and has additional marbling with some visible fat.

The five basic grades for veal, as determined by the USDA, are (1) prime, (2) choice, (3) good, (4) standard, and (5) utility. More than 93 percent of all graded veal is of prime or choice

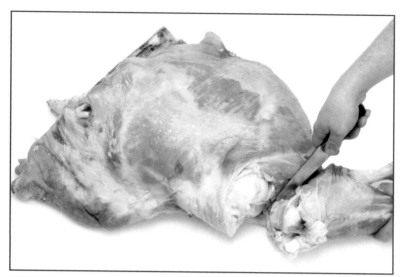

FIGURE 9–11A Removing the shank from the leg at the knee joint

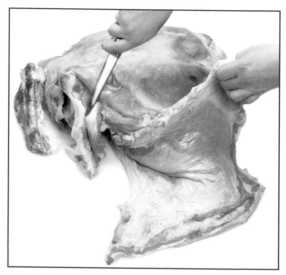

FIGURE 9–11B Removing the flank and butt tenderloin

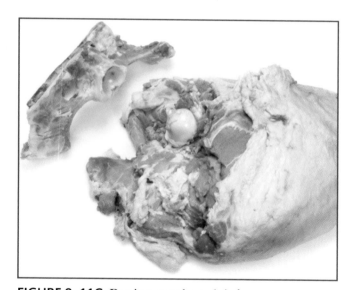

FIGURE 9–11C Boning out the pelvic bone

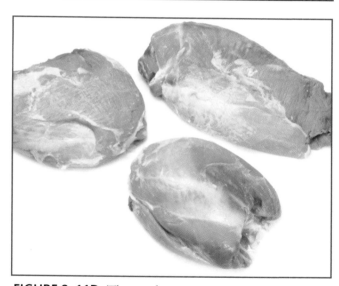

FIGURE 9–11D The top butt

quality. There is no need to hang or age veal because the meat is very tender to begin with and there will be no marked improvement in texture or flavor.

One of the most beneficial pieces of the animal is the leg. Learning how to break it down can be financially rewarding for the chef because the meat yielded can be very versatile in its use.

1. Remove the shank from the leg at the knee joint by cutting through the ball and socket joint (Figure 9–11A).
2. Trim the fat from around the pelvic bone, exposing the butt tenderloin and removing the flank.
3. Remove the butt tenderloin (Figure 9–11B).
4. Bone out the pelvic bone and turn the leg with the inside facing up (Figure 9–11C).
5. Find the natural seam and remove the top butt without cutting any flesh (Figure 9–11D).
6. Trim around the exposed bone at the shank end, removing the excess shank meat.
7. Cut under the bone, removing it cleanly from the meat.
8. Seam out the knuckle meat in one piece and then separate the inside round from the eye round.

PORK

Pork is the meat of hogs that are usually slaughtered before they are 1 year old. At this young age, their lean meat is very tender with a delicate flavor and does not require aging or hanging. More than two thirds of the pork marketed in the United States is cured in some way to produce charcuterie items, which remain a popular menu item. Pork is not graded with USDA quality grades because it is generally produced from young animals that have been bred and fed to produce more uniformly tender meat. The appearance is an important guide in buying fresh pork. Look for cuts with a relatively small amount of fat over the outside and with meat that is firm and grayish pink in color. The meat should have a small amount of marbling, which is a sign of the best flavor and tenderness.

The tenderloin of pork can be further broken down into small tournedos or flattened into escalope:

1. Trim all fat and sinew off the tenderloin.
2. Remove the chain or fillet string, and remove all silverskin by stripping with a sharp, thin-bladed knife.
3. Cut the tenderloin against the grain into eight pieces of equal weight (Figure 9–12).
4. Wet the plastic to ensure the meat does not stick and place the meat between the sheets of plastic. Flatten the meat with a mallet, striking it several times with glancing blows in different directions, to stretch the meat (Figure 9–13). Cover tightly for service.

LAMB

Lamb is meat from sheep slaughtered when less than 1 year old. Most are brought to market at about 6 to 8 months old. If "Spring Lamb" is written on a meat label, it means the lamb was produced between March and October, although lamb is available all the time. A lamb weighs about 120 pounds (54.4 kg) and yields approximately 72 pounds (32.6 kg) of retail lamb cuts, which include bone and fat.

Lamb is usually tender because it is from animals younger than 1 year old. Look for good marbling, which is the white flecks of fat within the meat muscle, which is fine textured and firm. The color of the meat should be pink and the fat should be firm, white, and not too thick.

There are two grades for lamb found at the retail level: prime and choice. The prime grade is extremely tender and very high in flavor, and has very even, clear marbling, which makes the cooked product moist and succulent. The choice grade has slightly less marbling than prime, but it is still of very high quality.

Note: Mutton is the meat of sheep slaughtered after its first year. The meat is generally tougher and more strongly flavored than lamb.

FIGURE 9–12 When cutting tenderloins for escalope, the length increases as the diameter of tenderloin gets smaller. *Note:* The dimension of the cut must change in order to maintain the same weight.

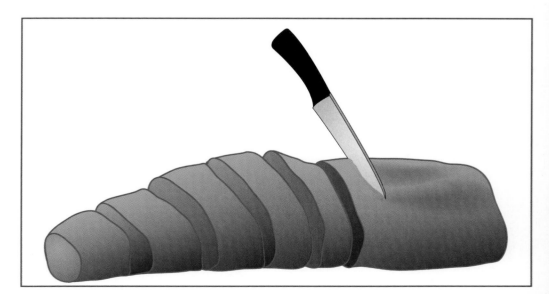

FIGURE 9–13A
Wetting the plastic
wrap to ensure that
the meat does not
stick

FIGURE 9–13B Placing the meat between durable
sheets of strong plastic wrap

FIGURE 9–13C Using a mallet to flatten the meat

Further aging of lamb is possible by following the guidelines under dry aging; it can be hung up to 15 days. Great care must be taken to ensure that the animal was held under the correct conditions before proceeding. When breaking down a saddle of lamb, one of the more unusual cuts is the rosette.

The following steps are used to prepare rosette of lamb:

1. Trim a saddle of lamb removing the loin and the fillet (Figure 9–14A).
2. Clean the fillet and remove the eye of meat from the loin, cleaning it of all silverskin.
3. Trim the fat cap of all sinew and flatten it with a mallet evenly and trim it into a rectangle (Figure 9–14B).
4. Place the fillet and the eye meat into the fat and wrap it tightly (Figure 9–14C).
5. Tie individual sections with twine and cut into rosettes (Figure 9–14D).

Lamb carcasses dry aging in cooler. Note the inspection stamps imprinted on the carcasses.

GOAT

Other than sheep and dogs, goat is thought to have been the earliest animal domesticated. Cave art from 10,000 to 20,000 years ago indicates that goats were common and played an important part in life. At the present time, goats provide the principal source of animal protein in many North African and Middle Eastern countries. Goat is also important in the Caribbean, Southeast

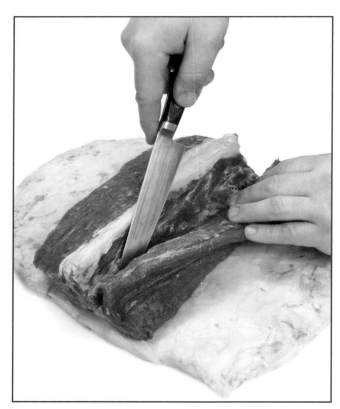

FIGURE 9–14A Removing the eye of the loin and the tenderloin from the fat

FIGURE 9–14B Flattening the fat to an even thickness

FIGURE 9–14C Rolling the fat around the meats

FIGURE 9–14D Cutting and tying the rosette

Asia, and other developing tropical countries. Three-fourths of all the goats in the world are located in the underdeveloped regions of the world. This is mainly due to the fact that they can survive in those harsh climates and can be sustained on very little food, tending to forage and survive on what they can find. Additionally, goat is not highly produced in the more economically developed regions of the world.

Goats younger than 1 year of age are called *kids* and are often slaughtered when 3 to 5 months of age and weighing from 25 to 50 pounds (11.3 to 22.7 kg). Kids do not store much body fat until they are about 1 year of age. Many goats are older and heavier when marketed, but most, except aged cull goats, are slaughtered when younger than a year of age. The meat of older goats is darker and less tender, but more juicy and flavorful than kid meat; it is usually served stewed. The meat from male goats is lighter in color and lower in fat than that of females; however, the meat from females is the more desirable for steaks and chops because it is more tender.

Goat can be hung to age for up to 7 days following the correct guidelines for dry aging. Great care must be taken to ensure that the animal was held under the correct conditions before proceeding. The saddle of a goat is ideal for stuffing, tying, and roasting or spit roasting.

The following steps are used to prepare a goat roast:

1. Remove the legs and shoulders from the animal, leaving the belly and ribs attached to the rack and saddle (Figure 9–15A).
2. Completely bone the belly, rack, tenderloins, and saddle, leaving the two halves connected (Figure 9–15B).
3. Remove all sinew and cartilage from the meat, laying it out meat side down.
4. Clean the skin off the outside, and clean silverskin off the fillets.
5. Pound the belly with a mallet to evenly distribute the meat (Figure 9–15C).
6. Season with appropriate seasonings, and lay the stuffing side-by-side with the tenderloins, down the center (Figure 9–15D).
7. Roll up the belly meat and secure with string to form an even shape (Figure 9–15E).

Tying meat is one of the basic techniques that become essential to a chef when working with a lot of meat, poultry, or game and must be mastered quickly. Tying helps the meat form the shape that is the most beneficial for carving and presenting the finished cooked product.

FIGURE 9–15A Removing the shoulders and the hind legs from the goat

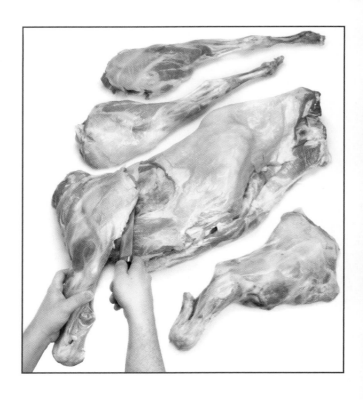

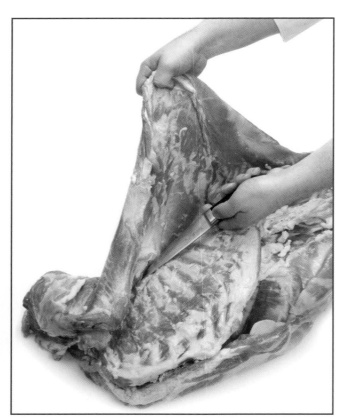

FIGURE 9–15B Boning the saddle and the loins

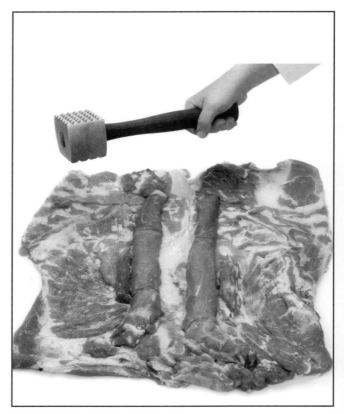

FIGURE 9–15C Flattening the belly meats

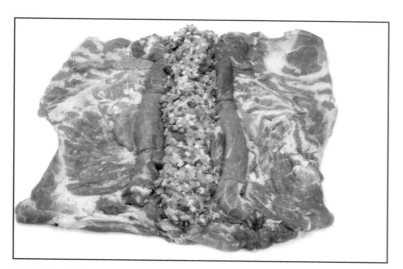

FIGURE 9–15D Laying the stuffing and the tenderloins into the cavity

FIGURE 9–15E Tying the roast

The following steps are used to tie a roast:

1. Wrap the loose end of the twine around the end of the goat roast (Figure 9–16A).
2. Create a loop and slide it down the roast to about 2 inches (5 cm) from the twine that you tied in a knot, and pull it tight until it ties the roast together where it is.
3. Make another loop, sliding it down the meat until it is 2 inches (5 cm) from the last loop and tighten it. Continue this way until the roast is completely tied (Figure 9–16B).

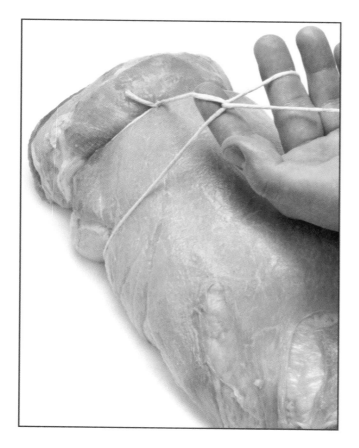

FIGURE 9–16A Wrapping the twine around and forming a slipknot

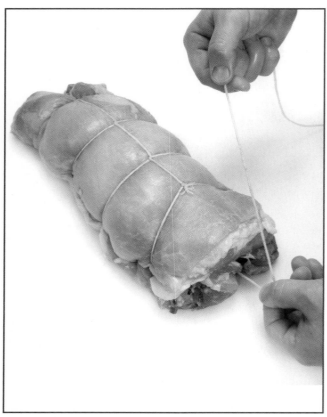

FIGURE 9–16B Looping the twine along the length of the loin, tightening as you go, to ensure an even shape

FIGURE 9–16C Securing the end of the twine around the final loop

4. At the end, turn the roast over and cut the twine long enough to slip it under each loop all the way back along the roast to where you started.
5. Turn over the roast and tie the twine to the first loop on the other side (Figure 9–16C).

Professional Profile

BIOGRAPHICAL INFORMATION

Name: Nabhojit Ghosh

Place of Birth: New Delhi, India

Recipe Provided: Seekh Kebab

CULINARY EDUCATION AND TRAINING HIGHLIGHTS

Upon graduating from high school in Calcutta, Chef Ghosh immediately began his education and training for the hospitality industry. After completing an extensive four-year program at Kolkata's Institute of Hotel Management and Catering Technology in 1985, he went onto the Oberoi School of Hotel and Management, specializing in chef training. (The school is run by the Oberoi Hotels Group of New Delhi to train and staff their large chain of hotels.) After graduating in 1987, Chef Ghosh worked as a demi-chef de Partie in the various restaurants of Calcutta's Oberoi Grand Hotel. He left Oberoi in 1993 to work for one year with Carnival Cruise Lines, out of Miami, Florida, traveling between the exotic islands of the Caribbean. Returning to the Oberoi group in 1994, Chef Ghosh became a senior sous chef based in Chennai, in southern India. In 1995, he opened Oberoi Airport Services in Calcutta and in 1996, he became a chef-instructor at the Oberoi Center of Learning and Development, located in New Delhi. Chef Ghosh continued his culinary journey in 1998, when he moved to Nepal, Kathmandu to open a Radisson Hotel as the executive chef. He now works for Taj Leisure Hotels, serving as the Executive Chef at the Taj Malabar on Willingdon Island near Cochin, India, a position he has held since 2003.

ADVICE TO A JOURNEYMAN CHEF

You must have a passion for cooking. This feeling will lead a person to everywhere he or she needs to be. Eventually, however, you must be particular in the job you choose. It must allow you the opportunity for creativity. This business requires a lot of tenacity. I am not time-bound; I love my work. But I must be physically and mentally capable for the work, for I will always stay until my last customer is served and satisfied.

Nabhojit Ghosh

RECIPE 9-1

CHEF NABHOJIT GHOSH'S SEEKH KEBAB

Recipe Yield: 35 ounces
(1 kg)

MEASUREMENTS		INGREDIENTS
U.S.	**METRIC**	
2.2 pounds	1 kg	Beef or lamb, finely ground
4 to 5 each	4 to 5 each	Scallions, minced
3 tablespoons	45 mL	Cilantro, minced
3 to 4 each	3 to 4 each	Green chilies, finely chop
1 teaspoon	5 mL	Red chili powder
2 teaspoons	10 mL	Dry pomegranate seeds, crushed
2 tablespoons	30 mL	Coriander seeds, crushed
2 teaspoons	10 mL	Garam masala
2 teaspoons	10 mL	Chat masala
2 teaspoons	10 mL	Kosher salt

PREPARATION STEPS:

1. Put ground meat in a bowl; add all ingredients and mix well. Make sure all ingredients are well blended with the minced meat.

2. Cover the meat and refrigerate for a few hours.

3. Light a charcoal grill and wait for a layer of ash to develop on top. (This reduces the temperature of the fire.)

4. Break handfuls of meat mixture and mold onto one end of separate metal kebab skewers. Put the skewers on a grill, over the hot coals.

5. Grill until brown on all sides, making sure to turn each side occasionally.

6. Serve with naan (bread) and chutney.

1. What is dry aging when referring to meat?

2. What is wet aging when referring to meat?

3. What are the common cuts of meat?

4. Draw a chart of the grades of beef.

5. What are steaks?

6. What are cutlets?

7. What are chops?

8. Where are the tenderloins located?

9. How do you prepare a crown roast?

10. What is a rosette?

A. **Group Discussion**

In small groups, discuss the pros and cons of dry aging meats for your own restaurant, bearing in mind the time and storage issues.

B. **Research Project**

In small groups, research 10 local white tablecloth restaurants to establish how much meat is being fabricated on site compared with how much is bought already cut and ready to cook.

C. **Group Activity**

In small groups, report to the class on some more meats that are commonplace in countries other than your own that may seem very unusual to you.

D. **Individual Activity**

Draw the skeletal structure of a shoulder of lamb to be used as a visual aid when boning.

CHAPTER 10

Fish and Shellfish

© 2005 CORBIS

After reading this chapter, you should be able to

■ explain the classification of fish and shellfish.

■ show how to select good fish for the kitchen.

■ describe how to handle fresh fish and shellfish.

■ eviscerate, clean, skin, and fillet fish.

■ clean and prepare shellfish for cooking.

■ prepare a variety of fish cuts.

Fish is the only food that is considered spoiled once it smells like what it is.

P.J. O'ROURKE

Market Forms of Fresh Fish and Shellfish

Fish and shellfish may be purchased in a variety of standard forms depending on the needs of the chef. Although it is easiest to identify the age and overall quality of seafood when it is in its whole and natural state, many times purveyors will process their products for a variety of reasons, including the preferences of their customers (Figures 10–1 and 10–2). The following forms are the most common and can be fabricated by either the fishmonger or garde manger chef.

■ Whole and Intact. The fish is literally untouched, with nothing removed. The gut is intact. It is in the same condition as when it was first landed (Figure 10–3).

■ Whole Drawn. The viscera or gut has been removed and the stomach cavity has been cleaned (Figure 10–4).

FIGURE 10–1 Fishing boats in Stornoway Harbour, Scotland

FIGURE 10–2 A fish processing plant. Courtesy of Nordicphotos/Alamy

FIGURE 10–3 Fresh caught herring, gut in

FIGURE 10–4 A salmon with the intestines removed

- Dressed Whole. The gut, all fins, and scales are removed; sometimes even the head and tail are removed (Figure 10–5).
- Filleted and Trimmed. The fish is removed from the bone and trimmed into fillets; the skin can also be removed at this stage (Figure 10–6).

FIGURE 10–5 A fish with the intestines, fins, and scales removed

FIGURE 10–6 A trimmed fillet of salmon

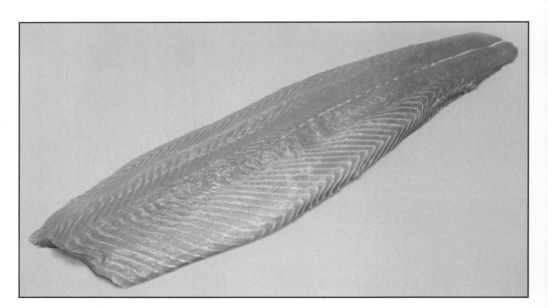

■ Portioned. The fillets are cut into portions to the customer's specification (Figure 10–7).
■ Alive and Well. Live shellfish should be purchased alive but also in as lively a state as possible (Figure 10–8).
■ Shucked or Shelled. Some shellfish are shucked or removed from their shells and cleaned as appropriate (Figure 10–9).

The Classification of Fish

The **classification** of fish is important; being able to categorize fish can be useful to the chef, especially when planning menus and creating new dishes. Some fish have certain qualities that others do not (e.g., fat content, yield of flesh, suitability for stock production). These decisions

FIGURE 10–7 Salmon portioned into darnes

FIGURE 10–8 Fresh mussels showing the closed shells

certainly affect how and what you buy; therefore, it is in the chef's interest to know the types of fish found in these categories.

■ Round White Fish. Commonly found swimming around or near the surface of the water. These fish have a vertical backbone with a fillet on either side. A round fish has one eye on each side of its head. The flesh is white in color.

■ Round Oily Fish. Commonly found swimming around or near the surface of the water. These fish have a vertical backbone that has one fillet on each side. A round fish has one eye on each

FIGURE 10–9 Shucked oysters

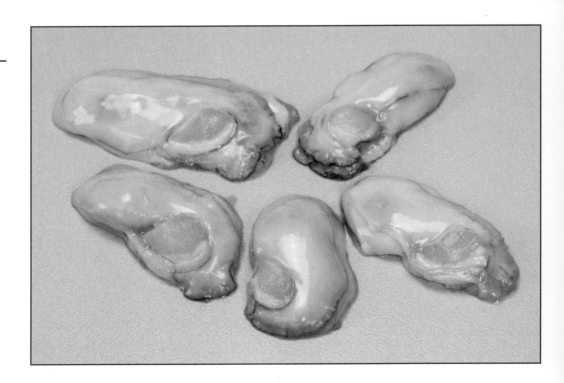

side of its head. Its flesh varies in color from shades of red to dark purple to off-white. The fat content also varies considerably: as high as 26 percent for the eel family, dropping to 12 percent for the salmon family, and as low as 6 percent for the herring family.

- Flat White Fish. Commonly found near or on the bottom of the sea. These fish have a horizontal backbone with two fillets on either side. A flat fish has both its eyes on the topside of the fish to look for predators swimming above it. Their skin is usually light colored underneath the body and dark colored on top to aid in camouflaging against predators.
- Nonbony Fish. These fish have cartilage instead of bone, which makes a difference in their handling and in their ability to make stock. This style of fish includes shark and rays.

Handling and Storage

Freshly caught fish or shellfish that is brought to the pan quickly is in a class by itself. Most chefs, and those fortunate to be around fresh seafood markets, understand the value of properly handling fish and shellfish. Seafood spoils more rapidly than most other food products and must be handled and stored with the utmost care. The fresh, rich flavor of seafood diminishes if processing, storing, or cooking is mishandled.

FRESH FISH

Fresh fish should be cleaned and gutted to preserve freshness. The removal of the intestines, liver, heart, and gills will eliminate the major sources of bacterial contamination. This is the most important step in retaining the freshness of fish. Occasionally, the gills are not removed when distributors clean fish. However, the complete removal of gills is necessary to preserve full freshness of fish. Red snapper, which has delicate meat, will spoil three to four times faster if the gills are not removed completely within a few hours of the catch.

Fresh fish should be refrigerated on ice at 35° to 40°F (1.5° to 4.5°C) as soon as it is received (Figure 10–10). Seal the fish in plastic wrap, if it has been skinned, before placing it on ice. Use

FIGURE 10–10 Fish sitting on ice at Stornoway Harbour, Scotland

a separate refrigerator, if available, or at least a section of the refrigerator to avoid contact with other foods. Fish should not be exposed to air unnecessarily because oxidation may alter the flavor. Fillets of fish lose their flavor more rapidly than whole fish and should be processed without delay. Purchase whole fresh fish if available. Eviscerate as soon as possible and process into fillets, steaks, sticks, or other forms shortly before cooking. This method of preparation helps preserve the full flavor of fish.

Note: Direct contact with ice can harm the flesh. Fish skin will insulate the flesh from damage.

FROZEN FISH

Fish to be frozen should be wrapped and sealed in moisture proof and vapor proof material. Do not freeze fish that are wrapped only in wax paper or polyethylene material. Fresh fish may be frozen in a block of ice or by glazing, both of which prevent moisture loss. To freeze fish in a block of ice, place fish in a container large enough to hold the fish and cover with water. Then freeze until solid.

Glazing is as effective as block freezing and takes less freezer space. To glaze fish (dressed, steaks, or fillets), place in a single layer on a tray, wrap, and freeze. As soon as the fish is frozen, remove from the freezer, unwrap, and dip quickly in ice-cold water. A glaze will form immediately. Repeat the dipping process three or four times. A thick coat of ice will result from each dipping. If necessary, return the fish to the freezer between dippings if the glaze does not build up after two or three consecutive dippings. Handle the fish carefully to avoid breaking the glaze. When glazed, wrap the fish tightly in freezer wrap or aluminum foil and return to freezer. Glazing may need to be repeated if the fish is not used within 1 or 2 months.

Commercially packaged frozen fish products should be placed in a freezer, in their original moisture-proof and vapor-proof wrapper, immediately after purchase, to maintain quality. Store at 0°F (18°C) or lower. At temperatures above this level, chemical changes cause the fish to lose color, flavor, texture, and nutritive value. Storage time should be limited in order to enjoy the optimum flavor of frozen fish. It is good practice to date the packages for easy rotation.

Recognition of Quality Points of Fish and Shellfish

Absolute freshness is essential if the best is to be enjoyed from any fish dish, both in flavor and nourishment. Stale fish are not only unappetizing but can also be the cause of digestive disorders or even food poisoning. The condition of fish and shellfish can be established by following some simple rules:

- The eyes should be bright and full, not sunken.
- The gills should be bright pinkish red in color.
- The flesh must be firm and springy, or elastic.
- Scales, if any, should be plentiful and firm, and should not come off when the fish is handled.
- The fish should have a pleasant, salty smell.
- The skin should be shining and have a good color.
- The flesh of white fish should be really white, not yellowish.
- The fish should feel heavy in relation to its size.

Likewise, there are sure indications of staleness that can be detected when evaluating fish:

- Unpleasant ammonia odor that increases with staleness
- Limp flesh that retains the imprint of one's finger
- Sunken eyes with loss of clarity
- Dull, discolored gills

The **quality points** of shellfish include:

- When inspecting all fresh shellfish, it is important to confirm that they are alive and healthy.
- Crustaceans, such as lobsters, should be lively and defensive when aggravated.
- Bivalves, such as mussels and clams, should generally have tightly closed shells. However, tapping on a slightly open shell and causing the shell to quickly close indicates a live shellfish. If the shell fails to react to the tapping and close itself, it is considered inedible and must be discarded.
- Univalves, such as snail and conch, are difficult to assess. However, poking the protected flesh in and around the aperture should cause a rapid withdrawal response by the shellfish.
- When cooking bivalves, the shells must open during boiling. Any shells that fail to open must be discarded, because the flesh is inedible.

Determining the Freshness of Fish

Understanding quality in fish is of paramount importance to a garde manger chef. It should not be assumed automatically that the fish received is of the highest quality. It is therefore of vital importance that chefs learn the signs that will allow them to make sound and safe judgments when deciding to accept or refuse fresh fish.

The maximum shelf life of some fish is 12 days; assuming that the fish has been correctly iced for the entire time it has been out of the water. Some fish, however, can only be safely held for far less time. Because it is nearly impossible to know with certainty how the fish was handled, the only sure means of telling is by observing the telltale signs of spoilage.

The appearance, smell, and touch of the fish are the most commonly used methods of determining freshness. Assessing certain parts—the gills, the eyes, the skin, the flesh, and the scales—also is favored for indicating the true condition of the fish.

Table 10–1 identifies the age of fish, while stored on ice, in relationship to its physical condition.

The Potential Yield of Fish

The yield attained when filleting fish is important information when calculating portions and margins of profit. Commercial processors calculate these percentages for their purposes, and can provide such information to their food service customers. Chefs need to understand the yield of fish in order to cost recipes accurately, including calculating the yield from the whole fish after

Table 10–1   Freshness Chart for Fish

	1–2 DAYS	3–4 DAYS	6–7 DAYS	9–10 DAYS	12 AND OLDER
Gills	Bright red, no bacterial slime	Darker red slime starting to appear	Thick slime, bloody briny odor	Bleached dark maroon slime, celery smell	Foul odor, very thick slime
Eyes	Bulging black pupil, clear cornea	Graying and flattening pupils	Cloudy and sunken pupils	Flat, cloudy, and bloodshot	Flat, cloudy, swollen, and bloodshot
Skin	Fresh sea slime present, nice sheen	Slime thickens, loss of sheen	Very little sheen, fishy smell, lines on skin not so distinct	Very dull and lifeless looking	Starting to yellow in color and becoming slimy
Flesh	Firm resilient and translucent	Beginning to appear less translucent	Soft to touch, no longer resilient to touch	Lost all clarity, waxy in appearance	Soft, darkening, and mushy
Scales	Flat to the body, plentiful, and very moist	Drying and reducing in number	Almost disappeared	Gone altogether	Nonexistent
Smell	Fresh sea smell	Still quite pleasant sea smell	Briny, yeasty, or malty	Distinctively fish and unpleasant	Bad fish smell, celery like

filleting, skinning, cleaning, and trimming into the desired portion size. Using this information, chefs then calculate how much whole fish to buy for their primary use, as well as the yield of by-products that can be used for stocks and sauces that accompany the fish. The following are some basic rules that should be understood to help in the planning of the menu:

- Yields vary according to the classification of fish, whether round or flat.
- Yields vary according to the species of fish.
- As the fish increases in size, there is a larger yield of fillet from the fish.
- Yields differ depending on the cut used. For example, a *darne* cut (which contains both bone and skin) when taken from a salmon will yield greater portion sizes than when the same fish is cut into *suprêmes* (containing no bone or skin.)
- The yield will drastically alter if the fish is processed by an inexperienced craftsperson.
- The yield will also lessen if the craftsperson is not acquainted with that species of fish.

YIELD CALCULATIONS

Weight loss from scaling is 2 percent.
Weight loss from skinning is 11 to 13 percent.
Weight loss from gutting is 13 to 16 percent.

The yields listed are based on the assumption that the fish were filleted, cleaned, and trimmed by an expert in both that field and that fish. Table 10–2 is a list of some popular fish from around the world; all the percentages are quoted for fillets with skins removed.

Note: As the fish increases in size, there is a corresponding increase in yield. This is the case whether the fish species is naturally larger or when the same species varies in size, such as snapper or salmon.

The Skeletal Structure of Round and Flat Fish

To be efficient and gain the most useable yield from a fish, it is important to understand the bone structure of the fish while learning to fillet (Figure 10–11). This information will help the chef see exactly where the knife should go in order to remove the flesh effectively.

Fabrication of Fish

For the fish to be of use to the garde manger, it must be fabricated to some level of purpose. Depending on the intended use of the product, **fabrication** of the fish will be handled in one or more of the following ways.

Table 10–2   Fish Yield Chart (Courtesy of Plitt Seafood)

Turbot—30 percent	Jumbo fluke—39 percent
John Dory—35 percent	Catfish—50 percent
Red grouper—42 percent	Chilean sea bass—53 percent
Black grouper—43 percent	Lake trout—55 percent
Walleye pike—44 percent	Mackerel—56 percent
Mahi-Mahi—48 percent	Halibut—57 percent
Rainbow trout—60 percent	Monkfish—60 percent
Tuna—69 percent	Salmon—71 percent

FIGURE 10–11 The skeletal structure of a flat and round fish

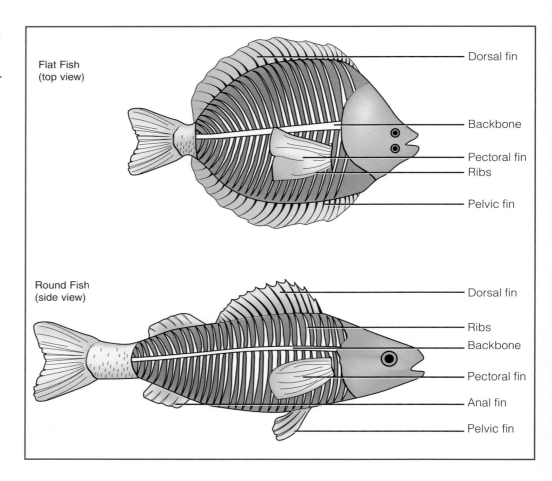

EVISCERATION

Evisceration is the removal of the intestinal tract and other organs found within the cavity of the fish. Elimination of these items aids in the extension of the fish's shelf life by removing many of the microbes and enzymes that hasten deterioration. Steps in the procedure include the following:

1. Wash and dry the fish with a clean cloth.
2. Cut off all fins and trim the tail (Figure 10–12A).
3. Remove the eyes and trim the mouth if the fish is to be served whole.
4. Remove the scales, if any, scraping with a knife or other scaling tool held at angle of 45 degrees from the tail to head (Figure 10–12B).
5. Remove gills, entrails, and roe, if any, from the gill slits (Figure 10–12C).

FIGURE 10–12A
Trimming the fins
and the tail

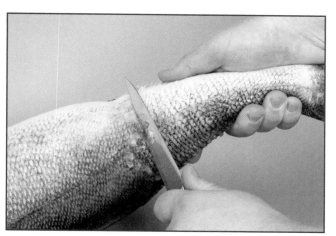

FIGURE 10–12B Scaling the fish

FIGURE 10–12C Removing the intestines and cleaning
the stomach cavity

For Round Fish
- Remove the gills and stomach from a round fish by making a small opening 2 inches (5 cm) in length from the vent to the belly.
- Remove the gut and the roe with care so as not to damage it and create spillage.

For Flat Fish
- Make the opening immediately behind the head of the fish where the gut is situated.
- Open the gill slit, on one side of the back of the head, with the fish on its side on the board. Hold the gills firmly with the fingers of the left hand and with the fish scissors cut the gill loose from both the left side, where it is attached to the throat, and the right side, where it is attached to the back of the gill slit.
- Repeat this operation on the other side of the fish until the gills are free.
- Using the tip of the knife and the thumb, draw out the entrails and the gills gently, taking care not to break the gut (which could spill and cause spoilage) or damage the flesh of the fish.

6. Wash the interior of the fish thoroughly under the cold water tap and scrape the backbone with a knife to remove all traces of blood or liver from the backbone. Rewash and dry the fish with a clean cloth. Ice down the fish and refrigerate.

7. Any roe removed from the fish should be placed at once into cold water and rinsed with cold water until the rinse water runs clear.

Note: Icing fish is an important step in the overall handling and storage of the fish, because it determines how long the fish will remain firm and fresh. Place the washed and well-dried fish on a perforated hotel pan. Set the perforated pan inside a deeper pan full of ice, and cover all with plastic wrap. Cover the fish with ice and store in the cooler (Figure 10–13). By following this procedure, the fish is surrounded by ice, but none of the ice or water from melting ice will be directly against the fish. Direct contact with ice can burn the flesh of the fish, while water can leach flavor from fish and soften its flesh.

FIGURE 10–13 The correct way to ice fish for refrigeration

SKINNING ROUND FISH

The steps in **skinning** round fish include:

1. Remove all dorsal, pelvic, and anal fins with a strong pair of scissors (Figure 10–14A).
2. Make an incision with the tip of a knife around the head, loosening the skin from the head (Figure 10–14B).
3. Loosen the skin along the length of the cut; then, starting at the head, pull the skin toward the tail in one stroke to remove the whole half-skin neatly (Figure 10–14C).
4. Repeat on the other side.
5. This technique is used when the fish is being served whole.

FIGURE 10–14A Trimming the fins neatly

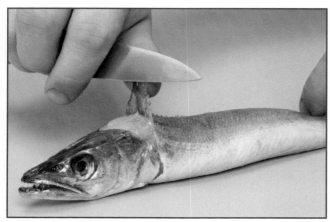

FIGURE 10–14B Loosening the skin around the head

FIGURE 10–14C Pulling the skin off the fish

SKINNING FLAT FISH

The steps in skinning flat fish include:

1. First, remove all fins from the fish using a sharp strong pair of scissors; cut through the skin across the middle of the tail and, with the tip of the filleting knife, scrape across the tail to loosen a few inches of skin (Figure 10–15A).
2. Raise the skin from the flesh, placing the thumb of the right hand between the flesh and the skin.
3. Hold the tail in the left hand. Pull gently with the right hand and remove the skin cleanly from the flesh (Figure 10–15B).
4. Follow along the fish with the thumb of the left hand, loosening the skin from the fins on either side of the fish.
5. The use of a little salt on the fingers, or better still, a clean cloth will ensure a firmer grip on the fish while skinning it.
6. If salt is used to enhance grip, always wash salt off skin before using it to make stock.
7. This technique is used when fish are being served whole.

FILLETING ROUND FISH

The steps in **filleting** round fish include:

1. Ensure the fish has been washed and dried well before starting.
2. Remove all fins and scales from the fish with sharp scissors and a scaling tool. This step becomes more important when dealing with fish that have a heavier bone and fin structure.
3. Make an incision along the backbone from the head end along to the tail, keeping your knife on the bone at all times (Figure 10–16A).
4. Continue to deepen the incision, scraping your knife over the fish ribs and holding the now freed flesh in your other hand until the fillet starts to come free (Figure 10–16B).
5. Follow the bones without cutting the flesh until you come to the other side of the fish and you have the whole fillet removed (Figure 10–16C).
6. Repeat the process on the other side. (*Note:* You will have two fillets from the boning of a round fish.)
7. Remove pin bones (Figure 10–16D).

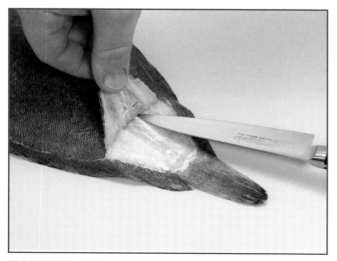

FIGURE 10–15A Loosening the skin with a knife

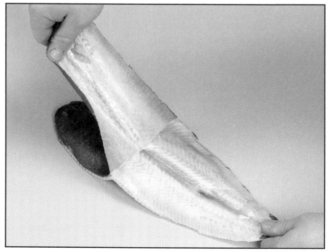

FIGURE 10–15B Pulling the skin off the fish from the tail to the head

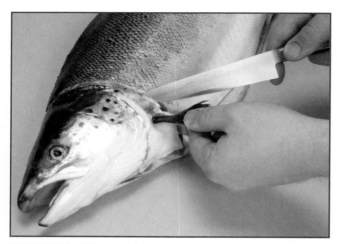

FIGURE 10–16A Making an incision around the head

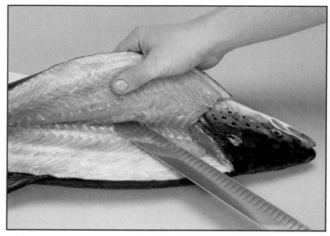

FIGURE 10–16B Continuing the cut over the bone, removing the fillet

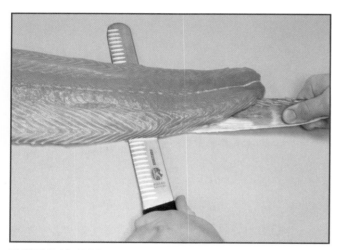

FIGURE 10–16C Skinning a salmon fillet

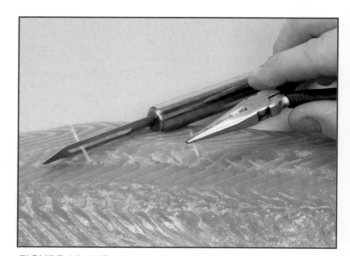

FIGURE 10–16D Removing the pin bones

FILLETING FLAT FISH

The steps in filleting flat fish include:

1. Remove all fins from the fish and dry it well.
2. Make an incision down the center of the backbone from head to tail (Figure 10–17A).
3. Separate the flesh from the bone by moving the knife down the length of the fillet and moving back the flesh as you go (Figure 10–17B). This will allow you to see as you go.
4. Continue to cut outward toward the fins, making sure your knife always stays on the bone until the fillet is completely removed.
5. Remove the other fillet from that side by turning the fish around and starting from the tail to the head (Figure 10–17C).
6. Remove both fillets from the other side to yield four fillets in all.

FILLETING LARGE FISH

When filleting larger fish such as tuna, Mahi-Mahi, and halibut, some difficulties may be encountered during the filleting process due to their bulk, weight, bone density, and awkwardness. The following are a few methods that will help you overcome these difficulties:

| TIP | There are two distinct techniques for filleting flat fish: (1) remove the individual fillets, yielding four or (2) remove the double fillet from one side and then the other. |

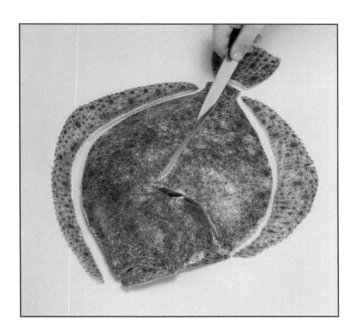

FIGURE 10–17A Making an incision down the backbone

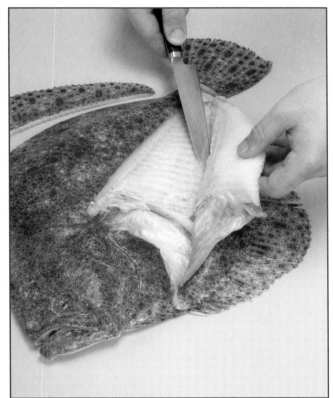

FIGURE 10–17B Cutting under the fillet against the bone to remove the fillet

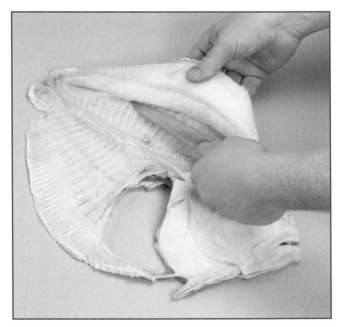

FIGURE 10–17C Repeating on the other side of the fillet

- Use a larger knife with a rigid blade and a very fine edge (Figure 10–18A).
- Always remove fins, scales, and any barbs.
- Use sharp, strong poultry scissors to aid you in trimming the fish.
- Try to support the fillet as it is being removed to prevent it from splitting (Figure 10–18B).
- Work on a very stable surface.

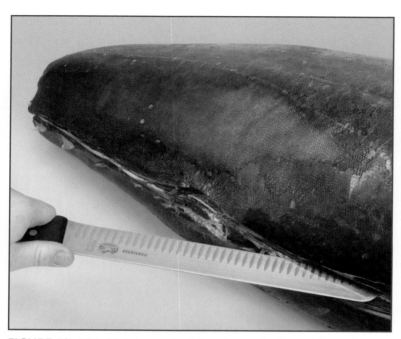

FIGURE 10–18A Making an incision down the lateral line of the fish to the bone

FIGURE 10–18B Cutting along the backbone to remove the fillet

- Wear good gloves for better grip.
- Make sure the fish is as dry as possible.

The Cuts of Fish

The following are some of the cuts associated with the innumerable shapes and sizes of fish. The basic fish cuts ensure consistency when writing menus and gives chefs, and more especially students, a reliable guide to follow when working with fish. These cuts offer a variety of ways of presenting an individual fish, enabling a very creative plate presentation. The cuts are designed to suit certain methods of cooking and certain plate shapes and sizes.

- Whole Fish. Some fish do not need to be cut. They just happen to be the correct portion size and ideal shape for the plate. Commonly, when whole fish are being cooked, lateral lines are cut into the flesh to the bone so that even cooking takes place. The term associated with this technique is *ciseler* (Figure 10–19).
- **Darne** (steak, sometimes called *côtelette*). A slice or steak cut from a large round fish on the bone. This fish should have been eviscerated and scaled before cutting. Being that this cut contains bone and skin, it has a certain sturdiness that allows it to be grilled without the fear of it breaking up.
- **Demi-Darne.** This is done when the darne is too big for the plate. The darne is simply cut in half through the spine. Salmon and cod work well for fabricating darnes. (See Figure 10–20 for a darne and demi-darne cut.)
- **Tronçon** (flat steak). A slice or steak cut from a large flat fish on the bone (Figure 10–21). All descriptions that apply to the darne also apply to the tronçon. Turbot, halibut, and brill make excellent tronçon.
- Fillet (fillets). Boneless, skinless portion of the fish achieved by completely boning fish (Figure 10–22). A flat fish yields four fillets and a round fish yields two. The term, when used on a menu, generally refers to small fish. Larger fillets are cut other ways. Sole, walleye, perch, and whiting make great fillets.

FIGURE 10–19 Lateral lines cut in a whole fish to help with even cooking

FIGURE 10–20 A darne and a demi-darne cut from a whole round fish through the bone

FIGURE 10–21 A tronçon cut from a large flat fish through the bone

FIGURE 10–22 Several fillets of fish, clockwise from top: Large round (yellow fin tuna), round (salmon), flat double fillet (brill), flat single fillet (turbot)

- ◼ Double Fillet. Removing two fillets at a time, leaving them attached, sometimes with the skin on (Figure 10–23).
- ◼ **Suprême** (medallion). Large fillets of fish, both round and flat varieties, cut into portions on a slant (Figure 10–24). Fillets of large Pacific fish are sometimes called *loins,* and when they are cut they are known as *steaks.*
- ◼ Boneless Darne. Produced by butterflying a suprême into the shape of a darne (Figure 10–25).
- ◼ **Goujon.** Fish cut into 3- by $\frac{1}{4}$-inch (7.5- by 0.5-cm) strips. This cut is a simple method of using the trim from filleting fish, although it can be cut from any boneless fillet (Figure 10–26).
- ◼ **Paupiette.** Fillets of fish that are rolled and stuffed with mousseline or forcemeat into a barrel shape (Figure 10–27). When preparing this cut, use a colored stuffing and carve the poached or baked paupiette for great presentation.
- ◼ **Plié.** Fillets that are folded, slightly flattened, and then folded in two (Figure 10–28). This is a method of reshaping or tidying up a fillet to fit the plate more neatly, although it can be stuffed.
- ◼ En Tresse (pleated or braided fillets). A fillet is cut into three strips, leaving them attached at one end. The strips are then braided or pleated for a fine presentation (Figure 10–29).
- ◼ **Tournedos.** Using a red flesh fish cut into circles approximately 3 inches (7.5 cm) in diameter, reserving trim for mousseline. Wrap with a slice of white fish, creating a tournedos shape, and pin with a toothpick (Figure 10–30). Poach, sauté, or pan roast.
- ◼ **Mignon.** Folded fillet as for a cornet or a triangular shape somewhat like a piping bag (Figure 10–31). Sometimes stuffed, small fillets are used.

FIGURE 10–23 Removing a double fillet from a halibut

FIGURE 10–24 Cutting suprêmes from large fillets

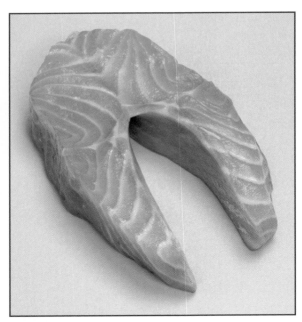

FIGURE 10–25 A boneless darne

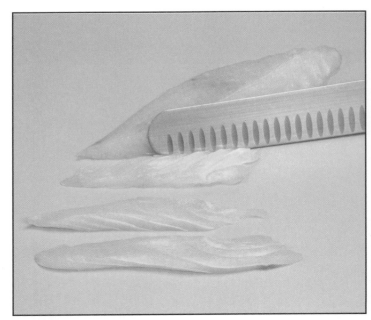

FIGURE 10–26 Goujons or fingers of fish

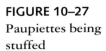

FIGURE 10–27
Paupiettes being
stuffed

FIGURE 10–28 Folding
and trimming fillets

FIGURE 10–29 Braiding small fillets

FIGURE 10–30 Making a tournedos

FIGURE 10–31
Stuffing a cornet or
mignon of fish

- **Délice.** A neatly folded trimmed fillet that is generally poached (Figure 10–32). This can be made with any small fillet.
- **Galette** (towered fish). Small fillets of equal size and width, or lightly beaten pieces of a larger fillet, layered on top of one another with a mousseline spread between each layer (Figure 10–33). Generally baked or classically poached in the oven.

FIGURE 10–32 Folding
a délice of fish

FIGURE 10–33
Layering a galette

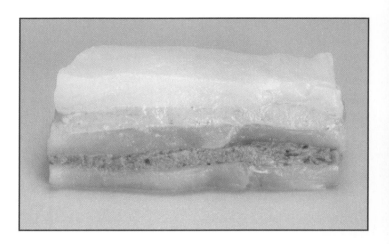

FIGURE 10–34 Stuffing a small whole fish

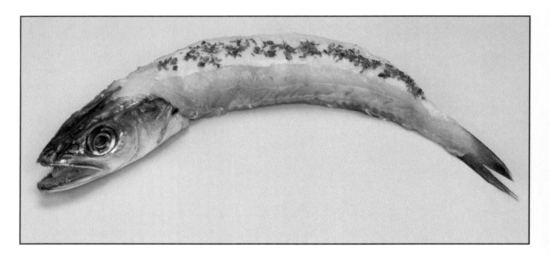

- **Farci** (stuffed whole). Applied to small whole fish. The fish is boned and skinned with head and tail still attached, then it is stuffed with mousseline and reshaped (Figure 10–34). This can have many variations depending on the size of fish and the cooking method.
- Skin-Wrapped Fillet. A large fillet of fish is used. The belly is cut from the skin, leaving a square piece of fish about 2 inches (5 cm) on all sides. Wrap the exposed fish with the flap of skin and tie well with string (Figure 10–35). This loin covered in skin can also be lightly stuffed before wrapping or cut into individual portions.
- **Tartare** of Fish. Fish flesh that has been neatly cut into small dice, and seasoned with spices and other flavoring agents (Figure 10–36). As with steak tartar, it is served raw.
- **Medusa**-Style. Filleted and skinned whole fish, with the head still attached. Cut each fillet into three strips (goujons) and dip in light batter to deep fry. Serve upside down to display goujons representing the snakes protruding from Medusa's head.
- Shaped Fish Fillets. Trimmed fish fillets, which are then cut accurately into circles, diamonds, triangles, or other shapes with the trim used for other dishes (Figure 10–37).
- Small Flat Fish, **Colbert**-style.
  1. Clean and prepare fish in the normal way.
  2. Skin and remove fins.

FIGURE 10–35 Wrapping a loin in its own skin

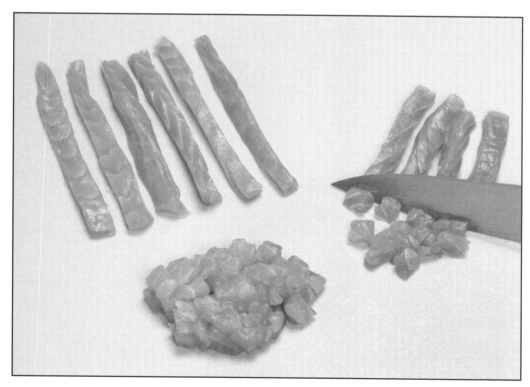

FIGURE 10–36 Fish cut tartare

FIGURE 10–37 Fish cut into different shapes

3. Make an incision along the backbone on one side, as if to fillet, within an inch (2.5 cm) or so from sides and ends of sole (Figure 10–38A).
4. Neatly fold back fillets.
5. Break backbone in two to four places, for easier removal after cooking (Figure 10–38B).
6. Wash, dry, season, flour, egg wash, and crumb (Figure 10–38C).
7. Deep fry whole fish.
8. When fully cooked, remove broken spine to expose cooked white fillets.
9. Place colored compound butters within and serve.

■ Small Round Fish, Colbert-style (Figure 10–39).
1. Skin fish or leave skin on, as desired.
2. Remove gills and eyes.
3. Wash and dry the fish.
4. Carefully remove the whole backbone, leaving 1 inch (2.5 cm) of the backbone and both fillets still attached to head and tail.
5. Season, flour, egg wash, and bread crumb the whole fish.
6. Apply vegetable spray to an oven-proof pan and stand coated fish as though it were swimming, but open to separate the two fillets.
7. Bake until done.

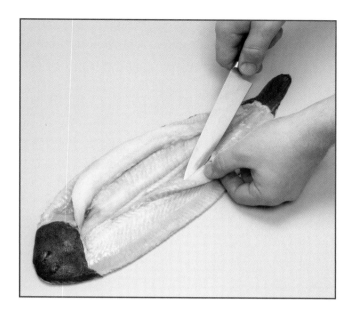

FIGURE 10–38A Opening the fillets along the spine

FIGURE 10–38B The fillets pinned back and the spine being broken

FIGURE 10–38C The whole fish dressed in breadcrumbs

FIGURE 10–39 Colbert-style with round fish

■ Curled Small Round Fish (Figure 10–40).
1. Skin as described.
2. Remove gill and eyes; force a knife through eye socket.
3. Wash, dry, and season.
4. Flour, egg wash, and breadcrumb.
5. Push tail through eye socket.
6. Secure, if necessary, with cocktail pick.
7. Deep fry or bake until done.

FIGURE 10–40 Curled round fish

- **En Colère** Small Round Fish (Figure 10–41).
  1. Prepare as Colbert round fish.
  2. Place tail between needle sharp teeth of fish.
  3. Bake until done.
- **En Lorgnette** Small Round Fish (Figure 10–42).
  1. Clean and skin as described.
  2. Remove backbone, leaving fillets attached to head.
  3. Season, flour, egg wash, and crumb fillets.
  4. Roll each fillet toward head, with the two fillets representing the glasses of a pince-nez, the head, and the handle.
  5. Other small fishes may be prepared in the same way.

Note: Other small fish, such as trout, may be prepared in the same manner.

Note: At times, the skin may be left attached to fillets for easier rolling and preparation.

Shellfish Through the Ages

The ancient Egyptians enjoyed **crustaceans** and shellfish in a variety of extraordinary dishes. Ancient Greeks, who settled along the shores of the Mediterranean, also loved to create seafood dishes. Accounts of the Roman's seafood banquets are legendary; they were also particular about where their seafood came from, having created artificial oyster beds in some of their marshy coastal towns. Marcus Caelius Apicius (born 25 BC), author of the oldest surviving cookbook, had many recipes for squid, shrimp, and lobster. The ninth volume of his book was devoted entirely to seafood.

Snails were eaten in Ancient Rome. In the last century BC, the Romans cultivated snails in breeding pens called *cochlearia,* which were later introduced into Gaul and Germania. In the heyday of the monasteries, snails were eaten on fast days. Sea snails were eaten as long ago as the Middle Stone Age.

In Greek mythology, gods and centaurs used to blow triton shells that can grow up to 2 feet, like bugles, and the shells are still used today as a musical instruments in Buddhist monasteries.

FIGURE 10–41 En colère small round fish

FIGURE 10–42 En lorngette small round fish

Helmet shells are used to make shell cameos, abalone shells are prized for their shimmering mother-of-pearl, and the cowrie was used as money in pre-Christian times.

MOLLUSKS

Generally speaking, there are two forms of mollusks commonly used in the culinary world, *bivalves* and *univalves*. The most striking feature of most mollusks is the shell, which not only provides protection and sturdiness but also replace muscle. The shell usually outlives the animal it houses and is often all the evidence we have of a particular animal. Seashells are still used in some cultures as everyday utensils, such as knives, spoons, and dishes.

All mollusks are invertebrates with bodies divided into four sections: the head, the muscular foot, the intestinal sac, and the "mantle," which encloses the mantle cavity and secretes the shell. Respiration in most mollusks is through gills, although in land snails the mantle cavity has developed into an air-breathing lung.

Bivalves

With around 8,000 species, bivalves are the second largest class of mollusks. The shell consists of two shells, or valves, joined by a dorsal hinge. When closed, the shells completely cover the shellfish. On one side is a hinge, formed by interlocking teeth on either valve. The two halves of the shell are closed by a muscle made up of two parts: one part is used to close the shell quickly, but this uses a lot of energy and soon becomes tired; the other half expends little energy and is able to keep the shell closed for weeks on end.

As a rule, bivalves are immobile. When they do move, it is by means of the foot, which is stretched as far forward as possible, and then anchored to the ground, so that the creature can drag itself slowly forward. Some species, however, are incapable of movement. These include edible mussels and oysters. Mussels, for instance, have byssus threads that anchor them to the rock. Other species dig or bore their way into peat, wood, or rock. A few bivalves, like scallops, are able to swim.

Clams and oysters in the shell should be alive, and the shells should be closed tightly or should close when the mollusks are tapped. The U.S. Public Health Service, in cooperation with the states, has a sanitation control program that covers the labeling and shipment of clams, mussels, and oysters. These shellfish may be harvested only from nonpolluted waters and processed for shipment in sanitary plants inspected by state shellfish inspectors.

Univalves

The largest groups of mollusks, at around 105,000 known species, are the snails. They are very common and are found throughout the world. They live in a wide variety of habitats—in the sea, in fresh water, and on land. The *queen conch* is an example of a widely consumed snail harvested from the sea, whereas the celebrated *escargot* is a land snail. Most have a spiral, univalve shell, but there are gastropods without shells and others with simple shells that, at first glance, look like the shell of a bivalve.

CRUSTACEANS

Crustaceans include shrimp, crabs, lobsters, crayfish, amphipods, isopods, ostracods, and barnacles. This group may have more than 50 million species and many have adapted to a life out of the water.

Shrimp, spiny and clawed lobsters, crayfish and common crabs, hermit crabs, robber crabs, spider crabs, mitten crabs, and stone crabs all belong to the great order of decapods. Decapods move by swimming, crawling, strutting, or raking through soft ground. The pincers are mainly used for feeding, for defense against enemies, and for holding onto the partner during mating.

Quality of Fresh Shellfish

The basic rule for evaluating shellfish is that it should arrive at your door alive and well, even kicking. The reality is that it is very difficult to truly know how alive and healthy every unit of shellfish remains. There are obvious exceptions, such as an open mussel or a "flying lobster" (Figure 10–43), but the really important and generally reliable guarantee is the chef's relationship with the supplier. A trustworthy source for shellfish is essential to have peace of mind when serving fresh shellfish.

Preparing Fresh Shellfish

Chefs must be vigilant to the freshness and quality of the shellfish they use. Freshness is the key to flavor and improved health. Shellfish, because of their vastly diverse skeletal structures, require different methods for preparation and cooking.

PREPARING LOBSTERS

Male and female lobsters offer different things to the chef, and it is important to understand the visual differences between the sexes (Figure 10–44).

The female lobster could contain the *coral* or egg sack in its body. As it is of great culinary importance for flavor, it should be retrieved during its processing (Figure 10–45).

Cooking Live Lobsters

Chefs use a court bouillon to poach lobsters, because seawater is rarely available. The great origin of all fish stocks is the ocean, with its slight sea salt content and water tasting of oxygen and the freshness of the sea. However, for practical purposes, court bouillon is generally used. Its ingredients vary, but commonly include water, acids such as lemon juice or vinegar, aromatic herbs, and vegetables. They are usually chosen as a complement to whatever style the lobster is to be cooked.

CLAW LOBSTERS The classic way of killing claw lobster is to cut through the nerve center by inserting a sharp, heavy knife where the head meets the body. This method can be clumsy and can damage the internal organs of the lobster, especially the egg sack in female lobsters. For males, this method works with little concern. However, the safest method for protecting the roe of the female lobster is to plunge the lobster into vigorously boiling court boullion just

FIGURE 10–43 A fresh lobster in the flying position, indicating its freshness

FIGURE 10–44 The sex organs of a male and a female lobster

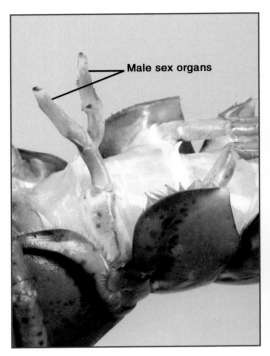

Male sex organs

Female sex organs

FIGURE 10–45 The egg sack from a female lobster

long enough to set the proteins of the flesh. (Allow the liquid to reboil when the lobsters are added, and simmer for 2 to 3 minutes). This method allows the partially cooked flesh to be removed with ease in order to access the egg sack. This method results in the lobster flesh being blanched, rather than cooked. It can then be further cooked, in a method so desired, like sauté.

To fully cook a lobster by boiling, be sure of the following:

- Allocate at least 1 gallon (3.8 L) of water per lobster.
- The flame must be set on high under the pot.
- The liquid must be fast boiling before the lobster is added.

To cook, plunge the lobster in head first and ensure that it is totally immersed. Hold it under the water with tongs for at least 2 minutes (for blanching). The lobster should die within 15 seconds. If the recipe calls for uncooked lobster, remove it after 2 minutes to prevent overcooking. Immediately shock and refresh the lobster in iced water, allowing it to get completely cold before proceeding. Preparing lobsters by the blanching technique allows lobster to be cooked further by a different method, thereby improving its flavor and lessening the chances of overcooking the lobster meat (Figure 10–46).

Lobsters that are going to be used for salad, appetizers, or other finished dishes should be cooked a further 6 minutes on the simmer, then shocked to prevent overcooking (Figure 10–47).

SPINY LOBSTERS AND CRAWFISH This much-valued shellfish has different handling qualities than its cold water cousin. There is no need to poach this lobster in order to get to its flesh; it can be removed in the raw state with ease. Splitting the shell down the back allows the chef access to the flesh which is available to be pried out using fingers, spoons, or small forks (Figure 10–48). The split lobster is often grilled with the flesh intact.

Cooking and Cleaning Crabs

There are many different types of crabs in the world, and they all have their own particular idiosyncrasies, so the following are some basic rules to follow for cooking and cleaning.

Boil the crabs in saltwater or seawater. If tap water is to be used, add ½ cup (118 ml) of kosher salt to every gallon (3.8 L) of water. Let the crabs boil for 5 minutes, and then cool them quickly under a running tap of cold water to set the meat.

Note: Freezing the claw lobster alive before it is boiled is considered more humane by some, because the lobster will gradually lose consciousness and die. It can then be plunged into boiling liquid.

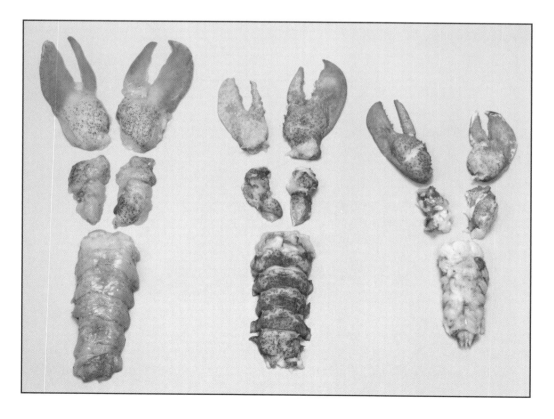

FIGURE 10–46 From left to right: Raw lobster meat, blanched lobster meat, and cooked lobster meat

FIGURE 10–47 Lobster with all the meat removed

FIGURE 10–48 Splitting a spiny lobster and placing the flesh still attached along the tail

Take the following steps to remove the meat from the crab (Figure 10–49A and B).

1. Hold the crab firmly in one hand and slide a sharp knife under the shell with the other.
2. Use the knife as a lever to prize off the shell.
3. Pull the shell away from the body.
4. Remove and discard the "feathery" internal gills.
5. Twist off all the legs and the claws.
6. To expose the body's white meat, cut it in half.
7. Open and separate the halves.
8. Dig out the white meat from the halves using the point of sharp knife.
9. Crack the claws and legs with a hammer or nutcracker and extract their meat.

Cleaning Shrimp

Cleaning shrimp can be done as follows:

1. The head may be removed by using thumbs or with a sharp knife. Placing a thumb under the head, between the body's shell and plate-like head, and quickly flicking the thumb as if

FIGURE 10–49A Removing the meat from a crab

FIGURE 10–49B Close-up of meat removal

to toss a coin will remove its head. Otherwise, placing the knife's tip in the same location and forcefully cutting will sever the head.

2. To peel the shrimp, straighten the shrimp's body and pull apart the shell starting between the legs, where it is weakest. Developing the proper touch, through repetition and practice, usually results in the shell and legs coming off in one piece (Figure 10–50A). If not, remove the shell and legs in pieces. When having difficulty loosening the shell, make a shallow cut between the legs to help get started.

3. To devein the shrimp, remove the contents of the alimentary canal (the digestive system) that runs from head to tail, down the back edge of each shrimp. To do so, make a shallow incision using sharp paring knife along the "back" (Figure 10–50B and C). When deveining un-peeled shrimp, cut through the shell using a sharp knife. Using the tip of the knife, scrape away the dark entrails.

Opening and Cleaning Scallops

The following are steps for opening and cleaning scallops:

1. To open the shell, hold the scallop with the flat shell uppermost.
2. Probe between the shells with a short knife to find a small opening. Insert the blade and run it across the roof of the shell (Figure 10–51A).
3. Separate the two halves of the shell and pull apart (Figure 10–51B).
4. Slide the blade under the grayish outer rim of the flesh, called the skirt, to free the scallop. Pull away the muscle with a small knife (Figure 10–51C). Use the trimmed scallop, whole or halved, for cooking.

FIGURE 10–50A Removing the shell from the shrimp

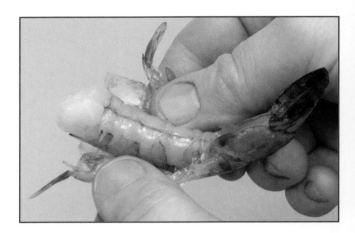

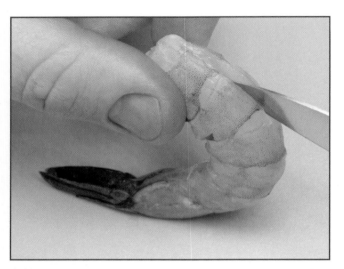

FIGURE 10–50B Cutting along the back and front of the shrimp

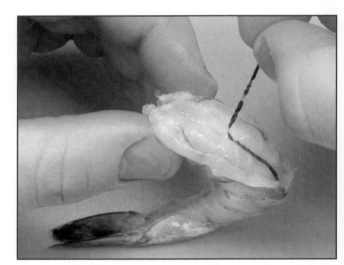

FIGURE 10–50C Removing the vein

FIGURE 10–51A Levering the knife into and between the shells

FIGURE 10–51B The separated shell showing the roe

FIGURE 10–51C Removing the scallop and roe

Shucking Oysters

The following are steps for shucking (opening) oysters:

1. Wrap the oyster in a towel and place it securely on a flat cutting surface with the small indentation in the end of the oyster facing outward. This indentation is precisely where you want to place the tip of the knife. Using a good deal of pressure, push the knife into the hinge. Twist the knife back and forth to pry the shell open (Figure 10–52A).
2. Once the knife has popped the hinge, pry the lid open wide enough to fit the top of your thumb inside (Figure 10–52B). Insert your thumb into the oyster to hold the lid open. Do not plunge the knife into the oyster once the hinge has popped.
3. At this point, slide the tip of the knife inside the oyster. Keep the tip of the knife slanted upward and slide it along the roof of the oyster. The tip should not scrape the roof. It should glide along the roof until it reaches the muscle that connects the two shells, then slice through the connective muscle. Cut the muscle from the top of the shell without piercing the oyster itself. This will allow you to lift off the top shell, exposing the oyster. Remove the oyster (Figure 10–52C).

Cleaning Mussels

When selecting mussels, never choose a mussel that is chipped, broken, or damaged in any way. Also, never choose a mussel that is open. The mussels should be tightly closed and stored in a cool area where they can breathe. Leave the mussels unwrapped so they can breathe, otherwise they may die before you cook with them.

The following are steps for cleaning mussels:

1. Just before cooking, soak mussels in fresh water. Soak them for about 20 minutes. As the mussels breathe, they filter water and expel sand. After about 20 minutes, the mussels will have less salt and sand stored inside of their shells.
2. Most mussels have what is commonly called "the beard," also known as byssal threads. The beard is comprised of many fibers that emerge from the mussel's shell.
3. To remove the beard, grasp it and give a sharp yank out and toward the hinge end of the mussel (Figure 10–53). This method will not kill the mussel. If you were to pull the beard

TIP It is necessary to exert a lot of pressure to open oyster shells, so it is important that the angle of the knife as well as the positioning of the towel holding the oyster is given special attention. Be careful.

TIP Oyster shells are brittle and splinter easily. If the oyster splinters and the knife is not angled so that it is pointing down toward the cutting board, it would be easy to lose control of the knife. Hold the oyster firmly on the cutting board to keep the oyster from sliding away.

FIGURE 10–52A Popping the hinge that connects the shell

FIGURE 10–52B Prying the shells apart

FIGURE 10–52C Removing the oyster

out toward the opening end of the mussel you can tear the mussel on the inside of the shell, killing it. Discard the byssal threads.

4. Remove the mussels from the water. Do not pour the mussels and water into a straining device because the sand has sunk to the bottom of the bowl. Pouring the mussels and water into a straining device would cause you to pour the sand back on top of the mussels. Place these mussels into another bowl full of clean cold water.

5. Once the mussels have been soaked, use a firm brush to brush off any additional sand or barnacles. Rinse the mussels under cool tap water, and set aside. Dry with a towel before cooking.

Cleaning Conch

The following are steps for cleaning conch:

1. Hold the shell on a flat surface or if you are experienced, in your hand, with the open hole side down. Tap a hole in the top of the shell between the third and fourth ring up from the pointy tip. The best tool for knocking a hole is a hammer with a flat edge. (*Note:* You must make the hole wide enough to insert a knife or machete, [about 2 inches/5 cm across the

FIGURE 10–53
Bearding and scrubbing mussels

shell and $\frac{1}{2}$ inch/1.25 cm wide] in order to cut the abductor muscle, which connects the meat to the shell.)

2. Insert your knife, cut the muscle, and pull the meat from the hole in the large opening by pulling it out by its black claw.

3. Cut around the hard conch meat to remove the loose hanging black and white viscera.

4. Now, with the face up to you and claw away from you, completely cut off the projecting proboscis (long snout) and eyes. Do not remove the claw yet because it can be used as a handle while working with the meat.

5. Make a small slit into the body of the meat from the base of the proboscis, below where the mouth was, to the ragged edge of the muscle and remove the esophagus. This leaves one piece consisting of a white upper body and a dark gray lower body.

6. Make a slit down the length of the gray portion of the muscle about an inch (2.5 cm) deep and work your thumb underneath to pull this outer gray layer off. Hold the claw firmly and use as a handle during this process.

7. Cut the claw from the meat, unless you choose to eat it scorched or raw, in which case you can use it as a handle to enjoy fresh from its shell.

Cleaning Squid

The following are steps for cleaning squid:

1. Remove the tentacles by cutting them off where they connect to the body.

2. Cut off the head with a sharp knife (Figure 10–54A). Cut off the fins.

3. Pull out the gladius (or pen) (Figure 10–54B). This is the clear, plastic-looking blade that runs the length of the body.

4. Pull out the rest of the innards (Figure 10–54C).

5. Rinse the tentacles, mantle (body cavity), and fins in fresh water, making sure that the inside of the cavity is very clean, that is, free of all membrane, stomach, and body fluids.

6. Cut as desired for cooking.

Cleaning and Preparing Octopus

Generally, octopus is tenderized when purchased. On receiving a freshly caught octopus, wash it in cold water and dry it well, then follow the steps as needed.

1. Tenderize the octopus by beating it on a large rock, holding the octopus alternate tentacles every four to five strokes until you have beaten it by all tentacles about 50 strokes in all. You can also beat it with a heavy mallet, but this does not work as well as the previous method.

FIGURE 10–54A Removing the head from the squid

FIGURE 10–54B Removing the pen from the body

FIGURE 10–54C Removing the intestines

2. Remove the tentacles just below the eyes, separating the head and body sections (Figure 10–55A).

3. Cut out the eyes and clean the body cavity thoroughly.

4. At the top of the tentacles where they join each other, locate the beak and remove it by pushing it through the joint and exposing it (Figure 10–55B).

5. Cut into sections for the chosen cooking method, washing well, especially the tentacles, which will contain some sand.

6. The skin on the octopus is difficult to remove when raw; by using a little salt to help you grip better, you can remove the skin with a little work (Figure 10–55C).

7. Parboiling for 5 minutes also helps with the skin removal.

8. Small octopus can be left whole. Split the body and remove the gut from the cavity, remove the eyes and the beak, and clean well.

FIGURE 10–55A Removing the tentacles just below the eyes

FIGURE 10–55B Extracting the beak from the head

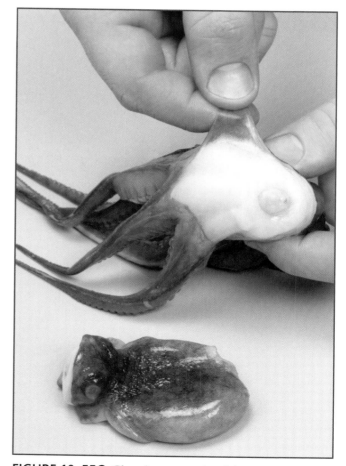

FIGURE 10–55C Cleaning away the skin

Cleaning and Preparing Fresh Abalone

The following steps are for cleaning abalone:

1. Similar to opening scallops, force a sharp blade tip in the thin part of the shell and separate the muscle from the shell (Figure 10–56A).

2. Remove the flesh and clean well, taking out the intestine and the sucker pad that appears as a dark heel on the side (Figure 10–56B).

3. Wash the flesh well and slice horizontally in two as evenly as possible (Figure 10–56C).

4. Place each piece into heavy plastic and beat with a mallet until very well pounded and limp.

FIGURE 10–56A Separating the shell and removing the abalone

FIGURE 10–56B Removing the sucker pad and the intestine

FIGURE 10–56C Cleaning and slicing the flesh

Removing the Roe from Fresh Sea Urchin

To Westerners, sea urchin is a very unusual shellfish that is gaining in exposure and popularity. To Asians, in countries like Japan, it is a delicacy enjoying widespread appreciation. Its attraction lies in its very soft and fragile roe, light yellow to light orange in color, found in the shell at the bottom of the fish by the mouth. It has fantastic sweet flavor, and can be eaten immediately raw or cooked in many ways. It is used to flavor risottos at the end of cooking, in chowders, or as a garnish. The roe is fragile, so great care must be taken when removing it. It is a little like getting a piece of china out of a well-nailed wooden box. It is wise to wear gloves when working with the urchin, because the spines can easily puncture human flesh. Take the following steps to remove the roe:

1. Turn the urchin over to expose the mouth, which has five teeth in its opening (Figure 10–57A).

2. With a pair of powerful pointed pliers, systematically break the shell upward in as large pieces as you can, trying not to create any splinters (Figure 10–57B).

3. Exposing the roe and the innards (the roe will resemble the segments of an orange), remove the roe from the shell using a spoon (Figure 10–57C). The innards will also come with the roe. Place the roe and innards on a very clean wire rack and discard the shell.

4. Carefully remove the larger and smaller specks of innards from the roe. Tweezers are an ideal tool for this procedure.

5. Carefully lower the roe into a bath of very cold water, and gently swill them around to clean. To avoid damaging the roe, do not run water directly on it.

6. Clean the roe with wet paper towels, removing any small specks that have lingered. Pat the roe dry, taking great care not to damage the tiny eggs.

FIGURE 10–57A
Locating the mouth of the urchin

FIGURE 10–57B Breaking the shell upwards carefully

FIGURE 10–57C Removing the soft roe

Cleaning Soft Shell Crab

The following steps are used to clean soft shell crab:

1. Pull back the pointed edge of the crab shell and remove the gills on both sides of the crab.
2. Cut off the head just behind the eyes and force out the green liquid that is found there.
3. Pull the intestinal vein out by pinching and drawing on the apron or tail flap, removing the tail flap in doing so.
4. Wash well and store for service.

Professional Profile

BIOGRAPHICAL INFORMATION

Name: Freddie Lightbourn

Place of Birth: New Providence, The Bahamas

Recipe Provided: Conch Salad

CULINARY EDUCATION AND TRAINING HIGHLIGHTS

Born on the sun-drenched island of New Providence in The Bahamas, Chef Freddie Lightbourn began his career at the original Poop Deck Restaurant in Nassau, The Bahamas. He worked part-time on weekends and during local holidays, until he finished high school, when he decided to take the restaurant business seriously and become part of the family-owned venture. The Poop Deck Restaurant is situated on the deck of the Bahamas Yacht Club and serves exceptional local Bahamian specialties. It has become an internationally renowned restaurant, attracting not only the many visitors to the beautiful islands but also many celebrities, heads of state, and foreign dignitaries. Its legendary conch chowder, minced lobster, and grilled hog snapper are famous throughout the Caribbean islands. Chef Lightbourn's training was virtually all hands-on, working with a group of talented local women who shared their secrets, passion, and love for food with him. The recipes they worked with had been handed down from generation to generation (and always cooked in a seasoned pot, never a new pot). His persistence to learn eventually paid off, earning him positions in the dining room and the management of the restaurant. In 1985, after being given part ownership of the restaurant, he saw the need to open a second Poop Deck Restaurant at a more pristine site. After much persuasion and debate, the beachfront property of The Poop Deck at Sandy Point opened in March 1999, on the western side of Nassau. Later, in 2002, Chef Lightbourn opened the Columbus Tavern on Paradise Island, which features Bahamian cuisine and has become a very popular local eatery. Chef Lightbourn has traveled extensively, to many countries, lecturing on Bahamian cuisine as well as judging local and international culinary competitions.

ADVICE TO A JOURNEYMAN CHEF

To be successful in any business, you must be very knowledgeable in all aspects of that business. If the business happens to be a restaurant, you better know and love people and food. I spent a lot of my spare time, as a young man, just watching the kitchen operate before I really understood why they were all rushing around. This allowed me to be better prepared for the kitchen when I eventually started to cook. Immerse yourself in the medium that you work with, and you will get a lot pleasure from it. One of the first lessons that I learned was to treat your guest as you would like to be treated yourself, and they will always return. Take pride in your food, and always remember every meal counts; food must always be consistent.

Freddie Lightbourn

RECIPE 10-1

CHEF FREDDIE LIGHTBOURN'S CONCH SALAD

Recipe Yield: 8 portions

| MEASUREMENTS | | INGREDIENTS |
U.S.	METRIC	
4 each	4 each	Conch, cleaned and skin removed (use only very fresh, live conch)
½ cup	118 mL	Onion, very finely diced
2 each	2 each	Celery stalk, very finely diced
1 each	1 each	Sweet red pepper, very finely diced
2 each	2 each	Large ripe tomato, very finely diced
½ cup	118 mL	Sour orange juice
1 teaspoon	5 mL	Goat pepper, very fine
1 tablespoon	15 mL	Salt

PREPARATION STEPS:

1. Carefully cut the conch flesh into small dices, always cutting across the grain.

2. Toss the conch with the other ingredients and allow to sit in the cooler for 1 hour.

3. Add more hot pepper, on the side, for additional heat and flavor.

1. How are fish available for purchase?

2. What are the classifications of fish?

3. What are the classifications of shellfish?

4. Describe the quality points of fresh fish.

5. Describe how to skin a flat fish.

6. Describe how to fillet a flat fish.

7. What advantages are there in leaving bone in small fish cuts?

8. How can you tell if a lobster is alive and well?

9. What are the signs that mollusks are beginning to deteriorate?

10. How do you ice fish properly?

A. **Group Discussion**

Discuss the forms in which fish are available for purchase, suggesting other possible ways the supplier could prepare them for the kitchen.

B. **Research Project**

In small groups, research how the stocks of wild fish are surviving in the wild. Report to the class on whether you think there are enough fish left in the wild or whether overfishing is jeopardizing the world's reserves of fish.

C. **Group Activity**

Bone a small flat or round fish and establish the percentage of weight loss due to cleaning, filleting, and skinning.

D. **Individual Activity**

Allow a small fish to deteriorate under controlled circumstances, observing daily how the fish changes in appearance and smell.

CHAPTER 11

Specialty Meats

After reading this chapter, you should be able to

- recognize the quality points of offal meats.
- describe how to preprepare fresh offal meats.
- describe how to prepare the meat for cooking.
- explain how to preparing the liver of different animals.
- select foie gras.
- describe the grades of foie gras.
- prepare foie gras for hot dishes.
- remove the veins from foie gras.
- prepare sweetbreads.
- prepare hearts.
- prepare kidney.
- explain how to use oxtail.
- demonstrate how to store and use blood.
- prepare the head.
- prepare tripe.
- prepare pluck.
- use caul fat.

Looks can be deceiving—it's eating that's believing.

JAMES THURBER

Offal/Specialty Meats/Organ Meats

The word **offal** actually comes from the Old English "off" and "fall," referring to the pieces that fall from an animal carcass during butchering. These cuts are given many names throughout the world including *offal cuts, specialty meats, organ meats, variety meats, bad meats,* and even *cheap meats.* However, they are perceived to hold a very important place in the world's cuisine. In fact, offal meat is eaten internationally on a regular basis and is included in the national dish of many countries. This is mainly due to dishes having been created in times when it was a necessity to eat all the internal parts of the animal, or else go hungry.

Often, people are still reluctant to eat brains, sweetbreads, and tripe, and many more are unaware from where they come, nevermind what they taste like. This confusion makes them somewhat of a "novelty meat." Some forms of offal have proven unpopular for social, economic, and religious reasons; however, the demand for offal meat has continued, especially the more alluring items such as liver. Today, with the ever increasing interest in culinary books and TV cooking shows, the acceptance of offal meats has grown; they are appearing ever more readily on international menus.

Recognition of Quality Points

All offal cuts should be purchased as fresh as possible, and the use of a very reputable butcher or supplier is the best guarantee of receiving a quality product. There are some very good frozen products on the market, but it is preferable with this specific classification of meat to use as fresh a product as possible. As a general rule of quality, an older or stale product will usually be much darker and deeper in color, with a somewhat dryer appearance. Most chefs, who are used to receiving and evaluating very fresh product, should be capable of noting the warning signs when the quality is lacking. Liver from a mature animal is also less tender, and has a stronger odor and flavor than that from a younger one. Individual quality points are given for all the offal meats discussed in this chapter.

Fabrication and Dressing

Always begin by checking whether the offal has a recommended soaking period, before the cleaning and dissection begins. This will aid in the butchering process and ensure that there is no spread of bacteria and that the offal does not acquire blood spotting.

LIVER

In the past, **liver** often had a much stronger flavor and was soaked in milk to absorb the intense flavors and to remove excess blood. Modern breeding and feeding methods, however, produce liver with a much milder flavor, especially in lamb and veal. Soaking pork, beef, and chicken livers is now really a matter of preference. It does, nevertheless, effectively remove some of the very strong flavor and some of the unwanted blood.

There is a school of thought that says soaking the livers in milk overnight removes any bitter flavor and keeps them tasting fresh and rich. When cooking liver, there are some precautions that should be observed in order to render it as palatable as possible. There is a very fine line between cooking it perfectly, and cooking until it is overdone and inedible.

- If liver is going to be cooked by any dry method of cooking, great care must be taken to ensure that it is not overcooked in any way.
- It will immediately harden, toughen, and change in flavor considerably.
- The thickness it is cut to is therefore of paramount importance because the thinner it is cut, the faster it will need to be cooked.
- Inaccuracy in the removal of the silverskin can result in the curling of the liver during cooking and result in poor presentation and toughness.

Checking the quality of the liver received in the kitchen is the first step to assuring success with the finished product. Each type of liver has its own quality points and must be carefully inspected. The following points are important to note:

- **Beef liver**
  - Beef liver should have a bright color with a moist but not slimy surface and a fresh smell.
  - Try to prepare it the day you buy it, or store it loosely wrapped in the refrigerator for no more than 1 day.
  - Beef liver is darker and has a stronger taste than all other livers.
  - The beef liver is generally the toughest of the livers.
  - Whole beef livers can weigh from 8 to 12 lb (3.6 to 5.4 kg)
  - The most suitable methods of cooking are braising and stewing.
- **Calf's liver**
  - Calf's liver should be a deep rose to reddish brown in color.
  - There should be no dark red or purple tinges.
  - There should be no blood spots or bruising.
  - Whole calf's liver weighs on average 7 lb (3.2 kg)
  - This is the finest quality of the livers and therefore the most expensive (excluding fois gras).
  - Grilling, frying, and sautéing are the preferred methods for cooking.
- **Lamb liver**
  - Lamb liver has a sharp and distinctive odor.
  - The color is a very light reddish brown and should be lively with a bright bloom. It should show no sign of dullness.
  - It resembles calf's liver and is sometimes mistakenly sold as calf's liver.
  - It is generally very tender.
  - A lamb's liver weighs on average about 2 lb (0.9 kg).
  - It can be broiled, grilled, sautéed, pan-fried, and deep-fried.
  - It dries very quickly if overcooked.
- **Pork liver**
  - Pork liver has a strong odor and flavor.
  - The color should be lively and have a light reddish brown tinge.
  - Whole pork liver weighs on average about 3 lb (1.4 kg).
  - It is primarily used in the prodution of patés and sausages, although it can be be sautéed and fried.

FIGURE 11–1A
Skinning the liver

FIGURE 11–1B Removing all the membranes from the liver

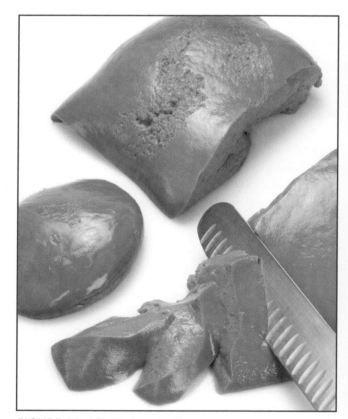

FIGURE 11–1C Slicing the liver

The general cleaning and preparation for beef, veal, pork, and lamb liver are similar because of their similar makeup shape.

1. Wash the liver well and pat dry.
2. Remove any tough membranes, tubes, and sinews from the liver.
3. Take great care not to damage the structure of the liver during the removal of the veins.
4. Skin the liver, removing the tough silverskin that surrounds it (Figure 11–1A and B).
5. Slice as desired for cooking (Figure 11–1C)

Chicken livers and **duck livers** are much smaller and have specific quality points that are important to the safety of the customer and the quality of the finished dish.

- Good fresh livers should be firm and well shaped.
- There should be no evidence of the gall bladder remaining; it is easily recognized by a green staining.
- The liver should be rich, dark, reddish brown with a bright bloom.
- The liver should be intact and not mashed or damaged in any way.

FOIE GRAS

In antiquity, the Egyptians appreciated the tasty flesh and liver of the geese that came to spend the winter on the edges of the river Nile. During their stay, the geese feasted on the lush land and built up fatty reserves that they stored to aid in their efforts for a return journey. It was much later in the first century BC when the Romans realized the improved qualities of the goose livers after the geese were fed fresh figs.

The Ashkenazi Jews of central Europe are credited with disseminating the method of cultivation for **foie gras**, a foodstuff not prohibited by their religion. Soon Alsace and the Southwest regions of France were producing pies and pots of foie gras. The foie gras became the food of kings at the table of Louis XV and then grew in stature at the table of Louis XVI.

Once the more modern gastronomes and chefs like Brillat-Savarin, Reyniere, and Carême started documenting some of the incredibly decorative and complex recipes, the foie gras fame was well established. In the late nineteenth century, Georges Auguste Escoffier (who simplified Carême's extravagant recipe) finally created step-by-step instructions on how to prepare the liver for the foie gras we know today.

The literal translation for foie gras from the French is "fat liver," and it is the term generally used for goose liver, although duck liver is also considered to be foie gras, when reared for that purpose. This specialty of Alsace in France is, in fact, the enlarged liver from a goose or duck that has been force fed and fattened over a period of 4 to 5 months. These specially bred fowl do not exercise and are overfed, resulting in a huge fatty liver.

This liver is much prized by chefs throughout the world and is considered a universal delicacy. Foie gras is an ingredient that, because of the aura surrounding it (and its high cost), has a reputation among chefs that special skills are required to handle it. This belief is by no means the case, although it has many stringent rules that must be followed in order to render it properly useable. But once mastered, the preparation of foie gras is very satisfying to a chef. There are specific quality checks that should be carried out when purchasing the fresh foie gras, and the inspection should include the ability to determine the grade that is being purchased (Table 11–1; Figure 11–2).

- The color should be a light yellow to amber.
- The lighter the liver, the less fat is contained in the liver.
- The liver should be firm and be resilient to touch.
- The liver should give slightly under thumb pressure and the thumb mark should remain visible.
- The higher the grade, the fewer blemishes the liver will have and the larger it will be.

FIGURE 11–2 The three grades of foie gras from left to right: A, B, and C

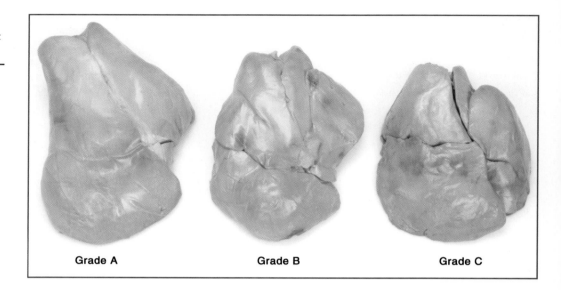

Grade A   Grade B   Grade C

- Grade A livers must weigh at least 1½ lb (0.68 kg) or more.
- Grade B livers weigh between ¾ lb and 1½ lb (340 g and 0.68 kg).
- Grade C livers weigh less than 1 lb (454 g).
- The size of the liver will determine how much vein is contained within.
- The finest livers should be relatively free of any bruises or blemishes.
- Surface blood spots, or small red pin dots, indicate a breakdown of capillaries or an excessive number of veins that will affect the flavor and texture of the finished dish.

Fresh foie gras is generally sold in individual vacuum-sealed packages and should be kept in the packages until they are to be used. They will keep refrigerated in its vacuum pack for up to 2 weeks from the time it is harvested. Try to use the foie gras livers within 1 week of having purchased them. Foie gras oxidizes when it comes into contact with air, so once removed from the vacuum pack, the liver should be used immediately or wrapped tightly in plastic film and used within 2 days.

It is always advisable to use fresh foie gras upon receiving it. Freezing and defrosting it will destroy the cell walls within the foie gras, allowing the moisture contained within the liver to evaporate. What is left is a dehydrated liver with a mealy, granular texture. It is also very important to use the cooler wisely when dealing with foie gras, because most cold preparations require

Table 11–1 Foie Gras Grades

GRADE	SIZE AND USE	SHAPE	TEXTURE	APPEARANCE
"A" This is the best quality liver and is the most prized.	1½ to 3 lb (0.68 kg to 1.4 kg) Usually hot and some cold dishes	Well-rounded and bulbous in appearance Both lobes visible	Firm to the touch; a thumbprint should remain slightly visible	Creamy off-white yellow amber color, relatively free of blemishes, few veins
"B" Of an inferior quality but very good for cold preparations	¾ to 1½ lb (340 g to 680 g) Generally cold presentations	May be flatter and not so compact or bulbous	May be softer or harder, depending on proportion of fat	May be darker, have a blemish or two, and have more veins
"C" Of an inferior quality but good for cold preparations	¾ to 1 lb (340 to 454 g) Always cold presentations	Slightly flatter and less rounded in shape	Soft or hard, depending on the fat content	More blemishes and slightly darker in color.

the removal of the veins from the liver, a task that can only be done with the whole liver at room temperature.

Drawing out residual blood left inside the veins of the liver can be somewhat helped by the immersion of the liver in a bath of milk for 2 hours, but due to modern processing techniques, this step is no longer required.

The debate between the duck and the goose will always be a matter of some conjecture; however, the only physical difference is that a goose liver is slightly bigger. The goose has two lobes of nearly the same size, whereas the duck has one that is one third larger than the other. Membranes, nerves, and veins are connected to the lobes, so their debris should be cleaned off carefully. If there are any green discolorations in the area, remove them with a sharp knife.

When *cleaning* foie gras, there are two different techniques: one for the livers that will be used for hot dishes and one for those that will be used for cold dishes. They differ in that the livers for hot dishes do not have the veins removed, whereas the liver used for cold dishes do.

When preparing the liver for *hot preparation* and service (e.g., roasting, searing, poaching, steaming, sautéing, and grilling), the liver is kept whole and cleaned, and then is cut into the desired size or left whole for cooking. Specific rules apply to this technique:

1. Working with a slightly chilled liver, separate the lobes by gently inserting your hands between the lobes and, with one lobe in each hand, pull them apart.
2. Use a sharp knife to cut the connective membranes and nerves between the lobes.
3. Trim away any visible membranes, veins, or green bile.
4. Cut the lobes into medallions of differing sizes, depending on the method of cooking.
5. Always use a sharp slicing knife dipped in hot water; slice the liver on a diagonal.
6. The size of the medallions will vary as the slices move along the lobe.

To avoid oxidization, the medallions of foie gras should be arranged in layers separated by white butcher paper or parchment paper, wrapped airtight with plastic film, and stored in the refrigerator until ready to use (Figure 11–3).

When preparing the liver for all *cold preparations* and service (e.g., torchon, mousse, terrine, forcemeat, sausage, cookies, and parfaits) the liver must first be cleaned and have all the veins removed before the cooking technique is started.

When preparing dishes for cold service, the liver is generally sliced and cut or puréed, and the residual blood, blemishes, veins, and other impurities when cooked will show as a dark, un-

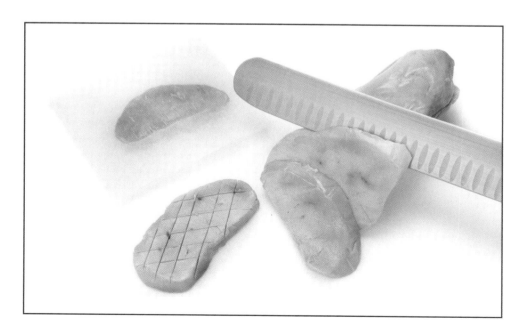

FIGURE 11–3 Slicing foie gras into medallions for hot preparations

appetizing mark. It is therefore imperative to remove all of these imperfections before starting to cook the liver. These blemishes are not inedible, but they are unattractive and they may affect the flavor of the finished dish.

During the cleaning process, it must be stressed that as little damage as possible should be done to the liver. The less damage that is done and the fewer pieces that are created, the greater the integrity of the finished terrine. To *remove the vein* from the liver (Figure 11–4) is a difficult and time-consuming task, but one that will yield a flawless finished product:

1. Bring the liver to room temperature by removing the foie gras from the vacuumed package, rinsing it well, and immersing it in a water bath of 95°F (35°C). After soaking for 1 hour, the liver will be pliable enough to clean.
2. Separate the two lobes and with the larger lobe lying upside down, find the area where the connecting membranes and veins have been cut.
3. Gently tug the membrane to reveal the location of the central vein of the lobe. As you pull, use your other hand to gently peel back the flesh of the liver, tracing the location of the vein.
4. Because of the temperature of the fat, the foie gras should be as moldable as soft clay.
5. Clean the foie gras without breaking it into pieces. The central vein reaches roughly two thirds down the middle of the large lobe before it forks into two separate directions, forming an upside down Y.
6. Remove it gently following it at all times; remove all membranes at the same time as the veins.
7. You should have a somewhat flattened but intact lobe.
8. Do the same for the other lobe.
9. Discard the membranes and veins. Using the cleaned lobes, continue with the recipe.
10. Refer to Figure 11–4 for removing veins and to recipes 11–1, 11–2, and 11–3 for cold preparations.

FIGURE 11–4
Removing the veins from foie gras for cold preparations

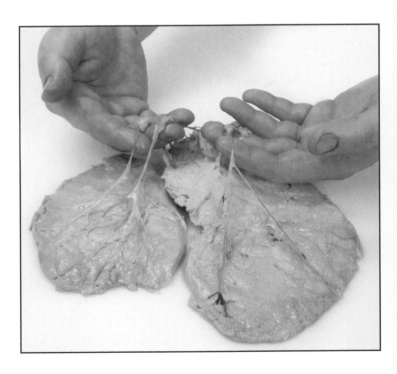

RECIPE 11–1

FOIE GRAS MOUSSE

Recipe Yield: 8 portions

MEASUREMENTS		INGREDIENTS
U.S.	**METRIC**	
1½ lb	680 g	Foie gras liver, whole
Pinch	Pinch	Sea salt
Pinch	Pinch	White pepper
2 tablespoons	30 mL	Brandy

PREPARATION STEPS:

1. Remove the veins from the liver and clean thoroughly.

2. Sprinkle the liver with salt, pepper, and brandy; chill for 3 hours (Figure 11–5A).

3. Place the liver into a resealable plastic bag and steam for 8 minutes (Figure 11–5B).

4. Remove the fat from the bag and chill well.

5. Remove the liver and pass it through a fine sieve.

FIGURE 11–5A Sprinkling the de-veined liver with seasonings and brandy

FIGURE 11–5B Plunging the sealed bag into boiling water

6. Beat the puréed liver, adding about half the reserved fat back into it (Figure 11–5C and D).

7. Reserve for further use as appetizers and canapés (Figure 11–5E).

8. Soften by taking to room temperature if you are required to mold it.

FIGURE 11–5C Puréeing the liver while adding the excess fat

FIGURE 11–5D The puréed liver

FIGURE 11–5E A plated foie gras parfait

RECIPE 11–2

INFUSED TORCHON WITH PORT AND RED WINE

Recipe Yield: 8 portions

MEASUREMENTS		INGREDIENTS
U.S.	**METRIC**	
1½ lb	680 g	Foie gras, whole
Pinch	Pinch	Sea salt
1 cup	237 mL	Port
1 bottle	1 bottle	Red wine (750 mL size)
1 sprig	1 sprig	Rosemary
½ ounce	14 g	Juniper berries, crushed
2 sprigs	2 sprigs	Thyme
2 each	2 each	Shallots, finely sliced

PREPARATION STEPS:

1. Bring the wine and port to a boil with the rosemary, thyme, shallots, and juniper.

2. Cook for 10 minutes and strain though fine cheesecloth.

3. Clean the foie gras liver of all veins and season well with the sea salt. Allow to rest, covered, in the cooler for 2 hours.

4. Place the foie gras in a bowl and bring to room temperature.

5. Bring the flavored wine port mixture to a boil, and allow the mixture to cool until it reaches 158°F (70°C). Pour cooled liquid over the liver (Figure 11–6A).

6. Allow to rest at room temperature for 1 hour and then chill in the cooler overnight.

7. Remove the liver from the liquid and allow it to come back to room temperature (Figure 11–6B).

FIGURE 11–6A Pouring the cooled liquid over the cleaned liver

FIGURE 11–6B Removing the liver from the liquid

8. Shape the liver, as for torchon, in plastic wrap. Roll tightly to form a tubular shape, like a large sausage. Remove plastic, and re-roll in cheesecloth and tie securely. Refrigerate overnight and serve as desired (Figure 11–6C, D, and E).

FIGURE 11–6C The room temperature liver being formed in plastic wrap

FIGURE 11–6D The shaped foie gras

FIGURE 11–6E The shaped foie gras, rolled in cheesecloth, and labeled for storage

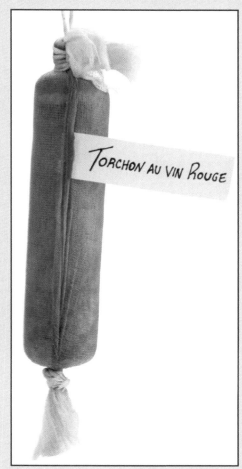

RECIPE 11–3

CURED FOIE GRAS

Recipe Yield: 8 portions

MEASUREMENTS		INGREDIENTS
U.S.	**METRIC**	
1½ lb	680 g	Whole foie gras liver, veins removed, lobes separated
1½ lb	680 g	Kosher salt
1¼ lb	567 g	White sugar
1 tablespoon	15 mL	Black pepper, finely ground

PREPARATION STEPS:

1. Lay out two large rectangles of cheesecloth on top of each other.

2. Shape the lobes of foie gras into two logs, similar to the shape of galantines, very tightly and tie off the ends very tightly (Figure 11–7A, B).

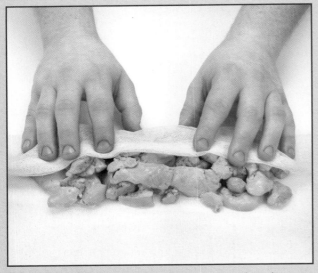

FIGURE 11–7A Rolling the foie gras into logs in cheesecloth

FIGURE 11–7B Tying ends securely

3. Completely submerge the tied logs in the salt, sugar, pepper mixture, ensuring that the logs are completely covered (Figure 11–7C).

4. Cover and chill for 30 hours.

5. Remove from the cure and clean the salt and sugar mix off the logs.

6. Keep the foie gras chilled until 1 hour before service (Figure 11–7D).

FIGURE 11–7C Covering the rolled foie gras with the cure

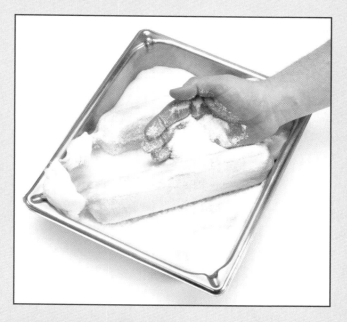

FIGURE 11–7D Presenting cured foie gras

Cooking Foie Gras

When *cooking foie gras* for hot dishes, specific rules should be followed for the different methods of cooking. The following recipes can be used as a guide for the cooking of foie gras.

PAN-ROASTED FOIE GRAS

Recipe Yield: 1½ lb (680 g)

MEASUREMENTS		INGREDIENTS
U.S.	METRIC	
1 each	1 each	Foie gras, Grade A, cleaned
Pinch	Pinch	Sea salt
Pinch	Pinch	White pepper
1 bunch	1 bunch	Herb of choice (e.g., rosemary, thyme, sage, or lemon balm)
1 tablespoon	15 mL	Whole spice of choice (e.g., peppercorns, coriander seeds, juniper berries, or allspice)
2 each	2 each	Shallots, sliced thin
4 each	4 each	Cloves garlic, bruised

PREPARATION STEPS:

1. Preheat the oven to 375°F (177°C). Season the foie gras generously with sea salt and white pepper.

2. Heat a sauté pan over high heat and sear the whole liver on both sides until golden in color. Add the herbs, spices, and vegetables, and baste well with the fat that has come from the liver searing.

3. Roast for 10 to 15 minutes, until the liver is cooked to medium rare and the flesh is firm to the touch.

4. Baste regularly during the roasting process.

5. Rest the liver for 6 to 8 minutes before carving.

> **TIP** For the **pan-roasted foie gras**, Grade A livers should be used, and the pan selected should be a nonstick type.

RECIPE 11–5

SEARED FOIE GRAS

Recipe Yield: 1½ lb (680 g)

| MEASUREMENTS | | INGREDIENTS |
U.S.	METRIC	
1 each	1 each	Foie gras, Grade A, cleaned and cut into 6 slices, each ³/₄-inch (0.6-cm) thick.
1 tablespoon	15 mL	Coarse salt
1 tablespoon	15 mL	Black pepper, freshly ground

TIP For **seared foie gras**, great care must be taken to recognize when the pan is ready for the foie gras slice. Because there is no temperature reading and because all pans react differently to heat, this technique must be practiced until perfected, preferably with the foie gras trim.

PREPARATION STEPS:

1. Season the foie gras slices with salt and pepper.

2. Heat a cast iron skillet over high heat and sear the foie gras for about 30 seconds on each side.

RECIPE 11–6

SAUTÉED FOIE GRAS

Recipe Yield: 1½ lb (680 g)

| MEASUREMENTS | | INGREDIENTS |
U.S.	METRIC	
1 each	1 each	1 Foie gras, Grade A, cut into ½-inch (1.3-cm) thick slices
1 tablespoon	15 mL	Coarse salt
1 tablespoon	15 mL	Black pepper, freshly ground
1 cup	237 mL	Any liquid of your choice (including wine, spirit, or stock)

TIP When **sautéing foie gras,** browning the liver well at the beginning of the cooking process will determine how good the flavor is at the end.

PREPARATION STEPS:

1. Season each slice of foie gras with salt and pepper.

2. Heat a sauté pan over high heat and sear the foie gras for about 30 seconds on each side, or just until a brown crust forms.

3. Remove the slices from the pan and dry on paper towels.

4. Deglaze the pan with the liquid for your sauce.

RECIPE 11–7

GRILLED FOIE GRAS

Recipe Yield: 1½ lb (680 g)

| MEASUREMENTS | | INGREDIENTS |
U.S.	METRIC	
1 each	1 each	Foie gras, Grade A, cut into ³/₄-inch (0.6-cm) thick slices
1 tablespoon	15 mL	Coarse salt
1 tablespoon	15 mL	Black pepper, freshly ground

TIP When **grilling foie gras,** check that the grill is very clean and that the grate has been slightly oiled before use. This will stop the liver from sticking and picking up debris.

PREPARATION STEPS:

1. Score the foie gras and season with salt and pepper.

2. Grill over wood or charcoal on hot clean grill bars, turning once.

3. Cook for about 1 minute per side, depending on the temperature of the grill.

4. The liver does contain a lot of fat, so be careful that there is no flare-up.

RECIPE 11–8

STEAMED FOIE GRAS

Recipe Yield: 1½ lb (680 g)

> **TIP** **Steaming foie gras** seems an unlikely way to prepare this noble liver, but the results are surprisingly good.

MEASUREMENTS		INGREDIENTS
U.S.	**METRIC**	
1 each	1 each	Foie gras liver, Grade A
1 tablespoon	15 mL	Coarse salt
1 tablespoon	15 mL	Black pepper, freshly ground
1 cup	237 mL	Liquid of choice (e.g., chicken, duck, or veal stock)
1 tablespoon	15 mL	Whole spices of choice (e.g., allspice, cinnamon stick, cardamon pods)
1 bunch	1 bunch	Whole herbs of choice (e.g., thyme, rosemary, lavender, tarragon, chervil, or parsley)

PREPARATION STEPS:

1. In the bottom of the steamer, bring the liquids and flavoring agents to a boil.

2. Season the foie gras generously with salt and pepper on both sides and place in the steamer basket.

3. Place the basket inside the steamer, cover, and steam the liver for 7 minutes.

RECIPE 11–9

POACHED FOIE GRAS

Recipe Yield: 1½ lb (680 g)

> **TIP** When **poaching foie gras**, make sure that the liquid comes to, and is maintained at, the correct temperature for best results.

MEASUREMENTS		INGREDIENTS
U.S.	**METRIC**	
1 each	1 each	Foie gras liver, Grade A, cut in 3-ounce (85-g) slices
1 tablespoon	15 mL	Sea salt
1 lb	15 mL	Black pepper, freshly cracked
2 cup	473 mL	Poaching liquid of choice (e.g., court bouillon or stock flavored with fruit juices, wines, aromatic vegetables, herbs, and spices)

PREPARATION STEPS:

1. Place the liquid in a saucepan and heat to 140°F (60°C)

2. Add the foie gras slices and poach for 10 minutes.

3. Remove the foie gras and set aside.

RECIPE 11–10

BUTTER-ROASTED FOIE GRAS

Recipe Yield: 1½ lb (680 g)

| MEASUREMENTS | | INGREDIENTS |
U.S.	METRIC	
1 each	1 each	Foie gras, Grade A, cut into 1-inch (2.5-cm) thick slices
1 tablespoon	15 mL	Sea salt
1 tablespoon	15 mL	Black pepper, freshly ground
2 tablespoons	30 mL	White bread crumbs
2 tablespoons	30 mL	Butter

> **TIP** For the **butter roasting of foie gras**, use only the best butter and baste the meat very well.

PREPARATION STEPS:

1. Season the foie gras slices with salt and pepper, and coat lightly with the bread crumbs.

2. In a very hot sauté pan, sear the foie gras slices for about 40 seconds on each side, until golden brown.

3. Add the butter to the pan and cook for an additional minute, using the butter to baste the foie gras.

4. Remove from the pan and drain on paper towels.

SWEETBREADS

Sweetbreads are without a doubt one of the most delicately flavored of the offal meats, and are much sought after for their subtle flavor and wonderful texture. They are, in fact, the small thymus glands from the neck and heart of young steers, calves, and lamb. As the animal grows and matures, these glands shrivel and disappear; so they are never found in older animals. The round lobe is found near the heart, and the longer, elongated lobe is found in the throat of the animal. These paired sweetbreads are generally sold together, and there is little difference in the flavor; no matter what their size. Generally the lamb sweetbreads, which are much smaller, are not as popular as the larger veal sweetbreads.

- Sweetbreads should be light, bright, and rosy in color.
- The larger they are in size, the more desirable they are to chefs.
- There should be no blood spots or bruising visible on the sweetbread.
- The outer membrane is removed either before or after cooking.

When preparing the sweetbreads for cooking, the following steps should be taken:

1. Soak in cold water for about 6 to 8 hours, changing the water often; this helps to whiten them and clean them of excess blood.
2. Blanch in simmering water with a little lemon juice or vinegar added for about 2 minutes to help firm their texture and prepare them for trimming.
3. Chill immediately and pat dry.
4. Carefully trim off all tubes, sinews, and any fat (Figure 11–8A).
5. Press very lightly between two boards to even their size (Figure 11–8B).
6. The sweetbreads can now be larded or studded as desired.
7. They are braised brown or white, sautéed, or fried.

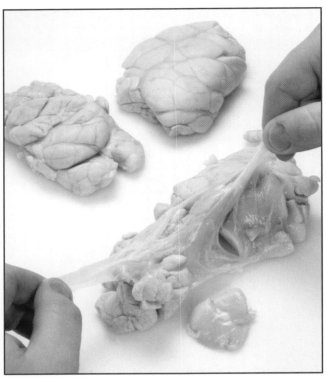

FIGURE 11–8A Cleaning the blanched sweetbreads of all sinew, tubes, and fat

FIGURE 11–8B Weighting the sweetbreads for even shape (shown here, between two cutting boards)

TONGUE

Fresh **tongue** is generally bought whole and intact with no bones or gristle attached (Figure 11–9). They have a finely knit texture and look much the same, apart from size, in all animal species. They are consumed cold or hot, and have a very distinctive flavor. Check the quality before purchasing:

■ There should be no throat bones or cartilage attached. They should be removed with a sharp knife, making sure not to cut the flesh of the tongue.

■ Wash well in cold water to remove any blood.

When cooking the tongues, the following are some important points to follow:

1. Tongues should be soaked in acidulated water for an hour or in plain water overnight if the tongues are to be salted.

2. Tongue has a thick outer layer of skin and requires a long, slow, moist cooking method to make them tender enough to eat.

3. Cook the tongue by poaching. When it is fully cooked, plunge it quickly in cold water. The skin is then split and peeled off—like a glove

4. Before serving, cut away any roots, small bones, or gristle that might still be present at the neck end.

5. Tongues can be pickled and pressed into shapes before cutting but this must be done after skinning and before they cool down.

HEART

The **heart** is a very active muscle and is very tough and elastic. Heart is the least expensive of all the variety meats and offers the bonus of rich nutritional benefits. The younger the heart is, the

FIGURE 11–9 Fresh beef tongues

more tender it is when cooked. Heart has a wonderful flavor and is highly underestimated as a pleasing meal; the effort involved in its preparation is definitely worthwhile.

Veal and lamb heart are the most highly prized and can be cooked a number of ways, with roasting and braising being the most common. Check for quality before purchasing:

- Hearts should have a fresh smell and red color, not brown or gray.
- Store loosely wrapped in the refrigerator for no more than 3 days.

When cooking the hearts, the following are some important points to follow:

1. All hearts should be thoroughly washed before cooking and the membrane inside that divides the two heart chambers should be removed, particularly if the heart is to be stuffed.
2. Open the heart without separating halves (Figure 11–10A).
3. Trim off excess fat and tubes.
4. Remove clots of blood and sprinkle with olive oil and lemon juice.
5. Marinade for 1 to 2 hours and season with salt and pepper.
6. Stuff with pork forcemeat or savory stuffing (Figure 11–10B).
7. Wrap in larding bacon cut paper-thin or in pig's caul.
8. Tie well with string and cook gentle by roasting or braising.
9. Or, the heart may be sewn closed for cooking (Figure 11–10C).

KIDNEYS

The **kidneys** are highly nutritious and have a very distinctive flavor. They are considered one of the great offal dishes and are a great delicacy. The kidneys of veal, lamb, and pork are the most tender, with veal being considered the most flavorful. Pork kidneys are normally used for pâtés, terrines, and stews. The beef kidney has a much more pronounced flavor and odor, and is used less for fine cooking, being reserved more for stew and pies. The following are some general

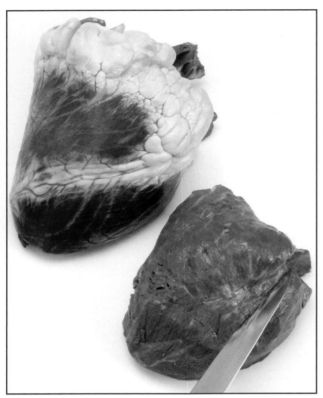

FIGURE 11–10A Opening the heart without separating the valves

FIGURE 11–10B Adding the stuffing

FIGURE 11–10C Sewing the heart closed

rules that should be observed when buying kidneys:

■ They should not appear limp or have a strong smell.

■ They are highly perishable and should be prepared promptly after they are purchased.

■ They should have a bright appearance and not be shriveled in any way.

■ Kidneys should be firm, pink or pale red rather than purple, and should not have a uric acid smell.

Lamb kidneys can be bought with the fat still surrounding them, which can be trimmed to within ½ inch (1.25 cm) of the kidney's surface in order to act as a natural basting agent during cooking. To prepare lamb kidneys for cooking, take the following steps:

1. Remove the skin from the kidney (Figure 11-11A).

2. Slit on bulging side and open without separating the two halves (Figure 11-11B).

3. Trim the sinew from the inside, skewer to hold open for grilling (Figure 11-11C).

FIGURE 11–11A Skinning the kidneys

FIGURE 11–11B Splitting a lamb kidney

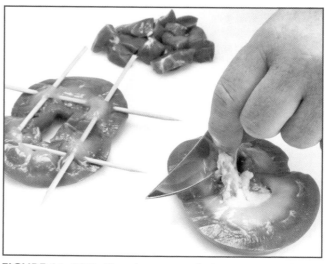

FIGURE 11–11C Cleaning away all the sinew and fat

PLUCK

In certain parts of the world, the word **pluck** can mean just the lungs on their own, or in any combination of the lungs, the liver, the spleen, and the heart. Pluck is generally used in the making of sausage, including the famous haggis of Scotland. In Europe, pluck is sometimes blanched and cut up into sauces as stews. When preparing the pluck, there are some important techniques that must be observed:

1. The lungs should be beaten vigorously to expel air, and the spleen should be skinned.
2. Pluck should be soaked in cold water with salt for 24 hours to remove the blood.
3. Blanch the spleen and lungs in salt water for 10 minutes. Slice it thinly and fry in clarified butter.
4. Or, lightly poach the spleen and lungs for $1\frac{1}{2}$ hours until tender and use for stuffing and sausage.

BRAINS

The most sought after **brain** is the calf's brain, although the lamb's, cow's, and pig's are used, but mainly in the production of sausage-style dishes and as fillings or stuffing. Care must be taken when choosing brains:

- Choose brains that are a clean, light pink color and free of blood clots and stains.
- Brains should be firm, plump, and pinkish-white.
- Chill well and use the same day as purchased.
  When cooking brains, the following important steps should be taken:

1. Before precooking, soak the brains in cold water and remove the outer membrane.
2. Brains should be soaked in cold water until all the blood is leached away, then remove the arteries and fibers.
3. Precooking is, in fact, a prerequisite to most methods of preparation and it also enhances the keeping quality of brains.
4. Simmer the brains for about 20 minutes in salted water, adding a tablespoon or two of lemon juice or vinegar, and other seasonings if you desire.
5. Or, simmer the brains in milk. This step will firm their mushy consistency for use in other recipes.
6. The brains are rested and cooled before the next method of cooking takes place.
7. Brains may be sautéed or cut in small pieces and deep-fried in batter or other coatings, fried, creamed, or scrambled with eggs.

BLOOD

Fresh **blood**, although difficult to find nowadays, is used in many international dishes and has specific techniques involved in its storage and use. It is used for its thickening properties, binding properties, flavor, and color. The following are considerations when using blood in cooking:

- Fresh blood straight from the animal is at great risk of spoiling unless dealt with immediately.
- A little lemon juice or vinegar should be added to stop the blood from clotting during refrigeration. One tablespoon (15 mL) of lemon juice per 2 quarts (1.9 L) of blood is the recommended ratio.
- Store the fresh blood for no more than 2 days.

HEAD

The head of the animal is used internationally for many interesting local dishes, and the ways of preparing them are complex and sometimes inexplicable (Figure 11–12). They are boiled whole for sausages in Louisiana; they are split and used in soups in Brazil and Argentina. The wool is

FIGURE 11–12 Skinned calf head

scorched and scraped, and the head is split and rubbed with the brain then made into soup in Scotland. The head is boned, rolled and stuffed, braised, chilled, and then sautéed in France; and, of course, the head is boned and used for sausage all over the world. When purchasing heads, there are some important points of quality that must be observed:

■ Check the neck for obvious signs of bruising and damaged flesh.

■ Ensure the windpipes have been removed.

■ Check the ears, nose, and mouth areas of anything that looks unclean.

The heads can be used in a variety of soups and stews, or poached to remove the flesh that is attached to the jaws, cheeks, and around the forehead; however, the classical technique is to bone stuff, roll, then poach the head. It is then carved, dressed in mustard and breadcrumbs, and sautéed for service, rendering this piece of meat tender, delicious, and well presented.

1. Make an incision down the center of the head from the top of the forehead to the nose of the animal.

2. Follow the head from top to bottom with your knife, removing the flesh from the forehead, around the eyes, along the snout, and along the jaw (Figure 11–13A).

3. Run your knife along the meat, keeping the flesh attached to the skin but following the contour of the skull until you have completely boned the head.

4. Lay the boned flesh out flat. Clean around the ears, nose, and tongue, removing any cartilage and sinew. Remove all large fat pockets that are visible, cleaning down to the flesh (Figure 11–13B).

5. Remove the tongue and cut into strips about 3 by $\frac{1}{2}$ inch (7.5 by 1.25 cm) and reserve (Figure 11–13C).

6. If the head was not previously skinned, remove the skin by using a long, firm ham knife and cutting under the flesh against the skin as if skinning a large salmon fillet.

7. Lay the flesh into the best rectangle you can make from it in between two sheets of strong plastic and lightly beat the head into a somewhat even looking sheet.

8. Roll this up with the tongue inside, seasoning well as you go (Figure 11–13D).

> **TIP** Take great care to keep your knife facing the bone at all times in order to remove as much flesh as possible.

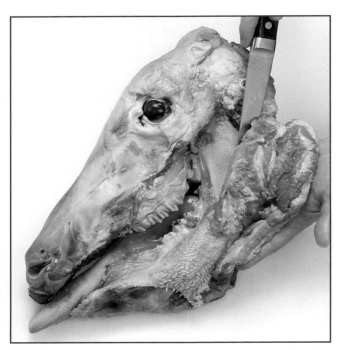

FIGURE 11–13A Removing the meat from the side of the head

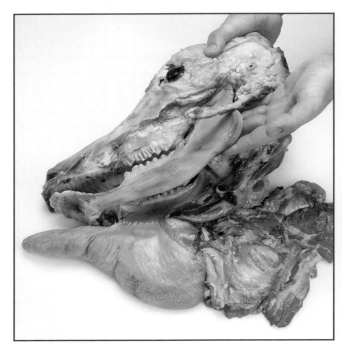

FIGURE 11–13B Laying the boned flesh out flat and cleaning

FIGURE 11–13C Removing the tongue and tongue skin and cutting into strips

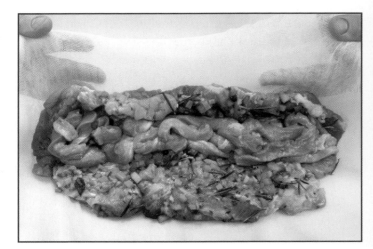

FIGURE 11–13D Rolling the meat up in cheesecloth with the tongue inside

9. Tie in muslin into a galantine shape and secure with butcher's twine.
10. Poach for 3 hours very gently until fully cooked.
11. Cool until you can handle it and then rewrap and tie tightly to create its final shape in fresh muslin, allowing it to chill in the liquid it was cooked in.

TRIPE

The **tripe** comes from the first and the second stomachs of beef. From the *rumen,* or first stomach, we get the plain tripe that tends to be rather tough. From the second, or the *reticulum,* we get the honeycomb that is the more tender and the more attractive of the two (Figure 11–14). Honeycomb is preferred, being easier to cook and having a little better flavor.

Today all tripe is chemically washed, soaked, scraped, and scalded or blanched to remove the inner stomach wall before leaving the packinghouse. Tripe is, however, very tough and will require 2 to 3 hours of cooking in a very good court bouillon to make it tender enough to eat.

1. Soak in acidulated water overnight and wash well the next day in plenty of running cold water.
2. Cut into strips about 4 inches (10 cm) long and 1 inch (2.5 cm) wide.
3. Braise or poach in rich court bouillon or good beef broth until tender.
4. Add to an appropriate sauce and serve.

TAILS

Oxtails are the **tails** of beef cattle. They are a great delicacy when handled well and make a truly remarkable meal. They contain a considerable amount of meat that has a rich, hearty flavor and a high quantity of gelatin. They are, however, extremely tough, contain a good deal of bone, and require long, slow, moist cooking. The tails of veal, lamb, and pork are also used in a similar manner to the oxtail. When choosing oxtails, there are some rules to follow:

■ Look for tails with an even distribution of meat and fat.
■ They should have an even coating of very white fat.
■ They should be skinned and trimmed of excess fat.

Tails can be bought cut into chunks through cartilage between segments of bone. They can also be boned without damaging skin. Season with salt and pepper and stuff. Roll and tie with cloth and string, like a galantine, for cooking.

FIGURE 11–14 Fresh honeycomb tripe at the market

CAUL FAT

Pig's **caul fat** is the lining of a pig's stomach, on which there is a certain amount of fat (Figure 11–15). The excess fat is removed and can be used as lard. The rest of the membrane is used as a protective wrapping for any meat item that needs to have a "natural plastic wrap" to hold it together during cooking. Soak it in salt water and then drain well before using. It can be very delicate, so handle with some care, especially when removing it to begin with; if it has been shrunk wrapped, it tends to be bound together rather tightly.

Caul fat will mostly disappear when cooked, except when it has been wrapped a number of times around a food item before cooking.

FIGURE 11–15 Sheet of caul fat showing the crepinette pattern

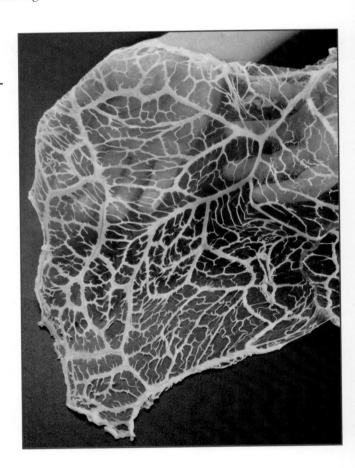

Professional Profile

BIOGRAPHICAL INFORMATION

Name: Bradley Ogden

Place of Birth: Traverse City, Michigan

Recipe Provided: Oven-Roasted Sweet Onions

CULINARY EDUCATION AND TRAINING HIGHLIGHTS

Bradley Ogden always has been inspired by the true quality and flavors of farm-fresh foods. Today, his enduring success can be attributed to bringing these exciting American ingredients from the farm to the table with dedication and innovative culinary flair.

Chef Ogden graduated with honors from the Culinary Institute of America and received the Richard T. Keating Award for the student most likely to succeed. Ogden spent four and a half years as chef at the renowned American Restaurant in Kansas City where, under the guidance of Joe Baum and Barbara Kafka, he was encouraged to explore the rich heritage of native foods while developing his signature expression of regional cuisine. He took that style to San Francisco as executive chef of the Campton Place Hotel, immediately earning countless accolades for his talents in the kitchen and his unique approach to American food.

Six years later, in 1989, Ogden opened The Lark Creek Inn, where he fully realized his vision of combining his cuisine with fine service, award-winning wines, and a relaxed, elegant setting. Today, The Lark Creek Inn continues to be a "must-stop" restaurant for both Bay Area residents and visitors alike. As chef and co-owner of the eight restaurants of the Lark Creek Restaurant Group, Ogden enjoys the reputation of being one of the country's most prolific and successful restaurateurs, while being at the vanguard of authentic American cuisine. Having established and nurtured close relationships with many American farmers, ranchers, and fishermen, he always brings farm-fresh ingredients with him wherever he goes.

In addition to publishing *Bradley Ogden's Breakfast, Lunch & Dinner* (1991), Ogden has shared his innovative cooking throughout California, co-owning Lark Creek restaurants in San Mateo and Walnut Creek; the venerable One Market Restaurant in San Francisco; Yankee Pier in Larkspur and San Jose; Parcel 104 in Santa Clara; and Arterra in San Diego. Each of these restaurants is co-owned by Michael Dellar, Ogden's long-time business partner, and his wife, Leslye Dellar.

On March 17, 2003, Chef Ogden's namesake restaurant opened at Caesar's Palace, in Las Vegas, Nevada. From its opening, *Bradley Ogden* received critical acclaim, including accolades such as "2004 Best New Restaurant"—James Beard Foundation; "Best of the Best Dining"—June 2004 *Robb Report* magazine; among "America's Best Restaurants"—October 2003 *Gourmet* magazine; and among "Top 10 Restaurants in Las Vegas"—February 2004 *Wine Spectator* magazine.

He received the International Food Service Manufacturers Association Silver Plate Award for the Independent Restaurant category. Ogden has been named Best Chef California by the James Beard Foundation, one of the Great American Chefs by the International Wine and Food Society, winner of the Golden Plate Award by the American Academy of Achievement, and Chef of the Year by the Culinary Institute of America. A popular television guest, Ogden has appeared on the *Today Show, Good Morning, America, AM/San Francisco, Dinner at Julia's,* and the Discovery Channel's *Great Chefs of the West* series.

RECIPE 11–11

CHEF BRADLEY OGDEN'S OVEN-ROASTED SWEET ONIONS

Recipe Yield: 8 portions

MEASUREMENTS		INGREDIENTS
U.S.	**METRIC**	
4 each	4 each	Sweet onions, such as Vidalia
4 each	4 each	Garlic cloves, peeled and sliced thin
6 each	6 each	Bay leaves, chopped
16 sprigs	16 sprigs	Parsley, fresh chopped
4 sprigs	4 sprigs	Oregano, fresh chopped
4 sprigs	4 sprigs	Sage, fresh chopped
1 teaspoon	5 mL	Salt, kosher
1 teaspoon	5 mL	Black pepper, fresh ground
½ cup	118 mL	Olive oil
½ cup	118 mL	Balsamic vinegar

PREPARATION STEPS:

1. Preheat oven to 500°F (260°C).

2. Cut whole onion into eight equal wedges.

3. Place wedges in a single layer in a shallow roasting pan.

4. Sprinkle onions with sliced garlic and chopped herbs.

5. Season with salt and pepper.

6. Drizzle with olive oil and vinegar; be sure to coat evenly.

7. Roast onions in hot oven until they are lightly caramelized and slightly tender (approximately 12 to 15 minutes).

8. Remove onions from oven and allow cooling.

1. What are offal meats?

2. What are the quality points of fresh beef liver?

3. What is foie gras?

4. How are foie gras preprepared for hot dishes?

5. How are foie gras preprepared for cold service?

6. Where are sweetbreads found?

7. How are lambs kidney cleaned?

8. What is tripe?

9. What is caul fat?

10. How is caul fat used?

A. **Group Discussion**

Discuss the use and popularity of offal meats in modern upscale restaurants. Discuss the revival of offal meats and whether they are here to stay.

B. **Research Project**

Research the history of foie gras, establishing why it is so revered amongst food products.

C. **Group Activity**

Remove the veins from a foie gras liver very carefully, recording their path through the liver in the form of a diagram.

D. **Individual Activity**

Report on the possible reasons why people are so intimidated by the thought of eating offal meats.

PART IV

Preserved Foods of the Garde Manger

CHAPTER 12

Methods of Preserving Foods

After reading this chapter, you should be able to

- describe the advantages of preserving food.
- use acids to partially preserve and flavor food.
- make carpaccio.
- make ceviche.
- make tartar.
- dry vegetables.
- dry fruits.
- dry herbs.
- can, jar, and pickle.
- make salsa and relish.
- make chutney.
- make chow-chow.

Mixed with fruit and flowers, sugar produces jams, marmalades, preserves, pastes, and candied fruits in a conserving process which lets us enjoy their flavors long after the time which nature had meant them to last.

JEAN ANTHELME BRILLAT-SAVARIN

The Advantages of Food Preservation in the Kitchen

Historically, the human race has preserved foods in jars, pots, tubs, barrels, and bottles longer than archeologists have been uncovering shards of pots. People were forced to capture the bounty of the harvest, and to store it for the cold months and lean times that lay ahead. Preserved foods have also played a significant role in our social and cultural history. According to Sue Shephard (2000):

> Food preserving helped make it possible for our nomadic ancestors to settle down in one place and build agrarian communities where they could live in reasonable confidence that they would not go hungry through the variable seasons and the many other difficulties that nature might throw at them. Food preserving also made it possible for some of our ancestors to travel, taking their food with them as they journeyed over long distances to explore unknown places, confident, if they could find no fresh food, that their portable provisions meant they would not starve. (p. 15)

Today, in the modern kitchen, it makes sense to try to capture the same freshness and quality, and preservation has become a very important part of kitchen philosophy (Figure 12–1). There are many advantages to preserving food in the kitchen, and the skill and knowledge involved can pay huge dividends. Some of the advantages include the following:

- Preservation captures the food at its very best quality.
- Seasonal foods are less expensive to purchase, saving on food cost.
- Preservation allows the chef to feature locally grown products on the menu at any time.
- It is convenient to have the preserved food on hand.
- The food can be marketed as having been made "in-house."
- The preserving agents enrich and add to the flavor of the food.
- Garde manger chefs can offer unique, custom-made products as signature items.

Using Acid to Partially Preserve and Flavor Food

Adding an **acid** ingredient to a piece of food not only extends its shelf life but also improves its flavor dramatically. In order to understand how to apply the acids to foods, the garde manger chef must first understand how to measure its effectiveness and which acids are the most effective.

The acidity of a food is indicated by its **pH value**. The pH scale ranges from 0 to 14, with pH 7 being neutral. Any pH registering below 7 falls in the acidic range, whereas those registering above pH 7 are in the basic range. The lower the pH reading, the more acid the food contains (Table 12–1).

FIGURE 12–1 An assortment of kitchen-made preserves. Courtesy of Getty Images, Inc.

Table 12–1   pH Value Scale

1	2	3	4	5	6	7	8	9	10	11	12	13	14
Less than 4.6 inhibits the growth of toxins.					7 is neutral.				Basic range.				

It is important that foods preserved by acidity have a pH level of 4.6 or less. At this level, the deadly toxins produced by the organism that can cause botulism are considerably inhibited. All foods that have a pH level greater than 4.6, including items such as meats and vegetables, are called low-acid foods. Most fruits and fruit products are high-acid foods with a lower pH level, and their juices are a good source for use in preservation.

Foods that are preserved with acid depend on one or more food-type acids, such as citric, malic, or acetic acid, in order to achieve stability (Table 12–2). When low-acid foods are used in recipes, they should be allowed to properly acidify before they have a chance to spoil. The rates at which they accept the acids are considerably altered by the portion size of the food. It is advisable to always cut the low-acid food items into smaller pieces, either by carving them thinly or cutting them into small strips.

When foods are allowed to remain in the **temperature danger zone** (between 41° and 135°F; 5° and 57°C), harmful bacteria begins to grow. The lag phase may last from 1 to 4 hours, which is followed by a period of accelerated bacterial growth known as the log phase (Figure 12–2). Dangerous bacteria grows in geometric proportions until it reaches the stationary phase where the bacterial growth planes out. For this reason, food should not be allowed to remain in the temperature danger zone for more than 2 hours. Foods higher in acid will experience a much slower rate of bacterial growth.

QUICK PICKLING LOW-ACID FOODS

Quick pickling involves the addition of acids and flavoring agents that are suitable to preserve or partially preserve foods. Internationally, this method of preparing food is done to enhance the

Note: When finishing a dish with flavored oil, the food should reach the correct pH of 4.6 or less before the oil is added.

Note: Temperature Danger Zone: The Food and Drug Administration (FDA) releases an updated food code every four years. At interim two-year periods, the FDA releases a supplement. As of the printing of this text, the FDA food temperature danger zone is between 41°F and 135°F (5°C and 57°C), a temperature range in which bacteria multiply rapidly. Additionally, initial temperature for cooling hot foods is from 135°F to 70°F within two hours, and then on down to 41°F within four more hours.

Table 12–2   ph Values of Commonly Used Foods for Preservation

Vinegar 2.5	Pineapple juice 3.3
Cider vinegar 3.10	Raspberries 3.2
Lime juice 1.8 to 2.1	Ketchup 3.6
Lemon juice 2.6	Honey 3.7 to 4.2
Orange juice 3.3	Cranberry juice 2.3 to 2.5
Worcestershire sauce 3.6	Apple juice 3.5
Tomato juice 4.0 to 4.6	

eating quality of the foods, but originally was created as a quick method of preservation that generally only lasted for a few days. According to Shephard (2000):

> We know from Herodotus that in the fifth century BC the Babylonians and Egyptians pickled fish such as sturgeon, salmon, and catfish, as well as poultry and geese (p. 76). . . . Salt's dehydrating properties could . . . be applied to almost any food. . . . In the hot southern regions of the Mediterranean, the Middle East and North Africa, fresh vine leaves were parboiled and preserved in jars of brine to make stuffed vine-leaf dishes such as dolma (p. 93). . . . Although the word "pickle" is used in many different preserving methods, preserving foods in containers filled with vinegar is always known as *pickling*. (p. 95)

Note: There is always the risk of food poisoning with the consumption of partially cooked or raw foods. Menu listings should be accompanied by a warning to the guests about their consumption, such as "Consuming raw or undercooked meats, poultry, seafood, shellfish, or eggs may increase your risk of foodborne illness."

Ceviche

Ceviche (also spelled *seviche* and *cebiche*) means *to saturate*. Ceviche refers to the marinating of fish and shellfish in lime or lemon juice, which in turn gives the fish a "cooked" appearance and texture (Figure 12–3). The origin of ceviche has been under debate for some time, but there is no doubt that it originated from one or more of the South and Central American countries. Locations in Mexico, Peru, Ecuador, and Chile still serve this popular and traditional dish. Good ceviche is only made with the freshest of fish and shellfish and with the sourest of limes, lemons, and sour oranges.

There are two basic ways of producing ceviche. They differ in that one classification of food needs to be lightly blanched in order to be completely edible, and the other does not.

FIGURE 12–2 Bacterial growth chart

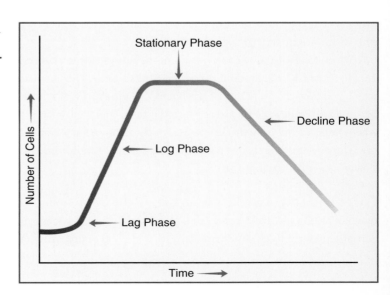

FIGURE 12–3 Appetizer of ceviche-style fish

ACID-COOKED Acid-cooked ceviche is generally made with all fish that can be safely eaten raw, including salmon, tuna, swordfish, mackerel, flounder, fluke, and scallops. They are first cut into small strips and then marinated with a combination of the chosen acids and finished with herbs, spices, and oils.

BLANCHED AND ACID-COOKED TO FINISH This technique involves lightly blanching the fish and shellfish in a richly flavored liquid, then finishing the cooking by adding the acids, flavorings, and oils when the fish is cold and allowing it to marinate. Shrimp, lobster, crabs, crawfish, octopus, squid, and prawns are all popular ingredients associated with this technique. If the chosen ingredients have tender flesh, try to marinate them without first blanching, because the flesh tends to get soft and mushy.

Note: All ingredients used for the production of ceviche should be checked for their suitability for consumption.

RECIPE 12–1

Turbot
Ceviche

Recipe Yield: 8 portions

MEASUREMENTS		INGREDIENTS
U.S.	**METRIC**	
3 cups	680 g	Turbot fillet cut into small cubes about $\frac{1}{2}$-inch (1.25-cm) square
1 cup	237 mL	Lime juice
$\frac{1}{2}$ cup	118 mL	Olive oil
1 each	1 each	Small red onion, finely diced
2 each	2 each	Tomatoes, peeled, seeded, and finely diced
1 each	1 each	Jalapeno pepper, roasted and finely diced
$\frac{1}{2}$ cup	113 g	Black olives, fine diced
1 fluid ounce	30 mL	Dry white wine
1 ounce	28 g	Gherkins, fine diced
1 teaspoon	5 mL	Chopped cilantro
To taste	To taste	Salt and white pepper

PREPARATION STEPS:

1. Marinate the turbot for at least 6 hours, but preferably overnight in 4 fluid ounces (118 mL) of the lime juice.

2. Rinse the fish in a colander and let it dry for 5 minutes.

3. Combine with the rest of the ingredients and toss lightly.

4. Season with salt and pepper to taste.

5. Serve chilled in attractive glasses.

RECIPE 12–2

SCALLOP CEVICHE

Recipe Yield: 8 portions

MEASUREMENTS		INGREDIENTS
U.S.	**METRIC**	
1½ pounds	680 g	Diver scallops, cut into ½-inch (1.25-cm) slices
½ cup	118 mL	Lime juice
¼ cup	59 mL	Olive oil
1 fluid ounce	30 mL	Balsamic vinegar
1 each	1 each	Serrano chile, roasted and fine diced
1 each	1 each	Avocado, fine diced
1 each	1 each	Tomato, skinned, seeded, and fine diced
1 teaspoon	5 mL	Fresh oregano, finely chopped
To taste	To taste	Salt and white pepper
½ each	½ each	Small red onion, very finely chopped

PREPARATION STEPS:

1. Put the cut scallops in a clean colander. Pour boiling water over them, and let them drain well; chill immediately.

2. Marinade the scallops with the lime juice, then drain well.

3. Toss the marinated scallops with the remaining ingredients and allow to rest chilled for 3 hours before service.

RECIPE 12–3

MAHI-MAHI CEVICHE

Recipe Yield: 8 portions

MEASUREMENTS		INGREDIENTS
U.S.	**METRIC**	
2 pounds	946 g	Mahi-Mahi, cut into ½-inch (1.25-cm) pieces
2 each	2 each	Lemons, juice of
2 each	2 each	Limes, juice of
2 each	2 each	Sour oranges, juice of
1 each	1 each	Jalapeno pepper, roasted and finely diced
1 each	1 each	Red onion, finely sliced
2 each	2 each	Garlic cloves, finely chopped
2 fluid ounces	59 mL	Olive oil
To taste	To taste	Salt and black pepper

PREPARATION STEPS:

1. Add the fish to the juices and chile.

2. Allow the fish to stand for about 6 hours in refrigerator.

3. Pour boiling water over the onions and drain.

4. Add all the ingredients together and season.

5. Let sit overnight before serving.

Carpaccio

The appetizer **carpaccio**, so named for the Renaissance painter Vittore Carpaccio, famous for the red colors he used in his paintings, is believed to have been invented in 1961 at Harry's Bar in Venice, Italy. The technique uses pieces of raw beef, sometimes seared on the outside. The beef is sliced paper thin and shingled onto a plate, covering the entire surface and turning it a bright red color. It was traditionally garnished with olive oil and lemon juice, or some kind of vinaigrette, with a few small garnishes.

Modern interpretations of this dish have changed over the years. Today, the technique still involves the very thin slicing of the product and the topping with certain acids like lemon, lime, and various vinegars. However, the ingredients that are used as the carved item have changed dramatically and include fresh fish as well as a variety of meats, poultry, game, vegetables, fruits, and even offal meats.

The original concept of trying to represent the red color on the plate has faded into obscurity, as modern chefs seek to experiment with other ingredients. But as we are exposed to a host of new and exciting ideas and products, it seems reasonable to accept the very thin slicing and covering of the bottom of the plate as one of many alternative representations of this wonderful dish (Figure 12–4).

FIGURE 12–4 Appetizer of Beef Carpaccio

RECIPE 12–4

BEEF CARPACCIO

Recipe Yield: 8 portions

MEASUREMENTS		INGREDIENTS
U.S.	**METRIC**	
24 ounces	680 g	Beef tenderloin, fresh, raw
4 tablespoons	59 mL	Balsamic vinegar, aged
4 tablespoons	59 mL	Freshly squeezed lemon juice
4 tablespoons	59 mL	Extra virgin olive oil
1 tablespoon	15 mL	Kosher salt
1 teaspoon	5 mL	Black pepper, fresh ground
4 cups	946 mL	Mixed green salad
4 tablespoons	59 mL	Olive oil
1 teaspoon	5 mL	Red wine vinegar
8 each	8 each	Parmesan cheese, fresh, thinly shaved

PREPARATION STEPS:

1. Wrap the beef into a nice even circular shape and lightly freeze until firm, about 1 hour.

2. With a sharp slicer, thinly shave the beef across the grain, and arrange onto plates in an attractive circular pattern.

3. Bring shaved meat to room temperature and drizzle with the balsamic vinegar and the lemon juice.

4. Drizzle with extra virgin olive oil and sprinkle with seasoning.

5. Toss the greens in the olive oil and the red wine vinegar and place a ball of greens in the center of the beef.

6. Sprinkle with the shaved parmesan and serve.

RECIPE 12–5

TUNA CARPACCIO WITH MANGO SALSA

Recipe Yield: 8 portions

MEASUREMENTS | INGREDIENTS

U.S.	METRIC	
1½ pound	680 g	Sushi grade tuna, center cut
1 each	1 each	Fresh lime, juiced
1 each	1 each	Fresh lemon, juiced
1 tablespoon	15 mL	Chives, finely chopped
2 each	2 each	Ripe mango, finely diced
1 each	1 each	Red onion, finely diced
1 teaspoon	5 mL	Mint, finely chopped
8 tablespoons	118 mL	Olive oil
2 teaspoons	10 mL	Kosher salt
1 teaspoon	5 mL	Black pepper, freshly ground

PREPARATION STEPS:

1. Slightly freeze the tuna loin, and then slice it very thinly.

2. Shingle it accurately over the bottom of the plate.

3. Drizzle with one-fourth of the lemon and lime juice. Sprinkle with salt and pepper and chopped chives.

4. Allow to rest in cooler for 30 minutes.

5. Make a salsa by combining the mango, onion, mint, the remaining juices, and half the olive oil; season and toss lightly, reserving for service.

6. For service, bring the fish back to room temperature and sprinkle with the remaining oil; sprinkle with salt and pepper.

7. Place a spoonful of mango salsa in the middle and serve.

Tartar

The word **tartare** has a checkered history. The dish has its origin as a culinary practice popular in medieval times among Mongolian tribes known as Tartars, who derived their name from Greek mythology, Tartarus. It is said that the dish originated from the fact that Attila the Hun's men ate raw meat cuts, which after being placed under their saddles were tenderized from a long day's ride. The low-quality, tough meat from Asian cattle that grazed on the Russian Steppe, was shredded to make it more palatable and digestible. At meal times, they added spices to their meat and consumed it completely raw. Today, the name is not only used for raw meat dishes but also for many finely diced raw fish dishes.

BEEFSTEAK TARTARE

Recipe Yield: 2 portions

MEASUREMENTS		INGREDIENTS
U.S.	METRIC	
1 pound	454 g	Beef tenderloin, fresh, trimmed of fat, chilled
1 each	1 each	Egg, coddled (optional)
1/2 teaspoon	3 mL	Salt, coarse
1/2 teaspoon	3 mL	Black peppercorns, fresh ground
1 ounce	29 mL	Vodka or gin (optional)
1/4 cup	59 mL	Onion, minced
1/4 cup	59 mL	Capers, drained
1 teaspoon	5 mL	White vinegar
1 teaspoon	5 mL	Mustard (optional)
4 slices	4 slices	Rye or pumpernickel bread

PREPARATION STEPS:

1. In a food processor bowl fitted with the metal blade, place one-third of the meat at a time; add egg and cover. Continue and process egg and meat for 7 to 9 seconds, or until all meat is finely chopped. Shape meat into a round on a platter or bowl. Press top flat.

2. Sprinkle meat with salt and pepper and worcestershire sauce.

3. Layer with onions and capers.

4. Sprinkle with vinegar, mustard, and gin.

5. Toss mixture together, until well blended, careful not to overmix.

6. Serve with dark bread.

RECIPE 12-7

SALMON TARTARE

Recipe Yield: 32 ounces
(907 g)

MEASUREMENTS		INGREDIENTS
U.S.	**METRIC**	
32 ounces	907 g	Salmon fillet, small dice
2 tablespoons	30 mL	Minced shallots
2 tablespoons	30 mL	Minced chive
3 tablespoons	44 mL	Minced capers
4 tablespoons	59 mL	Olive oil
2 each	2 each	Lemons, juiced
1 teaspoon	15 mL	Salt
1 tablespoon	15 mL	Black pepper, fresh ground

TIP It is imperative that the fish used in this recipe be freshly caught and cleaned at once, and that the preservation process take place immediately. This should all occur within a few hours of the fish being harvested, or else the flavor will be affected and the risk for spoilage will be increased.

PREPARATION STEPS:

1. Lightly toss all the ingredients together and chill for 1 hour.

2. Serve on toast points (Figure 12–5).

FIGURE 12–5 Appetizer of Salmon Tartare

USING HERRINGS, ANCHOVIES, SARDINES, AND SALMON These internationally popular fish have been marinated raw and stored for use as appetizers, entrees, and salads for many years. They are usually served in countries where the fish is available at its very freshest, and the method of preservation not only lengthens its shelf life but also allows people to enjoy the fish for longer periods of time. There are many regional and national dishes made this way, using a variety of different ingredients that are available in that locality. However, the basic method of preservation is very much the same.

The fish for these recipes should be fresh caught and cleaned at once, with the preservation process taking place immediately. The fish can be processed whole or in fillets, with or without skin, and sometimes even with the heads attached. The most important factor must always be the sanitation of the fish during the cleaning stage. The cavity must be completely cleansed, the scales must be removed, and particular attention must be paid to the removal of the gills.

The fish can also be lightly salted during the initial cleaning, which helps create an environment that inhibits bacterial growth (Figure 12–6).

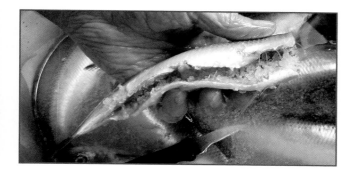

FIGURE 12–6A Applying coarse salt to herring

FIGURE 12–6B Pickled sardines for appetizer platter

RECIPE 12–8

PICKLED ANCHOVIES

Recipe Yield: 8 portions

MEASUREMENTS		INGREDIENTS
U.S.	**METRIC**	
24 ounces	680 g	Fresh anchovies, filleted and cleaned
8 tablespoons	118 mL	Kosher salt
4 each	4 each	Limes, juiced
2 tablespoons	30 mL	Red wine vinegar
1 cup	237 mL	Olive oil

PREPARATION STEPS:

1. Sprinkle the anchovies on both sides with all the salt and allow to sit for 30 minutes.

2. Wash the fish well and pat dry with paper towels.

3. Combine the vinegar and the lime juice and sprinkle it onto a layer of anchovies packed together in a glass dish.

4. Continue to add the fish in layers sprinkling with the mix until all the fish and the liquids are together, packed well into the dish.

5. Cover with the olive oil and chill for 36 hours before use.

6. Herbs and spices can also be added to this recipe.

FRUITS AND VEGETABLES

The quick **pickling** of fruits and vegetables is a cost-effective way of preserving highly perishable ingredients, and the finished product renders a flavorful addition to soups, appetizers, salads, and side dishes.

The selected vegetables or fruits are cooked, cooled, stored, and served in the pickling liquor, not only preserving them but also retaining their vibrant colors and flavors. These quick pickling methods can only improve the shelf life of the fruits and vegetables for a few extra days, but in the long run the overall savings involved are considerable.

Recipes for the pickling liquors vary and should always have some form of acid or citric juices present to aid in the pickling process.

Vegetable Pickling Recipes

Almost all vegetables can be pickled using the following recipes, but the vegetables that work the best are cauliflower, baby onions, baby beets, fennel, celery, squashes, pumpkin, mushrooms, artichokes, leeks, celeriac, sun chokes, Brussels sprouts, and cucumbers.

RECIPE 12–9

GREEK PICKLING LIQUOR

Recipe Yield: 1 quart
(946 mL)

Note: Recipe will pickle 24 ounces
(680 g) of vegetables.

| MEASUREMENTS | | INGREDIENTS |
U.S.	METRIC	
1 quart	946 mL	Water
1 cup	237 mL	White wine
3 each	3 each	Lemons, juiced
2 each	2 each	Sprigs thyme
6 each	6 each	Bay leaves
3 teaspoons	44 mL	Olive oil
1 teaspoon	5 mL	Black pepper, freshly ground
1 tablespoon	15 mL	Kosher salt

PREPARATION STEPS:

1. Bring the ingredients to a boil and simmer for 5 minutes.

2. Add the vegetables to be pickled.

3. Poach until just done.

4. Chill all together and store for service.

RECIPE 12-10

PORTUGUESE PICKLING LIQUOR

Recipe Yield: 3 pints
(1.42L)

Note: Recipe will pickle 24 ounces
(680 g) of vegetables.

| MEASUREMENTS | | INGREDIENTS |
U.S.	METRIC	
4 each	4 each	Garlic cloves, crushed
2 tablespoons	30 mL	Olive oil
1 quart	.95 L	Greek Pickling Liquor
1 pound	454 g	Tomato concassée

PREPARATION STEPS:

1. Stew the garlic in the olive oil until cooked

2. Toss with the tomatoes and add to the Greek Pickling Liquor (Recipe 12–9) before
   cooking the vegetables that are to be pickled

AMERICAN PICKLING LIQUOR

Recipe Yield: 1 quart (946 mL)

Note: Recipe will pickle 24 ounces (680 g) of vegetables.

MEASUREMENTS		INGREDIENTS
U.S.	**METRIC**	
2 cups	473 mL	White wine vinegar
2 cups	473 mL	Water
12 ounces	340 g	Sugar
1 ounce	28 g	Pickling spice
4 teaspoons	20 mL	Dill seeds
1 tablespoon	15 mL	Salt
1 each	1 each	Lemon, juiced

PREPARATION STEPS:

1. Simmer all pickling ingredients together for 5 minutes to blend flavors.

2. Briefly poach the selected vegetables in the simmering pickling liquid.

3. Allow vegetables to cool in the pickling liquid, then jar together and refrigerate.

PICKLED MUSTARD GREENS

Recipe Yield: 3 cups (710 mL)

MEASUREMENTS		INGREDIENTS
U.S.	**METRIC**	
1½ pounds	680 g	Mustard greens
1 bunch	1 bunch	Scallions (green onions)
1 quart	0.95 L	American Pickling Liquor

PREPARATION STEPS:

1. Cut the mustard greens and scallions into 2-inch (5-cm) lengths.

2. Wash well, then dry in the sun, food dryer, or low oven until slightly shrunk in size.

3. Transfer the greens to a bowl filled with the pickling liquid.

4. Using another bowl as a weight, keep the greens submerged in the liquid.

5. Cover the bowls and place them in a warm part of the kitchen for 3 days.

6. Transfer the greens and pickling liquid to canning jars, and refrigerate up to 2 months for use.

RECIPE 12–13

CORNICHONS

Recipe Yield: 6 pints
(2.84 L)

MEASUREMENTS		INGREDIENTS
U.S.	**METRIC**	
6 pounds	2.7 kg	Pickling cucumbers, 1 to 1½ inches (2.5 to 3.8 cm) in length
6 sprigs	6 sprigs	Fresh oregano (or marjoram)
6 sprigs	6 sprigs	Fresh tarragon (or chervil)
1 tablespoon	15 mL	Juniper berries (or white mustard seeds)
1 tablespoon	15 mL	Green peppercorns, whole
2 cups	474 mL	Water
4½ cups	1.07 L	Vinegar (red wine or champagne)
1½ tablespoons	22.5 mL	Sugar
1½ tablespoons	22.5 mL	Salt

TIP If extra crisp pickles are desired, eliminate the water bath processing. Cornichons will need to be refrigerated, instead.

PREPARATION STEPS:

1. Scrub the cucumbers under running water, using a vegetable brush. Reserve to drain.

2. Using six sterilized pint-sized canning jars, portion in each: 1 oregano sprig, 1 tarragon sprig, ½ teaspoon (2.5 mL) juniper berries, and ½ teaspoon (2.5 mL) green peppercorns.

3. Equally divide the cucumbers between the six jars. Tightly pack them in vertically, until they are all inserted.

4. In a nonreactive saucepan, bring the water, vinegar, sugar, and salt to a boil for 2 minutes.

5. Pour the boiling brine into the packed jars.

6. Clean jar rims and cover with cap and rings.

7. Process under boiling water for 6 minutes.

8. Remove jars and allow to cool. Store for 2 weeks before service.

Bread and Butter Pickles

Recipe Yield: Five 1½ pint jars

MEASUREMENTS		INGREDIENTS
U.S.	METRIC	
10 each	10 each	White onions, small, thinly sliced
18 each	18 each	Cucumbers, medium, thinly sliced
⅓ cup	79 mL	Salt
3 cups	711 mL	Distilled vinegar
2 cups	473 mL	Sugar
2 teaspoons	10 mL	Ground tumeric
2 teaspoons	10 mL	Celery seed
2 tablespoons	30 mL	Mustard seed, yellow
1 teaspoon	5 mL	Ground ginger
1 teaspoon	5 mL	Peppercorns

TIP Sweet, sour, and spicy seasoning may be customized by adjusting sugar, vinegar, or mustard seeds. (Black and brown mustard seeds are hotter than white mustard seeds.)

PREPARATION STEPS:

1. Combine sliced onions and cucumbers in a large bowl. Layer with salt and cover with ice cubes. Reserve and allow to stand 1½ hours. Drain and rinse well.

2. Combine remaining ingredients in a large saucepan and bring to a rolling boil.

3. Add drained onions and cucumbers, and return mixture to boiling.

4. Carefully pack the hot pickle mixture into sterilized, hot canning jars. Remove bubbles.

5. Cover and firmly secure the ring.

6. Process under boiling water for 10 minutes.

7. Remove and allow jars to cool. Check seals on jars.

Fruit Pickling Recipes

Once fruits have been cut, they tend to deteriorate rapidly and often they are wasted because there is not a good way of using them quickly. It makes sense for the garde manger chef to have a way of preserving them promptly so as to extend their useful life and use them profitably elsewhere on the menu. Common fruits that are used successfully in the quick pickling process include pineapples, melons, mangos, guavas, firm papayas, apples, plums, peaches, apricots, and pears.

RECIPE 12–15

Caribbean Fruit Pickle

Recipe Yield: 8 portions

Note: Recipe will pickle 24 ounces (680 g) of fruits.

MEASUREMENTS		INGREDIENTS
U.S.	**METRIC**	
2 cups	473 mL	Water
1 cup	237 mL	Apple cider vinegar
1 cup	237 mL	Brown sugar
1 cup	237 mL	White sugar
1 cup	237 mL	Apple juice
4 each	4 each	Star anise
3 each	3 each	Cinnamon sticks
4 tablespoons	59 mL	Ginger, grated fresh
10 each	10 each	Allspice berries
2 each	2 each	Limes, juiced

PREPARATION STEPS:

1. Boil all ingredients for 10 minutes, then pour the boiling mixture over the fruits.

2. Allow the fruits to cool in the liquor and chill for service.

RECIPE 12–16

Gingered Tomato Comfiture

Recipe Yield: 1 ½ pints (946 mL)

Note: Recipe will pickle 24 ounces (680 g) of fruits.

MEASUREMENTS		INGREDIENTS
U.S.	**METRIC**	
2 tablespoons	30 mL	Olive oil
1 tablespoon	15 mL	Grated ginger
1 each	1 each	Small red onion, diced fine
2 each	2 each	Garlic cloves, minced fine
1 each	1 each	Green onion, minced fine
4 cups	964 mL	Chopped tomato, peeled and seeded
As needed	As needed	Salt and pepper
1 cup	237 mL	Sugar

PREPARATION STEPS:

1. In the olive oil, sauté the garlic, onions, green onion, and ginger until soft.

2. Add the tomatoes. Cook for 25 minutes until a thick sauce consistency is achieved.

3. Add the sugar. Cook for another 5 minutes.

Drying Foods

Food **drying** is one of the oldest methods of preserving food. Compared with other methods, it is very simple to do. It also has the advantage of lasting a much longer time when stored properly. Dried foods keep well because the moisture content is so low. Organisms that cause spoilage cannot grow because the environment is not conducive to their growth. Although drying will not retain all the taste, appearance, and nutritive value of fresh foods that other methods of preservation can, it still has many advantages and positive considerations:

■ Drying provides a unique texture to the food that is useful when menu planning.

■ Dried food takes up a lot less space than canned or frozen foods.

■ Eating dried food has become a popular way of eating healthy snacks.

■ Because it is dehydrated, dried food weighs less and is preferred by hikers and climbers.

■ Very little equipment is required to dry food successfully.

■ Dried food lasts a long time and is therefore a good way of using surpluses when available.

METHODS OF DRYING

Over the centuries, several dependable methods of drying food have been used. Not much has changed with these methods, although technology continues to improve upon the drying machines. These machines are available to both the home cooking enthusiast and the commercial food service operator. In either case, the cost is relatively low.

Sun Drying

Sun drying is the ancient method used to dry food; it uses the warmth from the sun and the natural movement of the air to dehydrate the food. Bright sunshine, low humidity, and temperatures around 100°F (37.8°C) are necessary for the process to be most successful. It is a slow process and requires constant care and attention.

Sun drying is not as sanitary as other methods of drying, and the food must always be protected from insects, dust, animals, and birds. It is also not advisable to use this method in an area that is susceptible to any form of air pollution, including industrial areas, local airports, areas adjacent to freeways or traffic fumes, or close to any other obvious air pollutants.

If the conditions are ideal for sun drying foods, an alternative to unprotected air drying is the use of a natural draft dryer, which allows the free movement of air, while protecting the ingredients from birds and insects (Figure 12–7). It also is perfect for trapping the heat from the sun.

The following procedures are used to sun dry foods:

1. Place the prepared food on the drying trays covered with a net of fine cheesecloth.
2. Place the dryer in direct sunlight, raised from the ground and away from animals or insects.
3. Allow to dry for the recommended time.
4. Place into the shade for the final stage of drying, making sure it is well ventilated.
5. Bring the dryer indoors at night if the temperature drops by 20°F, because the dew and sudden change in temperature can lengthen the drying time.
6. The length of drying time depends on the conditions, which makes sun drying a slower and less dependable method of drying.

FIGURE 12–7 Diagram of a natural draft dryer

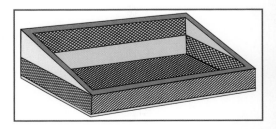

Oven Drying

Oven drying is the simplest method of drying food, because it requires almost no special equipment except what is already available in a commercial kitchen (Figure 12–8). It tends to be a lot faster than sun drying and is generally more reliable. The only drawback to oven drying is that only a small amount can be done at one time. However, if the food is dried on a daily basis for freshness and consistency, it meets a chef's needs very well. Because drying foods takes so much time, careful allocation of oven space is a wise precaution. One of the best times to dry foods is overnight when oven space is not at a premium. To achieve consistent results, there are some basic guidelines that should be followed:

- Use an oven set on the lowest possible setting and preheat to 140°F (60°C). Some ovens might run at this temperature with just the gas pilot lit.
- An accurate reading is very important, so the temperature should be checked with a reliable thermometer.
- Arrange the food on trays with a ½-inch (1.25-cm) gap between the food.
- Do not overfill the oven, because this will slow down the drying process. No more than four trays at a time should be used.
- Try to keep the oven door slightly ajar during the drying process to allow humidity to escape.
- Encourage airflow by using the oven fan or by placing a fan close to the oven door.
- If possible, rotate the trays for even drying.

FIGURE 12–8 Food drying in an oven

RECIPE 12–17

OVEN-DRIED TOMATOES IN OLIVE OIL

Recipe Yield: 1 cup (237 mL)

TIP See Chapter 13 for information on hot and cold smoking foods.

MEASUREMENTS		INGREDIENTS
U.S.	**METRIC**	
2 cups	473 mL	Tomatoes, split in half with core and seeds removed
½ cup	118 mL	Kosher salt
½ cup	118 mL	Sugar
8 cloves	8 cloves	Garlic, blanched and refreshed
1 cup	237 mL	Extra virgin olive oil

PREPARATION STEPS:

1. Sprinkle the salt and the sugar over the tomatoes and toss well together.

2. Spread the tomatoes onto a sheet pan and leave at room temperature for 4 hours.

3. Rinse the tomatoes briefly and dry well with paper towels.

4. At this stage they could be cold smoked for better flavor.

5. Allow to dry in a low oven for 6 to 8 hours or until completely dry.

6. Place the tomatoes into the oil with the garlic.

Using a Food Dryer

Machines that are specifically made to dry food come in all shapes and sizes, and they are a very reliable way of producing large quantities of dried food product (Figure 12–9). They tend to operate at lower temperatures, often around 125°F (52°C), thus taking a little longer to dry the food than a commercial oven. However, these ovens are well ventilated and tend to dry products more evenly and with a more consistent result.

1. Preheat the dryer to 125° F (52°C) and place the food evenly spaced on the racks.

2. Stack the racks in the drier and gradually increase the temperature to 140°F (60°C) until the food is completely dried.

DRYING INGREDIENTS

Any ingredient that has moisture can be dried. As mentioned earlier, the reasons for drying vary. In some cases, it aids in preservation, storage, transportation, flavor, or texture. In addition to protein products, the following ingredients are often dried (Figure 12–10).

Herbs

The growing and drying of herbs has gained popularity, both for the commercial kitchen and home enthusiast, as an inexpensive way of ensuring a dependable supply close at hand. For the best result, use only the young, tender leaves that are more flavorful and aromatic when dried. When picking herbs for drying, some important factors include the following:

- Cut the stalks when the leaves are mature.
- Use only the tender and leafy tops and flower clusters for drying.
- Avoid the leaves below 6 inches (15 cm) from the top of the stalk; they will not be as pungent as the top leaves.
- Always remove the dead or discolored leaves.
- Rinse carefully with cold water to wash off dust and dirt.
- Dry completely with paper towels.

FIGURE 12–9 Food drying in a food dryer

FIGURE 12–10 Assorted dried fruits and vegetables

When air drying herbs, there are several successful ways of achieving good results:

1. Tie a bunch of the picked herb together by the stems.
2. Place the bunched herbs into a ventilated brown bag and tie shut, making sure the herb does not touch the sides of the bag.
3. Hang it in a warm, dry, airy room or attic. Herbs will dry in 1 to 2 weeks.
4. Alternatively, pick the leaves and dry them on a tray in a warm, dry, airy place away from direct sunlight.

When drying in an oven, it is important that the oven be used solely for the purpose of drying at that time. It should be done at an appropriate time in the kitchen when no one can interfere with the drying process.

1. Place the clean, fresh leaves in a single layer on the racks.
2. There should be at least 1 inch (2.5 cm) around the edge of the racks and between the herbs so that the air can circulate freely.
3. Place in an oven at 130°F (54°C).
4. Keep the oven door propped open slightly for ventilation and to control the heat. Drying will be complete in 2 to 4 hours.

The careful storage of the herbs after they have been so carefully dried will ensure the best quality and utilization of the herbs:

1. When the leaves are dry, shake them from the stems.
2. Crush the leaves at this point, but keep in mind that crushed or ground herbs do not retain their flavor as long as whole herbs.
3. Store dried herbs in small airtight containers away from direct light.

Vegetables

Virtually all vegetables can be dried successfully in the kitchen, especially when cut small and used as garnitures for finished salads and appetizers. Always start with fresh, mature produce that is very ripe and ready to eat. When selecting vegetables for drying, there are several factors that the chef should consider:

- Harvest or buy only the amount you can dry at one time.
- Wash all dirt off the vegetables and cut out any bad spots.
- Cut the vegetables into pieces of a suitable size or shape for the display.
- Thin pieces dry faster than thick ones.
- Almost all vegetables need to be blanched in boiling water for a short time before drying.
- Blanching also protects certain nutrients and reduces the drying time somewhat.
- Tomatoes can be split and salted for 1 hour before drying to remove a lot of the liquid and to season the tomato well.
- Spread the prepared vegetables in thin layers on the drying trays, spaced well apart.
- Mature beans, peas, and soybeans may be fully or partly dried on the vine.
- Avoid drying green vegetables because they lose a lot of their flavor during the drying process.
- Vegetables with a strong odor should not be dried with any other vegetables, because they will impart their overpowering flavor.

Fruits

When drying fruits, choose only those that have been harvested very recently. They should be fully ripened, naturally sweet, and flavorful, and they should be of the same high quality that would be served at the table. Proper sanitation and avoiding the possibility of cross contamination should be of the utmost importance. All fruits need some kind of pretreatment before drying. The following points should be considered:

- Ensure that the fruit is washed well, removing any leaves or dirt.
- Discard any bruised, overripe, or otherwise damaged fruit.

- Remove the unwanted parts of the fruits such as pits, seeds, stones, and cores.
- The skins can be left on the fruits and will help support the shape of the fruits when sliced while adding a nice splash of color.
- Fruits that have a waxy skin such as plums and cherries should be blanched to remove them. Score or crack the skins and plunge the fruit into boiling salted water, refreshing in an ice bath immediately. This loosens the skin, which can then be removed easily and quickly.
- Drain this fruit well on absorbent towels as soon as they are peeled.
- Oxidation or discoloration can occur with some light-colored fruits, and if it is not stopped, it can damage the texture, flavor, aroma, and aesthetic qualities of the fruit.
- This darkening is caused by a chemical reaction on contact with air, but it can be checked with the use of an antioxidant.
- Ascorbic acid diluted in water (2 teaspoons [10 mL] per cup for apples, and 1 teaspoon [5 mL] per cup for other light-colored fruits) will keep fruits white during drying.
- Do not soak fruits in water before drying; the fruits absorb the water, which slows down the process.
- Arrange the fruits in a single layer, spaced $1/2$ inch (1.25 cm) apart, on the racks for drying.
- The length of drying time depends on the size that the fruit is cut and the method that is used to dry.

Preserving by Hot Pickling, Jarring, or Canning

This method of preservation has been called by many names in many different countries. For all intents and purposes, it is the art of preserving foods in a container by cooking the jar, sealing the lid, and preserving the food in the condition in which it was prepared. In this chapter, we refer to the process as *canning*, as it is called in the United States (Figure 12–11).

Because microorganisms, such as molds, yeasts, and bacteria, are found in their natural state in the air and in the foods we use, it is important to try to control their growth. One method of controlling bacterial growth is by canning foods to extend their shelf life, thereby enabling the food to be kept beyond its natural storage period.

It is best as a general guideline for the canning of the foods to follow the rules as set forth by the manufacturer of the jars that are being used. However, there are some basic rules that the chef should follow in order to achieve the safest and best results:

1. Follow a tried and tested recipe, and use the equipment as directed.
2. Use only the best quality ingredients, preserving them at the peak of their ripeness.
3. Sanitize all jars lids correctly.

FIGURE 12–11 Finished canned goods

4. Fill the hot jars with the warm prepared recipe, leaving the recommended headspace.

5. Remove any air bubbles.

6. Wipe the rim and threads with a clean, damp cloth, making sure that they are very clean.

7. Center the heated lid on the jar, and screw the band down evenly and firmly until it is fingertip tight.

8. Process under boiling water for at least 15 minutes, or as needed to kill any toxins.

9. After processing, remove jars from the boiling water and canner. Stand the jars upright on a towel to cool.

10. Do not retighten bands or check for a seal while jars are hot.

11. After 24 hours, check the lids to determine whether they have downward curve, confirming the seal.

12. Press the center of the lid to ensure it has been vacuum-sealed. It should not be flexible to the touch.

13. Wipe the jars clean with a damp cloth and dry.

14. Label the jars, and store in a cool dry place. Use within 1 year.

Clostridium botulinum

Clostridium botulinum, the organism that thrives in an oxygen-free environment, may grow in a can when strict sanitary procedures are not practiced. *Clostridium botulinum* bacteria are present in the soil and water. Spores can be associated with any food that comes into contact with the soil or grows in water. Improperly (home or commercial operations) canned low-acid foods, such as vegetables, meat, poultry, and fish, are usually the cause of a botulism foodborne illness.

During the canning process, oxygen is removed from the container by heating the jar and its contents. If the proper temperatures to destroy the spores are not reached, the spores will have the suitable environment to grow into vegetative cells and eventually produce the deadly toxin. Proper canning methods must be used for preserving low-acid foods. Pressure processing is necessary to obtain the temperatures required to destroy the *Clostridium botulinum* spore. When the bacterium grows, it can produce a gas that causes canned items to bulge. Never taste food from leaking, bulging, or damaged cans; from cracked jars or jars with loose or bulging lids; from containers that spurt liquid when opened; or from any canned food that has an abnormal odor or appearance.

Although many garde manger chefs regularly use canning in their commercial food service operations, some local health code regulations may forbid the use of noninspected canned products. It is recommended to check with the local health department regarding this practice.

Note: To process the jars: The toxin can be destroyed by submerging the filled jars in boiling water for 10 minutes at sea level. Add 1 minute of processing time for every 1000 feet above sea level.

RECIPE 12–18

SAGE BOURBON JELLY

Recipe Yield: 1 quart
(946 mL)

MEASUREMENTS		INGREDIENTS
U.S.	**METRIC**	
3 cups	710 mL	Bourbon
2 ounces	57 g	Powdered pectin
1 tablespoon	15 mL	Lemon juice
$\frac{1}{4}$ teaspoon	2 mL	Vegetable oil
$4\frac{1}{2}$ cups	1.1 L	Sugar
6 tablespoons	88.5 mL	Sage, chopped fresh

PREPARATION STEPS:

1. Bring the bourbon, pectin, lemon juice, and oil to a boil.

2. Stir in the sugar and bring back to a boil a second time.

3. Boil hard for 1 minute stirring constantly and add the sage

4. Remove from the heat and reserve to cool for service, or it can be canned in Mason jars for future use.

RECIPE 12–19

SWEET ONION MARMALADE

Recipe Yield: $\frac{1}{2}$ cup (118 mL)

MEASUREMENTS		INGREDIENTS
U.S.	**METRIC**	
2 cups	473 mL	Red onions, sliced thin
2 tablespoons	30 mL	Vegetable oil
2 tablespoons	30 mL	Balsamic vinegar
2 tablespoons	30 mL	Red wine vinegar
$\frac{1}{2}$ cup	118 mL	Brown sugar
As needed	As needed	Salt and white pepper

PREPARATION STEPS:

1. Fry the onions in the oil until golden brown, at least 10 minutes in a thick-bottomed sauté pan.

2. Deglaze with the vinegar and reduce until almost dry.

3. Add the sugar and allow to dissolve.

4. Season with salt and pepper.

5. This can be canned in Mason jars for future use.

RECIPE 12-20

TAMARIND FIG JAM

Recipe Yield: 1½ pints
(710 mL)

MEASUREMENTS		INGREDIENTS
U.S.	**METRIC**	
½ cup	118 mL	Tamarind pulp
½ cup	118 mL	Black mission figs, dried, finely diced
¼ cup	59 mL	Golden raisins
1 teaspoon	5 mL	Cumin, ground
1 teaspoon	5 mL	Coriander seeds, ground
1 teaspoon	5 mL	Curry powder
1 teaspoon	5 mL	Ginger, ground
½ teaspoon	2 mL	Cayenne pepper, ground
1 cup	237 mL	Water
1 tablespoon	15 mL	Champagne vinegar
½ cup	118 mL	Brown sugar

PREPARATION STEPS:

1. Simmer all the ingredients for 20 minutes.

2. Purée and pass through a fine strainer to remove seeds.

3. This can be canned in Mason jars for future use.

CHOW-CHOWS

Chow-chow, or *pickalilly,* is a relish or a pickled relish that is used in the southern states of America. Its roots can be traced back to either China or India, and it consists of almost anything that can grow in a summer garden. Recipes can include cabbage, cucumbers, zucchini, okra, corn, various peppers, tomatoes, carrots, onions and even some squashes.

Chow-chow is prepared similar to the Korean favorite kimchee, the fermented Korean pickle. In ancient times, and still now, Korean winters were long and severe. The harsh conditions forced people to preserve vegetables. The word *kimchee* in Korean means "sunken vegetable." Chinese cabbages and radishes were "sunk" into salted water and then seasonings such as chile pepper were added. Later, the special flavor of salted fish was added. Kimchee can be described as fermented vegetables, because many kinds of bacterial reactions contribute to build its flavor. Most important is lactic acid, which aids digestion.

The basic recipe for chow-chow involves the fine chopping of the available vegetables and seasonings, then adding a pickling liquid and jarring.

RECIPE 12–21

CHOW-CHOW PICKLING LIQUID

Recipe Yield: 1 quart
(946 mL)

MEASUREMENTS		INGREDIENTS
U.S.	**METRIC**	
3 cups	710 mL	White vinegar
1 cup	237 mL	Water
2½ cups	591 mL	Sugar
1 cup	237 mL	Brown sugar
¼ cup	59 mL	Salt
2 teaspoons	10 mL	Allspice
2 each	2 each	Cinnamon
1 teaspoon	5 mL	Mustard seeds
1 teaspoon	5 mL	Black peppercorns
1 teaspoon	5 mL	Celery seeds

PREPARATION STEPS:

1. Bring the ingredients to a boil and simmer for 20 minutes.

2. Strain and use for jarring.

RECIPE 12–22

THAI PICKLED CUCUMBERS

Recipe Yield: 1 quart
(946 mL)

MEASUREMENTS		INGREDIENTS
U.S.	**METRIC**	
3 cups	710 mL	Cucumbers, thinly sliced
⅓ cup	79 mL	Onion, finely chopped
½ cup	118 mL	Rice wine
2 teaspoons	10 mL	Sugar
¼ teaspoon	1 mL	Salt
4 each	4 each	Red chile peppers, seeded, chopped
1 tablespoon	15 mL	Cilantro leaves

PREPARATION STEPS:

1. Combine all ingredients and allow flavors to blend for 4 hours.

2. Serve chilled or at room temperature as a condiment (Figure 12–12).

FIGURE 12–12 Thai Pickled
Cucumbers

A wide variety of fruits and vegetables may be pickled, including the commonly used vegetable, the cucumber. In addition, the flower and buds of different plants may also be pickled. The caper bud is harvested from the bush (Figure 12–13A) before it flowers and is then submerged in a pickling solution similar to the Chow-Chow Pickling Liquid. The pickled capers (Figure 12–13B) are often used as a condiment ingredient in Mediterranean recipes.

FIGURE 12–13A Capers on the bush

FIGURE 12–13B Pickled capers

SALSA AND RELISHES

The terms **salsa** and *relish* may be used interchangeably. The only difference is that the word *relish* is of French origin and the word *salsa* is of Spanish origin. They are both condiments intended to add flavor to other foods, and both can be either raw or cooked.

RECIPE 12–23

PEACH TOMATO CHERRY SALSA

Recipe Yield: 1 quart
(946 mL)

| MEASUREMENTS | | INGREDIENTS |
U.S.	METRIC	
2 cups	473 mL	Peaches, very ripe, medium dice
1 cup	237 mL	Tomato concassée
½ cup	118 mL	Dried cherries
⅓ cup	79 mL	White sugar
⅓ cup	79 mL	Brown sugar
⅓ cup	79 mL	Apple cider vinegar
1 tablespoon	15 mL	Jalapeño, finely diced
1 tablespoon	15 mL	Ginger, freshly grated
1 tablespoon	15 mL	Lime juice
1 tablespoon	15 mL	Kosher salt
1 tablespoon	15 mL	Mint, finely chopped

PREPARATION STEPS:

1. Place all ingredients in a glass bowl and chill overnight.

2. Ladle relish into hot jars, filling to ½ inch (1.25 cm) from the top.

3. Apply the lids and process for 15 minutes in the boiling water.

4. Remove the jars and cool, ensuring that they have sealed.

RECIPE 12–24

BLACKENED JALAPEÑO SALSA

Recipe Yield: 1½ pints
(710 mL)

MEASUREMENTS		INGREDIENTS
U.S.	**METRIC**	
1 each	1 each	Red onion, ½-inch (1.25-cm) thick slices
8 each	8 each	Garlic, fresh peeled
4 teaspoons	20 mL	Oregano (Mexican oregano preferred), dried
24 each	24 each	Jalapeño chiles, stemmed, seeded, and halved
6 each	6 each	Anaheim chiles, stemmed, seeded, and halved
½ cup	118 mL	Olive oil
4 each	4 each	Limes, freshly juiced
1 teaspoon	5 mL	Kosher salt
1 cup	237 mL	Water

PREPARATION STEPS:

1. Heat a cast iron frying pan over high heat. Add the onion slices and char on both sides, separating them into rings. Remove onions and reserve.

2. Add the garlic and quickly stir to lightly brown.

3. Add the oregano to the garlic and toss together for 1 minute. Remove and reserve with the onions.

4. Add the oil to the hot pan. Add the chiles and sear until evenly char-browned. Remove chiles from the heat and combine with the onion mixture.

5. Combine all of the ingredients in a food processor, and quickly pulse until mixture is roughly chopped.

6. Add water and pulse quickly until mixture is salsa consistency.

7. Store in a covered nonreactive container, under refrigeration, until service.

RECIPE 12-25

MEXICAN SALSA CRUDA

Recipe Yield: 1½ pints
(710 mL)

MEASUREMENTS		INGREDIENTS
U.S.	**METRIC**	
2 each	2 each	Tomato, fresh diced
½ cup	118 mL	Spanish onion, freshly minced
¼ cup	59 mL	Cilantro, freshly minced
6 each	6 each	Serrano chiles, finely chopped (with seeds)
1 teaspoon	5 mL	Salt
1 each	1 each	Lime, juice of
½ cup	118 mL	Water, cold (as needed)

TIP Salt, lime juice, and liquid hot sauce may be used to adjust seasoning.

PREPARATION STEPS:

1. In a large bowl, combine all of the ingredients, except the water.

2. Add water to achieve desired consistency.

3. Adjust seasoning.

4. Store in a covered nonreactive container, under refrigeration, until service.

RECIPE 12-26

AMERICAN APPLE KETCHUP

Recipe Yield: 2½ pints
(1.18 L)

MEASUREMENTS		INGREDIENTS
U.S.	**METRIC**	
12 each	12 each	Apples, tart, large and firm
As needed	As needed	Water
2 each	2 each	White onions, medium-sized, finely minced
½ cup	118 mL	Prepared horseradish
1 cup	237 mL	Sugar
1 teaspoon	5 mL	White pepper
1 teaspoon	5 mL	Ground cloves
2 cups	474 mL	White pickling vinegar
2 teaspoons	10 mL	Cinnamon
1 tablespoon	15 mL	Salt

PREPARATION STEPS:

1. Pare, core, and quarter the apples.

2. Place the apples in a saucepan and barely cover with water. Slowly cook over medium heat until the apples are soft and the water has almost evaporated.

3. Rub the apples through a sieve or food mill, without their skins. The resultant pulp should measure about 1 quart (0.95 L). Return the apple pulp to a nonreactive saucepan (stainless steel or tempered glass).

4. Add the remaining ingredients and heat to the boiling point. Reduce heat to low and slowly simmer for 1 hour and 20 minutes.

5. Place in sterilized canning jars while warm and process for 15 minutes.

6. Cool, inspect lids for a good seal, then store properly.

RECIPE 12–27

HOT AND SPICY TOMATO KETCHUP

Recipe Yield: 7 cups
(1.66 L)

Note: Alternatively, the ketchup may be stored in clean jars and refrigerated for several months.

MEASUREMENTS		INGREDIENTS
U.S.	**METRIC**	
24 ounces	680 g	Stewed tomatoes (strained, reserving the drained liquid)
18 ounces	510 g	Tomato paste
1 cup	237 mL	Sugar
4 teaspoons	20 mL	Salt
1 tablespoon	15 mL	Celery salt
2 to 3 teaspoons	10 to 15 mL	Cayenne pepper (adjusted for desired heat)
½ teaspoon	2.5 mL	Cloves
1 teaspoon	5 mL	Cinnamon, ground
1⅓ cups	316 mL	Cider vinegar
As needed	As needed	Water
3 tablespoons	45 mL	Worcestershire sauce
4 teaspoons	20 mL	Angostura Bitters

PREPARATION STEPS:

1. Strain the stewed tomatoes, pressing out any excess moisture. Reserve the strained liquid in a large measuring container.

2. Combine the strained stewed tomatoes, tomato paste, sugar, salt, and celery salt in a nonreactive saucepan. Add cayenne pepper, adjusting amount for desired heat.

3. Stir in the cloves and cinnamon. Gradually stir in the vinegar.

4. Measure the strained tomato liquid, and add enough water to make $2\frac{1}{2}$ cups liquid. Add to the saucepan with tomatoes and seasonings.

5. Add the Worcestershire sauce and bitters.

6. Bring mixture to a boil, stirring frequently, over medium heat. Reduce to low simmer, stirring mixture occasionally until thick and rich, about 18 minutes.

7. Place in sterilized canning jars while warm and process for 15 minutes under boiling water.

8. Cool, inspect lids for a good seal, then store properly.

9. Allow ketchup to stand for 2 days before service.

RECIPE 12-28

SWEET 'N SPICED BEER MUSTARD

Recipe Yield: 1½ quarts (1.42 L)

MEASUREMENTS		INGREDIENTS
U.S.	METRIC	
1⅓ cups	316 mL	Beer (strong flavored beer or ale)
2 cups	473 mL	Mustard powder, dry
¼ cup	59 mL	Molasses
¼ cup	59 mL	Honey
¼ teaspoon	1.3 mL	Ground turmeric
½ teaspoon	2.5 mL	Salt
2 tablespoons	30 mL	Candied ginger, minced
1 cup	237 mL	Dried apricots, minced

PREPARATION STEPS:

1. Combine the beer and mustard powder, blending well. Reserve for 30 minutes.

2. Add the remaining ingredients, blending well.

3. Portion into clean jars; cover. Refrigerate; may store for several months.

4. Or, the mustard may be canned and processed. Carefully portion into four hot, sterilized, pint-sized canning jars. Clean rims and cover; process under boiling water for 10 minutes.

5. Remove and allow to cool, checking lids for proper seal.

RECIPE 12-29

COUNTRY-STYLE GRAIN MUSTARD

Recipe Yield: 1 quart (946 mL)

Note: Brown mustard seeds are the hottest, followed by black, and then white (yellow-colored) is the least spicy. Select the seeds according to desired potency.

MEASUREMENTS		INGREDIENTS
U.S.	METRIC	
¾ cup	177 mL	Black or brown mustard seeds
¾ cup	177 mL	Yellow (white) mustard seeds
1½ cups	356 mL	Brown or yellow mustard powder
1½ cups	356 mL	Cold water
¾ cup	177 mL	Vinegar (white wine or apple cider preferred)
2 tablespoons	30 mL	Salt

PREPARATION STEPS:

1. Use a mortar and pestle, spice grinder, or food processor to grind the mustard seeds to desired texture. (There is no correct texture for the grain of the mustard seeds.)

2. In a medium-sized bowl, combine the ground seeds, mustard powder, and water. Blend evenly and allow to stand 10 minutes.

3. Add the vinegar and salt. Blend evenly.

4. Store in covered container and refrigerate a minimum of 12 hours before serving.

> **TIP** This mustard can be used as is or as a base for other flavored condiments.

> **TIP** Using kitchen-made mustards and ketchups (catsups) is an excellent way for a garde manger chef to elevate otherwise pedestrian foods such as cold cuts, burgers, and bratwurst (Figure 12–14).

FIGURE 12–14 Assortment of mustards and ketchups (catsups)

CHUTNEYS

Chutney is a relish made from fresh fruits and spices, that is suspended in a cooked sugar and vinegar solution. During the Colonial era, the British enjoyed the discovery of chutney while colonizing India, and the recipe was taken back to Britain. Over time, the process for making chutney spread to their other colonies in Africa and the Caribbean.

Chutneys have changed with travel. Although the ingredients have varied, the nature of what goes in chutney, as well as how it is made, has not. Chutneys are still made from a wide variety of fruits, which may include mangoes, apples, tamarind, sour sops, pears, pineapples, and papaya. And they are still scented with sweet onions and some dried fruits, such as raisins, cherries, or sultanas simmered with vinegar, brown sugar, and spices for about 2 hours and then jarred for future use.

Chutneys are served with almost every meal in India, with some usually as side dishes for their curries. They have, however, become very popular worldwide in the last decade as side dishes and sauces to accompany fish, poultry, roasted meat, and game entrées.

RECIPE 12–30

MANGO CHUTNEY

Recipe Yield: 1½ quarts
(1.42 L)

MEASUREMENTS		INGREDIENTS
U.S.	**METRIC**	
½ cup	118 mL	Sugar
1 cup	237 mL	White vinegar
3 cups	710 mL	Mango
1 cup	237 mL	Pineapple
1 each	1 each	Red onion, finely diced
1 tablespoon	15 mL	Tamarind pulp
½ cup	118 mL	Green pepper
2 tablespoons	30 mL	Jalapeño, roasted and finely diced
1 teaspoon	5 mL	Cinnamon
¼ teaspoon	1 mL	Allspice
2 tablespoons	30 mL	Ginger, freshly grated
1 tablespoon	15 mL	Orange zest
1 teaspoon	5 mL	Salt
1 cup	237 mL	Dried cherries, finely diced

PREPARATION STEPS:

1. Boil the vinegar and sugar for 3 minutes.

2. Add all the other ingredients and cook slowly for 30 minutes until the consistency resembles that of porridge.

3. Jar in 6 to 8 jars and cook for 15 minutes in the jars.

RECIPE 12–31

BANANA CHUTNEY

Recipe Yield: 8 portions

MEASUREMENTS		INGREDIENTS
U.S.	**METRIC**	
12 each	12 each	Bananas, finely chopped
1 cup	237 mL	Raisins
1 cup	237 mL	Currants
1 teaspoon	5 mL	Cinnamon
2 cups	473 mL	Apple cider vinegar
1 cup	237 mL	Brown sugar
½ cup	118 mL	White sugar
1 tablespoon	15 mL	Curry powder
1 each	1 each	Red onion
1 teaspoon	5 mL	Salt
2 teaspoons	30 mL	Lime juice

PREPARATION STEPS:

1. Add the vinegar to the sugars and bring to a boil.

2. Add all the other ingredients and simmer for 15 minutes or until the consistency resembles that of porridge.

3. Jar and cook for 10 minutes in the water bath.

RECIPE 12–32

PINEAPPLE MINT CHUTNEY

Recipe Yield: 2½ cups (691 mL)

MEASUREMENTS		INGREDIENTS
U.S.	**METRIC**	
1¾ cups	414 mL	Brown sugar
2 tablespoons	30 mL	Honey
1½ cups	355 mL	White vinegar
1 tablespoon	15 mL	Ginger, freshly grated
½ teaspoon	2 mL	Salt
2 pounds	907 g	Pineapple, fresh, diced
2 tablespoons	30 mL	Mint, fresh, chopped

PREPARATION STEPS:

1. Simmer all ingredients, except the mint, in a thick-bottomed pan for 50 minutes.

2. Add the mint and cook for a further 10 minutes.

3. Cool and jar if necessary.

RECIPE 12–33

CRANBERRY
GINGER
CHUTNEY

Recipe Yield: 1½ quarts
(1.42 L)

MEASUREMENTS		INGREDIENTS
U.S.	**METRIC**	
1½ cups	356 mL	Water
1½ pounds	0.68 kg	Whole cranberries, fresh or frozen
2 each	2 each	Granny Smith apples, peeled and cut in medium dice
1½ cups	356 mL	Sugar
2½ tablespoons	37.5 mL	Ginger, peeled and freshly minced
1 tablespoon	15 mL	Orange zest, finely grated
3 tablespoons	45 mL	Cider vinegar

PREPARATION STEPS:

1. Combine all ingredients in a large nonreactive (stainless steel or glass) saucepan. Stir well to evenly distribute ingredients.

2. Bring mixture to a boil, stirring until sugar is dissolved.

3. Gently boil sauce until slightly thickened, about 8 to 10 minutes (less if chunky texture is desired.)

4. Cover and chill for 2 hours before service.

Professional Profile

BIOGRAPHICAL INFORMATION

Name: Stuart Richard Sturgeon

Place of Birth: Westmeston, Sussex, England

Recipe Provided: Marinated Salmon

CULINARY EDUCATION AND TRAINING HIGHLIGHTS

Chef Sturgeon began his culinary career when he attended Brighton Technical College from 1960 to 1964, earning City and Guilds Certificates 147 and 151 as part of the London National Apprenticeship for Cooks. His training continued at the famed Dorchester Hotel in London, England, where he gained invaluable experience during his 4-year apprenticeship, working primarily under Swiss-born chefs. This unique experience of being part of a brigade of 140 chefs, in arguably one of the world's most influential culinary cities, molded his superior work ethic and professional standards—a hallmark of his career. Chef Sturgeon's training was organized under the classical brigade system, including the parties (stations) of fish, sauce, and larder. From London, he moved to the Marine Hotel in Troon, Scotland, as sous chef, later becoming the executive chef in 1980. He has spent time in a work-exchange in Arbois, France, and as an adjunct instructor at the Glasgow College of Food Technology. Chef Sturgeon has presided over four British Open Championships played over the Royal Troon Golf Course, adjacent to the hotel. His property is the hotel of choice for the top players on the professional golf circuit. Chef Sturgeon has cooked for Her Royal Highness Queen Elizabeth II and many other members of the royal family, among many celebrities and world leaders.

ADVICE TO A JOURNEYMAN CHEF

Seek professional advice as to where you are likely to obtain the very best training in your area. It may be at a large hotel with a larder department or a full-time college course, or a combination of both. Working and going to school is where you are likely to combine learning larder skills while obtaining practical experience with butchers, fishmongers, delicatessen outlets, cheese merchants, and fruit and vegetable traders. Take time to learn and perfect basic skills, and once these are achieved—move on. Select a workplace that allows you to learn and perfect even more skills. Each career move should be for just these reasons. Once confident that you possess the skills and ability to become a garde manger chef—GO FOR IT—and enjoy a career which will give you and your customers a great deal of satisfaction and pleasure.

Richard Sturgeon

RECIPE 12–34

CHEF RICHARD STURGEON'S MARINATED SALMON

Recipe Yield: 7 to 8 lbs

MEASUREMENTS		INGREDIENTS
U.S.	METRIC	
		Salmon
7 to 8 pounds	3½ kg	Salmon fillets, whole
2 ounces	57 g	Sea salt
2 ounces	57 g	Sugar
1½ ounces	43 g	Peppercorns, freshly ground
2 ounces	57 g	Dill, coarsely chopped
		Sauce
12 ounces	340 g	Arran Wholegrain Mustard
3 ounces	85 g	Coleman's Dry Mustard
3 each	3 each	Egg yolks
2 ounces	57 g	Caster sugar
1½ each	1½ each	Lemons, freshly juiced
1 pint	473 mL	Vegetable oil
2 ounces	57 g	Chopped dill

Note: Lemon wraps are made by encasing lemon wedges in commercially available decorative netting, similar to cheesecloth, and tied with ribbon. Their purpose is to prevent lemon seeds from being released onto the fish.

PREPARATION STEPS:

1. Mix salt, sugar, and pepper together. Rub this mixture into the meat side of the salmon fillet and sprinkle with chopped fresh dill.

2. Place one fillet on top of the other, skin side out, and in opposite directions. Place salmon in the refrigerator at 40° to 50°F (4.4° to 10°C) for 3 days, turning twice a day. Baste with the brine that forms.

3. For the sauce, dissolve Coleman's Mustard in lemon juice, and add Arran Mustard, sugar, and egg yolks.

4. Whisk in oil a little at a time.

5. Add dill and season to taste.

6. Serve as you would smoked salmon and garnish with lemon wraps, sprig of fresh dill, and mustard sauce.

REVIEW QUESTIONS

1. Why is preservation necessary in a modern kitchen?

2. What are the advantages of using preserved foods in a busy kitchen?

3. Explain the term *ceviche.*

4. Describe the different methods of preparing ceviche.

5. Where did the term *carpaccio* come from?

6. What is carpaccio?

7. What is tartar?

8. What is chutney?

9. How do you oven dry foods?

10. What does pH value mean?

A.   Group Discussion

As a group, discuss the importance of preservation of foods in the modern kitchen.

B.   Research Project

Choose a country other than your own and research the forms of food preservation that are used by the native peoples. Report on how the foods have changed due to modern practices.

C.   Group Activity

Create a chutney recipe using local products and produce a small sample for evaluation by your peers.

D.   Individual Activity

Study a menu from a local restaurant, concluding where preserved foods could have been used as a financial saving.

ACTIVITIES
AND
APPLICATIONS

CHAPTER 13

Curing, Sausage Making, and Smoking

After reading this chapter, you should be able to

- cure food.

- use the ingredients that are used to cure food.

- make a shellfish cure.

- cure fish.

- make lox.

- explain the difference between wet and dry curing.

- explain the use of brine.

- explain how to use all types of sausage casings.

- demonstrate how to grind meat for sausage.

- explain how to season meat for sausage.

- demonstrate how to stuff sausage.

- explain how to smoke food.

- discuss the difference between hot and cold smoking.

- make bacon.

- discuss how to barbecue food.

- define American barbecue.

- define Jamaican barbecue.

- define South African barbecue.

- define Indian barbecue.

- define Brazilian barbecue.

The herring is a lucky fish
From all disease inured,
Should he be ill when caught at sea;
Immediately—he's cured

SPIKE MILLIGAN

A Perspective in the Curing of Meats and Seafood

For thousands of years, food has been preserved with salt and smoke in order to make it more suitable for long-term storage. However, these foods were very salty or so strongly flavored with smoke that they were almost inedible. This condition resulted in the foods being soaked in fresh water, and the oversmoke became an acquired taste to those who ate it. How salt curing was discovered remains a mystery; however, it could have been as simple as someone noticing that food washed in sea water lasted longer than food washed in fresh water. Or it may have been observed that an animal appeared still edible after having been killed and naturally preserved in a salt-water lake or brine pool.

In their book on garde manger, The Culinary Institute of America (2004) writes:

Records of various curing methods have been tracked back as far as 3000 BC, when it is believed the Sumerians salted meats as a way to preserve this valuable but perishable food. Historical evidence shows that the Chinese and the Greeks had been producing and consuming salted fish for many years before passing their knowledge on to the Romans. (p. 4)

It would make sense, in a primitive home, that freshly cleaned fish left hung to dry near the warming fire would have had a longer shelf life (Figure 13–1). Not only would this dry smoking process have been a great discovery, but the ancient people probably thought it tasted better as well. Although the use of salt and smoke to cure and preserve is part of our ancestors' ancient history, the scientific reasons have only been discovered in modern times.

Throughout history, meat, game, poultry, and fish have been common food for most people in the world, and many groups of people have survived because they have been able to cure their

FIGURE 13–1 Fish hanging in an ancient home rafter. Courtesy of Cephas Picture Library/Alamy

food. Tribes, such as the Laplanders and Eskimos, have subsisted almost exclusively on a meat diet for many generations. Today, curing food is as important as it was for these peoples and has become a vast industry of its own. In this chapter, we show how curing can be applied to the modern garde manger, allowing the chef to become more versatile with menu and dish creations.

The Curing of Foods

Curing is the addition of salt, sugar, and **nitrite** or **nitrate** to any protein to preserve, flavor, and color. Salt penetrates the cells of the flesh by a process called *osmosis,* which dehydrates the meat. This results in the lower moisture content that inhibits the ability of bacteria to thrive and re-produce. The primary reason for curing meat is to retard the growth of harmful microbes, such as those causing botulism, particularly while the meat is in the temperature danger zone (Figure 13–2). Therefore, it is very important that meats and sausages that are cooked and smoked at low temperatures be properly cured.

There are several methods of applying the curing process to the flesh. When choosing the method for curing, several simple guidelines should be observed:

- The shape and size of the piece of meat that is being cured is important, because this will determine how long the curing process will take.
- The regularity of the shape determines how to choose the curing method. It would be very difficult to reach every nook and cranny of a whole chicken with a dry cure, whereas a wet brine will reach every part of the bird, thereby ensuring an even and complete cure.
- The delicacy of the food that is to be cured should be evaluated. It would not take a very strong solution or dry mix to cure a tender sea scallop compared to a dense ox tongue.
- When curing only the surface of the product, in order to emphasize the color more obviously, the cure must be light and briefly used.
- The introduction of dry flavoring agents into the cure, such as herbs and spices, would sug-gest the use of a dry rub as the best choice among the alternatives.
- The use of liquid flavoring agents, such as boiled and chilled wines or spirits, would suggest the use of wet brine as the best choice among the alternatives.
- The strength of the cure has to be determined, because certain foods need stronger cures to penetrate them.

FIGURE 13–2 FDA Temperature
Danger Zone Chart

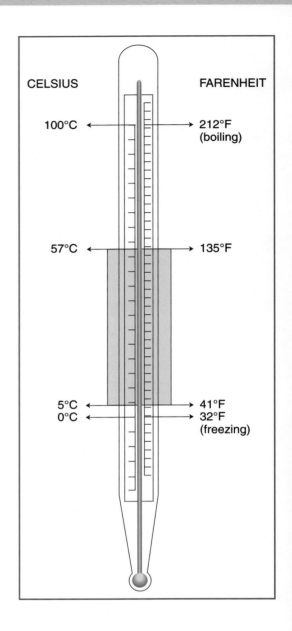

The Basic Curing Methods

The basic curing methods can be applied in many different ways to the food product. It is important to choose the correct method for the food selected:

■ **Dry curing** involves the application of the dry curing mix directly on the meat. This method results in products with very low moisture content and takes the longest amount of time to completely cure. Dry curing is used in curing hams, bacon, salt pork, salmon lox, and other small cuts of meat.

■ **Dry sugar curing** involves the addition of sugar and nitrate and/or nitrite to the salt that is then applied directly to the surface of the meat. This is always done in the cooler and can be applied to foie gras, fish fillets, and smaller cuts of meat, poultry, and game.

■ **Brine curing,** or pickling, involves the mixing of the salt, nitrite or nitrate, and/or sugar with water. The meat is cured with this brine by soaking it for a specific time in the cooler. Larger cuts of meat and poultry such as hams and turkeys are cured this way, although smaller products including whole chickens and fish may be soaked in a curing brine solution. It is not advisable to reuse brines because their strength has been considerably weakened when used, and it would not be of the correct strength to cure again successfully. Additionally, they will always contain contaminants from the previously cured products. There are three ways of accelerating the brining process, including stitch pumping, artery injection, and needle injection (Figure 13–3). They are generally applied to larger pieces of meats that the chef wants to cure in a shorter amount of time.

- *Stitch pumping* is done by inserting a needle with multiple holes on the end into the meat and injecting the brine into the flesh so that it cures from the inside as well as the outside. This method is not foolproof because it is difficult to know exactly where the liquid reaches inside the meat. An electric pump will aid the injecting and give a more accurate result.
- *Artery injection* involves the use of a needle with a single hole. The brachial or femoral artery is located, and the brine is pumped into it to flood the arterial system and cure from the inside as well as from the outside.
- *Needle injection* is the most common way of commercially curing meat today. A machine that contains multiple needles automatically injects the meats with the brining solution.
- Sausage cure method is the method for making cured sausage that differs because the meat is ground when the curing salt and spices are mixed into it, resulting in the cure getting directly to the meat.
- Combination cure method combines the dry rub cure with injection of brine solution and is generally used for hams and other larger pieces of meat. The advantage is that this method shortens the curing time required and reduces the chance of spoilage because the cure process takes place inside and outside the ham.

Ingredients Used in Curing and Sausage Making

The ingredients that are used to cure food must first be understood in order for them to be used correctly. They all have their own specific job to do during the curing process, and they need to be measured accurately at all times for effective results. The combination of curing ingredients used can also have flavoring agents added to them in order to further enhance the overall flavor.

- Salt is the most important product used in the curing of meat and the making of sausage, so it must be of the highest grade possible. In order to test the quality of the salt, dissolve 2 tablespoons (30 mL) of salt in a pint-sized (473-mL) glass container partially filled with water.

FIGURE 13–3 Injecting brine into turkey with a brining needle

Check the cloudiness of the resulting liquid. Clear liquid suggests good-quality salt, whereas cloudy liquid suggests poor-quality salt containing lots of impurities. Do not use iodized salt for curing; kosher flake salt or canning salt should be used. The salt will enhance flavor, act as a microbial inhibiter, and assist in the binding qualities of dry-cured meat and sausages. It is also responsible for the "setting up" or "stiffening" of the meat.

Note: According to Dr. Janet Starr Hull, PhD, CN, "Sodium nitrite and sodium nitrate are two closely related chemicals used for centuries to preserve meat. While nitrate itself is harmless, it is readily converted to nitrite. When nitrite combines with compounds called secondary amines, it forms nitrosamines, extremely powerful cancer-causing chemicals. The chemical reaction occurs most readily at the high temperatures of frying. Nitrite has long been suspected as being a cause of stomach cancer."

■ Natural sugars, such as maple sugar, brown sugar, honey, and molasses, are used primarily in the curing process as a flavoring agent and to counteract the harshness of the salt used in the recipe. Caramelized white sugar is used as a flavoring agent, to enhance the browning process of the sausage during cooking, and as nourishment for the beneficial bacteria that are active during the curing process. Modified sweeteners, such as corn syrup solids, dextrose, and crystalline fructose, are also used in the sausage-making process for their binding qualities and lighter sweetening power.

■ Sodium nitrite is usually blended with salt to help control the level and distribution of the nitrite in the cured item or sausage. It improves the flavor of the food and forms the cured pink color of the food after cooking. **Sodium nitrite** also retards rancidity and helps reduce microbial growth.

■ **Prague powder #1**, also known commercially as InstaCure #1 or **tinted cure mix**, is a combination of 1 ounce (28 g) of sodium nitrite and 1 pound (457 g) of salt thoroughly mixed together. It is used for curing all meats that are to be cooked, smoked, or canned. One ounce of this mix cures 25 pounds (11.34 kg) of meat or sausage; 1 level teaspoon cures 5 pounds (2.27 kg) of meat or sausage.

■ **Prague powder #2**, also known commercially as InstaCure #2 or tinted cure mix #2, is a combination of 1 ounce (28 g) of sodium nitrite, 0.64 ounce (18 g) of sodium nitrate, and 1 pound of salt thoroughly mixed together. It is used for curing products that do not require cooking, smoking, or refrigeration. One ounce (28 g) of this mix cures 25 pounds (11.3 kg) of meat or sausage; 1 level teaspoon cures 5 pounds (2.27 kg) of meat or sausage. This mixture acts like a time-release mechanism, breaking down into sodium nitrite and then into nitric oxide to cure the meat over extended periods of time.

■ Fresh ground spices and dried herbs are used in the making of a rub or dry cure mix that will be packed onto a piece of meat for curing. In the production of sausages, they are incorporated into the meat by adding them to the water so as to prevent uneven distribution throughout the entire mix. Whole spices can be roasted and added to brines for extra flavor, or they may be crushed into dry rubs.

■ Water used in the production of brines should always be boiled and cooled before being used in order to clean and purify it. The water used in the production of sausage should contain a large quantity of ice, to ensure that it is very cold. The amount used is generally 1 pint (473 mL) per 10 pounds (4.5 kg) of meat

■ **Powdered dextrose** is a sweetener used in sausage making and is only 70% as sweet as sugar. It is used in dry-cured or semi–dry-cured sausages, and aids in the fermentation process, giving the sausage the required amount of tang. It also helps reduce the harshness of salt and helps with the browning process during the smoking and cooking of the sausage.

■ **Soy protein concentrate** and nonfat dry milk are ingredients that are used in the production of sausages to help retain the juices of the meat and to bind the meat well together. They also contribute to the weight of the sausages and give them a nice plump look when cooked. Their addition makes the sausage easier to slice and allows the sausage to hold its shape when sliced. They are rarely added to fresh sausage, because they give a bland and greasy look to the meat, but they are excellent in sausages that are to be cooked or smoked.

■ **Corn syrup solids** are used to add flavor and help the fermentation process. They also are very good binding agents for sausages that are to be cooked at low temperatures. They help hold the color in the sausages and are especially used when the sausages are to be displayed under fluorescent light (which bleaches out the color in the sausage).

Curing and Brining Formulas with Time Charts

Accurately following recipes and formulas for the production of sausage is imperative, not only to enhance flavor but also, and more importantly, to ensure health and safety. The ingredients that are used in the making of sausage are powerful, so they need accurate measurement in order to render the finished product edible.

Note: It is very easy to spoil a batch of sausage by inaccurate weighing of salts or spices.

PREPARING AND CURING SHELLFISH

Shellfish such as shrimp, scallops, oysters, and conch cure well. As a result, they take a light smoke that gives them a wonderful flavor when cooked and then served either hot or cold. It is important to note some prebrining rules:

- Clean the shellfish well and ensure that they are very fresh.
- Prepare brine made of 1 cup (237 mL) of salt to 1 gallon (3.8 L) of water, and presoak the shellfish in the brine for 30 minutes.
- Rinse in fresh, clean water and dry thoroughly

Note: This cleans any blood or debris from the evisceration process and ensures a bacteria-free product with which to start.

RECIPE 13–1

SHELLFISH BRINE FOR HOT SMOKING

Recipe Yield: 1 gallon (3.8 L) for 10 pounds (4.5 kg) of shellfish

MEASUREMENTS		INGREDIENTS
U.S.	**METRIC**	
16 ounces	454 g	Kosher salt
8 ounces	227 g	Sugar, brown or white
½ ounce	14 g	Prague powder #1
½ ounce	14 g	White pepper
1 ounce	28 g	Mixed spice of choice or herbs of choice
1 gallon	3.8 L	Hot water
1 teaspoon	2 g	Garlic powder (optional)
1 teaspoon	2 g	Onion powder (optional)
1 fluid ounce	30 mL	Lemon juice (optional)

PREPARATION STEPS:

1. Dissolve all the ingredients in the water and allow the brine mixture to cool.

2. Pour enough brine over shellfish to completely submerge them. Use a plate or plastic wrap to keep them completely below the surface.

3. Remove and discard the brine, patting the shellfish dry.

4. Allow the shellfish to air dry for 2 hours before smoking.

PREPARING AND CURING FISH

To prepare the large oily or game fish for curing and smoking:

1. Scale the fish well, cleaning the skin thoroughly.
2. Fillet the fish and remove any pin bones.
3. For large fillets, periodically remove 2-inch (5-cm) slices of the skin down the back of the fillet so the dry cure can act from both sides.
4. Small white or oily fish can be filleted with the skin still attached, split down the back with the bone and head still attached, left whole with the bone slit to the skin from the inside, or filleted and skinned.
5. Prepare a brine of 1 cup of salt and 1 gallon (3.8 L) of water.
6. Place the fillets in the brine for 30 minutes to soak out any blood diffused through the flesh. We call this a *prebrine.*
7. Rinse in fresh water and dry thoroughly.
8. The ingredients of the brines and cures can be altered to create differing flavor with other ingredients; however, the quantities of cure mix and salt should remain the same.
9. The fillets are ready for the dry cure. (See Recipe 13–2.)

RECIPE 13–2

DRY CURE FOR LARGE FISH FILLETS (TO BE HOT SMOKED)

Recipe Yield: 26 ounces (0.7 kg) for 10 pounds (4.5 kg) of fish fillet

MEASUREMENTS		INGREDIENTS
U.S.	**METRIC**	
16 ounces	454 g	Kosher salt
8 ounces	227 g	Sugar, brown or white
½ ounce	14 g	Prague powder #1
½ ounce	14 g	White pepper
1 ounce	28 g	Mixed spice of choice or herbs of choice
		Cheesecloth as needed

PREPARATION STEPS:

1. Prebrine the fish fillets; wash and dry well.
2. Mix all ingredients together well.
3. Rub all over the fish, making sure that all the mix is used. Wrap in cheesecloth to hold shape.
4. Placing weights on the fillets at this time will retain good shape for service.
5. Cure for 6 to 8 hours, turning once. If the fillet exceeds 1½ inches (3.75 cm) in thickness, extend the curing time 2 hours per inch (2.5 cm).
6. Wash thoroughly and air dry for 6 hours before hot smoking.

RECIPE 13–3

BRINE FOR SMALL OILY OR GAME FISH (TO BE HOT SMOKED)

Recipe Yield: 1 gallon (3.8 L) for 10 pounds (4.5 kg) of fish fillet

MEASUREMENTS		INGREDIENTS
U.S.	**METRIC**	
16 ounces	454 g	Kosher salt
8 ounces	227 g	Sugar, brown or white
½ ounce	14 g	Prague powder #1
½ ounce	14 g	White pepper
1 ounce	28 g	Mixed spice of choice, or herbs of choice
1 gallon	3.8 L	Distilled water, hot
1 teaspoon	2 g	Garlic powder (optional)
1 teaspoon	2 g	Onion powder (optional)
1 fluid ounce	30 mL	Lemon juice (optional)

PREPARATION STEPS:

1. Prebrine the fish fillets; wash and dry well.
2. Mix all ingredients. Add hot water. Stir to dissolve and then cool.
3. Pour enough brine over fish to completely submerge them. Use a plate or plastic wrap to keep them completely below surface.
4. Cure the fish for 2 to 3 hours if fillets, and 3 to 4 hours if bone is still present.
5. Wash the fish in fresh water and dry well.
6. Discard the brine.
7. Air dry the fish on racks for 2 hours before hot smoking.

PREPARING AND CURING LOX

The production of **lox** is as diverse as the countries that claim it as their own. The recipes here are very simple, with a minimum number of ingredients, thus allowing the fish to speak for itself. The quality of the salmon used is of paramount importance; it should be the freshest possible for these recipes. Prepare the salmon fillet, as with all large fillets, taking care not to lose any of the fat bellies of the fish. They are one of the most important areas of flavor in a carved slice of lox. (See Recipes 13–4, 13–5, and 13–6.)

RECIPE 13–4

DRY CURE FOR LOX I

Recipe Yield: 26 ounces (0.7 kg) for 10 pounds (4.5 kg) of fish fillet

Note: Alternative recipes do not include the use of the cure mix Prague powder #1. However, we choose to use it in order to promote safety in the consumption of the fish.

MEASUREMENTS		INGREDIENTS
U.S.	METRIC	
16 ounces	454 g	Kosher salt
8 ounces	227 g	Brown sugar
8 ounces	227 g	White sugar
½ ounce	14 g	Prague powder #1
½ ounce	14 g	White pepper
1 ounce	28 g	Dill weed
		Cheesecloth as needed

PREPARATION STEPS:

1. Trim and shape fillets.

2. Mix all ingredients together well.

3. Rub all over the fish, making sure that all the mix is used. Wrap in cheesecloth to hold shape.

4. Place weights on the fillets to retain good shape for service.

5. Cure for 6 to 8 hours, turning once. If the fillet exceeds 1½ inches (3.75 cm) in thickness, extend the curing time 2 hours per inch (2.5 cm).

6. Wash thoroughly and air dry for 6 hours before cold smoking.

7. Smoke cold for 6 hours.

8. Chill and allow the fish to stiffen or set (within 24 hours).

9. Brush with oil and wrap in parchment. Do not wrap in plastic wrap because this encourages molding.

RECIPE 13–5

DRY CURE FOR LOX II

Recipe Yield: 32.3 ounces (0.9 kg) for 10 pounds (4.5 kg) of fish fillet

MEASUREMENTS		INGREDIENTS
U.S.	METRIC	
16 ounces	454 g	Kosher salt
8 ounces	227 g	Brown sugar
8 ounces	227 g	White sugar
½ ounce	14 g	White pepper
		Cheesecloth (optional)

PREPARATION STEPS:

1. Trim and shape fillets.

2. Mix all ingredients together well.

3. Rub all over the fish, making sure that all the mix is used.

4. Wrap the cure-packed fish well in cheesecloth.

5. Cure for 24 hours in the cooler.

6. Wash and dry well, discarding the cure.

7. Cold smoke the fish for 6 hours and brush with oil, wrapping in parchment paper for cooling and setting.

RECIPE 13–6

DRY CURE FOR LOX III

Recipe Yield: 32.3 ounces (0.9 kg) for 10 pounds (4.5 kg) of fish fillet

MEASUREMENTS		INGREDIENTS
U.S.	METRIC	
32 ounces	907 g	Kosher salt
1 ounce	28 g	White pepper

PREPARATION STEPS:

1. Trim and shape fillets.

2. Sprinkle part of the salt all over the fish, and allow it to sit at room temperature.

3. After 1 hour, sprinkle more salt all over the fish, and leave for an additional 30 minutes.

4. Continue to do this over a 4-hour period in 30-minute intervals, until all the salt has been used. There should be a considerable amount of liquid releasing from the fish.

5. After 4 hours, wash the fillets in lots of cold water and dry well.

6. Allow the fish to air dry for 2 hours.

7. Cold smoke for 6 hours and brush with vegetable oil.

8. Wrap in parchment paper and chill until set up and firm to the touch.

CURING AND BRINING MEATS, POULTRY, AND GAME

When dry curing flat pieces of meat, trim the meats to a square shape before curing begins to ensure ease in carving the finished product (Table 13–1). (See Recipes 13–7 and 13–8.)

Table 13–1   Time Chart for Curing Meats

THICKNESS	TIME	PRODUCT
¼ inch (0.64 cm)	2 to 3 hours	Lean meats such as loins and tenderloins
1 inch (2.5 cm)	6 to 8 hours	Lean meats such as loins and tenderloins
2 inches (5 cm)	10 to 12 hours	Lean meats such as loins and tenderloins
2 inches (5 cm)	6 to 10 days	Pork bellies for the making of bacon. One day per pound
4 to 6 inches (10 to 15 cm)	30 days	Bone-in loins of pork
6 inches (15 cm)	45 days	Bone-in hams

RECIPE 13–7

DRY CURE FOR MEATS

Recipe Yield: 26 ounces (0.7 kg) for 25 pounds (11.3 kg) of meat

MEASUREMENTS		INGREDIENTS
U.S.	**METRIC**	
16 ounces	454 g	Kosher salt
8 ounces	227 g	White sugar (or 4 ounces; 113 g of powdered dextrose)
2 ounces	57 g	Prague powder #1

PREPARATION STEPS:

1. Combine all ingredients well and store in an airtight jar.

2. Rub all over the meat, then stack the meats on top of each other.

3. Chill and turn every 2 days, until cured.

4. Curing times depend on the thickness and the texture of the meats (Table 13–1).

The brining of meat, poultry, and game must be taken very seriously, because there is a distinct possibility of causing food poisoning due to inaccurate brining times and procedures. The meat has to remain in the brine for the appropriate length of time for the curing process to work all the way through to the center of the meat. When the cut of meat is large, as in the case of hams and barons of beef, the amount of time a general brine will take to penetrate might be longer than the amount of time the meat bone and marrow in the center will take to spoil. It is therefore advisable to inject the brine into product of this size in order to create curing from the inside out, as well as from the outside in. Hams such as those shown being salted (Figure 13–4) and the prosciutto hanging in a drying room (Figure 13–5) are generally salted externally as well as injected with a brine cure. When brining, it is important to consider each piece of food separately and to adjust the recipe or method of application of the brine to suit the piece of meat (Table 13–2). Some very important considerations include the following:

- Foods that have a thick or tough skin, which create a barrier to the brines
- Products with a considerable layer of fat on the exterior
- The density of the meats

Note: Injection can be done by any of the methods mentioned under the Basic Curing Methods and should be administered accurately to approximately 10 percent of the weight of the piece of meat.

FIGURE 13–4 Salting hams

FIGURE 13–5
Prosciutto hanging in a drying room

- Large pieces of meat that contain bone in the center
- Very large and tough poultry and game birds
- Products that contain bone with marrow
- Ensuring that the meats are completely submerged in the brine

Table 13–2   Time Chart for Brining Meats

FOOD	SOAKING TIMES (WITHOUT PUMPING)	WITH PUMPING TO 10 PERCENT OF WEIGHT
Chicken breasts	24 to 28 hours	Do not pump
Whole chickens	24 to 28 hours	12 hours
Game birds	24 to 28 hours	12 hours
Large game meats	7 to 8 days	3 days
Pork, boneless	5 to 6 days	2 to 3 days
Turkeys	5 to 6 days	2 to 3 days
Hams, bone in	20 to 22 days	6 to 7 days
Corned meats (brisket)	7 to 8 days	4 to 5 days

RECIPE 13-8

Brine for Meats, Poultry, and Game

Recipe Yield: 1 gallon (3.8 L) for 25 pounds (11.3 kg) of meat, poultry, or game

MEASUREMENTS		INGREDIENTS
U.S.	**METRIC**	
8 ounces	227 g	Kosher salt
4 ounces	113 g	Powdered dextrose (or 8 ounces; 227 g white sugar)
2½ ounces	71 g	Prague powder #1
½ ounce	14 g	White pepper
1 ounce	28 g	Mixed spice of choice or herbs of choice
1 gallon	3.8 L	Distilled water, ice cold

PREPARATION STEPS:

1. Dissolve and mix all ingredients together well.

2. Brine for required amount of time.

3. Wash the meat in fresh water and dry well.

4. Discard the brine.

5. Air dry the meats on racks for 2 hours before hot smoking.

Sausage Making

The word *sausage* originally comes from the Latin word *salsus,* which means "to salt" or "to preserve." This process was necessary for people to allow their meat products to preserve longer. Sausage making became a crucial method for using the ground or chopped scraps and trim that were left over after the slaughter of an animal too large to consume in one or two meals. Trimmings were always enclosed in the intestines, appendix, gullet, or bladder of the slaughtered animal, in a manner distinct by region, creating a very provincial delicacy that has developed into a culinary art form.

In some parts of the world, especially in colder climates, the sausage would survive during the winter without refrigeration, allowing people to eat them regularly during the bleak months. However, during the summer, people had to develop ways of preserving meats. Hence, the process of smoking and drying sausages allowed people living in even the fairest of climates to have sausage available year-round, without the need for continuous cooling. Sausages became known by the name of the towns from whence they originated, such as bologna coming from the northern Italian town of Bologna.

CLASSIFICATIONS OF SAUSAGE

Although the culinary world is blessed with thousands of sausages, there are basically six classifications of sausage:

- Cooked sausage is made with fresh meats and then fully cooked and sold as is. It is generally eaten immediately after cooking or is refrigerated and reheated before eating. They are also commonly eaten cold. Examples include braunschweiger, veal sausage, and liver sausage.
- Cooked smoked sausage is basically the same as cooked sausage, but it is smoke-cooked or it is cooked and then smoked. It can be eaten hot or cold and is generally reheated in different ways before being eaten, although it can be eaten cold. Examples include wieners, kielbasa, and bologna.
- Fresh sausage is made from fresh meats that have not been previously cured. This sausage must be refrigerated and thoroughly cooked before eating. Examples could include boerewors, sweet Italian sausage, and fresh breakfast sausage.
- Fresh smoked sausage is fresh sausage that is smoked and refrigerated and cooked thoroughly before eating. Examples include mettwurst and andouillie sausage.
- Dry sausage is the most complex of all sausages and is made from a variety of meats. The drying process has to be carefully controlled and the sausage is ready for eating once it has been dried. This sausage keeps for very long periods under refrigeration. Examples include all styles of salamis and summer sausage.

SAUSAGE-MAKING EQUIPMENT

It is very important to ensure that all equipment is sanitized correctly before proceeding with the sausage making. The pieces of equipment involved contain many edges, ridges, and holes that are particularly difficult to inspect. However, these are the areas that must be vigilantly maintained and inspected at all times.

It is also to the sausage maker's advantage that the equipment be very cold before proceeding. Submerging the grinder head, blade, and die in a large pail of ice water for a few minutes before their use, prevents the sausage meat from warming and the separation of fats within the mix. The undesired warming of the sausage meat can result in a difficult stuffing process.

The meat grinder that is used for producing sausage can be a freestanding unit, an attachment to a mixing machine, or a small table attachment. However, there are some basic rules that chefs should always observe:

- The grinder should always be secured and completely stationary for use.
- Always refrigerate the grinder parts before use.
- Never freeze the grinder parts because food will stick.
- After washing lightly, oil the attachments before chilling.
- Clean the grinder well, being careful to reach the inside of the grinder body and the holes on the blades.
- Do not overfill the grinder or force food through.
- Assemble the parts correctly; the most common mistake is facing the blades in the wrong direction.
- Never use a metal spoon to push the meat through; always use the plastic or wooden implement provided. Never try to use an implement that is too small for the opening.
- Soak or wash immediately; do not allow the meat particles to dry on the machine parts because it will be much more difficult to wash.

Assembly of the Meat Grinder

The following are basic steps in the assembly of the meat grinder (Figure 13–6):

1. Attach the chilled grinder head to the machine, making sure that the pin fits into the hole on the machine. This will lock the grinder head in place and prevent it from spinning.
2. Slide the auger (worm) into the opening, ensuring it is properly lined up and fully inserted.
3. Place the knife blade, with flat part facing outward, onto the end of the auger.
4. Secure the chilled grinder die onto the auger, lining up the pin on the bottom of the auger with the hole in the base of the die.
5. Attach the collar and turn to secure.
6. Add the feeding tray.

Note: There are several dimensions of holes available in grinder dies, including fine, medium, and coarse. These are selected for their desired effect on the texture of the sausage.

FIGURE 13–6 Meat grinder assembly

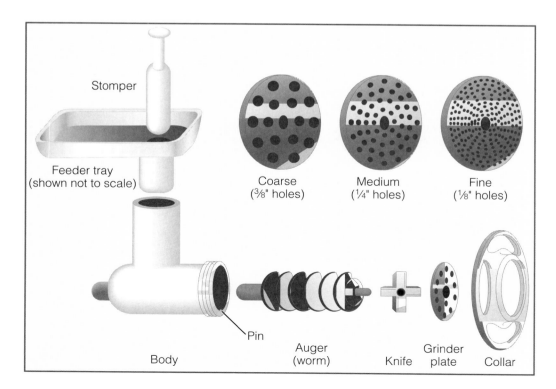

Stomper

Feeder tray
(shown not to scale)

Coarse
(⅜" holes)

Medium
(¼" holes)

Fine
(⅛" holes)

Pin

Body

Auger
(worm)

Knife

Grinder
plate

Collar

FIGURE 13–7 A variety of sausage horns

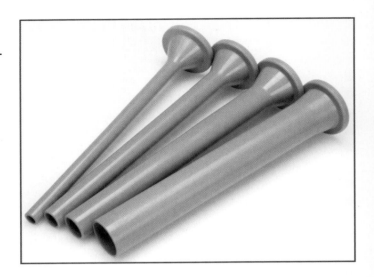

The **sausage horns** used in the production of sausage come in at least four different sizes and are specifically sized to fit the sausage casings that are in general use (Figure 13–7).

■ Sausage stuffers come as individual units or are attached to machines, and are by far the best way to stuff sausage (Figure 13–7). Rolling and tying in plastic wrap or using a piping bag can be successful; however, these methods are not very efficient nor do they result in a good finished product. When using a sausage stuffer, some basic rules should be followed for safety and efficiency:
   1. Sanitize all equipment well, ensuring that the smaller pieces are thoroughly cleansed.
   2. Check that the air pressure valve is working and in position before use.
   3. Understand the possibility that the stuffer crank handle may slip and reverse one full crank because of the pressure buildup in the chamber.
   4. Make sure the stuffer is well secured and stationary during the stuffing.
   5. Always assemble correctly with accurate guidance and instruction.
   6. Do not operate the stuffer on your own. Always have one person cranking the machine while another handles the sausage.
   7. Soak or wash all parts immediately after use. Do not allow the meat particles to dry on the machine parts because they will be much more difficult to wash.

■ Assembling a sausage stuffer can be the key to the making of good sausage; it can also lead to very serious injury if not done correctly (Figure 13–8). Be sure of the following:
   1. Chill all parts of the stuffing machine before use.
   2. Secure the correct sausage horn on the bottom of the meat chamber.
   3. Pack the prepared sausage into the meat chamber.
   4. Place the "O" ring and the air pressure valve onto the piston.
   5. Place the crank handle on the low gear.
   6. Apply the skin onto the horn and prepare to crank.
   7. Release the air as you go to avoid a buildup of pressure.

Other Equipment Used in Sausage Making

■ Large plastic, nonreactive tubs are very useful when making sausage because they hold a great deal of product and serve as a curing tub for meats that need time to cure once they have been cut or ground. They also become important when storing sausage meats, because it is very difficult to tell flavors apart when they are not well labeled. Always ensure that sausage meat, whether in casing or in bulk, is labeled well—a common fault with inexperienced sausage makers.

Note: When placing bulk ground sausage meat into these tubs, it is imperative that the meat be flattened well and all possible air pockets removed from the mass, which can otherwise become an area for bacterial spoilage when storing or curing for long periods of time.

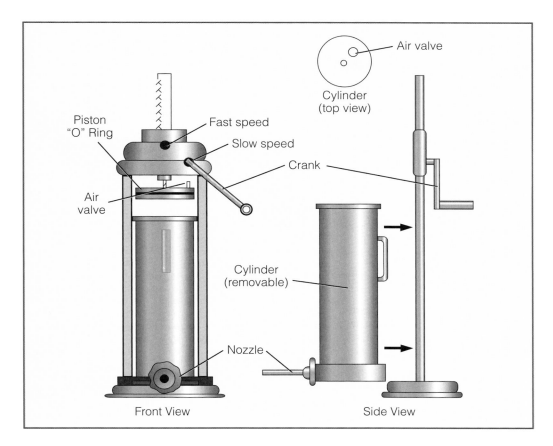

FIGURE 13–8 Assembly of a sausage stuffer

- The sausage knife with its three to four sharp prongs (Figure 13–9) is used to eliminate holes under the skin in the sausage when they have been stuffed. If the sausage has been stuffed tightly enough, when the sharp prong hits the air hole, the pressure of the sausage meat should fill that hole instantly. This prevents possible spoilage in that area of the sausage.
- Hog staplers or hog ring pliers (Figure 13–10) are used to close larger sausages that require their filling to be under pressure during the stuffing. They ensure a tight seal at the end of the sausage and allow the sausages to remain firm and well packed.

SAUSAGE CASINGS AND SKINS

Sausages are encased in a variety of the interior tube-shaped membranes found naturally within lambs, sheep, pigs, and cows. The fresh skins are by far the best when making sausage; however, they are time consuming to use and can break if not handled expertly. As a result, many other manmade sausage skins are available to sausage makers, making their job less labor intensive.

Natural Casings

Generally, sausages are made with the small intestines of the animal, although some regional specialties are made with the large intestine, stomach, bladder, or bung. The advantages of **natural casings** are that the sausage has a fresh, homemade appearance and that it is a semipermeable porous membrane that allows the skin to accept smoke more readily than manmade skins. All fresh casings must be washed thoroughly when removed from the animal. They are then preserved in salt, both on the inside and the outside, in 100-yard lengths that are sold as a hank, bundle, or cap, or by the ounce. These hanks, which are smaller in length as the intestine gets larger in diameter, can be purchased salted, preflushed, or preflushed on tubes. The more that is done to the skins before you receive them, the more expensive they are going to be per hank—an important consideration for the chef.

FIGURE 13–9 Sausage knives

FIGURE 13–10 Hog ring pliers and staplers used for securing the end of large sausage

Salted hanks must first be untangled and cut into appropriate lengths (Figure 13–11A). Then they are flushed clean of the salt on the outside and the inside of the sausage skin. It is advisable then to soak the skin overnight, immersed in cold water, to completely reconstitute it and avoid the cooked sausage from having resistance to the consumer's teeth when biting into it. Skins that are larger in diameter, such as beef round, are generally easier to work with because there is no need to untangle them. However, the **flushing** and soaking procedure is the same (Figure 13–11B).

After prepping the sausage skin, it can be placed on the sausage horn for stuffing (Figure 13–12). Take great care not to tangle the skins when feeding them onto the horn. The use of running water helps a great deal with this technique.

Preflushed casings are basically ready to use, but great care must be taken when feeding the skins onto the sausage horns because they tangle easily.

Some sheep and lamb skins come packaged in a liquid preservative that keeps the casings soft and pliable for immediate use. Preflushed sausage skins are also available untangled and fed onto plastic tubes (Figure 13–13). They can be easily slipped onto the sausage horn with no labor involved in the preparation. This is a convenient way of purchasing sausage skins, but they are more expensive.

Natural casings come in a variety of sizes and from a variety of animals and should be chosen for the specific type of sausage that is to be made. They are generally labeled according to their millimeter size and their animal type (e.g., 32- to 35-mm hog casing). The millimeter size is the internal diameter of the skin and has two dimensions to emphasize an uneven measurement throughout the length of skin. Although the measurement varies between the two dimensions, the difference is difficult to notice with the naked eye when the sausage is stuffed. The following list details the various types of casings, or skins, used in sausage production.

■ Lamb and sheep casings can be difficult to use, but with care and patience they give great results. They are used for breakfast sausage, frankfurters, and fresh pork sausage, and are usually the skins used when making fish and shellfish sausage. As with all natural products, all

FIGURE 13–11A Hanks of sausage skins

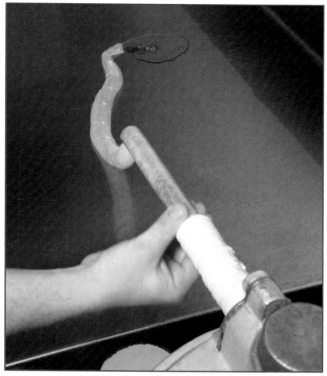

FIGURE 13–11B Flushing fresh sausage skins

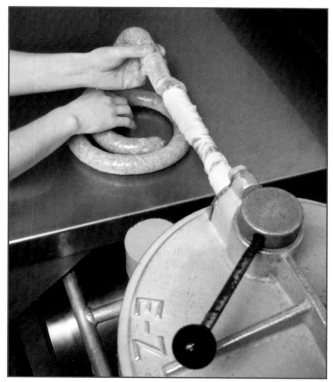

FIGURE 13–12 Filling 31-mm/35-mm hog casing from a preloaded sausage horn

FIGURE 13–13 Preflushed sausage skins

Note: One pound (2.2 kg) of meat stuffs about 6 feet (0.45 kg) of medium-size 22- to 24-mm lamb casing. One hank of medium-sized 22- to 24-mm lamb casing uses about 55 to 60 pounds (25 to 27 kg) of sausage meat.

Note: One pound of meat will stuff about 2 to 3 feet (0.6 to 0.9 m) of 32- to 35-mm hog casings.

Note: One hank of 32- to 35-mm hog casing holds about 110 pounds (50 kg) of sausage meat.

measurements are approximate because of the fluctuations in size and breakages in the skins. The most commonly used lamb and sheep skins are 22- to 24-mm and 24- to 26-mm casings, and are bought in hanks of 100 yards (91 m). There are other areas of the animal that are used for very specialized sausages. The stomach, bung, and bladder of lamb and sheep are also used (Figure 13–14).

■ Pig or hog casings are undoubtedly the most popular casing and can be used for almost any sausage of that general diameter. They are versatile, easy to handle and give consistent results when making fresh sausage. The most commonly used skins are 32- to 35-mm hog casings and come in 100-yard (91-m) hanks. Other areas of the hog's intestine used for sausage making include the hog middle (or chitterlings), which are used for a savory hot entrée, and the hog bung end, which is normally used for salamis and liverwurst (Figure 13–15).

■ Beef casings, or sausage casings from the beef carcass, are all much larger and are used for the larger style sausages. They are easy to handle because there is little to no tangling involved with such large casings. They are bought in a variety of lengths or as individual skins because of the variance in their holding capacity (Figure 13–16).

FIGURE 13–14 Location of sausage skins in a sheep

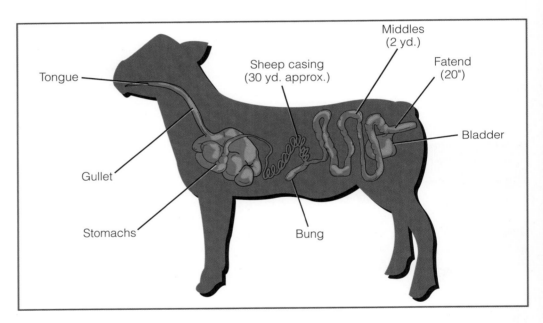

FIGURE 13–15 Location of sausage skins in a hog

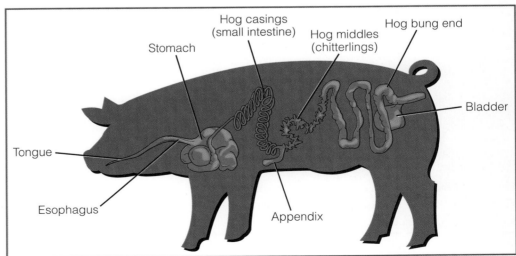

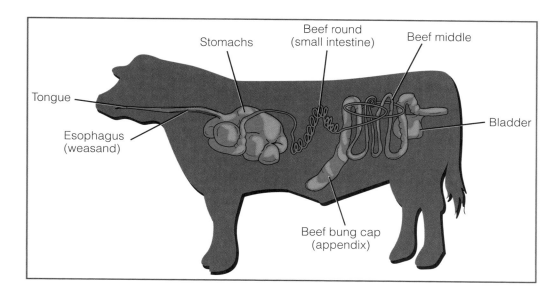

FIGURE 13–16 Location of sausage skins in a cow

- Beef rounds are from the small intestines and are named for the familiar ring shape they create when stuffed. They are sold in 100-foot (91-m) lengths and can vary in diameter from 35 to 45 mm, allowing them to stuff an average of 80 to 90 pounds (36 to 40 kg) of sausage meat. They are used predominantly for the production of ring bologna, mettwurst, liver sausage, blood sausage, and Polish sausage.
- Beef middles are sold in "sets" of 18- to 20-inch (45- to 51-cm) lengths, adding up to their entire length of 57 feet (17.9 m). They vary in width, but a complete set stuffs approximately 80 to 90 (36 to 40 kg) pounds of sausage meat.
- The beef bung, or appendix, and beef bladder are generally sold as individual casings because they are so large. These are specialized skins and are rarely used in restaurants, although there are always new and innovative ways of using them. The bungs come in a variety of diameters ranging from $3\frac{1}{2}$ inches to 5 inches (8.9 to 12.7 cm). The most common size is $4\frac{1}{2}$ inches (11.4 cm) and stuffs between 8 and 10 pounds (3.6 and 4.5 cm) of sausage. They are used to make sausages such as capocollo, large bologna, and cooked salamis.
- Bladders can be large, medium, or small and can stuff between 7 and 14 pounds (3.2 and 6.4 kg) of sausage. Their most famous use is for the production of the mortadella (Figure 13–17).

FIGURE 13–17 Bolognese butcher holding large mortadella

Consider the advantages of fresh sausage skins before opting for the easy selection of collagen or synthetic casings:

■ They do look more like a natural product. In most cases, in a restaurant setting, this is a very good thing.

■ They cook well and taste extremely good by almost any method of cooking.

■ They have better flavor.

■ They accept other flavors (other than smoke) well.

■ They can be stuffed with a greater variety of food products.

■ They shrink equally with the meat during the drying stage.

Collagen Casings

The use of collagen casings became necessary because there were just not enough natural sausage skins to keep up with the ever-growing popularity of sausage products. They also met the need for regular and precise portion control that fresh skins could not deliver and that the industrial catering of the modern world demands.

The **collagen casing** is formed when the hides of beef cattle are split and the flesh side of the skin, or the corium layer, is (1) ground and swollen in acid, (2) sieved and filtered, and then (3) extruded into perfectly even-sized skins of various sizes that are uniform in diameter for their entire length. They are considered to be a fresh product and, because they are edible, should be refrigerated after being soaked. There are some major advantages to using collagen casings:

■ Collagen casings are available in the same variety of sizes of natural casings and they are available curved to imitate the curving of a beef round.

■ They are generally used for smoked sausages, but not as commonly for fresh sausages.

■ They accept smoke well, giving a wide variety of shades of a mahogany color.

■ They accept smoke so quickly that the smoking period can be greatly shortened.

■ They are available in a colored form to imitate smoking.

■ They are much easier to use because there is no tangling or preparation beforehand.

■ They are available in an accordion style, which makes it easier to load greater quantities of skins onto the sausage horn.

■ They are slightly stronger than fresh skins and are less likely to break during stuffing.

■ They need little time to soak before they are ready for use, making them extremely convenient (Figure 13–18).

Synthetic Casings

Synthetic casings come in a variety of sizes and are available in a variety of shades of brown to imitate the smoking process. **Synthetic casings** are available in two forms, including plastic casings and fibrous casings (Figure 13–19). They are both inedible and are used predominantly for their strength. Synthetic casings are able to withstand more pressure during the stuffing process, expelling more air and creating a very compact sausage. They are easy to store and once they have been soaked for use, they require no special handling or refrigeration.

Some of the fibrous casings contain proteins, or in some cases spices, that are lined on the inside of the skin that attach to the exterior of the sausage during the drying process.

Synthetic casings are versatile and very useful when making sausage, especially when the outer skin is not important for the finished product. The following are the major advantages:

■ They are strong and durable when stuffing large, tightly packed sausage, ensuring ease when slicing and good portion control.

■ They are convenient, cleaned and ready to use. They can then be peeled and thrown away.

■ They coat the outside of the sausage during the drying process.

■ They are easy to store and do not need refrigeration.

■ They are not meat based, making them appropriate for fish, poultry, or vegetable sausage.

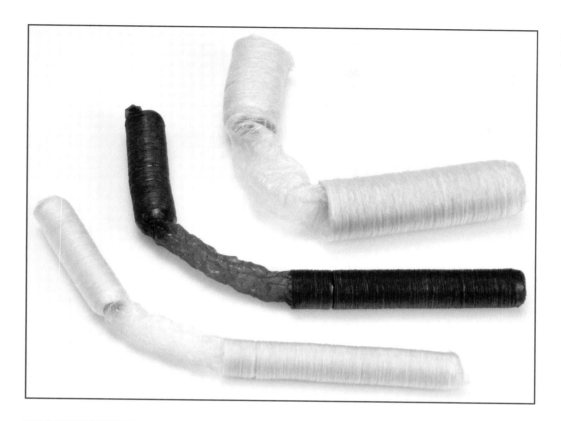

FIGURE 13–18 A variety of collagen casings

FIGURE 13–19 A variety of synthetic casings

Other items can be substituted for sausage skins, but they are not as effective as the natural, collagen, or synthetic skins. Caul fat, the thin membrane attached to the liver and surrounding the stomach of a pig, can be used. It is, however, easily torn and only suitable for individual small sausages that do not have to be tightly rolled. Plastic wrap can also be used with some success in poaching, but there are several limitations. For example, plastic wrap cannot be used in smoking or grilling.

PREPARING AND SELECTING THE MEATS

Although most sausage is made from a mixture of pork, beef, and fat, sausage is a great way to use up tough pieces of meat, no matter where the meat, poultry, game, fish, or shellfish came from. However, most other meats do not contain the appropriate quantities of suitable fats for the making of sausage. This is not to say that other meats cannot be tried, but it may require the blending of multiple meats and fats.

The basic rule of thumb for most sausage is that it should contain at least 25 percent of its weight in fat. The chef can use this guideline to mix and match the meats, often substituting pork fat and some pork meat (for that matter), into the sausage recipes that are being created.

For example, if pheasant sausage is to be served as an accompaniment to a dish being prepared with pheasant breast, the tough pheasant legs and thigh can be ground with a quantity of pork fat and meat to accomplish the needed ratio. When the protein used is of a finer consistency and structure, as is the case in fish and shellfish, the fats used should be more refined and better flavored, like butter and heavy cream. In this case, there may also be a need for panada because the proteins themselves will not sufficiently bind the sausage.

Table 13–3 provides an approximate guideline for the creation of sausage using a variety of proteins. It is important to alter the recipes as needed for each individual protein, to suit the flavor and consistency desired. The addition of fresh herbs, spices, and other flavoring agents can also be chosen as appropriate.

Cutting and Grinding

The strictest rules of sanitation and safety must be observed during the process of cutting and grinding the meats for the production of sausage:

- Sanitize the table, cutting board, meat tubs, grinding equipment, and knives.
- Use only the best quality and the most suitable cuts of meat.
- Bone the meat carefully, removing all the bones and their splinters.
- Make sure that the proportion of fat to meat is correct (Figure 13–20).
- Remove the gristle, sinew, blood spots, and excess fat from the meat.
- Make the meat the appropriate size for the grinder.
- Do not store the cut meat for too long after it is cut.
- Make sure that the meat is very cold before grinding.
- Always measure the amount of spice to meat accurately.

Once the meat is ready for grinding, prepare the machine by checking that it has been set up accurately and is well chilled.

Grinding the Meats

Several precautions should be observed during the grinding process: including:

- Always grind starting with the large-holed die. Second and sometimes third grinds are made with sequentially smaller-holed die (Figure 13–21A).

Table 13–3   Suggested Alternative Sausage Formulas

FOOD GROUP	RATIO OF PORK FAT (PERCENT)	RATIO OF PORK (PERCENT)	BREAD PANADA	BUTTER/ CREAM
Poultry	20	5		
Game birds	25	10		
Meat and game	25	8 to 10		
Fish			15 to 20	5
Shellfish			10 to 15	5 to 8

FIGURE 13–20 Meat cut for sausage, showing the ratio of fat to lean meat

- Have the meat and the grinder well chilled.
- Make sure that the fat is well distributed within the mix.
- Feed the meat through easily with limited pressure; do not force the meat.
- Have the correct size container to receive the ground product. The container should be at least twice the capacity of the amount being ground so that the meat can be easily mixed in the container.

Adding Water and Spices

When adding the spice to the ground sausage, it is important to distribute the spices evenly throughout the sausage mixture. To facilitate the addition of the spice to the sausage, the spice should be mixed well with ice-cold water first before being added (Figure 13–21B). The addition of the ice water, at a rate of 1 pint (473 mL) per 10 pounds (4.8 kg) of meat, not only aids in the even distribution of the spice but also gives the sausage lubrication for the stuffing process. Added water also results in a moist texture in the cooked sausage and assists in the cooling of the sausage mix during stuffing.

The water also makes the mixing process easier. When the integration is complete, the sausage should squirt out from in between the fingers when a handful is squeezed together (Figure 13–21C). It is important when the sausage has been blended well that it be stuffed immediately into the sausage casings or skins. Placing the sausage mixture in the refrigerator before being stuffed will result in the meat becoming too firm to stuff, requiring additional mixing. The flavors of the raw ingredients are sufficiently developed, by resting in the skin, after be stuffed. Resting the meat in bulk, before being stuffed, does not appreciably improve the flavor of the finished product and only impedes the stuffing process.

Note: It is prudent to pan fry a patty of sausage before stuffing the mixture. Testing the binding qualities and flavor, before stuffing, allows the garde manger chef an opportunity to adjust the ingredients, if needed.

Stuffing and Storing

Once the sausage meat has been seasoned and lubricated, it can be stuffed into the casings, which should be standing by, already flushed, soaked overnight, and on the horn.

Once the stuffer is in place and has been assembled correctly, the following precautions should be adhered to:

1. When placing the meat into the stuffer, pack it down well, eliminating as many air pockets as possible.
2. Ensure that the casing to be stuffed is very wet inside and outside. This will aid in the free flow of the stuffing.

FIGURE 13–21A Grinding the sausage meat

FIGURE 13–21B Adding the spices and water to the meat for sausage production

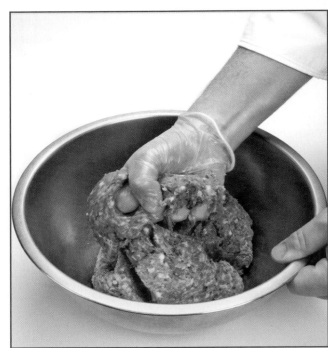

FIGURE 13–21C Mixing the sausage

Note: If the skin bursts, tie the tear off, and start the process again. A tear or break in the skin is not considered a failure; it's a common occurrence in making sausage.

3. Crank slowly at first, until the meat reaches past the open end of the skin.
4. Tie the skin into a strong knot, and begin to stuff.
5. Eliminate any air pockets that develop using the sausage prong.
6. Always ensure the casing is in front of the horn, and free from twists or obstructions (Figure 13–22A, B, and C).
7. Link the sausage in appropriate lengths, and store them in the cooler at once; or place them on smoke sticks and load them straight into the smoker.
8. Twist the sausage at the desired lengths, or tie the lengths of sausage with pieces of twine.

FIGURE 13–22A Stuffing a fresh skin

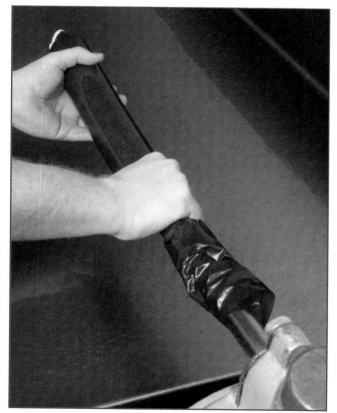

FIGURE 13–22B Stuffing a synthetic casing

FIGURE 13–22C Stuffing a collagen casing

SMOKING FOOD

Man has been smoking food since ancient times, probably unwittingly at the beginning. It was widespread practice for our ancestors to hang their "kill" in the rafters of their timber, straw, or stone dwellings, to prevent wild animals from running off with it, and to preserve the meat through air drying. The prevalent practice of building fires in the center of these lodgings for cooking and heating purposes allowed the smoke to swirl upward and envelope the hanging meats. Chimneys were not often built, and the smoke was usually trapped under the sod roofs. The "blackhouses" of Scotland get their name from the soot that completely coated the walls and ceilings of those ancient drystone and thatched-roof croft houses.

It is likely that this practice led to the discovery that foods exposed to the smoke from the fire remained in better condition, and for a longer time. And probably, just as important, they improved the taste of the otherwise unseasoned and possibly rancid meat. Once realized, smoking meats and seafood became a common method of preservation, helping to provide meat, fish, and other food for the long dark winter months.

In modern times, the smoking process still involves the same basic steps of brining and salting, or somehow curing the food; air drying it; and then smoking it slowly over a smoldering fuel source. The process still effectively extends the product's shelf life, but only to a point. **Smoking** is now primarily used to impart a pleasant taste and color to the food, as well as to enhance the natural flavors. The methods used to smoke food have changed dramatically as technology has progressed, and the demand for smoked products has risen—regardless of the season of the year.

The smoke itself consists of numerous tiny droplets of various natural chemicals, such as aldehydes, phenols, ketones, and carbolic acid. These chemicals tend to condense on the food being smoked and form a tacky film called the **pellicle**. The chemicals settle on the surface as well as penetrate into the meat and through the porous skin, and all the way into the casing of the sausage. This action gives the meat a smoky flavor and allows the natural chemicals to kill or stop the formation of bacteria, yeast, and molds that would lead to spoilage. The phenols in the smoke also help prevent the oils and fats within the product from turning rancid.

SMOKING EQUIPMENT

The most primitive of smokers were simple, enclosed vessels or rooms through which air could flow freely. This enabled the smoke to continually generate, and the heat within the chamber could be controlled. They tended to last for years, and the person operating them had an intimate knowledge of their idiosyncrasies and knew exactly how to produce the best results for anything that was going to be smoked.

Note: To have the most versatile smoker, it is necessary to have the option of operating each element of the machine independent of the others. Separate functioning allows the elements to be used together simultaneously, or in any combination that the recipe or technique demands.

Today, a host of machines are available for smoking, and they all have their individual procedures for being used, generally according to the manufacturer's instructions (Figure 13–23). Although it would be impossible to write about them all, a set of guidelines on the important facts of smoking, as noted in the following section, can offer general insights into the process of smoking foods.

UNDERSTANDING THE ELEMENTS INVOLVED IN SMOKING

The first step is to break the smoking process down into its elements or components, and then to understand the difference between **cold smoking** and **hot smoking**.

■ The smoking chamber can be made from almost anything and comes in many shapes and sizes, although there are some very important elements to keep in mind when choosing one (Figure 13–24):

   ■ The chamber should be of the appropriate size for the quantity of product that is generally smoked by the establishment.

   ■ It should have the ability to hang smoke sticks, as well as receive small and large sheet pans or screens for supporting resting products.

FIGURE 13–23 Sausage hanging in a large smoker

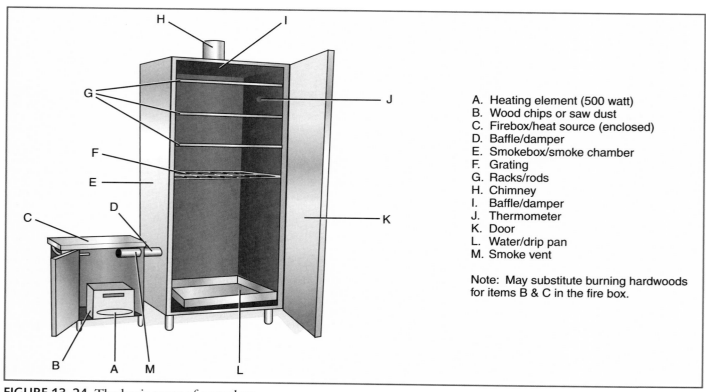

A. Heating element (500 watt)
B. Wood chips or saw dust
C. Firebox/heat source (enclosed)
D. Baffle/damper
E. Smokebox/smoke chamber
F. Grating
G. Racks/rods
H. Chimney
I. Baffle/damper
J. Thermometer
K. Door
L. Water/drip pan
M. Smoke vent

Note: May substitute burning hardwoods for items B & C in the fire box.

FIGURE 13–24 The basic parts of a smoker

■ It should have easy access for loading, with a solid well-sealed door.

■ It should have effective dampers for releasing moisture and smoke.

■ The dampers control the flow of smoke and can affect the color of the finished product. There should be a draft control at the base of the chamber to work with the damper and to create the flow of air. The dampers should be easy to operate and have the ability to stay wide open or shut, or any point in between, as required.

■ The heat source can be wood, peat, pellet, gas, or electricity. It must work dependably, and should have accurate and reliable temperature gauges and controls so as to monitor the cooking process properly. It does not have to be the smoke source as well, and often works in conjunction with a smoke source.

■ The smoke source can be exterior to the food chamber and heat source where it is more versatile and controllable. It should be able to be used for both cold smoking and hot smoking. The smoke source should be fitted with a fan to allow for a controlled flow of smoke, enabling it to produce the correct amount of smoke for the correct length of time. It also allows the chamber to be used for cold smoking because the heat source can be turned off independently.

■ The humidity control allows for a greater range of products including dried and semidried sausage and meats that have to be very carefully monitored. Sometimes, if a great quantity is being smoked, it is better to do this drying process in a separate air-controlled room that draws the moisture out and controls the relative humidity. The ideal conditions for this drying process are between 45° and 55°F (7.2° and 12.8°C) with a humidity of 70 to 72 percent.

■ The fuel to be used will affect the eventual flavor of the product and should be carefully selected (Table 13–4). A combination of hard woods, such as oak, alder, and mesquite, generally provides the best results for smoking. Fruit woods, such as cherry and apple, contain a lot of tar compounds and should be used sparingly or in combination with hardwoods. Soaking the dried wood well, before use, helps to give a colder smoke and a much better final product. Soaked rice, tea leaves, spices, specialty woods, dried fruit skins, peat moss, and plant leaves are examples of fuel alternatives used throughout the world.

Table 13–4   Various Fuels Used in Smoking

	TYPE OF FOOD	STRENGTH	SOAKING	FLAVOR
All fruitwoods (peach, apple, and cherry)	Poultry, game, pork, salmon	Mild	30 minutes	Sweet scented
Pecan wood	Turkey, goose, salmon	Dark color with delicate pecan flavor	1 hour	Delicate pecan flavor
Alder wood	Salmon, trout, swordfish, fish roe, chicken, pork	Medium to mild; good color	1 hour	Wood smoke flavor
Mesquite wood	Beef, duck, lamb, some strong fish	Medium; more delicate than hickory	1 to 2 hours	Sweet delicate flavor
Hickory wood	Hams, bacon, turkey, beef ribs, any pork cut	Strong, smoky	1 to 2 hours	Strong, pungent, and smoky
Grapevine	Fish, delicate poultry, game	Very delicate	30 minutes	Light and slightly fruity
Spices	All foods; should be added to wood as a flavor enhancer	Smolder only for a delicate taste	30 minutes	Light and slightly fragrant
Air-dried peat	Oily fish, especially herring, mackerel, and salmon	Smolder only for delicate flavor and color	No soaking	Rich peaty aroma with good coloring qualities

Understanding the elements of smoking and knowing what the chef requires a smoker to do, make choosing a smoker much easier.

COLD AND HOT SMOKING

There are two methods of smoking food: cold smoking and hot smoking. As their names suggest, cold smoking is done without any heat being present and hot smoking is with the addition of heat. Cold smoking merely imparts a smoke flavor and color without increasing the internal temperature of the food product, allowing it to remain raw if desired. Hot smoking raises the internal temperature of the product and helps render the food cooked.

- Cold smoking is done by controlling the flow of smoke from the smoke generator, preferably located on the outside of the chamber and blowing it over the food while having the heat source switched off. Or, the food can be trapped in between two layers of ice on pans, and then the smoke and heat source can be directly below the ice, thus preventing the heat from negatively affecting the food. The time and amount of smoke differs by recipe and needs to be carefully monitored.
- Hot smoking is accomplished by controlling the heat and the application of smoke over a long period of time, allowing the food to cook and smoke evenly together, rendering a fully smoked and fully cooked end product.

HOW TO SMOKE FOODS

In order to use the smoker to its best advantage, it is important to understand some basic rules that will render a superior end product:

- Use tested recipes.
- Follow the recipe's recommended times, temperature, and humidity readings accurately for the best results.
- Ensure that the products to be smoked are allowed sufficient time to air dry, usually 3 to 5 hours depending on their size, before adding them to the smoker. Spots of moisture on the outside of the sausage will result in a mottled surface on the finished product. It also does not allow the desired brown color to develop on the outside as it should.
- When air drying sausages, they should be hung on the smoke sticks on which they will eventually be smoked.
- Make sure that the sausage stays cool during the drying process or else bacteria will develop.
- When the product goes into the smoker, continue the drying process by initially opening the dampers fully to allow the excess moisture to escape and not settle on the product.
- Excess heat at the beginning of the smoking process will create sweating, and moisture will build up on the surface of the sausage. Opening the dampers and minimizing the heat initially can avoid this.
- Excess heat at this stage can cause the fat to run before the skin has dried and coagulated.
- Cook the sausage to an internal temperature of 152°F (66.8°C) to avoid spoilage.
- When the sausage has been cooked, shower it well with cold water to avoid shriveling and the development of wrinkles on the surface. Wrinkling happens quickly, so showering has to be done immediately.
- If the sausage does shrivel, lightly poach the sausage in hot water for a short time to remove the wrinkles somewhat, but the showering process has to be repeated.
- Once the sausage has been removed and showered, it should be allowed to hang at room temperature to develop even more color. The longer it hangs, the more brown it becomes; this technique is called "blooming."

> **TIP** Soak all wood chips for at least 30 minutes before use. Just before putting the food on the grill, add the chips directly to the charcoal. One or two handfuls is about the right amount; add more for a stronger flavor.

Making the Products

The following recipes represent the international diversity that exists in the world of sausage making (Figure 13–25A, B, and C). Also included are some recipes for other cured and smoked pieces of meat. The recipes are identified by their country of origin.

FIGURE 13–25A Cured meats and sausages in an Italian charcuterie

FIGURE 13–25B Meat market sausages

FIGURE 13–25C A selection of dried and cured sausages

RECIPE 13–9

ITALIAN DRY-CURED PEPPERONI

Recipe Yield: 10 pounds
(4.5 kg)

MEASUREMENTS		INGREDIENTS
U.S.	**METRIC**	
5 pounds	2.3 kg	Boneless pork butt (approximately 25 percent fat content), cubed
5 pounds	2.3 kg	Lean beef, cubed
9 tablespoons	133 mL	Salt
3 tablespoons	45 mL	Powdered dextrose
2 teaspoons	10 mL	Prague powder #2
2 ounces	57 g	Corn syrup solids
1 teaspoon	5 mL	Allspice
5 teaspoons	142 mL	Anise seed, ground
1 tablespoon	15 mL	Cayenne pepper

PREPARATION STEPS:

1. Grind the well-chilled meats through the medium plate.

2. Add all the remaining ingredients, and regrind through the small plate.

3. Stuff immediately into medium lamb casings (24 to 26 mm).

4. Hold the sausage at 70°F (21.1°C) at a humidity of 75 percent for 48 hours.

5. Chill and hold for 20 days in the cooler to allow hardening before use.

RECIPE 13–10

CAJUN ANDOUILLE SAUSAGE

Recipe Yield: 10 pounds
(4.5 kg)

MEASUREMENTS		INGREDIENTS
U.S.	**METRIC**	
10 pounds	4.5 kg	Boneless pork butts (approximately 25 percent fat content), cubed
8 tablespoons	118 mL	Salt
2 tablespoons	30 mL	Thyme, dried
1 cup	237 mL	Fresh garlic, finely chopped
1/2 cup	118 mL	Black peppercorns, cracked
2 tablespoons	59 mL	Cayenne pepper
2 cups	473 mL	Iced water
1 teaspoon	5 mL	Prague powder #1

PREPARATION STEPS:

1. Grind the well-chilled meat through the large die (through a die with 1/4-inch [0.6-cm] holes, if possible.)

2. Mix well with all the other ingredients and stuff into beef middles or beef rounds.

3. Air dry for 2 to 3 hours and smoke for 4 to 5 hours at 175° to 200°F (80° to 93.3°C) until an internal temperature of 152°F (21.1°C) is achieved.

4. Flush with cold water and bloom for 2 hours at room temperature.

5. Chill overnight before use.

GERMAN BRATWURST

Recipe Yield: 10 pounds
(4.5 kg)

MEASUREMENTS		INGREDIENTS
U.S.	**METRIC**	
5 pounds	2.3 kg	Boneless pork butt (approximately 25% fat content), cubed
5 pounds	2.3 kg	Veal trim, cut
2 cups	473 mL	Whole milk, ice cold
1 ounce	28 g	Salt
1 tablespoon	15 mL	Black pepper, ground
1 tablespoon	15 mL	Powdered dextrose
½ teaspoon	2 mL	Nutmeg, ground
1 teaspoon	5 mL	Coriander, ground
1 teaspoon	5 mL	Mace, ground

PREPARATION STEPS:

1. Grind the pork and veal through a medium die.

2. Add all the remaining ingredients to the milk and mix well.

3. Mix the ground pork and veal well with the milk and seasonings.

4. Stuff into 32- to 35-mm hog casings and chill well.

RECIPE 13–12

TRADITIONAL SCOTTISH HAGGIS

Recipe Yield: 10 pounds (4.5 kg)

> **TIP** The lungs are sometimes hard to find, so lamb stew meat can be substituted.

MEASUREMENTS		INGREDIENTS
U.S.	**METRIC**	
8 pounds	3.6 kg	Sheep pluck (including the heart, liver, and lights or lungs)
1 pound	0.45 kg	Toasted pinhead oatmeal
10 ounces	283 g	Lamb suet, finely chopped
3 cups	710 mL	Onion, finely diced
2 tablespoons	30 mL	Salt
1 tablespoon	15 mL	Allspice, ground
1 tablespoon	15 mL	White pepper, ground
½ teaspoon	2 mL	Nutmeg, ground
1 cup	237 mL	Lamb stock

PREPARATION STEPS:

1. Soak the pluck in cold salted water for 2 to 3 hours.

2. Gently simmer the pluck for 2 to 3 hours, making sure it is always completely covered with water.

3. Chill well and grind through the medium plate.

4. Combine well with all the other ingredients and stuff into beef bungs.

5. Close the bung by sewing the opening with butcher's needle and thread. Gently poach in a stockpot for 1 hour.

6. Chill and use for service.

FIGURE 13–26 Ode to a Haggis

> **TIP** This sausage is now completely cooked and ready to eat; it should be gently heated in the original skins or it can be molded into desired shapes and heated for service. Serve with mashed potatoes, mashed cooked rutabagas, and sometimes skirlie (traditional oatmeal and onion stuffing). Haggis is normally presented to the dining room on a silver platter. It is carried by the chef who is led by a bagpiper playing a traditional Scottish tune. The haggis is then addressed with a recital of a famous Robert Burns poem ("Ode to a Haggis," 1786), and served to the guests with a dram of Scottish whiskey (Figure 13–26).

Ode to a Haggis
by Robert Burns

Fair fa' your honest, sonsie face,
Great chieftain o' the puddin race!
Aboon them a' ye tak your place,
Painch, tripe, or thairm:
Weel are ye wordy of a grace worthy
As lang's my arm.

The groaning trencher there ye fill,
Your hurdies like a distant hill,
Your pin wad help to mend a mill
In time o' need,
While thro' your pores the dews distil
Like amber bead.

His knife see rustic Labour dight,
An' cut you up wi' ready sleight,
Trenching your gushing entrails bright
Like onie ditch;
And then, O what a glorious sight,
Warm-reekin, rich!

Then, horn for horn, they strech an' strive:
Deil tak the hindmost! on they drive,
Till a' their weel-swall'd kytes belyve,
Are bent like drums;
Then auld Guidman, maist like to rive,
'Bethanket!' hums.

Is there that owre his French ragout
Or olio that wad staw a sow,
Or fricassee wad mak her spew
Wi' perfect sconner,
Looks down wi' sneering, scornfu' view
On sic a dinner?

Poor devil! see him owre his trash,
As feckless as a wither'd rash,
His spindle shank, a guid whip-lash,
His nieve a nit;
Thro' bluidy flood or field to dash,
O how unfit!

But mark the Rustic, haggis-fed,
The trembling earth resounds his tread.
Clap in his walie nieve a blade,
He'll make it whissle;
An' legs, an' arms, an' heads will sned,
Like taps o' thrissle.

Ye Pow'rs wha mak mankind your care,
And dish them out their bill o'fare,
Auld Scotland wants nae skinking ware
That jaups in luggies;
But if ye wish her gratefu' prayer,
Gie her a Haggis!

RECIPE 13–13

FRESH POLISH KIELBASA

Recipe Yield: 10 pounds
(4.5 kg)

MEASUREMENTS		INGREDIENTS
U.S.	**METRIC**	
10 pounds	4.5 kg	Boneless pork butt (approximately 25 percent fat content), cubed
2 cups	473 mL	Ice water
5 tablespoons	74 mL	Salt
3 tablespoons	44 mL	Fresh garlic, finely diced
1 tablespoon	15 mL	Black pepper, ground
1 teaspoon	5 mL	Marjoram
1 tablespoon	15 mL	Sugar

PREPARATION STEPS:

1. Grind the well-chilled meat through the medium die, and place ground meat in a large bowl.

2. Add the seasonings to the water and combine with the meat.

3. Mix well until spices are evenly distributed.

4. Stuff into 32- to 35-mm hog casings and chill immediately.

RECIPE 13–14

SOUTH AFRICAN BOEREWORS

Recipe Yield: 10 pounds
(4.5 kg)

MEASUREMENTS		INGREDIENTS
U.S.	**METRIC**	
6 pounds	2.7 kg	Lean beef
4 pounds	1.8 kg	Boneless pork butt (approximately 30 percent fat content), cubed
1½ ounces	43 g	Coriander seeds, freshly roasted, coarsely ground
1 ounce	28 g	Salt
1 ounce	28 g	Black pepper, ground
1 teaspoon	5 mL	Nutmeg, ground
6 fluid ounces	177 mL	Red wine vinegar
1 cup	237 mL	Iced water

PREPARATION STEPS:

1. Grind the well-chilled meats through the large die.

2. Mix the seasonings with the water and the vinegar.

3. Add all the ingredients together and stuff into 32- to 35-mm hog casings.

4. Chill well before use.

RECIPE 13-15

SPANISH CHORIZO

Recipe Yield: 10 pounds (4.5 kg)

MEASUREMENTS		INGREDIENTS
U.S.	**METRIC**	
10 pounds	4.5 kg	Boneless pork butt (approximately 25 percent fat content), cubed
5 tablespoons	74 mL	Salt
5 tablespoons	74 mL	Spanish paprika
10 each	10 each	Fresh garlic cloves
1 tablespoon	15 mL	Oregano
2 teaspoons	10 mL	Black pepper
3 tablespoons	44 mL	Cayenne pepper
1 cup	237 mL	Iced water
1 cup	237 mL	Vinegar

PREPARATION STEPS:

1. Grind the well-chilled meat through the medium die, and place ground meat in a large bowl.

2. Add the water, vinegar, and all the seasonings and mix well with the meat.

3. Stuff into 38- to 42-mm hog casings and chill well for use.

4. Use the fresh sausage within a few days.

RECIPE 13–16

CANADIAN SMOKED COUNTRY VENISON SAUSAGE

Recipe Yield: 10 pounds
(4.5 kg)

| MEASUREMENTS | | INGREDIENTS |
U.S.	METRIC	
5 pounds	2.3 kg	Venison meat, cubed
5 pounds	2.3 kg	Boneless pork butt (approximately 30 percent fat content), cubed
2 cups	473 mL	Iced water
6 tablespoons	89 mL	Salt
2 teaspoons	10 mL	Prague powder #1
3 ounces	83 g	Corn syrup solids
2 cups	473 mL	Soy protein concentrate
1 teaspoon	5 mL	Nutmeg, grated
1 tablespoon	15 mL	Ginger, ground
1 teaspoon	5 mL	Black pepper, ground
2 teaspoons	10 mL	Garlic powder

PREPARATION STEPS:

1. Grind all the well-chilled meat through the small die.

2. Mix well with all the other ingredients.

3. Stuff into 32- to 35-mm hog casings and link into 8-inch (20-cm) links.

4. Dry at room temperature for 1 hour on smoke sticks.

5. Place into the smoker, with the dampers wide open, for 1 hour at 120°F (48.9°C).

6. Close the dampers to one-fourth, and apply heat—increasing it to 160°F (71°C) while applying a heavy smoke.

7. Continue to heat in increments of 10°F (−12°C) every 30 minutes until an internal temperature in the sausage has reached 152°F (21°C).

8. Shower with cold water until the internal temperature has reached 100°F (38°C).

9. Bloom for 1 hour in the kitchen and chill immediately overnight for service.

RECIPE 13–17

ENGLISH BREAKFAST SAUSAGE

Recipe Yield: 10 pounds
(4.5 kg)

MEASUREMENTS		INGREDIENTS
U.S.	**METRIC**	
10 pounds	4.5 kg	Boneless pork butts (approximately 30 percent fat content), cubed
2 teaspoons	10 mL	White pepper
1 teaspoon	5 mL	Ginger, ground
2 teaspoons	10 mL	Dried sage
1 teaspoon	5 mL	Mace, ground
8 ounces	227 g	White bread crumbs
3 ounces	85 g	Salt
1 cup	237 mL	Iced water

PREPARATION STEPS:

1. Grind the well-chilled meat through the small die.

2. In a mixing bowl, add the other ingredients, mixing well.

3. Stuff into 32- to 35-mm hog casings.

4. Link in 4-inch (10-cm) links and chill for service

5. Or, the sausage meat can be shaped into a log and sliced into patties.

RECIPE 13–18

PASTRAMI

Recipe Yield: 25 pounds
(11.3 kg)

MEASUREMENTS		INGREDIENTS
U.S.	**METRIC**	
25 pounds	11.3 kg	Beef briskets, whole
5 ounces	142 g	Prague powder #1
20 each	20 each	Garlic cloves, crushed
10 ounces	283 g	Salt
6 ounces	170 g	Powdered dextrose
5 quarts	4.7 L	Iced water
2 ounces	57 g	Black pepper, ground
2 ounces	57 g	Coriander seeds, ground

PREPARATION STEPS:

1. Prepare a brine by dissolving the salt, dextrose, and Prague powder in the iced water and adding the garlic.

2. Allow to sit for 2 hours and then pump the brisket to 15 percent of their weight.

3. Submerge the briskets in the brine for 4 to 5 days and holding in the cooler.

4. Dry the briskets well and rub all over with the spices.

5. Place in the smoker at 130°F (54.4°C) for 1 hour with the dampers wide open until the meat dries well.

6. Increase the temperature to 220°F (104.4°C) with the dampers one-fourth open and allow the brisket to smoke and cook to an internal temperature of 175°F (80°C).

7. Cool the meat at room temperature for 1 hour and chill well overnight before service.

DRY-CURED BACON

Recipe Yield: 25 pounds
(11.3 kg)

MEASUREMENTS		INGREDIENTS
U.S.	**METRIC**	
25 pounds	11.3 kg	Pork belly, skin removed and trimmed to rectangular shapes
12 ounces	340 g	Salt
16 ounces	454 g	Brown sugar
4 ounces	113 g	Powdered dextrose
2 ounces	57 g	Prague powder #1

PREPARATION STEPS:

1. Mix all the dry ingredients very well and rub liberally onto the pork bellies.

2. Pack them together, removing any air pockets; cover and chill for 2 days.

3. The bacon should be turned and repacked in the natural brine that has developed.

4. Brine for 7 days.

5. Wash well with cold water.

6. Dry the bacon thoroughly and hang in the smoker with the dampers open at 130°F (54.4°C) until the bacon is very dry.

7. Now cook at this temperature with the dampers one-fourth open, applying smoke until an internal temperature of 130°F (54.4°C) is achieved.

8. Chill well overnight before slicing.

Barbecue

Barbecue, or barbeque, could have originated when early man first pulled a charred piece of animal flesh from a fire and discovered the wonders of cooking over wood. It might have occurred in the Caribbean where native Indians used gratings of wood over a slow fire to cook pieces of meat and fish. The true origin may never be decided. Wherever it came from, barbecue simply means to slowly cook the seasoned food over burning coals or aromatic woods. This can be done by rotating the food on a spit, placing the food on a grill or down in a pit, or placing the food in any other suitable apparatus. The food is often cooked enclosed with the fuel source to impart the smoke flavor onto the food. It is generally sauced during or after it has been cooked.

In attempting to spell an Arawak Indian word representing a method of cooking, the Spanish came up with *barbacoa,* the Caribbean name for the sticks used to hold meats over the fire. The French have an old expression for *barbe a queue,* meaning "whiskers to tail," referring to the roasting of the whole animal on a spit over coals. When the word was first used in the English language during the seventeenth century, it meant a wooden framework that was used to sleep on, or on which to store things. It had nothing to do with food. However, with the European explorers entering the New World and with the later conquest of the American territories, the barbecue method of cooking is considered an American original. Over the years, methods have been further refined, and barbecue is now regionalized throughout the United States.

FUEL

There are many fuels from which a chef may choose to prepare barbecue; each has its own merits and characteristics.

- Charcoal briquettes burn easily and evenly, although most of them have some form of combustion-enhancing chemicals added that will affect the flavor of the food. It is therefore advisable to look for briquettes that are "all-natural."
- Lump hardwood charcoal, which is also known as chunk hardwood, is all-natural with no additives at all, adding a wonderful flavor and aroma to food. It burns easily but is much hotter than briquettes, so great care must be taken when using the direct grilling method.
- Hardwoods are the ideal fuel for giving the food its characteristic flavor and aroma; however, hardwoods are unpredictable, causing uneven burning while losing heat quickly. It takes a lot of practice to master the idiosyncrasies of hardwood cooking, but the results are worth the effort. Always use hardwoods, because the high levels of resin in softwoods will create soot and harsh smoke flavors that can be acrid. Cherry, hickory, maple, mesquite, and pecan are good hardwoods for cooking.

COOKING BARBECUE

As any experienced barbecue enthusiast will say, taming the fire is the biggest challenge to producing great barbecue. The often-quoted expression for cooking "Q" is "low and slow." Allowing the flavors of the smoke to slowly penetrate the meat, while natural enzymes and marinades gradually break down the resistant flesh—the control of time and temperature—is vital to successful barbecue. The following methods are used to cook barbecue:

- Grilling by direct heat is when the food is placed on the grill over direct heat. Generally, this is done with smaller pieces of tender meats, firm-fleshed fish, shellfish, poultry, game, and vegetables. The fuel used can be briquettes, lump hardwood charcoal, wood, or gas and the meat can be marinated or rubbed. The temperatures are normally between 350° and 550°F (177° and 288°C).
- Grilling by indirect heat is a technique used to grill large pieces of meat, racks of ribs, fish containing bone, poultry containing bone, and larger pieces of food that would burn before being fully cooked over direct heat. The food is placed on the cooking rack, sometimes over a

bath of water, with the coals spread out around the food or to one side (never directly underneath). Temperatures are normally at the 350°F (177°C) range, and fuel is added periodically to continue the cooking process. The food can be marinated, rubbed, or have sauce added during cooking.

■ Slow smoking is when the food is covered in a smoker or any enclosed container and smoke is applied, maintaining a temperature of between 200° and 220°F (93° and 104°C) for a long, slow cooking period. Slow smoking is used for tougher, larger pieces of meats such as briskets, racks of ribs, pork butts, whole pigs, shanks, and poultry with or without bone. Larger, firm-fleshed fish as well as shellfish can also be slow cooked this way.

■ Pit barbecuing is the use of a large structure, often for barbecuing larger pieces of meat or whole animals. The pit should always be able to be closed in order to create the smoke and the heat needed to cook the food correctly. The pit can be a hole dug in the ground, or it can be freestanding cement, or a brick or metal oven that in some cases has a special lining of clay to create more flavor (Figure 13–27). Another great favorite is the large metal drum that has been specially adapted for barbecuing by splitting it down the middle and hinging the top to the bottom, creating a base for the fire and a lid to create the heat and smoke. This slow method of smoking is applied to all tougher and larger pieces of meat with a constant temperature of no more than 200° F (93°C).

Barbecue Seasonings

In addition to the important smoke flavor that is imparted to the meat during the cooking process, the taste of barbecue varies greatly by the seasonings used to flavor it. Seasonings can be imparted dry or in a liquid state before, during, or after the cooking process. Seasonings can also help tenderize the final product. They provide the unique characteristic to each area's barbecue.

■ Cures and brines are used to flavor the foods that are to be smoked or grilled. These change the texture of the meats as well as render a moister finished product. The recipes listed near the beginning of this chapter, under Curing and Brining, would be appropriate for barbecuing.

■ Marinades are highly flavored liquids that are used to tenderize and flavor food. In the case of barbecue, they are sometimes also used to baste the meat during the cooking process. Marinades contain acids, oils, and essential flavoring agents that can include herbs, spices, vegetables, fruits, chiles, canned or jarred oriental sauces, and flavored vinegars.

FIGURE 13–27 A simple, but effective, barbecue pit using cinder blocks and a steel grate. Courtesy of Profimedia. CZ s.r.o/Alamy

RECIPE 13-20

LIME GINGER SWEET SOY MARINADE

Recipe Yield: 2 cups (473 mL) to marinate 4 pounds (1.8 kg) of fish or shellfish

MEASUREMENTS		INGREDIENTS
U.S.	**METRIC**	
¼ cup	59 mL	Lime juice, fresh
¼ cup	59 mL	Lemon juice, fresh
¼ cup	59 mL	Sweet soy sauce
⅓ cup	79 mL	Dry white wine
1 tablespoon	15 mL	Asian chile sauce
2 tablespoons	30 mL	Fresh ginger, grated and diced
½ cup	118 mL	Green onion, minced
2 tablespoons	30 mL	Garlic, minced
½ cup	118 mL	Olive oil
1 tablespoon	15 mL	Lemon grass, minced

PREPARATION STEPS:

1. Combine all ingredients and reserve in a nonreactive container (such as glass or stainless steel) for marinating.

2. Combine well with fish or shellfish of choice, and marinate for 3 hours before barbecuing.

RECIPE 13-21

AMERICAN SOUTHWESTERN MARINADE

Recipe Yield: 1 quart (946 mL) for 4 racks of ribs

MEASUREMENTS		INGREDIENTS
U.S.	**METRIC**	
½ cup	118 mL	Ancho chile paste
½ cup	118 mL	Olive oil
2 tablespoons	30 mL	Ground cumin
2 tablespoons	30 mL	Fresh coriander seeds, ground
¼ cup	59 mL	Apple cider vinegar
¼ cup	59 mL	Lemon juice
¼ cup	59 mL	Molasses
¼ cup	59 mL	Brown sugar
2 tablespoons	30 mL	Fresh garlic cloves, minced
2 tablespoons	30 mL	Soy sauce
1 cup	237 mL	Tomato ketchup
3 tablespoons	44 mL	Serrano chile, chopped

PREPARATION STEPS:

1. Combine all the ingredients and rub well over the meat, poultry, or ribs.

2. Allow to rest and chill for 24 hours.

3. Smoke slowly until very tender, basting with any extra marinade periodically during cooking.

4. Use for briskets, beef or pork ribs, and chicken.

■ Dry rubs are combinations of dry ingredients that are rubbed well into the meats before cooking. They are generally made from a combination of herbs, spices, sugars, salts, and dehydrated zests, barks, fruits, and vegetables. They are used for briskets, ribs, poultry, and any larger cuts of meat. Adding them to a small tender piece of meat and grilling it can be called *encrusting*. The addition of too much salt to these rubs will change them into dry cures depending on how long the rub is applied (Figure 13–28A and B).

FIGURE 13–28A Coating meat with a dry rub

FIGURE 13–28B Rubbing the seasoning into the meat

RECIPE 13–22

ALL-PURPOSE RUB

Recipe Yield: ½ cup (118 mL) for 5 pounds (2.2 kg) of meat

MEASUREMENTS		INGREDIENTS
U.S.	**METRIC**	
2 tablespoons	30 mL	Salt
2 tablespoons	30 mL	Paprika
2 tablespoons	30 mL	Black pepper, fresh ground
2 tablespoons	30 mL	Garlic powder

PREPARATION STEPS:

1. Mix well and store in an airtight container for use.

2. Rub generously over the meats and smoke slowly until tender.

3. Use for all meats.

RECIPE 13–23

FISH AND SHELLFISH RUB

Recipe Yield: 1 cup (237 mL) for 4 to 5 pounds (1.8 to 2.2 kg) of fish or shellfish

MEASUREMENTS		INGREDIENTS
U.S.	**METRIC**	
½ cup	118 mL	Brown sugar
2 tablespoons	30 mL	Salt
1 tablespoon	15 mL	Allspice, ground
1 tablespoon	15 mL	Coriander, ground
1 teaspoon	5 mL	Clove, ground
1 tablespoon	30 mL	White pepper, ground
¼ teaspoon	1 mL	Nutmeg, ground
1 tablespoon	30 mL	Dried lemon peel, ground

PREPARATION STEPS:

1. Combine all the ingredients in an airtight container for use.

2. Rub well into the fish or shellfish and allow to rest for 2 hours.

3. Grill over indirect heat with heavy smoke or directly over hardwood coals.

RECIPE 13–24

CAJUN RUB

Recipe Yield: ¾ cup (1.77 mL) for 5 pounds (2.2 kg) of meat

MEASUREMENTS		INGREDIENTS
U.S.	**METRIC**	
¼ cup	59 mL	Chile powder
2 tablespoons	30 mL	Oregano, dried
2 tablespoons	30 mL	Thyme, dried
2 tablespoons	30 mL	Garlic powder
2 tablespoons	30 mL	Black pepper, ground
2 tablespoons	30 mL	Brown sugar

PREPARATION STEPS:

1. Combine ingredients and rub well into chicken or pork.

2. Smoke or grill as desired.

- Pastes are similar to the dry rubs but they are combined with thick mustard, bottled oriental sauce, peanut butter, or ketchup. These pastes are rubbed into the meats before cooking, giving the exterior of the meat, when finished, its own sauce-like moisture that has a pleasant mouth feel and burst of flavor.

- Glazes, or "mops," are any of the sauces that are glazed or basted onto the foods while they are cooking. They give a rich shine and color to the finished dish, much like a lacquer or glaze, which gives food the distinct look of having been on a grill. They can be marinades, clarified butter, barbecue sauces, or a specially made sauce for glazing (Figures 13–29 and 13–30).

- Dipping sauces or sops can be any sauce or style of ketchup that is served by the plate of food to be used as a dipping sauce or for sopping up the meats. They are generally the barbecue sauces themselves, although some are specifically made as table sauces.

- Barbecue sauces are rich tasting, mouth-watering sweet, sour, and spicy sauces that are used to accompany the barbecued meats basted on them during cooking, or thinned and used as a flavorful marinade. There are many recipes for this sauce throughout the United States, and each region claims theirs to be the best. Most barbecue sauces are tomato based, but they can also be mustard based (northern Florida) or vinegar based (North Carolina). The barbecue sauce recipes included in this text represent a cross-section of America's best (see Chapter 5 for Recipes 5–33 to 5–37).

AMERICAN BARBECUE

Traveling in the United States in search of barbecue can be a lot of fun, because each region proudly claims and protects its own special version of "Q." All over the world, people have been marinating fire-roasted meats with sauces to tenderize and enhance the flavor since early records were kept. However, the primary difference between American and ancient European, Asian, and African marinades and sauces is the addition of the tomato from the New World.

In the state of Tennessee, barbecue pork is cooked in a pit, serving it pulled with spicy sauces. In the city of Chicago, the barbecue is much like the kind found in the southeast and, along with the blues, was brought by the migration of African Americans to the area. Sauces are

FIGURE 13–29 Applying a mop

FIGURE 13–30 Chinese barbecue

FIGURE 13–31 A Southern barbecue shack. Courtesy of Mark Sykes/Alamy

heavy and sweet, and are normally livened up by generous applications of pepper that become the focus of the dish.

Throughout the American southeast, pork is the favored meat (Figure 13–31). The reason may possibly date back to the Colonial period when the pigs in this area were allowed to roam loose and to grow fat on apples and nuts. They were then recaptured and eaten later, saving the farmer the effort and the expense of sheltering and feeding the livestock. In Virginia and North Carolina, the sauces are thin and vinegary, and provide a sharp contrast to the rich pork. The rest of this region tends to use thick, sweet, tomato-based sauces seasoned with lemon juice. The meat itself is sauced during cooking over a low flame and is served mixed with or topped by the sauce. In Florida, the sauce is made with mustard and spiced with datil peppers, in contrast to the rich tomato, brown sugar, and vinegar sauces of the mid-Atlantic states.

Texas is known for its beef barbecue, in particular beef ribs and the method of dry rubbing the meat before smoking and cooking. These are served with hot and sweet sauces that are used to dip or sop the meat in. Other parts of the southwest use the same technique, but do not serve sauces as an accompaniment. The larger pieces of meat are generally carved or pulled.

On the coast of California, barbecuing fish and vegetables has become commonplace (Pacific Northwest salmon has been cooked over open fire and coals for centuries). The Eastern Seaboard has also developed its own seafood grills, and some would argue that the shellfish cooked in open pits and clambakes are developed forms of barbecue.

The following recipes are categorized under the techniques involved in the production of barbecued foods. The words used can have multiple meanings and are common American terms throughout the barbecuing regions of the United States.

Serving American Barbecue

There are basically four ways of serving American barbecue (13–32):

- Whole, wherein the protein is small and tender enough to be served as an individual portion, including, for example, pork spare ribs, chicken legs or breasts, salmon steaks, or ribeye steaks.
- Chopped, wherein the meat comes from a larger piece, often a whole hog, and is simply chopped up with large cleavers. The customer can choose to add a table sauce for additional seasoning. This product could also be from a pork butt, shoulder, or turkey leg.
- Pulled, wherein the meat is shredded off the bone or pulled from larger pieces of meat that have been smoked and barbecued. This meat often comes from a whole hog, shoulder, butt, or shank.
- Carved, wherein the meat is carved into thin slices from larger pieces of meat and served with a sauce for sopping. This is often from a whole brisket or from a salmon side.

JERK (JAMAICAN BARBECUE)

Jerk is a unique Jamaican way of preserving meats and then cooking them over a fire pit or an open barbecue grill. The origins of the word are unknown, but this method of preparing food dates back to the native Arawak Indians who traditionally cooked wild pigs in this way. The technique was further developed and perfected by escaping slaves, called Maroons, who lived in the Blue Mountains while fighting the British troops.

FIGURE 13–32 The four types of American barbecue (clockwise from left): carved, chopped, whole, and pulled

Today, jerk seasoning is applied to pork, chicken, and fish while they are cooked in the colorful and rustic jerk huts that line the sides of the road. The flavor of the food, when cooked in this way, is spicy, sweet, and hot—which seems to be a perfect reflection of the lifestyle of the native people.

The jerk flavor is created by a special combination of scallions, onions, wild local thyme, cinnamon, nutmeg, and the careful use of the native fiery Scotch bonnet pepper. Additionally, this combination has one more key ingredient—the Jamaican pimento, known as allspice. The source for the allspice is very important to the success of the final flavor; it should be bought locally to guarantee the authentic, unique pungency of Jamaican jerk.

Another unique addition to the jerk flavoring is the use of the pimento wood for the fuel during the barbecuing process. Using both the fruit and the wood from the pimento tree, in the same recipe, adds its own unusual flavor and intensity to the final dish. There are many recipes for jerk available throughout the world, but the true native flavor is unique to Jamaica. The distinctive flavor, developed in its native environment, is very difficult to replicate outside of the region.

There are two methods of applying the jerk to meat: dry or wet, in the form of a marinade.

RECIPE 13–25

DRY JAMAICAN JERK RUB

Recipe Yield: 1 cup (237 mL)

MEASUREMENTS		INGREDIENTS
U.S.	**METRIC**	
2 tablespoons	30 mL	Onion flakes
2 tablespoons	30 mL	Onion powder
4 teaspoons	20 mL	Thyme, ground
4 teaspoons	20 mL	Salt
2 teaspoons	10 mL	Jamaican allspice, ground
1/2 teaspoon	2 mL	Nutmeg, ground
1/2 teaspoon	2 mL	Cinnamon, ground
2 teaspoons	10 mL	Sugar
2 teaspoons	10 mL	Black pepper, ground
2 teaspoons	10 mL	Cayenne pepper, ground

PREPARATION STEPS:

1. Mix the ingredients well and store in an airtight container until use. It will keep for up to 1 month.

2. Rub over the meats and chill for 24 hours before barbecuing.

RECIPE 13–26

JAMAICAN JERK MARINADE

Recipe Yield: 1½ cups (355 mL)

MEASUREMENTS		INGREDIENTS
U.S.	**METRIC**	
1 each	1 each	Onion, finely chopped
1/2 cup	118 mL	Green onion, minced
4 teaspoons	20 mL	Fresh thyme, minced
4 teaspoons	20 mL	Salt
2 teaspoons	10 mL	Jamaican allspice, ground
1/2 teaspoon	2 mL	Nutmeg, ground
1/2 teaspoon	2 mL	Cinnamon, ground
2 teaspoons	10 mL	Sugar
2 teaspoons	10 mL	Black pepper, ground
2 teaspoons	10 mL	Cayenne pepper, ground
3 tablespoons	44 mL	Soy sauce
1 tablespoon	5 mL	Vegetable oil
1 tablespoon	15 mL	Jamaican rum

PREPARATION STEPS:

1. Blend all ingredients well in a food processor.

2. Store in an airtight jar.

3. Marinate the meat for 24 hours before barbecuing.

CHURRASCO (BRAZILIAN BARBECUE)

Churrasco, or Brazilian barbecue, was introduced to Brazil when cattle were first brought to the country in the early sixteenth century. It quickly became the traditional staple food of the gauchos or cowboys of southern Brazil before becoming nationally recognized as a popular style of barbecue. It has now become very fashionable, and there are excellent churrascaria restaurants specializing in Brazilian barbecue all over Brazil, and around the world. In this style of restaurant, the waiters go from table to table with different types of meats on skewers from which they slice portions onto your side plate.

Originally, the metal-skewered meats were cooked in trenches over hot coals. The meat was seasoned with coarse salt, and each gaucho had his own churrasco knife that he used to cut pieces of meat from the spit. This old style of churrasco is still practiced by the people of southern Brazil, although the meat is now turned on motorized skewers over charcoal grills.

The array of meats used for churrasco can range from tenderloin to chicken hearts, and in most churrascarias, their service is preceded with a host of salads and appetizers. Brazilian sausages, beef steaks, pork tenderloins, and all other red meats are typically only seasoned with coarse salt, whereas all white meats are marinated in a mixture of garlic, salt, and lime juice. The salt can be applied to the meats by sprinkling or by brushing it on using a bunch of parsley dipped in a salt solution. The meats are carved tableside and then returned to the rotating grill for further cooking, with the exposed sides seasoned or brushed again.

RECIPE 13–27

BRAZILIAN WHITE MEAT MARINADE

Recipe Yield: $^1/_2$ cup (118 mL)

MEASUREMENTS		INGREDIENTS
U.S.	**METRIC**	
$^1/_4$ cup	59 mL	Lime juice
$^1/_4$ cup	59 mL	Garlic, minced
2 tablespoons	30 mL	Salt
1 tablespoon	15 mL	Vegetable oil

PREPARATIONS STEPS:

1. Mix the ingredients well together and use immediately.

2. Marinate the meats for 3 to 4 hours before barbecuing.

TANDOOR (INDIAN BARBECUE)

The **tandoor** oven originated in the northern regions of India, and has spread throughout the country and the world. It resembles a large clay vase or jug, out of which one might expect a snake charmer to musically produce a serpent. Heavily insulated by the clay from which it is made, the ovens were originally shaped like a vase to retain the precious heat and conserve fuel. In order to further insulate the ovens, they were surrounded by a thick layer of plaster, or otherwise were sunk neck-deep into the ground, where the cooks used them from a seated position. A modern version of the oven is set on wheels with a metal exterior and can be used alongside other standard kitchen equipment (Figure 13–33A and B).

The principle of the tandoor is to barbecue, bake, and roast simultaneously; this is done by impaling marinated food items on 3- to 4-foot (0.9- to 1.2-m) long metal skewers and placing the skewer tips into the hot coals with the other end resting on the neck of the clay tandoor

FIGURE 13–33A Using a modern tandoor oven

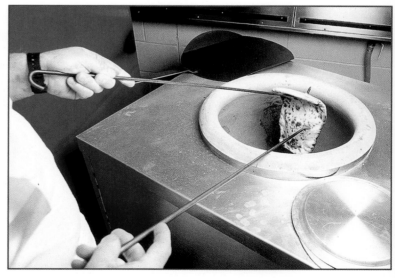

FIGURE 13–33B Closeup of a modern tandoor oven

oven. Basting with marinades and the Indian version of clarified butter, called *ghee,* the food items roast and barbecue in the well of the oven.

The wood or charcoal (or recent alternative of gas fuel) used to heat the tandoor is ignited on the very bottom of the vase, intensely heating the clay sides of the oven up to near the halfway point. A very hot glow is achieved toward the bottom, with the heat gradually dissipating past the narrowing neck of the oven. The oven itself is made with specially molded clay and needs to be seasoned before each use. The process of seasoning involves lighting a small quantity of coals (or just the pilot light, if it is gas) to distribute the heat evenly at first, gradually raising the heat as time goes on for 2 to 3 hours. This warming process is sufficient when the tandoor oven is used daily, but sporadic usage requires 4 to 6 hours of seasoning to avoid cracking the clay while ensuring uniformity of heat distribution.

Traditionally in India, marinating was used to preserve foods for a longer time (necessary in such a hot climate) as well as imparting more flavor to the food. Another important reason for marinating was to tenderize the normally tough meat. Sometimes the rind of papaya was added to the yogurt and spice marinade to assist in the tenderizing process. Fresh garlic, ginger, saffron, and different spice mixes were the main flavorings, giving the final product an interesting fragrance and an eclectic color. Some chefs enhance the color by adding orange food coloring to the marinade recipes.

Additional flavors are achieved from the tandoor oven itself. A pleasant odor is emitted from the clay as it heats the food while the classic tandoor fragrance, created from the marinade dripping onto the coals, provides a sensational flavor. Reasonable results can be achieved without a tandoor oven by barbecuing the seasoned meats over a charcoal fire; however, the intense encompassing heat of the tandoor creates unique results when used.

RECIPE 13–28

TANDOOR MARINADE

Recipe Yield: 2 cups (473 mL)

| MEASUREMENTS | | INGREDIENTS |
U.S.	METRIC	
1 each	1 each	Onion, finely diced
8 each	8 each	Garlic cloves, finely diced
1 ounce	28 g	Ginger, fresh grated
1½ cups	355 mL	Plain yogurt
3 tablespoons	45 mL	Lemon juice
1 tablespoon	15 mL	Cumin, ground
1 tablespoon	15 mL	Coriander, ground
1 tablespoon	15 mL	Garam marsala
½ tablespoon	7 mL	Black pepper, ground
½ tablespoon	7 mL	Cayenne pepper, ground
1 teaspoon	5 mL	Saffron

PREPARATION STEPS:

1. Combine all the ingredients well and use immediately.

BRAAI (SOUTH AFRICAN BARBECUE)

Braai, or South African barbecue, comes from the Afrikaans word *braaivleis*. The term means "meat grill," and it is a form of barbecue that has become a social tradition among the people of South Africa. **Braai** is commonly cooked outside on an open barbecue, usually with family and friends. Recently, specialized indoor braai equipment has become popular additions in home construction.

This rich cultural event is completely interracial and has become part of the South African tradition of sharing food and beverages around an open fire with slowly sizzling food on the grill. Depending on the part of the country, braai can be anything from meats to seafood. However, a traditional barbecue meal consists of large marinated steak, lamb cutlets, and the beloved boerewors (see Recipe 13–14), a spicy farmer's sausage much prized on the barbecue. A spicy marinade called piri-piri (Peri-Peri, Recipe 13–29) can be used for the meats and poultry for the braai; however, its use results in very spicy meat. The recipe should be adjusted to personal taste.

RECIPE 13–29

Peri-Peri Marinade

Recipe Yield: ¾ cup (177 mL)

MEASUREMENTS		INGREDIENTS
U.S.	**METRIC**	
3 each	3 each	Red chile, very finely diced
¼ cup	59 mL	Vegetable oil
¼ cup	59 mL	Lemon juice, fresh
2 tablespoons	30 mL	Lime juice, fresh
2 tablespoons	30 mL	Garlic, fresh minced
1 tablespoon	15 mL	Paprika
1 tablespoon	15 mL	Cayenne pepper
1 tablespoon	15 mL	Salt

> **TIP** When barbecuing, the meats should be basted with a mixture of equal quantities of butter and red wine that are thoroughly mixed together.

PREPARATION STEPS:

1. Combine all ingredients in a large mortar.

2. Grind into a paste with a mortar and pestle and reserve for service.

3. Adjust the cayenne pepper for heat control.

4. Marinate the meats for 30 minutes before grilling.

RECIPE 13–30

Manyeleti Red Meat Marinade

Recipe Yield: 2 cups (473 mL)

MEASUREMENTS		INGREDIENTS
U.S.	**METRIC**	
½ cup	118 mL	Red wine
½ cup	118 mL	Vegetable oil
½ cup	118 mL	Soy sauce
½ cup	118 mL	Tarragon vinegar
4 each	4 each	Garlic cloves, minced
¼ teaspoon	1 mL	Black pepper, freshly ground

PREPARATION STEPS:

1. Heat the red wine to burn off the alcohol. Cool.

2. Combine with other ingredients and blend well.

3. Marinate meats at room temperature for 3 hours and then grill.

4. Alternatively, meats may be marinated overnight under refrigeration.

Professional Profile

BIOGRAPHICAL INFORMATION

Name: William Christian Franklin, CMC AAC

Place of Birth: Sevierville, Tennessee

Recipe Provided: Smoky Mountain Pork Sausage Confit

CULINARY EDUCATION AND TRAINING HIGHLIGHTS

Chef Franklin began his food service career in the Bavarian-style village of Leavenworth, Washington, in 1965. His first position, in the afternoons and evenings after school, was as a butcher's assistant and all around "gopher" in a small grocery market. In 1970, he enlisted in the U.S. Navy with the intention of learning a trade such as electrician or mechanic. However, the galley found him first. After 4 years as a Navy cook, he was discharged in Hawaii and worked as sous chef at the Sheraton Waikiki under the guidance of German Master Chef Gordy Dambauch. Upon finishing his degree work with the University of Hawaii, he attended The Culinary Institute of America where he graduated in November 1977. From there, chef Franklin's career has included serving as the executive chef for the U.S. Navy club systems, Director of Culinary Training for U.S. Air Force clubs, Dean of Faculty for the National Cooking Institute, General Manager of Wright Patterson Air Force Base Officer's Club, Executive Chef of the Denver Athletic Club, and Corporate Executive R & D Chef for Nestle Foodservice. Additionally, he and his wife were owner-operators of their own restaurant. Chef Franklin returned to the CIA in January 1988 to become the 37th American Culinary Federation Certified Master Chef in the United States.

ADVICE TO A JOURNEYMAN CHEF

Surround yourself with friends and associates that encourage you to stretch yourself and lift you up. Seek to develop skills that scare you, such as public speaking. Learn and practice attributes that have stood the test of time in all of humanity such as honesty and integrity. Seek ways to emulate these throughout your life regardless of current trends or popularity. Be a friend to all. Work is noble; learn to embrace and enjoy it. Learn from seniors by listening to them and not by making their same mistakes. Most of all, don't take yourself too seriously; life is just too short.

William Christian Franklin

RECIPE 13–31

CHEF WILLIAM CHRISTIAN FRANKLIN'S SMOKY MOUNTAIN PORK SAUSAGE CONFIT

Recipe Yield: 16 balls or 8 portions

MEASUREMENTS		INGREDIENTS
U.S.	**METRIC**	
1 pound	455 g	Fresh, medium ground pork (approximately 30 percent fat content)
1 teaspoon	5 mL	Salt
$\frac{1}{2}$ teaspoon	2.5 mL	Black pepper, ground
$\frac{1}{4}$ teaspoon	1.2 mL	Dried red chile flakes
1 teaspoon	5 mL	Ground sage, dried
$\frac{1}{2}$ teaspoon	2.5 mL	Ground thyme
1 pound	455 g	Fresh fatback or belly, minced

PREPARATION STEPS:

1. In a chilled bowl over ice, combine all ingredients and blend evenly.

2. Using a #40 scoop, form balls using the entire product. Store chilled.

3. In a heavy skillet over medium-low heat, render the fat back and remove the solids, if any.

4. Reduce heat to approximately 300°F (148.8°C).

5. Fry the balls in shallow oil until evenly browned or until they reach 150°F (65.5°C) internal temperature.

6. Immediately place balls snugly into a sterile canning jar and finish by pouring hot lard to completely cover the meat.

7. Cover the jar with a sterile lid and ring cap. Close tightly and set aside to chill, forming a vacuum seal.

PRESENTATION SUGGESTIONS:

Traditionally, for the sausage, once cooked and covered with the very hot fat, the jars would seal themselves. They were then stored in the cool smokehouse and used throughout the winter in various dishes. The most common use was warmed sausage in unpasteurized milk gravy and hot buttermilk biscuits. Often the jar was set near a warm part of the wood-fired range so the sausage balls could be more easily removed from the fat. The fat would then be used to heat the sausage. This same fat would be used to make the roux for the milk gravy.

CULINARY HISTORICAL NOTE:

This process was a late Fall ritual of the western foot hills of the Smoky Mountains because as the fatted family hog was butchered in late November or early December. It was a family event and everyone had responsibilities.

The process (maybe grim for some) is worth mentioning.

On a farm, you learn early that in order to sustain life, life is taken. So all food is sacred. At our farm, the animals to be slaughtered were given extra portions of their favorite foods the evening before—a small measure of gratitude.

The hog was, more often than not, shot with a single .22 caliber shot. Because pigs are not skinned, we had to get the outside as clean as possible. This was achieved by pouring boiling water from a large cauldron, a little at a time, over sections of the carcass. This process of scalding made it easier to scrape the hide clean of hair and dirt with our sharp knives.

To hang this carcass, we used a double-pulley system with ropes mounted at the top on a recently cut 15-foot timber tripod. Once up, the hog was rinsed again with boiling water to minimize contaminants. The throat was then cut and the blood collected in a steel tub to be set in the creek and cooled. Starting from the elevated rear end, the hog was eviscerated, with the entrails falling into another waiting tub. With the coveted offal gathered and the still-warm carcass steaming in the cool air, it would be quartered from the bottom up. The hams and hocks would be the last items hanging. The hams, loins, and bellies were off to the smokehouse for a layer of salt. The balance was not wasted. The hide was for crackling, the head (less the brains and ears) was rendered for a type of potted meat called souse meat. Excess fat was used for soap and the lean trimming was ground for the sausage or confit. I remember well that as a young boy, my treat was the pig's bladder tied to a reed so I could inflate it as if I had a prize balloon!

William Christian Franklin

FIGURE 13–34 Confit of sausage

1. Explain why we cure food.

2. List five ingredients that are used in the curing and sausage-making process.

3. Briefly describe the production of lox.

4. Describe the uses of Prague powder #1.

5. Describe the uses of Prague powder #2.

6. What are collagen casings and how are they made?

7. Why is it necessary to use only ice-cold water in the sausage making process?

8. Explain the differences between hot and cold smoking.

9. Discuss the reasons for using injection brining.

10. Define barbecuing.

A. **Group Discussion**

Discuss the historical development of sausages and cured meats, reporting on how important they were as food groups for a people's development and survival.

B. **Research Project**

As a group, research a country's regional sausage and cured meat specialties; report on your more unusual findings.

C. **Group Activity**

Analyze a dry-cured sausage recipe, explaining the use and function of the individual ingredients.

D. **Individual Activity**

Create a sausage recipe of your own using proteins other than pork and beef.

CHAPTER 14

Pâtés, Terrines, and Mousselines

After reading this chapter, you should be able to

- explain the historical nature of charcuterie, from its early development to modern trends.

- identify equipment used to make charcuterie products.

- list the basic elements of charcuterie production.

- explain differences in how charcuterie products are cooked.

- recite techniques used to make various pâtés, terrines, and mousseline recipes.

KEY TERMS

bain marie	forcemeat	pâtés
binding agents	galantines	pâté en croûte
charcuterie	glaze	quenelles
extenders	mousselines	terrines
farce	mousse	timbales
food for carriage	panadas	wrappings

Whenever you eliminate the inedible, whatever remains, however unpalatable, must be food.

UNKNOWN

A Brief History of Charcuterie

As a derivative from the French words *chair* and *cuit* **charcuterie** translates to mean "cooked flesh." The original products of the charcutier were, in fact, cooked pork. In Chapter 13, the teaching of sausage making is separated from the other items commonly considered as charcuterie. This chapter encompasses the remainder of the topic of charcuterie, from its humble beginnings to its present-day form. "No matter how unpretentious and simple, nor how fancifully adorned, terrines, pâtés, and galantines are basically composed of a **forcemeat** cooked in some sort of wrapping," writes Victoria Wise (1986).

Around 400 BC, Sparta's King Agesilus received a gift of fatted geese from Egypt for making, what would more than 1,300 years later be called, *pâté de foie gras.* Known to the ancient Egyptians and popular among wealthy and noble Romans, this pâté is still made from the enlarged livers of force-fed geese.

Pâtés, terrines, and timbales have existed since the early history of the Francs and Gaules and of the Roman Empire. It was in the thirteenth century that the wealth of buffets and *tables of entremets* began in earnest. In these particular times, not only were chefs involved in the creations, but sculptors, painters, and architects were also part of the formation of these extravagant banquets. The buffets were generally considered entertainment, as well as art and food, until the sixteenth century. As times changed, so did the abundance of artful creations and sumptuous banquets.

Charcuterie has existed in small kitchens and in large kitchens, and the traditions that started more than 3,000 years ago continue today.

From Necessity to Haute Cuisine

Whether as hunters and gatherers or as an agrarian society, mankind has had to find sustenance in the world since the dawn of civilization. The limited means of preservation gave rise to experimentation, as man sought ways of using an entire animal for nourishment. The worn expression "necessity is the mother of invention" appropriately explains man's quest for survival.

Every portion of an animal's flesh and blood was considered part of the human diet. All meat and seafood was precious and could not be left to spoil, so the means of preserving and preparing food became of vital importance. Large joints of the animals were roasted over wood fires, while trimmings were often mixed with roots, stale bread, and herbs, and then boiled as a crude form of soup. This style of food preparation satisfied the majority of townspeople and farmers, but did not adequately meet the needs of traveling armies and wandering souls.

From those early times, the human race has learned that specialization within a craft yields superior results. Throughout time, and even in our modern world, there have been tragedies that have occurred due to the improper handling of food products. According to T. G. Mueller (1987), "In the Middle Ages, strict separation between fisheries, slaughterhouses, butchers, and prepared-meat shops was enforced in order to control disease. Not until the sixteenth century did charcutiers, as these cooks were called, legally obtain the right to butcher their own pigs and sell both raw and cooked products."

Over time, as the need of **food for carriage** became essential, portable food items such as pork pies, blood puddings, savory pasties, country style pâtés, and terrines were developed, although their use was not strictly limited to carriage. Along with cheese, bread, and fruit, these foodstuffs, which we now refer to as the "fast food" of its era, became the provisions for military campaigns and religious pilgrimages, and for those toiling far from home.

These forcemeat (called **farce** by the French) products were made of finely ground or coarsely chopped meat, poultry, or fish. They were seasoned and then bound with a combination of egg, cream, and bread stuffing. As the skills of chefs advanced and consumers had more money, forcemeats were layered with strips of meat, sweetbreads, liver, truffles, and mushrooms to vary their tastes and textures. The edible wrappings of pastry, caul fat, salt pork, grape leaves, and poultry skins encased the meats. The bundles were then placed on hot stones near the fires to slowly bake, or they were roasted on spits above the open flames. Eventually, earthenware crocks were also used as vessels for baking.

The elegance of these charcuterie products evolved so that they soon became mainstays of both the privileged and impoverished classes. Even though a series of crop failures in the early 1300s caused widespread famine and the bubonic plague of 1348 spread across Europe with catastrophic devastation, the balance of resources had completely tipped. According to Albala (2003), "It appears that because grains were so cheap, the average peasant diet included a much higher proportion of meat" (p. 3). Several centuries later, these products endure and have gained in global acceptance. They have become entrenched in world cuisine, and embraced by members of all socioeconomic and ethnic groups. What was once the food of the masses during medieval times has again become a food of choice for many.

The internationally celebrated French chef Paul Bocuse earned Michelin's coveted three-star rating in 1965 at his nine-table bistro north of Lyons, with menu items such as Stuffed Bass in Pastry, trout mousse, mallard gallantine, and other regional specialties. In 1971, the "dame of American regional cooking" Alice Waters opened Chez Panisse in Berkely, California, offering a three-course menu of pâté maison, duck with olives, and almond tart. Today, trendy magazines like *Art Culinaire* feature items such as terrine of salmon, red mullet, and foie gras; calve's head terrine; pigeon with duck confit; and marinated foie gras terrine with pig trotters. Charcuterie is enjoying continued international interest everywhere, from take-away stands on Edinbugh's Royal Mile to the bistros and haute cuisine establishments on the Champs-Elysées in Paris (Figure 14–1).

THE NEW AGE REVOLUTION OF PÂTÉS AND TERRINES

Many innovations have contributed to the revolution that is occurring with charcuterie products. Once considered a risk to human health by both physicians and nutritionists, the consumption of pâtés, mousselines, timbales, and terrines faltered some during the late 1970s and 1980s. Health-conscious consumers have now embraced a lighter eating style that emphasizes using fat-free products, light creams, and foods that are lower in animal fats. As a result, the use of fish, shellfish, and vegetable pâtés and terrines has also gained in popularity. New innovations in the types and sizes of molds being used make the smaller portions more acceptable to the consumer. Still, the consump-

FIGURE 14–1 Menu containing charcuterie style foods

Charcuterie Menu

Chicken liver parfait with candied walnuts and eau de vie prunes — $9

Morcon, San Danielle ham, lomo, chorizo & salchichon — $8

Chilled scallop galantine with cantaloupe and honey dew coulis — $11

Charcuterie plate with rosette de Lyon salami, duck rillette, pork pepper pate, garlic salami — $13

Foie gras terrine with cherry tomato puree — $18

tion of charcuterie products has never been higher in modern times, as garde manger chefs develop new styles and flavor profiles that are blended with the traditional methods.

Traditional garde manger styles have always been based on Western European food styles, typically German (well known for charcuterie), French (farces), Scandinavian (cured and marinated fish), and Swiss (the best of German and French styles). In melting-pot North America, contemporary garde manger styles are influenced by a variety of Asian cultures—Chinese, Japanese, Korean, Thai, and Vietnamese—as well as those "south of the border," principally Mexican. A familiarity with all of these influences is essential to understanding the creative possibilities.

David Larousse (1969)

Equipment Used in the Making of Pâtés and Terrines

As with all things modern, both the past and the present are represented in the charcutier's equipment. The basic equipment involved was for two purposes, either to grind the ingredients into some edible paste suitable for its purpose or as a container in which to bake or mold the final product.

GRINDING, BLENDING, AND WHIPPING

The early fabrication of pâtés, terrines, timbales, and mousselines required that the meat, the seafood, and the occasional vegetable be coarsely ground. The use of mortar and pestles, knives, grinding stones, and clubs were all employed to shred and pummel the food so that it could be seasoned and bound together in some vessel or wrapping. The pot was then boiled, baked, or roasted to cook the charcuterie product.

FIGURE 14–2 Clockwise from top left: A blixer, a food processor, a paco jet, and an immersion blender

As centuries passed, chefs became more concerned about the texture of the product and its overall appearance to the guest. They devised ways of adding moisture, binding agents, and contrasting colors and textures, while seeking the means to aerate the mixture. The result has provided a range of textures and styles for the consumer. Today, modern garde manger chefs use meat grinders, food processors, sieves, paco jets, and blixers among other tools, to grind, blend, smooth, and whip air into their meat, seafood, and vegetable pastes (Figure 14–2).

MOLDING

The means of shaping the final charcuterie product has evolved over the centuries from the very free-formed to the ornate. The human hand provided the first mold, not unlike the humble meatball. Knives and straight-edged tools were later used to rub off portions of paste, similar to how spaetzle noodles are scraped into vats of boiling water or broth, while the use of spoons provided the means of forming quenelles of forcemeat. The Japanese used seaweed, Eastern cultures used banana leaves, the French used caul fat, the Britons used fish heads, and Latin Americans used gourds to hold their fish and meat pastes for cooking.

These customs still widely exist, although chefs have been using metal collapsible-sided pâté en croûte molds, ceramic and clay terrines, copper timbales, and other manmade vessels for more than 100 years (Figure 14–3).

New age pâté and terrine molds feature various shapes and materials. Everything from half-moon to triangular-shaped molds are used and are made from PVC, terra cotta, enamel, glass, and ovenable and flexible rubber and plastics (Figure 14–4).

FIGURE 14–3 Traditional molds

FIGURE 14–4 Modern molds

Elements of Production

The finished quality of any prepared foodstuff is directly affected by the condition of its ingredients when first received by the chef and by the competent manner in which it is handled afterward for production. The wholesome condition of all commodities, including fish and shellfish, meats, poultry, and game, as well as the seasonings, binding agents, and **wrappings**, has much to do with the end results of the prepared charcuterie product.

The chef should only use products that are fresh and nourishing to ensure a quality food item is produced. Ingredients that are not indigenous or locally available, possibly such as truffles, morels, or other specialty products, must be evaluated for their value-added benefit. Nature's flavors need only be enhanced by the skillful work of the chef. However, that being said, the garde manger must use proper culinary techniques so that the final product is not compromised. An incompetent cook can destroy even the best foodstuffs.

Virtually all charcuterie goods contain elements that fall under the following categories: seasonings; pastry dough; panadas; stocks, glazes, and gelatins; and forcemeats. Beyond the ingredients themselves, the methods and international styles used by the chef create the primary differences in products.

SEASONINGS

The use of seasonings and flavoring agents can make all the difference between two or more forcemeats. Combinations of herbs, spices, and seasonings develop the distinctive flavor characteristics of the farce, whether it is mild and delicate or robust and bold. Some combinations are simple and forthright, such as chopped garlic blended with Kosher salt to yield garlic salt. Other formulas are in ratios, such as four spices (quatre epices) that is made from eight parts white pepper to one part each of nutmeg, cloves, and cinnamon or ginger. Others use extensive blends of spices and herbs to create spicy and aromatic flavor profiles. (See Table 14–1 for a list of the principle herbs and spices used in charcuterie.)

Salt is a foundation spice that plays a significant role in producing quality farces. In addition to its ability to act as a natural curing agent and as a flavor enhancer, salt affects the albumen and protein contained in meats, aiding in their ability to bind. Too much salt can toughen the final product and render it bitter and inedible; too little salt can lessen the flavor potential of the ingredients.

The garde manger chef must be accurate in measurements and ratios. It is always important to follow the formulas precisely and to conduct quality control tastings after each batch is prepared. The time to solve an ingredient problem is during the preparation phase, before the item has been cooked. The spice mix needs to be blended with the farce before taste testing occurs, because each mixture will taste quite different separately.

Ready-made seasoning mixes may be purchased from both retailers and commercial distributors. It is important to buy them in smaller quantities so they may be used up quickly and replaced frequently. Spices tend to last longer than herb blends, in that the flavoring oils in the herbs dramatically weaken after a few months. Mixes should be stored in cool and dark areas, because sunlight and extended exposure to heat also destroys the flavors of the seasonings.

Many chefs like to mix their own blends, allowing for unique combinations and fresher ingredients (Figure 14–5).

Table 14–1   Principle Herbs and Spices in Charcuterie

BEEF AND PORK	POULTRY	FISH AND SHELLFISH	GAME	OFFAL
White pepper	Cumin	Dill	Juniper	Clove
Black pepper	Coriander	Tarragon	Star anise	Allspice
Mace	Cardamom	Basil	Allspice	Coriander
Nutmeg	Paprika	Fennel	Black pepper	Ginger
Coriander	White pepper	White pepper	Marjoram	Cayenne
Marjoram	Garlic	Lemon grass	Black pepper	Garlic
Rosemary	Sage	Mint	Allspice	Mace
Sage	Basil	Oregano	Nutmeg	Paprika
Ginger	Tarragon	Coriander	Caraway	Rosemary
Allspice	Clove	Parsley	Chili	Thyme
Garlic	Savory	Cilantro		White pepper
Bay leaf		Italian parsley		Cinnamon
		Lemon balm		
		Aniseed		

FIGURE 14–5 Electric spice/coffee grinder with assorted spices

All spice mixes listed in this chapter are used by the following ratio:

■ 2 ounces (57 g) spice mix to 18 ounces (500 g) kosher salt

■ ⅓ ounce (9 g) of the above mix for approximately 1 pound (453 g) of forcemeat

In addition to herbs and spices, flavoring agents such as wines and other alcoholic liquids, vinegars, nuts, onions, citrus fruits, mushrooms, seaweed, and aromatic vegetables can be added to the farce for further flavor. Sauce reductions, extracts, and flavor concentrates can also be added. In all cases, items should be thoroughly chilled before being added to raw meat.

RECIPE 14–1

AMERICAN SPICE MIX

Recipe Yield: 8 ounces
(227 g)

MEASUREMENTS		INGREDIENTS
U.S.	METRIC	
1 ounce	28 g	Ground mace
1 ounce	28 g	Ground white pepper
1 ounce	28 g	Ground ginger
1 ounce	28 g	Ground nutmeg
½ ounce	14 g	Cayenne pepper
½ ounce	14 g	Ground fennel
½ ounce	14 g	Cinnamon
1 ounce	28 g	Ground dried morel
½ ounce	14 g	Ground thyme
¼ ounce	7 g	Bay leaf

PREPARATION STEPS:

1. Combine all the ingredients.

RECIPE 14–2

BRAZILIAN SPICE MIX

Recipe Yield: 6 ounces
(170 g)

MEASUREMENTS		INGREDIENTS
U.S.	METRIC	
1 ounce	28 g	Ground allspice
1 ounce	28 g	Ground black pepper
1 ounce	28 g	Ground cumin
1 ounce	28 g	Dried malagueta pepper, ground
½ ounce	14 g	Ground clove
½ ounce	14 g	Ground ginger
½ ounce	14 g	Ground coriander

PREPARATION STEPS:

1. Combine all the ingredients.

RECIPE 14–3

BRITISH SPICE MIX

Recipe Yield: 6 ounces (170 g)

MEASUREMENTS		INGREDIENTS
U.S.	METRIC	
1 ounce	28 g	Ground white pepper
1 ounce	28 g	Ground mace
1 ounce	28 g	Ground nutmeg
1/2 ounce	14 g	Ground ginger
1/2 ounce	14 g	Ground coriander
1/2 ounce	14 g	Ground cayenne
1/2 ounce	14 g	Ground sage

PREPARATION STEPS:

1. Combine all the ingredients.

RECIPE 14–4

CARIBBEAN SPICE MIX

Recipe Yield: 8 ounces (227 g)

MEASUREMENTS		INGREDIENTS
U.S.	METRIC	
1 ounce	28 g	Ground white pepper
1 ounce	28 g	Ground black pepper
1 ounce	28 g	Ground mace
1 ounce	28 g	Ground ginger
1 ounce	28 g	Ground nutmeg
1 ounce	28 g	Garlic powder
1 ounce	28 g	Ground allspice

PREPARATION STEPS:

1. Combine all the ingredients.

RECIPE 14–5

EAST ASIAN SPICE MIX

Recipe Yield: 8 ounces (227 g)

MEASUREMENTS		INGREDIENTS
U.S.	METRIC	
2 ounces	57 g	Ground cinnamon
2 ounces	57 g	Ground fennel
1 ounce	28 g	Ground white pepper
1 ounce	28 g	Ground clove
2 ounces	57 g	Sesame seeds

PREPARATION STEPS:

1. Combine all the ingredients.

RECIPE 14–6

FRENCH SPICE MIX

Recipe Yield: 8 ounces
(227 g)

MEASUREMENTS		INGREDIENTS
U.S.	**METRIC**	
½ ounce	14 g	Paprika
½ ounce	14 g	Ground nutmeg
½ ounce	14 g	Ground mace
1 ounce	28 g	Ground clove
1 ounce	28 g	Ground ginger
¼ ounce	7 g	Ground cayenne pepper
½ ounce	14 g	Marjoram
½ ounce	14 g	Thyme
½ ounce	14 g	Basil
½ ounce	14 g	Rosemary

PREPARATION STEPS:

1. **Combine all the ingredients.**

RECIPE 14–7

ITALIAN SPICE MIX

Recipe Yield: 6½ ounces
(182 g)

MEASUREMENTS		INGREDIENTS
U.S.	**METRIC**	
1 ounce	28 g	Ground white pepper
1 ounce	28 g	Ground mace
1 ounce	28 g	Ground nutmeg
1 ounce	28 g	Ground clove
1 ounce	28 g	Ground garlic
½ ounce	14 g	Cayenne pepper
½ ounce	14 g	Ground marjoram
½ ounce	14 g	Ground basil

PREPARATION STEPS:

1. **Combine all the ingredients.**

RECIPE 14–8

SOUTH AFRICAN SPICE MIX

Recipe Yield: 8 ounces
(227 g)

MEASUREMENTS		INGREDIENTS
U.S.	**METRIC**	
1 ounce	28 g	Ground mace
1 ounce	28 g	Ground allspice
1 ounce	28 g	Ground nutmeg
1½ ounces	43 g	Ground coriander
½ ounce	14 g	Ground cumin
1 ounce	28 g	Thyme
1 ounce	28 g	Fresh rosemary
1 ounce	28 g	Ground white pepper
½ ounce	14 g	Fresh marjarom

PREPARATION STEPS:

1. Combine all the ingredients.

RECIPE 14–9

SOUTH ASIAN SPICE MIX

Recipe Yield: 10 ounces
(283 g)

MEASUREMENTS		INGREDIENTS
U.S.	**METRIC**	
1 ounce	28 g	Anise seed
1 ounce	28 g	Ground black pepper
2 ounces	57 g	Ground cardamom
1 ounce	28 g	Ground cumin
1 ounce	28 g	Saffron
2 ounces	57 g	Ground turmeric
2 ounces	57 g	Garam masala

PREPARATION STEPS:

1. Combine all the ingredients.

PASTRY AND DOUGH

Pastry and dough crusts have been used for several hundred years as a means of protecting and containing the forcemeat filling. They are generally of the same simple ingredients and are made sufficiently tough to be handled without breaking. The raw dough works best when kept chilled and allowed to relax. Most dough will shrink when cooked, so it is essential not to stretch or tear the dough when handling it.

The following represent the most commonly used dough formulas when making **pâté en croûte** and savory pies.

RECIPE 14–10

SHORT PASTE

Recipe Yield: 14 ounces
(397 g)

MEASUREMENTS		INGREDIENTS
U.S.	**METRIC**	
8 ounces	227 g	All-purpose flour
2 ounces	57 g	Butter
2 ounces	57 g	Lard
2 fluid ounces	60 mL	Water, cold
Pinch	Pinch	Salt

PREPARATION STEPS:

1. Rub the fats into the sifted flour and salt.

2. Add water.

3. Bind gently into a dough, and rest the dough for 30 minutes before use.

4. This dough can be covered and refrigerated for 1 week or frozen and kept for use.

RECIPE 14–11

BRIOCHE DOUGH

Recipe Yield: 12 pounds
(504 kg)

MEASUREMENTS		INGREDIENTS
U.S.	**METRIC**	
5 pounds	2.27 kg	Bread flour
1⅓ ounces	38 g	Instant dry yeast
2 pounds	907 g	Eggs, beaten
16 fluid ounces	480 mL	Milk, whole
8 ounces	227 g	Sugar
1½ ounces	43 g	Salt
3 pounds	1.36 kg	Butter, chipped

TIP Use the brioche dough on the following day, because it is a rich dough and very sticky when warmed by hands. The dough is generally easier to handle when well chilled.

PREPARATION STEPS:

1. In a commercial mixing bowl, combine the yeast and flour.

2. Add the eggs, milk, sugar, and salt. Mix on low speed, using a dough hook, for about 6 minutes.

3. With the mixer on medium (2) speed, slowly add the butter chips. Clean the mixer bowl sides, as needed. Continue to mix another 12 minutes, until the dough becomes elastic and pulls fairly cleanly from the side of the bowl.

4. Place the dough on a parchment-covered sheetpan, and cover tightly with a bun-pan bag.

5. Refrigerate the dough overnight.

RECIPE 14–12

SUET PASTRY

Recipe Yield: Approximately
1 pound (454 g)

MEASUREMENTS		INGREDIENTS
U.S.	**METRIC**	
8 ounces	227 g	All-purpose flour
1 teaspoon	5 mL	Baking powder
4 ounces	113 g	Beef suet, chopped
2 fluid ounces	59 mL	Water
Pinch	Pinch	Salt

PREPARATION STEPS:

1. Sieve flour and baking powder, mixing well.

2. Add suet, salt, and water to form soft dough.

3. Chill for 1 hour and rest for another hour before use.

RECIPE 14–13

HOT WATER PASTRY

Recipe Yield: Approximately
1 pound (453 g)

MEASUREMENTS		INGREDIENTS
U.S.	**METRIC**	
8 ounces	227 g	All-purpose flour
¼ pint	119 mL	Water
2½ ounces	71 g	Lard
½ teaspoon	2.5 mL	Salt

PREPARATION STEPS:

1. Add the salt to the flour and sieve well.

2. Rub 1 ounce (28 g) of the lard into the flour.

3. Bring the water to a boil and add the rest of the lard.

4. Partially cool the water and lard, and form into dough with the flour; knead well and store for service.

5. It is recommended that this paste be made just before use.

Note: This is only a guide for those who are unfamiliar with the production of these types of dishes. More or less panada can be added as the specific meat, poultry, game, fish, or shellfish requires.

PANADAS AND SECONDARY BINDERS

In addition to the proteins found in the forcemeats, there are several other binding agents that are commonly used in charcuterie. The **panadas** are the largest group of **binding agents,** encompassing several categories of farinaceous items. Starch is used as a thickener and binding agent in many preparations. The panadas can be made from well-cooked grains, including wheat, corn, and rice, or from tubers, such as cooked potatoes or tapioca. Another popular method of making a panada is by soaking bread in milk. (See Figure 14–6A, B, and C for steps in making panadas.) Starches are inert and insoluble in cold milk or water. When the liquid is slowly heated, the starch swells to thicken and bind, and then becomes gelatinous.

The protein found in eggs, gelatin, and milk or cream also serves to bind the farce as it coagulates when heated. Additionally, milk and the egg white both have the ability to hold an emulsion, thereby allowing air molecules to be trapped in the mixture and later be expanded by heat.

In addition to their role as binding agents, panadas serve as **extenders** in the forcemeat. Cooked rice, bread, or potatoes are much less expensive than meats, fish, and liver. (See Figure 14–7 of four panadas.) Their addition to the charcuterie product allows the chef to stretch the budget a little further while meeting the production needs of the operation. The panadas also aid in the slicing of the final product and to the textural interest to the dish.

The Use of Panadas

The use and makeup of panadas vary slightly. Table 14–3 lists the different types of ingredients used in forming the panadas and gives their suggested uses.

RATIOS AND RECIPES FOR THE USE OF PANADA Different panadas are selected for a variety of reasons, including but not limited to available ingredients, texture, appearance, flavor, and slicing ability. (See Table 14–2 on the use of panadas.) They may be enriched through the addition or substitution of rich (fatty) ingredients, such as cream, crème fraiche, glace, butter, eggs, and foie gras.

FIGURE 14–6A Adding the eggs to the milk

FIGURE 14–6B Mixing the bread into the wet ingredients

FIGURE 14–6C Mixing ingredients by hand

FIGURE 14–7 Clockwise from left: Egg panada, rice panada, flour panada, pure forcemeat

Table 14–2   The Use of Panadas

PANADA/BINDING INGREDIENTS	USES FOR PANADA/BINDING AGENT
Pure meat panada uses only the meat of the pâté or terrine that is being made with the addition of neutrally flavored meats such as pork or veal.	Normally used in heavy meat and game pâtés and terrines. Use only well-cleaned and sinew-free meat and pure pork fat.
Flour panada consists of a either a pâte a choux or a very thick béchamel sauce.	Used for poultry and white meat pâtés and terrines.
Bread panada is bread well soaked in milk, half and half, or single or heavy cream, and is added to the forcemeat. It is normally used in conjunction with egg whites.	Generally used in forcemeats made from fish, shellfish, finer small poultry and game birds, vegetables, and offal. Used for parfaits and timbales.
Whole or separated eggs are used as an enriching and binding agent. They are neutral in flavor and using only the white can substitute for the use of fat, making a much healthier product.	Generally used with vegetables, fish, shellfish, and fine white meats. Very good for parfaits and timbales.
White rice is cooked in a flavorful liquid.	Same as for bread and egg panadas.

RECIPE 14–14

PURE FORCEMEAT

Recipe Yield: 5 pounds
(2.3 kg)

MEASUREMENTS		INGREDIENTS
U.S.	METRIC	
2 pounds	900 g	Meat, poultry, or game (as the main ingredient), cubed
2 pounds	900 g	Pork fat, chopped
1 pound	450 g	Lean pork or veal, cubed

PREPARATION STEPS:

1. Chill meats and purée in a food processor.

2. Fold together in a large mixing bowl.

3. This ratio will bind into a pâté or terrine.

RECIPE 14–15

BREAD AND EGG PANADA

Recipe Yield: 12 ounces

Note: This panada binds 1 pound (453 g) of fish, shellfish, light white meat, or vegetable forcemeat and will accept at least 12 fluid ounces (354 mL) of heavy cream.

MEASUREMENTS		INGREDIENTS
U.S.	METRIC	
6 ounces	168 g	White bread, crusts trimmed, torn or cubed
4 fluid ounces	118 mL	Heavy cream
2 each	2 each	Egg whites

PREPARATION STEPS:

1. Soak the cream and egg white with bread.

2. Store mixture in a cold place for 30 minutes.

3. Store together to a form a smooth paste.

RECIPE 14–16

FLOUR PANADA

Recipe Yield: 16 ounces
(453 g)

Note: Two ounces (56 g) of this
panada binds 8 ounces (227 g) of
fish, shellfish, or any light white
meat. One egg white can be
added, removing ½ ounce of
panada. This panada will accept 8
fluid ounces (200 mL) of heavy
cream.

| MEASUREMENTS | | INGREDIENTS |
U.S.	METRIC	
8 fluid ounces	237 mL	Milk
2 ounces	56 g	Butter
4 ounces	112 g	All-purpose flour
2 each	2 each	Eggs

PREPARATION STEPS:

1. Heat milk and butter. Add flour as for choux pastry.

2. Add eggs, one at a time, beating well.

RECIPE 14–17

RICE PANADA

Recipe Yield: 8 portions

Note: This mixture binds 8 ounces
(227 g) of fish, shellfish,
vegetable, or any light white
poultry or game bird. This panada
will accept 4 fluid ounces (100
mL) of heavy cream.

| MEASUREMENTS | | INGREDIENTS |
U.S.	METRIC	
2 ounces	56 g	White rice, cooked
1 each	1 each	Egg white

PREPARATION STEPS:

1. Blend these ingredients together to form a smooth paste.

STOCKS, GLAZES, AND GELATINS

Stocks are the foundation to sauces and glazes. A properly made stock, rich in gelatin and flavor, will add greatly to the final product. In charcuterie, stocks are generally used as either a poaching liquid for galantines, a flavoring agent for the farce, a gelatin for binding a terrine or mousse, or as a **glaze** that serves to color, flavor, and protect the surface of the final product. However, aspic is also a highly susceptible breeding ground for bacteria. Extreme caution must be used when preparing the equipment and ingredients for usage.

So that the final pieces are both attractive and satisfying to the tastes, the chef must understand the importance of using only quality ingredients, seasoned properly. Also, the final clarity of the stock, glaze, or gelatin is of the utmost importance (Figure 14–8). Cloudy gelatin is of no use to the garde manger, because it only serves to mask and detract from the appearance of the products.

Because food is subject to being discolored and dried by the warm air that passes over, glazes can be both practical and attractive. Decorated pâtés, terrines, timbales, and galantines enhance the beauty of a buffet presentation. The skills of the garde manger can be illuminated, while the establishment gains enhanced stature from the artful presentations.

Trends in decorating food platters and charcuterie products have come and gone over the decades; Chapter 17 explains the process in greater depth.

FIGURE 14–8 Cubed, gelatinized, clarified stock

RECIPE 14–18

Brown Venison Stock

Recipe Yield: 10 quarts
(9.46 L)

MEASUREMENTS		INGREDIENTS
U.S.	METRIC	
10 pounds	5 kg	Venison bones, backs, and necks, split and cut small and brushed lightly with oil
2½ gal	11 L	Water
1 pound	453 g	Onion, small dice
1 pound	453 g	Carrots, small dice
1 pound	453 g	White of leek, small dice
1 each	1 each	Garlic bulb, split in half
4 each	4 each	Sprigs of thyme
4 each	4 each	Sprigs of rosemary
½ pound	226 g	Tomato paste
1 ounce	28 g	Whole peppercorns
20 each	20 each	Juniper berries
8 each	8 each	Bay leaves

PREPARATION STEPS:

1. Roast the venison bones to an even, light golden brown.

2. Deglaze roasting pans with water, and reserve for the stock.

3. Repeat the process for the vegetables.

4. Add the bones and the deglazing liquid to the rest of the water, and bring slowly to a boil.

5. Skim off all fat and simmer for 3 hours.

6. Add the vegetables and continue to simmer for 1 more hour.

7. Add herbs and spices and finish simmering for 30 more minutes.

8. Strain though a fine chinois at least twice. Chill rapidly and reserve for service.

ASPIC CLARIFIED WITH MEAT

Recipe Yield: 1 quart (.95 L)

MEASUREMENTS		INGREDIENTS
U.S.	METRIC	
2 pints	0.95 L	Very fine white or brown stock, very cold (any flavor that suits the pâté you are using this for)
3 each	3 each	Large egg whites
6 ounces	150 g	Ground shin of beef, completely free of all fat and sinew (meat from other animals can be used but must come from the areas of that animal that contain the highest levels of gelatin.)
4 ounces	100 g	Onion, fine dice
4 ounces	100 g	Carrots, fine dice
4 ounces	100 g	Celery, fine dice
4 ounces	100 g	Leek, fine dice
4 ounces	100 g	Fennel, fine dice
4 each	4 each	Sprigs of thyme
2 each	2 each	Sprigs of rosemary
1 ounce	25 g	Black peppercorns, lightly crushed
6 each	6 each	Bay leaves
½ cup	118 mL	Any suitable wine or fortified wine
8 each	8 each	Leaves of gelatin, soaked in cold water
1 ounce	25 g	Salt

PREPARATION STEPS:

1. Thoroughly combine all vegetables, herbs, and spices with the egg white.

2. Vigorously work this into the meat, mixing well with your hands.

3. Mix into the cold stock.

4. Bring to a boil as quickly as possible in a thick-bottomed pan.

5. When the stock begins to boil, reduce the heat slightly to control the floating raft of meat and egg. Prevent the raft from breaking apart by controlling the heat and creating a hole on the side of the raft, allowing the stock to simmer gently.

6. Add the wine and simmer gently; avoid disturbing the raft that has formed.

7. Cook for 1 hour.

8. Decant the aspic through cheesecloth at least twice.

9. Chill quickly and store for further use.

10. Use melted aspic to fill pâtés and glaze and to lightly coat slices of finished pâtés or terrines. Aspic serves to enhance presentations and to give the pâté moisture.

RECIPE 14–20

ASPIC CLARIFIED WITH EGG WHITE

Recipe Yield: 1 quart (0.95 L)

The stock that is used in this recipe should be very high in natural gelatin because that part of the clarification cannot be substituted with the egg white. The egg white is only being used to help further clarify the stock. It is suggested, therefore, that this stock be fortified using a neutrally flavored, highly gelatinous bone such as veal shin during its original cooking.

MEASUREMENTS		INGREDIENTS
U.S.	**METRIC**	
2 pints	0.95 L	Very fine white stock, very cold (any flavor depending on the pâté you are using this for)
5 each	5 each	Egg whites, lightly beaten to soft peaks
4 ounces	100 g	Onion, fine dice
4 ounces	100 g	Carrots, fine dice
4 ounces	100 g	Celery, fine dice
4 ounces	100 g	Leeks, fine dice
1 ounce	25 g	White peppercorns
4 each	4 each	Bay leaves
4 each	4 each	Thyme sprigs
1 ounce	25 g	Salt
½ cup	118 mL	Any suitably flavored wine or fortified wine
8 each	8 each	Leaves of gelatin, soaked in cold water

PREPARATION STEPS:

1. Combine all vegetables, herbs, and spices with the egg white, mixing well with your hands or a spatula.

2. Combine this mixture well with the stock and salt.

3. Bring to a boil as quickly as possible in a thick-bottomed pan.

4. On the point of boiling, control the heat. The egg white will have set and will begin to float. Add the wine.

5. Simmer the stock for 30 minutes, being very careful not to disturb the raft that has formed.

6. Decant the now clarified stock through cheesecloth at least twice.

7. While the stock is still warm, dissolve the gelatin leaves in the stock.

8. Chill quickly and store cold until needed.

FORCEMEATS

Forcemeat originated with the use of pork and included mainly the lesser cuts of flesh mixed with a percentage of fat and seasonings. Today, forcemeats include the use of fish and shellfish, beef, game, poultry, and vegetables. They come in a variety of textures, from the very smooth to the coarsely ground. The farce can be of one animal and color, or blended to create a variety of textures, flavors, and appearances.

Before grinding and blending the ingredients, both the meat and grinding heads must be well chilled to prevent fat separation. Additionally, maintaining a temperature below 40°F (4.4°C) reduces the risk of food-borne illness by keeping the food out of the danger zone. Generally, a traditional meat grinder is used for the task, and its metal grinder head and grinder plates may be submersed in ice water or placed in the freezer before the grinding process begins.

Often, the forcemeat will require several successive passes through the grinder, starting with larger holes, and repeating the process using progressively smaller holes in the grinder plates. This process is called *progressive grinding.* The seasonings and panadas are sometimes ground along with the meat product to aid in their equal distribution. It is important that the forcemeat mixture is thoroughly combined with the desired seasonings and appropriate binding agents before being placed in its mold.

However, before placing the mixture into the mold, it should be tested for quality. A patty can be quickly pan-fried to check for seasoning and texture. Alternatively, a dumpling can be poached in plastic wrap to check for seasoning and the correct binding consistency. Loose mixtures may be tightened by the addition of egg whites or starch. Tough samples can be improved upon by incorporating additional fat or cream. If additional ingredients are added, a second test is recommended.

Cooking Times and Methods

When we were in culinary school, we were taught to check the temperature of our roasts and finished forcemeats by using kebab skewers. We were directed to plunge the slender metal rods into the cooked products, slowly count to 10, and then place the metal skewers softly against our chins, just under the lip. Other chefs, presumably more concerned for sanitation than our tender chins, would instruct us to lay the heated skewers across the underside of our wrists. The sensitivity to heat was evident in both locations, and one could quickly learn to gauge temperatures by this method.

Today there are pocket thermometers that provide accurate readings of the internal temperatures of cooked products and laser guns that accurately detect external temperatures (Table 14–3). Time and temperature are the two most important variables for chefs to understand in cooking. To be a garde manger chef, one must master the ability to tame the flame!

Table 14–3   Temperature Chart for Pâtés and Terrines

INGREDIENT TEMPERATURE	COOKING TEMPERATURE	COOKING TIME	INTERNAL TEMPERATURE	WATER BATH
Pâté en croûte	425°F (218°C)	15 minutes		
	350°F (177°C)	25 minutes	155°F (68°C)	None
Pâté (no pastry)	350°F (177°C)	40 minutes	155°F (68°C)	None
Pâté de foie gras	270°F (132°C)	35 minutes	140°F (60°C)	175°F (79°C) max
Terrines	350°F (177°C)	55-65 minutes	155°F (68°C)	Simmering
Parfaits	350°F (177°C)	45 minutes	155°F (68°C)	Simmering
Timbales	350°F (177°C)	15 minutes	155°F (68°C)	Simmering

Pâtés or terrines with a solid core should be cooked rare to an internal temperature of 141°F (61°C).

When baking terrines, it is important that the temperature at which they are baked be carefully regulated. The use of a water bath, or **bain marie**, can serve to insulate the terrine from temperature extremes. The filled mold is placed in a roasting pan or other high-sided ovenable container. Simmering water is carefully added to the pan until it reaches two-thirds of the way up the side of the terrine mold. The bain marie and terrine are placed in an oven where the water bath temperature should be maintained between 170° and 175°F (77° and 79°C).

Cooling and Storage

The cooling procedure is an equally important part of the production process. As products change temperature, they also change in molecular ways. As proteins cool, they tend to solidify their structures while allowing the moisture to return to the cells. Heat causes pressure that repels moisture. Roasting or poaching a galantine will cause the juices to gather in the center of the meat, furthest from the heat. As the galantine cools, the moisture returns to the outer cells that were left partially dehydrated by the cooking process.

Cooked foods must not be allowed to remain within the temperature danger zone (between 41° and 135°F [5° and 57°C]) for more than 2 hours. It is of great importance that the chef use proper care by sufficiently cooling, wrapping, and labeling the products for storage. The use of product identification and date tags helps the staff properly rotate the stock while giving the chef of critical information.

Production of Pâtés, Terrines, Mousselines, Parfaits, and Timbales

RECIPE 14–21

VEAL FORCEMEAT

Recipe Yield: 8 portions

Note: Used as a binding agent for any meat, poultry, or game pâtés and terrines.

MEASUREMENTS		INGREDIENTS
U.S.	**METRIC**	
1 pound	453 g	Lean veal
½ pound	226 g	Lean pork
1 pound	453 g	Pork fat
1 ounce	28 g	Pâté spice of choice
2 ounces	56 g	Salt

PREPARATION STEPS:

1. Cut all meat and fat into 6- by ½-inch (15- by 13-mm) strips.

2. Ensure all ingredients are well chilled.

3. Season meat with salt and spice (Figure 14–9A).

4. Grind meats individually through a medium die and then through a fine die (Figure 14–9B).

5. Mix all of the ingredients together well.

6. Pass through a fine sieve (Figure 14–9C).

7. Store chilled until needed (Figure 14–9D).

FIGURE 14–9A Seasoning the measured meats

FIGURE 14–9B Grinding the meats through medium and fine dies

FIGURE 14–9C Passing through a fine sieve

FIGURE 14–9D The finished forcemeat

RECIPE 14–22

SHRIMP FORCEMEAT, FOOD PROCESSOR STYLE

Recipe Yield: 8 portions

Note: This can be used for terrines, mousselines, parfaits, timbales, quenelles, and stuffing for fish cuts. It can also be used in a host of hot and cold appetizer dishes.

MEASUREMENTS		INGREDIENTS
U.S.	**METRIC**	
4 ounces	113 g	Bread and Egg Panada (Recipe 14–15)
1 pound	453 g	Shrimp, peeled and deveined
½ ounce	14 g	Onion, softened in butter and cooled
1 ounce	28 g	Pâté salt
8 fluid ounces	240 mL	Heavy cream

PREPARATION STEPS:

1. Ensure all ingredients are very cold.

2. Combine shrimp, onion, and panada. Sprinkle with pâté salt and place into a chilled food processor bowl fitted with a sharp blade (Figure 14–10A).

3. Blend well, scraping down the side at least two to three times until the mixture is well blended.

4. Ensure that the blending does not take too long, because friction causes heat, which can create fat separation.

5. Remove from bowl. Chill well and pass through a fine sieve (Figure 14–10B).

6. Over an ice bath, beat in the heavy cream (Figure 14–10C). Store well, chilled, for further use (Figure 14–10D).

FIGURE 14–10A Adding the ingredients into a food processor

FIGURE 14–10B Passing through a fine sieve

FIGURE 14–10C Beating in cream over ice bath

FIGURE 14–10D Finished fine forcemeat

RECIPE 14–23

TROUT FORCEMEAT WITH BREAD PANADA

Recipe Yield: 8 portions

Note: This can be used to create terrines, mousselines, parfaits, timbales, quenelles, and stuffings for fish cuts.

MEASUREMENTS		INGREDIENTS
U.S.	**METRIC**	
6 ounces	170 g	White breadcrumbs
4 fluid ounces	118 mL	Heavy cream
2 each	2 each	Egg whites
1 pound	450 g	Trout fillet, skinned and trimmed, diced small
1 ounce	28 g	Pâté salt of choice
1 ounce	28 g	Shallots, softened in butter and cooled
12 fluid ounces	355 mL	Heavy cream

PREPARATION STEPS:

1. Soak breadcrumbs in 6 ounces (170 g) of heavy cream (Figure 14–11A).

2. Add egg white, mixing well. Set aside in refrigerator to soak for 10 minutes.

3. Combine the trout bread and mixture with the shallot and season all with the pâté salt (Figure 14–11B). Chill well for 30 minutes.

4. Grind through the fine die (Figure 14–11C). Chill well.

5. Repeat the grinding a second time, and pass through a fine sieve.

6. Over an ice bath, add the remaining cream, beating continuously (Figure 14–11D).

7. Reserve until needed.

FIGURE 14–11A Mixing the egg white, cream, and bread

FIGURE 14–11B Combining the trout with the panada

FIGURE 14–11C Grinding through a fine die

FIGURE 14–11D Adding the remaining cream over an ice bath

PÂTÉS

Historically, **pâtés** were pastes of light colored meat, poultry, or game birds that were spread in molds lined with short paste and strips of fat or bacon. The molds were baked in an oven, and topped with liquid gelatin while they cooled. After being cooled, these *pâté en croûtes* were unmolded and sliced. Today, the term *pâté* is commonly, and perhaps wrongly, applied to many shapes and forms that encompass several other categories of forcemeat mixtures including terrines, mousselines, parfaits, and galantines. Most pâtés are still made from mixtures of game birds, poultry, pork, fish, or liver, and are often baked in dough.

RECIPE 14–24

DUCK PÂTÉ

Recipe Yield: 12 portions

MEASUREMENTS		INGREDIENTS
U.S.	**METRIC**	
1½ pounds	680 g	Short paste (Recipe 14–1)
6 ounces	170 g	Lean duck thigh, cleaned of all sinew and fat
6 ounces	170 g	Lean pork
6 ounces	170 g	Pork fat
1 ounce	28 g	French style spice mix (salt added)
4 ounces	115 g	Duck breast, cut into 1- by 3-inch (25- by 76-mm) strips
2 ounces	60 g	Duck fat (for searing)
2 fluid ounces	59 mL	Brandy
4 fluid ounces	118 mL	Duck glace
½ ounce	14 g	Orange zest
2 ounces	56 g	Pine nuts
2 ounces	56 g	Ham, diced
1 ounce	28 g	Truffles, diced
1 ounce	28 g	Thyme, finely chopped
1 ounce	28 g	Egg wash
1 ounce	28 g	Pink peppercorns
10 fluid ounces	290 mL	Duck and Madeira aspic

PREPARATION STEPS:

1. Line a pâté mold with pastry and reserve for filling.

2. Cut duck thighs and lean pork into strips and season with salted spice mix.

3. Grind the meat twice, once through a medium die and then through a fine die.

4. Grind the fat twice through a fine die.

5. Mix the two together over ice until well combined.

6. Pass through a fine sieve and chill well.

7. Sear duck breast quickly in hot duck fat.

8. Remove the meat and excess fat from the pan, and deglaze with the pink peppercorns, brandy, and the duck glace.

9. Add the orange zest to this liquid. Chill and add all to the chilled forcemeat.

10. Fold the nuts, ham, truffle, and thyme into the forcemeat.

11. Start to layer the pâté by adding one-third of the forcemeat into the lined mold and topping with the duck breast. Continue layering with more forcemeat and so on, until the mold is full.

12. Cover with the pastry. Create vent holes in the top to allow the steam to escape during the cooking process.

13. Egg wash and bake for 15 minutes at 425°F (218°C) and then for 25 minutes at 350°F (177°C).

14. Test for an internal temperature of 140°F (60°C).

RECIPE 14–25

Ceann Cropic

Recipe Yield: 8 portions

TIP	This dish is best served warm, with toasted buttered brown bread and the cheeks and other retrievable meat from the head. It can also be left to get cold and carved, spooned, or shaped into quenelles for service with tossed green salad.

MEASUREMENTS / INGREDIENTS

U.S.	METRIC	INGREDIENTS
16 ounces	454 g	Cleaned cod liver
1 teaspoon	15 mL	Salt
4 ounces	113 g	Onion, very finely chopped, lightly sautéed in butter until fully cooked with a golden color
14 ounces	397 g	Oatmeal, very fine
1 each	1 each	Large cod's head, gills removed and washed thoroughly, approximately $^2/_3$ pound (304 g)

PREPARATION STEPS:

1. Sprinkle the cleaned liver with the salt and allow to rest at room temperature for 1 hour.

2. Add the onion into the liver, combining well. Add the oatmeal slowly, mixing in four stages and allowing a resting time of 10 minutes between additions.

3. When all the oatmeal is added, press the mixture into the cod head, packing well to avoid breakage.

4. At this stage, the whole head can be wrapped in cheesecloth to further secure the pâté. However, it is not necessary if the cooking stage is done with great care.

5. Immerse the whole head in salted water and bring to a very slow boil. Allow to sit in cooking liquid for 35 minutes before removing the pâté with great care.

6. Keep the pâté intact in the head at all times.

TERRINES

Terrines were originally mixtures of ground pork, game, poultry, or livers that were baked in earthenware crocks or molds, known as terrines, which were lined with fat. Today, they are made from a variety of ingredients including fish and shellfish, game, poultry, meat, or vegetables. The mixtures are bound with eggs, cream, or gelatin; spread or laid into any vessel of glass, ceramic, enamel, or earthenware; and gently poached in a water bath. Terrines are served cold, either still in the mold, or released and sliced for service.

RECIPE 14–26

VEGETABLE TERRINE

Recipe Yield: 12 portions

Note: The terrine must be pressed well and then sliced carefully while in plastic wrap. Liquid clear gelatin may be painted between layers of vegetables if extra strength is desired.

Note: The powdered gelatin needs to be bloomed in $\frac{1}{2}$ cup (118 mL) of balsamic vinegar and $\frac{1}{2}$ cup (118 mL) olive oil heated to 110°F (43° C).

MEASUREMENTS		INGREDIENTS
U.S.	**METRIC**	
8 ounces	227 g	Green zucchini, cut lengthwise in $\frac{1}{4}$-inch (6.4-mm) slices
8 ounces	227 g	Yellow summer squash, cut lengthwise in $\frac{1}{4}$-inch (6.4-mm) slices
1 pound	454 g	Eggplant, peeled and cut lengthwise in $\frac{1}{4}$-inch (6.4-mm) slices
8 ounces	227 g	Red onion, sliced into $\frac{1}{4}$-inch (6.4-mm) circle
8 ounces	227 g	Red peppers, skinned and roasted
8 ounces	227 g	Yellow peppers, skinned and roasted
8 ounces	227 g	Roma tomato flesh, blanched and seeded
1 cup	237 mL	Olive oil
1 ounce	28 g	Salt
1 tablespoon	15 mL	Oregano, finely chopped
1 ounce	28 g	Powdered gelatin
$\frac{1}{2}$ cup	119 mL	Balsamic vinegar, reduced from 1 cup
1 ounce	28 g	Red pepper flakes
$\frac{1}{4}$ cup	58 mL	Basil leaves, blanched and chilled

PREPARATION STEPS:

1. Brush all the vegetables with the olive oil and sprinkle with the salt and oregano. Grill on a hot grill evenly on both sides. Chill immediately.

2. Brush grilled vegetables with gelatin, olive oil, the balsamic vinegar, and red pepper flakes.

3. In a terrine mold lined with plastic wrap, layer the grilled vegetables with the basil leaves alternately until it is full; cover with the remaining plastic wrap and weight down with a light weight. Chill overnight.

4. Carefully remove from the mold and slice to order.

RECIPE 14–27

SALMON AND CRAWFISH TERRINE

Recipe Yield: 8 portions

TIP	Any leftover farce can be used for appetizers or stuffing.

MEASUREMENTS / INGREDIENTS

U.S.	METRIC	INGREDIENTS
2 pounds	907 g	Salmon fillet, skin-on, well scaled
6 ounces	170 g	White breadcrumbs
4 fluid ounces	120 mL	Heavy cream
2 each	2 each	Egg whites
2 ounces	56 g	Shallots, softened in butter
1 ounce	28 g	Salt
1 teaspoon	7 g	Cayenne
Pinch	Pinch	Nutmeg
1 fluid ounce	30 mL	Lemon juice
12 fluid ounces	350 g	Heavy cream
¼ ounce	7 g	Tarragon
2 ounces	59 mL	Carrot, small, diced, blanched, and chilled
2 ounces	56 g	Green beans, cooked and diced
4 ounces	113 g	Crawfish, cooked
2 ounces	60 g	Clarified butter

PREPARATION STEPS:

1. Trim the belly area of the fish by removing the skin up to 2 inches (51 mm) from the edge. Leave a 2-inch (51-mm) square of the flesh, attached to the skin, the whole length of the fillet. The removed salmon flesh should add up to 1 pound (454 g).

2. Grind this flesh twice through the fine die and reserve.

3. Meanwhile, soak the breadcrumbs with 4 fluid ounces (120 mL) of the cream and the egg whites.

4. Mix the ground salmon, soaked breadcrumbs, shallot, salt, cayenne, nutmeg, and lemon juice thoroughly together.

5. Pass this mixture through a fine sieve and chill well.

6. Beat the mixture over ice, adding the rest of the cream.

7. Carefully fold in the tarragon, carrots, beans, and crawfish.

8. Brush an oval 3-pound (1.9-kg) mold, or a length of PVC piping, with clarified butter and place the salmon skin with the salmon piece still attached into the mold. The salmon piece attached to the skin should be hanging over the edge so when it is flipped over it becomes the bottom of the terrine. Fill with the forcemeat and cover with the extra flap of skin with salmon attached.

9. Brush the top of the terrine with clarified butter. Place a light weight on top for 30 minutes, in the refrigerator, to flatten the shape of the top.

10. Place the terrine into a bain marie of simmering water and then into a 350°F (177°C) oven for about 40 minutes.

11. Cool with a light weight on the top to retain a nice even shape for presentation.

MONKFISH LIVER TERRINE

Recipe Yield: 8 portions

MEASUREMENTS		INGREDIENTS
U.S.	METRIC	
1½ pounds	680 g	Monkfish liver
4 ounces	113 g	White breadcrumbs
2 fluid ounces	60 mL	Heavy cream
2 each	2 each	Egg whites
2 ounces	56 g	Shallots, minced, softened in butter without color
Pinch	Pinch	Ginger
Pinch	Pinch	Allspice
Pinch	Pinch	Cayenne pepper
½ ounce	14 g	Salt
4 ounces	118 g	Lobster meat, diced
2 ounces	56 g	Scallops, diced
½ ounce	14 g	Basil, chopped
2 fluid ounces	60 mL	Madeira wine, brought to a boil and flambéed, then chilled
2 ounces	56 g	Clarified butter

PREPARATION STEPS:

1. Line a mold with buttered parchment paper, leaving as much as will cover the top hanging over the edge.

2. Slice the liver into ½-inch slices. Overlapping each other, cover the whole base of the terrine.

3. Grind the remainder of the liver (at least 12 ounces [340 g]) through the fine die twice.

4. Soak the breadcrumbs in the cream and the egg white.

5. Combine the liver with the wine, bread mixture, adding the shallots, ginger, allspice, cayenne, and salt. Beat all together well and pass through a fine sieve.

6. Carefully fold in the lobster, scallops, and basil. Fill the terrine with this mixture, carefully covering the top with the parchment. Brush with clarified butter.

7. Place terrine in a bain marie, in lightly simmering water. Place bain marie in oven for 40 minutes at 325°F (163°C). Chill quickly, with a light weight on the top, to retain good shape for presentation.

RECIPE 14–29

TERRINE OF FOIE GRAS

Recipe Yield: 8 portions

Note: It is essential that the very best liver be used in this recipe. Before using the liver, it should be very carefully cleaned of all skin and blood vessels. This should be done with the liver at, or as close to, room temperature as possible. Great care should be taken so as not to damage the liver.

TIP	Terrine slices may be coated with duck port aspic to improve appearance and extend shelf life.

MEASUREMENTS		INGREDIENTS
U.S.	**METRIC**	
1¾ pounds	894 g	Fattened goose liver (foie gras)
½ ounce	14 g	Salt
¼ ounce	7 g	White pepper
20 fluid ounces	591 mL	Port (good quality), brought to the boil, then chilled
8 fluid ounces	237 mL	Brandy, flambéed
2 ounces	57 g	Clarified butter (for coating parchment)

PREPARATION STEPS:

1. Clean and remove all veins from the liver. Break into manageable size pieces, about 2 inches (5 cm) square.

2. Place in a nonreactive container, and season with salt and pepper. Drizzle in the brandy and port. Knead the liquid into the liver without damaging it. Marinate for 24 hours, turning often.

3. Butter a mold thoroughly. Press in the liver to form a terrine, making sure there are no air pockets.

4. Cover with a lid, and place in a bain marie of water that is exactly 176°F (79°C). Place the bain marie into an oven, preset at 275°F (135°C) for about 35 minutes, or until an internal temperature of 140°F (60°C) is reached. Chill quickly and store until needed.

RECIPE 14–30

SMOKED GUINEA FOWL TERRINE WITH TRUFFLES

Recipe Yield: 18 portions

MEASUREMENTS		INGREDIENTS
U.S.	**METRIC**	
10 each	10 each	Pork fat, very thinly sliced
1 pound	454 g	Guinea fowl leg and thigh meat, trimmed of all sinew
½ pound	227 g	Lean pork
½ ounce	14 g	Pâté spice with salt
½ pound	227 g	Pork fat
¼ pound	113 g	Foie gras
6 ounces	177 g	Breadcrumbs
4 fluid ounces	118 mL	Heavy cream
2 each	2 each	Egg white
2 fluid ounces	59 mL	Armagnac
½ ounce	14 g	Sage, finely chopped
½ ounce	14 g	Rosemary, finely chopped
12 fluid ounces	355 mL	Heavy cream
2 ounces	58 g	Carrots, small dice, blanched and chilled
2 ounces	58 g	Pistachio nuts
½ ounce	14 g	Black truffle, finely diced
2 ounces	58 g	Fresh morels, small dice
1 pound	454 g	Smoked guinea fowl breast, cold smoked, cut in half lengthwise
½ ounce	14 g	White pepper, ground

PREPARATION STEPS:

1. Line the 3-pound (1.4-kg) terrine mold with the pork fat (Figure 14–13A). Chill and reserve for filling.

2. Grind the guinea fowl leg and thigh meat with the lean pork, pâté spice, pepper, and salt. Grind it twice through the fine die.

3. Grind the fat with the foie gras through the fine die.

4. Soak the breadcrumbs in 4 fluid ounces (118 mL) heavy cream and egg white.

5. Beat the ground meat, ground fat, and the panada together well (Figure 14–13B). Pass through a fine sieve.

6. Beat the Armagnac (brandy), sage, and rosemary into the mix and chill well.

7. Over an ice bath, beat in the rest of the heavy cream and fold in the carrot, pistachio, truffle, and morel (Figure 14–13C).

8. Place one half of the mix into the mold. Place the cold smoked guinea breast end to end down the middle (Figure 14–13D).

9. Top with the rest of the mix. Fold over the pork fat to seal.

10. Place in a bain marie of water that is just on the point of simmering. Bake at 325°F (163°C) for 45 to 50 minutes until internal reading shows 155°F (68°C).

TIP Terrine slices may be coated with guinea fowl Marsala aspic to improve appearance and extend shelf life.

11. Remove and chill quickly with a light weight on the top to improve the presentation of the slice (Figure 14–13E). (A finished plate of guinea fowl terrine is shown in Figure 14–14.)

FIGURE 14–12 Ingredients ready for the terrine: Whole bird, truffle, morel, pistachio, carrot, foie gras, pork loin, salt pork fat, spices, terrine mold

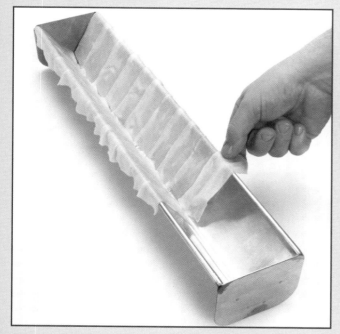

FIGURE 14–13A Lining the terrine mold with very thin pork back fat

FIGURE 14–13B Clockwise from top left: The panada, the ground meat, and the ground fat with the foie gras

FIGURE 14–13C Folding the carrot, truffle, pistachio, and morel through the forcemeat

FIGURE 14–13D Placing the cold smoked guinea fowl breast down the center of the first layer of forcemeat

FIGURE 14–13E Placing weights on the top of the terrine after cooking in order to create a clean shape when cold

FIGURE 14–14 A plated dish of guinea fowl terrine

RECIPE 14-31

TERRINE OF DUCK CONFIT

Recipe Yield: 12 Portions

MEASUREMENTS		INGREDIENTS
U.S.	**METRIC**	
4 pounds	1.8 kg	Duck legs
4 ounces	113 g	Salt
20 each	20 each	Black peppercorns
2 each	2 each	Rosemary stalks
4 each	4 each	Thyme sprigs
6 each	6 each	Bay leaves
10 each	10 each	Juniper berries
6 cups	1.4 L	Duck fat, melted
1 cup	237 mL	Duck glace
½ cup	119 mL	Fresh peas
½ cup	119 mL	Dried cherries, chopped
½ cup	119 mL	Carrot, small dice, blanched and chilled
1 each	1 each	Orange, zest of

PREPARATION STEPS:

1. Salt the duck legs and chill for 24 hours.

2. Remove all the excess salt and liquid from the duck legs and combine with the peppercorns, rosemary, thyme, bay leaves, juniper berries, and melted duck fat.

3. Bake covered in a 350°F (177°C) oven for about 3 hours or until they are fork tender.

4. Remove from the fat and strain the fat though a fine strainer.

5. Separate the small quantity of liquid from the fat and add it to the duck glace.

6. Line a 3-pound (1.4-kg) terrine mold with parchment paper.

7. Remove all the skin, bone, and fat from the legs. Shred the meat into a clean bowl.

8. Combine the meat, duck glace, peas, cherries, carrots, and orange zest. Place into the terrine mold, pressing down to pack the mixture well.

9. Cover and chill well over night before removing and carving.

RECIPE 14-32

PIG'S HEAD AND TONGUE TERRINE

Recipe Yield: 8 portions

MEASUREMENTS		INGREDIENTS
U.S.	**METRIC**	
1 each	1 each	Pig's head, split and trimmed with the tongue attached
2 each	2 each	Onions, whole
2 each	2 each	Carrots, whole
2 each	2 each	Celery stalks
1 each	1 each	Leek, split
1 ounce	28 g	Peppercorns
6 each	6 each	Bay leaves
4 ounces	113 g	Long grain rice
2 ounces	57 g	Carrot, fine dice
2 ounces	57 g	Rutabaga, fine dice
2 ounces	57 g	Celery, fine dice
1 ounce	28 g	Butter
½ ounce	14 g	Parsley, very finely chopped
1 ounce	28 g	Salt
½ ounce	14 g	White pepper, finely ground
½ ounce	14 g	Marjoram, finely chopped
1 ounce	28 g	Cider vinegar

PREPARATION STEPS:

1. Clean the head of all blood and fat. Soak in salted water for 1 hour and run under the cold water for 1 hour.

2. Cover the head with cold water and bring to a boil. Simmer for 2 hours. Add the whole carrots, onions, celery stalks, leeks, peppercorns, and bay leaves. Simmer for 1 more hour.

3. Remove the head and strain the stock.

4. Cook the rice in 1 cup (237 mL) of the stock.

5. Reduce the stock to 1 cup (237 mL) and reserve.

6. Sauté the fine dice carrots, rutabaga, and celery until soft.

7. Remove the skin from the tongue, and the flesh from the head.

8. Dice the tongue small, and shred the meat (approximate yield 24 ounces [680 g] finished vegetable and meat).

9. Prepare a terrine mold by lining with parchment paper (Figure 14–15A).

10. Combine the meat with the sautéed vegetables, reduced stock, salt, pepper, parsley, marjoram, cider vinegar, and the cooked rice and pack into the terrine mold (Figure 14–15B, C, and D).

11. Place a light weight on the top to improve the shape and chill overnight.

FIGURE 14–15A The mold lined with plastic wrap

FIGURE 14–15B The ingredients

FIGURE 14–15C Packing the mold with the ingredients and topping with stock

FIGURE 14–15D Covering the mold with the plastic wrap

MOUSSELINES

A mousseline is a mixture similar to a mousse. Their differences are in two areas: (1) the use of raw or cooked ingredients and (2) the method of binding. **Mousselines** use raw ingredients that require cooking, sometimes with panadas or egg whites, that also aid in the binding process. **Mousses** are made from cooked ingredients that require only binding, usually with gelatin, by refrigeration. Mousselines are often used as fillings or formed into oval-shaped dumplings called **quenelles**.

RECIPE 14–33

LOBSTER QUENELLES

Recipe Yield: 8 portions

TIP For the making of quenelles (oval-shaped dumplings), take two similar-shaped spoons and create the shape of the inside of the spoons, using the mousseline. Use warm water to keep the mixture from sticking to the spoons. This mousseline can be used for timbales, terrines, parfaits and for stuffing of other fish cuts.

MEASUREMENTS		INGREDIENTS
U.S.	METRIC	
1 pound	454 g	Lobster meat
1 ounce	28 g	Shallots, softened in butter, without color
½ ounce	14 g	Pâté spice
Pinch	Pinch	Cayenne pepper
1 fluid ounce	30 mL	White wine reduction
4 ounces	113 g	Bread and egg panada (Recipe 14–15)
8 fluid ounces	237 mL	Heavy cream

PREPARATION STEPS:

1. Finely mince the lobster meat and the shallots and place in food processor.

2. Season with the pâté spice and the cayenne pepper.

3. Add the white wine and the panada and continue to mix well in the food processor.

4. Pass through a fine sieve and beat the cream into the mixture over an ice bath.

STUFFED LOBSTER TIMBALE

Recipe Yield: 12 portions

MEASUREMENTS		INGREDIENTS
U.S.	**METRIC**	
1½ pounds	680 g	Lobster mousseline, made with lobster leg meats
1 ounce	28 g	Shallots
½ ounce	14 g	Butter
1 cup	237 mL	Shitake mushrooms, diced
8 ounces	227 g	Lobster tail meat
2 ounces	59 mL	Brandy
1 cup	237 mL	Sauce Americaine

TIP Sauce Americaine is made by thickening lobster stock with a blond roux, and the addition of tomato purée and heavy cream.

PREPARATION STEPS:

1. Line a small timbale mold with greased parchment paper (Figure 14–16A).

2. Place the mousseline in a piping bag. Partially fill the mold across the entire bottom ½ inch (1.25 cm) of the timbale. Then fill ½ inch (1.25 cm) more all the way around the sides, leaving a hollow center (Figure 14–16B). Chill until ready to fill.

3. Sauté the shallots in the butter until soft. Add the mushrooms and sauté 2 minutes. Add the lobster and flambé with the brandy.

4. Add the sauce and chill well for stuffing.

5. Place the prepared stuffing into the hollow of the timbale, being careful not to touch the edges (Figure 14–16C).

6. Place some more mousseline on the top to seal the liquid stuffing inside. Be careful not to trap any air in the timbale (Figure 14–16D).

7. Poach the timbale in a simmering water bath for 15 minutes in a 325°F (163°C) oven.

8. Remove and present split open to show the stuffing and allow the sauce to flow out onto the plate.

FIGURE 14–16A Lining a timbale mold with greased parchment paper

FIGURE 14–16B Piping the mousseline into the mold

FIGURE 14–16C Placing the lobster filling into the timbale

FIGURE 14–16D Closing the timbale with more mousseline

PARFAIT

Parfaits are considered to be the grandest and finest of all pâtés and terrines. They are generally bound with a fine panada of bread or rice, and enriched with fats such as fois gras, butter or heavy cream. Parfaits are only made using the most tender, and generally expensive, cuts of meat, fish, poultry, or game. Parfaits are poached in water baths, as with terrines, and can be made in terrine molds or timbales. They are served both hot and cold, and are delicate to handle.

RECIPE 14–35

SMOKED SCALLOP AND WATERCRESS PARFAIT

Recipe Yield: 16 portions

MEASUREMENTS		INGREDIENTS
U.S.	**METRIC**	
1 pound	454 g	Large scallops, cured and cold smoked
½ pound	227 g	Scallops, diced
1 ounce	28 g	Salt
½ ounce	14 g	White pepper
Pinch	Pinch	Cayenne pepper
6 ounces	170 g	White breadcrumbs
2 each	2 each	Egg whites
4 fluid ounces	118 mL	Heavy cream
12 fluid ounces	355 mL	Heavy cream
½ cup	119 g	Roasted red pepper, finely diced
3 ounces	85 g	Watercress, blanched, squeezed dry, chilled

PREPARATION STEPS:

1. Grind the scallops through the fine die twice. Add the salt, pepper, and cayenne. Mix well and chill.

2. Soak the breadcrumbs in the 4 ounces (113 g) of cream and the egg white to form a panada.

3. Combine the panada with the scallop mixture in a food processor.

4. Pass half this mixture through a fine sieve and beat in 6 ounces (177 mL) of the cream over ice. Fold in the red pepper and chill until needed (Figure 14–17A).

5. Pass the other half of the mixture through a fine sieve with the watercress and beat in the remaining cream over ice (Figure 14–17B).

6. Prepare a mold by lining it with parchment paper.

7. Place the two mousselines into the mold, alternately creating a split color on the bottom and on the top (Figure 14–17C). Use the freezer to firm up the mixes to get accurate separation of colors.

8. Cover the top with parchment. Poach in a simmering water bath placed in a 325°F (163°C) oven for 40 minutes.

9. Chill quickly with a light weight on the top to improve the shape.

FIGURE 14–17A Folding the peppers into one half of the mousseline

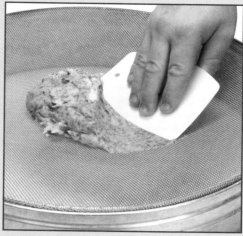

FIGURE 14–17B Passing the watercress and the mousseline through a fine sieve

FIGURE 14–17C Layering the different mousselines into the mold

MOUSSE

HAM MOUSSE

Recipe Yield: 16 portions

MEASUREMENTS		INGREDIENTS
U.S.	**METRIC**	
1½ pound	510 g	Lean ham, fully cooked but cold, diced small
13 fluid ounces	384 mL	Chicken velouté or béchamel, cooked but cold
15 each	15 each	Leaves of gelatin, bloomed in cold water
2 cups	473 mL	Heavy cream
½ ounce	14 g	Salt
½ ounce	14 g	White pepper

PREPARATION STEPS:

1. Grind the ham through the fine die twice and bind with the seasoning and velouté or béchamel sauce. Mix well.

2. Pass the ham and sauce mixture through a very fine sieve. Reserve.

3. Melt the gelatin leaves. Reserve.

4. Lightly whip the cream into soft peaks. Reserve.

5. Fold the softened aspic (without excess water) into the ham, distributing it evenly.

6. Fold in the whipped cream until well distributed.

Note: The mousse is now ready for use. Place the mousse into a mold or use it quickly for stuffing or piping for appetizers.

RECIPE 14–37

SMOKED SALMON MOUSSE

Recipe Yield: 10 portions

MEASUREMENTS		INGREDIENTS
U.S.	**METRIC**	
1 pound	453 g	Side of smoked lox (cold smoked salmon)
12 ounces	340 g	Cream cheese
2 ounces	57 g	Clarified butter
1 teaspoon	5 mL	Lemon juice
½ ounce	14 g	Chopped dill

PREPARATION STEPS:

1. Very carefully carve the salmon into equal slices, as thin as possible.

2. Line a small terrine mold with plastic wrap, leaving some plastic wrap over the edge for enclosing the top (Figure 14–18A).

3. Pass any trim from the carving of the salmon through a fine sieve, with the cream cheese, clarified butter, lemon juice, and dill. Hold at room temperature.

4. Shingle a layer of the thin salmon onto the plastic wrap, then alternate layers of cheese and salmon. Continue layering until the terrine mold is full (Figure 14–18B).

5. Fold over the final piece of plastic wrap to cover the terrine, and place a light weight on top to retain a nice shape (Figure 14–18C).

6. Chill 24 hours and carve to serve (Figure 14–18D).

FIGURE 14–18A Lining the plastic-lined terrine mold with the thinly sliced salmon

FIGURE 14–18B Layering the cream cheese mixture with the salmon to fill the terrine

FIGURE 14–18C Pressing the plastic wrap down on the top to create a flat surface

FIGURE 14–18D A finished plate of the salmon terrine

RECIPE 14–38

ROASTED RED PEPPER MOUSSE

Recipe Yield: 8 portions

MEASUREMENTS		INGREDIENTS
U.S.	**METRIC**	
10 ounces	283 g	Red peppers, roasted
8 fluid ounces	237 mL	Tomato sauce, roux method
1 fluid ounce	30 mL	Balsamic vinegar, boiled and cooled
8 each	8 each	Leaves of gelatin, bloomed in cold water
6 fluid ounces	177 mL	Heavy cream, whipped to soft peaks
1 ounce	28 g	Shallot, finely chopped
1 teaspoon	5 mL	Basil, finely minced
½ ounce	14 g	Salt

PREPARATION STEPS:

1. Blend the red peppers in a high-powered blender with the tomato sauce and vinegar.

2. Pass through a very fine sieve.

3. Add the melted gelatin and fold in the whipped cream.

4. Add the shallots, basil, and salt, and place in oiled molds to chill for 24 hours.

GALANTINE

Galantines were originally made only from poultry, as their French name "galine" suggests. Today, **galantines** are formed from poultry, and sometimes portions of small game or fish, that have had their bones completely removed, leaving the skin and flesh intact. The meat is often left attached to the skin, and usually flattened with a mallet for even distribution. Ground force-meats, sausage, panadas, vegetables, and seasonings are then layered upon the flattened flesh, and the whole galantine is rolled tightly into a cylindrical shape. It is then encased in plastic wrap, aluminum foil or cheesecloth to secure its uniformity. The galantine is cooked either by slow roasting in an oven, or by poaching in a flavored liquid. It is generally served cold, in slices, with accompanying pickled vegetables or fruit sauces.

RECIPE 14–39

LAMB BREAST GALANTINE

Recipe Yield: 8 portions

MEASUREMENTS		INGREDIENTS
U.S.	METRIC	
2 each	2 each	Lamb breast
2 pounds	907 g	Greek Feta Stuffing (Recipe 14–46)

PREPARATION STEPS:

1. Prepare the lamb breast by laying it out skin side down and removing all sinew and bones from that side.

2. Turn them over and remove the very fine skin on the other side.

3. Lightly beat the lamb breast with a mallet between two sheets of strong plastic until the breast is all the same width.

4. Now consider the breasts as a whole, gauging whether the meat within the whole pieces is evenly distributed over the entire breast. To redistribute the meat evenly, lightly beat again forming the whole back into one.

5. Place the stuffing onto the breasts, using 1 pound (457 g) for each, and spread it out evenly.

6. Roll and tie the breasts tightly lengthwise, in cheesecloth, until it almost is as firm as it would be when cooked and cold. This stage is of great importance, because the final presentation will be affected greatly by shoddy workmanship in the rolling and tying procedure.

7. Poach in very rich lamb stock for $2^{1}/_{2}$ hours in a 325°F (163°C) oven completely covered in stock.

8. Cool in the stock. When they are cool enough to handle, do one of two things: (1) very carefully unwrap the galantine and wrap in strong plastic, reshaping as needed or (2) just wrap the galantine in strong plastic still in the cheesecloth. (This is another opportunity to reshape the galantine to perfection before it finally solidifies with refrigeration.)

9. Chill well overnight before carving.

RECIPE 14-40

PIG'S TROTTER GALANTINE

Recipe Yield: 12 portions

MEASUREMENTS		INGREDIENTS
U.S.	**METRIC**	
4 pounds	1816 g	Pig's trotters (approximately 2 pounds each)
1½ pounds	680 g	Lean pork
1 teaspoon	5 mL	Salt
1 teaspoon	5 mL	White pepper
1 ounce	28 g	Pâté spice
12 ounces	340 g	Pork fat
½ ounce	14 g	Thyme, finely chopped
Pinch	Pinch	Mace
2 ounces	57 g	Garlic, roasted
Pinch	Pinch	Allspice
4 ounces	113 g	Ham, cooked and diced
3 ounces	85 g	Pickled tongue
2 ounces	57 g	Pork fat, diced
3 ounces	85 g	Pistachio nuts

PREPARATION STEPS:

1. Blanch the trotters and clean the skin of any hair bristles.

2. Bone out the interior of the trotters and clean the insides thoroughly.

3. Grind the lean pork, salt, pepper spice, pork fat, thyme, mace, garlic, and allspice twice through a fine die.

4. Pass the mix through a fine sieve and chill well.

5. Add the ham, tongue, diced pork fat, and pistachio nuts, then stuff into the trotters evenly.

6. Sew up the end well with strong string.

7. Wrap well in several layers of cheesecloth and prick several times with a needle.

8. Poach carefully in for 3 to 4 hours in veal stock. Cool in the stock overnight.

CHICKEN GALANTINE (3 WAYS)

Recipe Yield: 8 portions

MEASUREMENTS		INGREDIENTS
U.S.	METRIC	
3 each	3 each	Whole chickens (reserved for methods 1, 2, or 3)
½ pound	227 g	Lean pork
1 ounce	28 g	Pâté spice
2 ounces	57 g	Shallots, softened in butter without color
½ pound	227 g	Pork fat
½ pound	227 g	Chicken breast, cut in long strips 1 inch (2.5 cm) in diameter
Pinch	Pinch	Mace
½ ounce	14 g	Sage, fine chopped
2 fluid ounces	59 mL	Calvados (apple brandy)
6 ounces	170 g	White breadcrumbs
4 fluid ounces	100 mL	Heavy cream
2 each	2 each	Egg whites
12 fluid ounces	355 mL	Heavy cream
3 ounces	85 g	Ham, cooked and diced
3 ounces	85 g	Ox tongue, cooked and diced
2 ounces	57 g	Pistachio nuts
1 tablespoon	15 mL	Sage, fine chopped
2 ounces	57 g	Red peppers, roasted and diced
½ pound	227 g	Chicken breast

PREPARATION STEPS:

1. Prepare whole chicken according to method 1, 2, or 3 as listed following these steps for the filling.

2. Grind the chicken breast meat with the lean pork, pâté salt, and shallots through a fine die twice; hold chilled for mixing.

3. Grind the pork fat through a fine die twice and chill well.

4. Marinate the chicken strips with the mace, ½ ounce (14 g) of sage, and half of the calvados; chill well.

5. Soak the breadcrumbs with the 4 fluid ounces of heavy cream, the egg whites, and the remaining calvados.

6. Beat together the meat, the fat, and the panada until smooth and pass through a fine sieve.

7. Over an ice bath, thoroughly beat in the remaining heavy cream.

8. Fold in the ham, tongue, pistachio nuts, sage, and peppers; keep well chilled until required.

Note: All three methods are gently poached in the oven, completely covered with a rich chicken broth.

METHOD 1

1. Very carefully remove the complete skin from the chicken, beginning with a cut through the skin right down the spine, and then working toward the breast. Use fingers to loosen the skin while working.

2. The skin will come off the chicken easily. Try not to damage the skin in any way, keeping it as whole as possible. Reserve.

3. Roll out the forcemeat between two sheets of plastic 1 inch (25 mm) thick, and as long as the trimmed skin.

4. Place the chicken strips down the center of the forcemeat. Roll into a tight log.

5. Place this stuffed log in the skin, and then wrap tightly in cheesecloth, tying well with string to form a firm galantine.

METHOD 2

1. Completely bone the chicken, starting at the spine, removing all the flesh still attached to the skin.

2. Place the boneless chicken skin side down on a generous piece of cheesecloth. Add a log of forcemeat, formed as in Method 1, down the center.

3. Roll the galantine up tightly in the cloth to form a very firm log. Tie securely with string.

METHOD 3

1. Cut the skin of the chicken down the spine; remove all of the spinal bones, the rib bones and the breast bones, opening the bird out to expose the boned breasts from the inside.

2. Make a ball of forcemeat and fill the cavity created by the boning. Sew up the cavity to reform the original chicken.

3. Reshape the bird, tying the whole bird as if trussing it for roasting.

4. This galantine should look realistically like the bird you are using.

RECIPE 14–42

TORCHON OF FOIE GRAS

Recipe Yield: 8 portions

Note: Although this food item is not wrapped in any form of skin, it closely resembles a galantine.

U.S.	METRIC	INGREDIENTS
1½ pounds	680 g	Grade A foie gras liver
1 tablespoon	15 mL	Fleur du sel (sea salt)
½ teaspoon	3 mL	White pepper, finely ground
½ teaspoon	5 mL	White sugar
1 quart	946 mL	Whole milk
2 quarts	1.9 L	Veal stock

PREPARATION STEPS:

1. Clean the goose liver. (Refer to Chapter 11 for cleaning and deveining a goose liver.)

2. Ensuring that no part of the outside of the liver has been broken during the cleaning of the liver, season the liver on both sides with the salt, sugar, and pepper (Figure 14–19A). Wrap well and hold chilled for 24 hours.

3. Remove the liver and manipulate it into a log of about 3 inches (76 mm) wide and 8 inches (203 mm) long (Figure 14–19B). Wrap it well in plastic to help with the shaping. Chill well.

4. Remove the log from the plastic when it has set up, and wrap it tightly in cheesecloth to form a galantine shape. An assistant can help hold cheesecloth tight (Figure 14–19C).

5. Secure the ends well with string, ensuring a very tight wrap and tubular shape.

6. Plunge the torchon into the boiling veal broth for 90 seconds. Remove and place into an ice bath (Figure 14–19D).

7. Unwrap as soon as it is cold. Reshape and wrap in a clean towel. Hang in the refrigerator until ready for service (Figure 14–19E).

FIGURE 14–19A Seasoning the foie gras

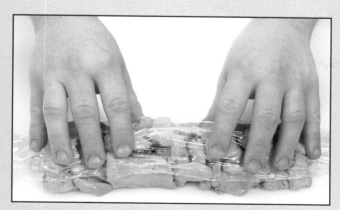

FIGURE 14–19B Rolling the foie gras in plastic wrap to create a cylindrical shape

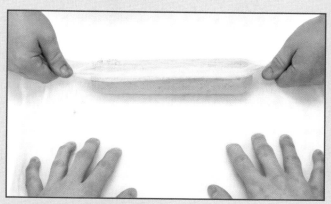

FIGURE 14–19C Rolling the foie gras in the cheesecloth once the shape has been established

FIGURE 14–19D Plunging into boiling veal stock with ice bath waiting to receive it

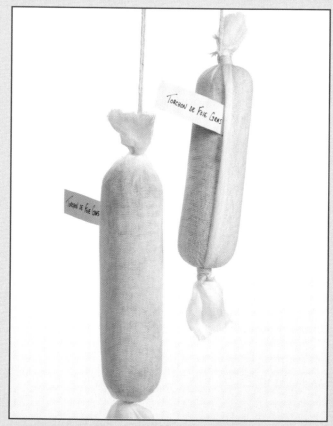

FIGURE 14–19E The finished torchon, rewrapped and hanging in the cooler for service

INTERNATIONAL STUFFING

The use of different stuffings, flavored with their own distinct herbs, spices, and other ingredients, can differentiate one national dish from another. Although most meats and poultry are universally available around the globe, their stuffings and methods of cookery are often the distinguishing characteristic. The following recipes are representative of those used in various countries or global regions, utilizing indigenous ingredients.

RECIPE 14–43

AUSSIE KANGA STUFFING

Recipe Yield: 8 portions

MEASUREMENTS		INGREDIENTS
U.S.	**METRIC**	
1 pound	454 g	Ground kangaroo meat
8 ounces	227 g	Fine white breadcrumbs
½ pint	237 mL	Beer (such as Foster's preferred)
2 each	2 each	Whole eggs
½ pound	227 g	Pork, ground
1 tablespoon	15 mL	Garlic, minced
1 tablespoon	15 mL	Ginger, minced
1 tablespoon	15 mL	Curry powder
1 cup	237 mL	Onion, minced
1 cup	237 mL	Cooked lentils
1 tablespoon	15 mL	Salt

PREPARATION STEPS:

1. Combine all ingredients and use immediately.

RECIPE 14–44

Brazilian Stuffing

Recipe Yield: 8 portions

Portion Size:

Note: Manioc flour is ground, roasted cassava root.

| MEASUREMENTS | | INGREDIENTS |
U.S.	METRIC	
½ pound	227 g	Manioc flour
¼ pound	113 g	Biscuit crumbs
½ pound	227 g	Black beans, cooked
½ pound	227 g	Rice, cooked
¼ pound	113 g	Bacon, cooked, diced fine
1 each	1 each	Onion, large, finely diced, cooked golden brown in batter
1 tablespoon	15 mL	Malagueta pepper, finely diced
1 tablespoon	15 mL	Garlic, diced
1 each	1 each	Orange, zest of
¼ pound	113 g	Plantains, diced, deep-fried golden brown
1 tablespoon	15 mL	Salt
2 each	2 each	Eggs

PREPARATION STEPS:

1. Combine all the ingredients and use immediately.

RECIPE 14–45

CHINESE STUFFING

Recipe Yield: 8 portions

MEASUREMENTS		INGREDIENTS
U.S.	METRIC	
1 pound	454 g	Jasmine rice, just cooked and still warm
8 ounces	227 g	Chinese sausage, small dice
2 ounces	57 g	Onion, small diced
2 ounces	57 g	Celery, small diced
2 ounces	57 g	Green pepper, small diced
2 ounces	57 g	Dried black Chinese mushrooms, reconstituted, diced small
1 tablespoon	15 mL	Ginger, minced
1 tablespoon	15 mL	Garlic, minced
2 ounces	57 g	Green onion, small diced
2 tablespoons	30 mL	Sweet soy sauce
1 tablespoon	15 mL	Toasted sesame oil

PREPARATION STEPS:

1. Combine all ingredients well and use immediately. Do not allow the rice to get cold or the stuffing will start to set up, becoming very difficult to work with (Figure 14–20).

FIGURE 14–20 Chinese stuffing being carved from a roasted bird

RECIPE 14–46

Greek Feta Stuffing

Recipe Yield: 12 portions

MEASUREMENTS		INGREDIENTS
U.S.	**METRIC**	
1 pound	453 g	Ground lamb (with at least 15 percent fat present)
8 ounces	227 g	Fine white breadcrumbs
6 ounces	170 g	Rice, cooked
4 ounces	113 g	Pine nuts, toasted to a golden brown
4 ounces	113 g	Onion, finely chopped, cooked in butter to a golden brown
1 ounce	28 g	Garlic, finely chopped and lightly sautéed
1 tablespoon	15 mL	Oregano, finely chopped
2 each	2 each	Whole eggs
1 cup	237 mL	Rich lamb stock
6 ounces	170 g	Feta cheese, crumbled
1 tablespoon	15 mL	Flat leaf parsley, finely chopped
½ cup	119 mL	Roasted red peppers, finely diced

PREPARATION STEPS:

1. Combine all of the ingredients together, mixing well. Chill for further use.

2. Ensure that the ingredients are well incorporated.

Professional Profile

BIOGRAPHICAL INFORMATION

Name: Grant Achatz

Place of Birth: St. Clair, Michigan

Recipe Provided: Sponge of Atlantic Shellfish with Celery Branch, Licorice, and Pear

CULINARY EDUCATION AND TRAINING HIGHLIGHTS

Grant Achatz enrolled at the Culinary Institute of America immediately upon graduating from his Michigan high school. Soon afterward in 1996, he secured a position with legendary chef Thomas Keller at the celebrated French Laundry in Napa Valley, California. He completed 4 years in the renowned kitchen, the last 2 as its sous chef. During his time in Napa, Chef Achetz also worked as an assistant winemaker at La Jota Vineyards, furthering his breadth of knowledge. In 2001, he took his first executive chef position at Trio in the Chicago suburb of Evanston, where he also held the position of partner. This experience heralded a stream of accolades, including *Food and Wine Magazine's* "Best New Chefs, 2002"; The James Beard Foundation's "Rising Star of the Year 2003"; four-star ratings from *The Chicago Tribune* and from *Chicago Magazine* in 2003; a coveted Fifth Mobile Star in 2003; and recognition of the "Meal of the Year" by *USA Today* in 2003. In late 2004, Chef Achatz left Trio to continue his culinary evolution and fulfill a boyhood dream; he opened his own restaurant, Alinea, in the Lincoln Park area of Chicago in early 2005, to the rave reviews of the culinary community.

ADVICE TO A JOURNEYMAN CHEF

The importance of learning basic techniques cannot be stressed enough. Only the foundation of proven techniques and a developed palate can provide the platform to execute creative and original cuisine. Take the time to seek out and understand the tasks that ground the craft, then you will be able to creatively process your surroundings.

Grant Achatz

RECIPE 14–47

Chef Grant Achatz's Sponge of Atlantic Shellfish with Celery Branch, Licorice, and Pear

Recipe Yield: 8 portions

U.S.	METRIC	INGREDIENTS
8 each	8 each	Pears, D'anjou, ripe
8 each	8 each	Gelatin sheets
2.2 pounds	1 kg	Mussels, rinsed, beards removed
24 each	24 each	Clams, littleneck
12 each	12 each	Clams, razor
16 each	16 each	Scallops, Nantucket
		Sachet
1 each	1 each	Bay leaf
1½ teaspoon	8 g	Fennel seeds
1⅛ teaspoon	6 g	Licorice root, whole, crushed
1½ teaspoon	8 g	Star anise
2 teaspoons	10 g	Black peppercorns, whole
3½ ounces	100 g	Shallots, peeled and sliced thinly
1 tablespoon	15 g	Garlic cloves, peeled and crushed
7 ounces	200 g	Fennel bulb, sliced thinly
5⅓ ounces	150 mL	White wine
5⅓ ounces	150 mL	Vermouth, dry
7 ounces	200 mL	Pernod
To taste	To taste	Salt, kosher
6 stalks	6 stalks	Celery branch, leaves reserved
1 each	1 each	Fennel bulb, fronds reserved
4 teaspoons	20 g	Sugar
As needed	As needed	Nonstick spray
1 bunch	1 bunch	Chervil sprigs
5 teaspoons	25 g	Black licorice powder

PREPARATION STEPS:

1. Peel and core six pears (reserve two) and place in a sous-vide bag. Cook for 1 hour at 175°F (79°C) or until pears are very tender.

2. Purée pears in blender until very smooth; pass through a chinois and reserve.

3. Bloom gelatin in ice water.

4. Prepare the sachet.

5. In a large rondeux, place the shellfish, sachet, shallots, garlic, sliced fennel, and wines. Cover and rapidly bring to a simmer just until shellfish opens up. Pour into colander, reserving all liquid that pours off.

6. Remove all shellfish from shells and trim accordingly. Strain the liquid through the cheesecloth and dissolve gelatin in the liquid.

7. Place liquid gelatin mixture in a small mixer with whip attachment and begin to whisk on high. Whisk until liquid produces a meringue type consistency of stiff peaks. Adjust seasoning with salt.

8. Cut the celery branch into small shapes and blanch for 2 minutes in hot salted water; shock in ice water.

9. Cut the fennel bulb into julienne and simmer in water, sugar, and salt until tender and glazed.

10. With the remaining two pears, peel and cut into medium dice, batons, and demispheres.

11. In an 80-mm ring mold sprayed with nonstick spray, spoon in 2.6 ounce (75 g) of pear purée and smooth it out. Spoon in a heaping mound of shellfish sponge and place into a refrigerator to set.

12. Remove the ring and arrange shellfish, glazed fennel, celery, and pear pieces around the molded sponge.

13. Place chervil sprigs, fennel fronds, and celery heart leaves on top of shellfish and vegetables.

14. Sprinkle licorice powder over sponge and serve.

1. Why does the equipment used in the making of forcemeats have to be cold?

2. Describe the term *panada*.

3. What is a forcemeat?

4. Describe how forcemeat is used.

5. What is mousseline?

6. How is gelatin applied to a vegetable terrine?

7. Why is gelatin applied to sliced pâté?

8. Describe the term *pâté*.

9. Describe the production of a timbale.

10. Describe the production of torchon.

A. **Group Discussion**

In groups, discuss the use of pâtés and terrines in the modern kitchen. Report on some modern ways that the techniques can be adapted to modern menus.

B. **Research Project**

In groups, research how pâtés have changed in appearance and flavor over the past 200 years.

C. **Group Activity**

Write a recipe for an original terrine using ingredients available in your own area. Create a name and method of presentation for the terrine. Experiment with your terrine, ensuring that the recipe works and has good flavor.

D. **Individual Activity**

Create an original spice mix and experiment with the flavor using plain forcemeat to determine the ratio needed to season a pâté.

CHAPTER 15

Kitchen-Made Cheeses and Creams

After reading this chapter, you should be able to

- communicate how cheese was first discovered.

- discuss the various benefits to the making of cheese.

- explain that cheese is a widespread food product and that similar varieties are produced globally.

- describe the steps involved in producing commercially made cheeses.

- demonstrate processes involved in producing kitchen-made cheeses.

- describe the nature of equipment used to make cheeses, and its proper care and sterilization.

- explain the function of ingredients in cheese making.

- show how to adjust the flavor and appearance of cheese.

Blessed are the cheese makers.

MONTY PYTHON

A Brief History of Cheese

In some ways, cheese is like fire. There are many theories as to its first discovery, but these remain conjecture. What is known, however, is that in order to have cheese, early mankind would have had to use the milk of available sheep, goats, cattle, camel, or oxen. It is likely that human beings and animals shared the same watering holes and that their fates, in this way, became entwined. Archeologists suggest that people began domesticating goats and sheep in the region of the Mediterranean somewhere between 6,000 and 10,000 years BC.

The Bible contains numerous references to cheese; however, cheese even preceded the events depicted in Genesis. As documented in the *History of Food,* Toussaint-Samat (2001) wrote:

> If we go back to the time of the Sumerians, 20 centuries before Abraham, we encounter another stock breeder. This one is anonymous, but his existence is well attested by the careful accounts he kept. . . . These accounts, engraved in cuneiform on clay tablets are now in the State Museum of the Middle East. They tell us that the breeder's herd of cattle increased fivefold within eight years, and production rose from. . . 8 litres of cheese a year . . . to 63.3 litres of cheese. (p. 114)

Additionally, the Iraq Museum in Baghdad has a Sumerian fresco of 2500 BC that depicts a peasant milking a cow while the calf is tethered to the cow's leg. And early records of cheese have been found on earthenware vessels dating back to 2300 BC in ancient Egyptian tombs. Around 2000 BC, dairymen of the Lake Constance Stone Age community developed a pottery colander for draining whey. So it is safe to say that cheese has existed for several thousand years.

Conventional wisdom, or at least cheese lore, holds that an Arab nomad who used a sheep's stomach to carry milk was the first to accidentally discover cheese. He probably spent several days in the sun, perhaps even crossing a desert, while the natural enzymes (**rennet**) contained in the sheep's stomach curdled the milk. Far from his home and regular source of sustenance, he was likely driven by hunger to test the coagulated mass left in the bottom of the skin after drinking the sour-tasting whey. The combination of movement and heat produced the curds and whey. Thus began the craft that has benefited from centuries of fine tuning and revered interest.

FIGURE 15–1 Herder milking a cow

During a culinary study tour to South Africa, the authors visited an interesting store in Johannesburg. The shop's two floors were filled with handwoven grain baskets, woodcarvings, Zulu spears, ornate clothing, wooden drums, weavings, and other marvelous handmade treasures. Among the artifacts was a herder's cheese jug that is still used in parts of Africa. The herder milks the cow (Figure 15–1), adds the milk to the jug, and then goes about his business of tending his herd of goats or cows (Figure 15–2). Over the course of time, the movement and heat cause the liquid to slightly solidify in the jug. Chef Garlough couldn't resist purchasing the ves-

FIGURE 15–2 African tribesman cheese jug

sel as a rustic reminder of how cheese is still crudely made in parts of the world. Even today, the container continues to emit a slight odor of cheese when the plug is removed.

Early Varieties

Cheese is simply an artificial form of coagulated milk. Defined as a food product made from the pressed curd of milk, cheese is made of milk proteins, milk fat, and water. One could easily argue that cheese was not truly invented by anyone; it was developed more by the act of circumstance than by man's ingenuity.

The differences in early cheeses had more to do with conditions that were beyond the control of man than they did with man's intentions. The kind of milk produced by his animals (sheep, goats, asses, cattle, buffalo, yak, reindeer, camels), the kind of rennet available in the region, the atmospheric conditions present in the caves, cellars, and storehouses where the cheese matured, the type of bacteria floating about, and the length of time available to ripen the cheese all created regional differences in early cheese production (Figure 15–3).

As this form of dairy chemistry evolved, mankind again stumbled on another twist that resulted in additional varieties being produced. Farmers discovered that certain thistles, fig bark, and herbs could be soaked in water to make extracts that could be substituted for rennet, thereby introducing the use of herbs into the production of cheese and making the first vegetarian-styled cheeses. The Greeks were known to stir their goat and sheep's milk with the green branches of fig trees to coagulate the liquid, and the Romans used safflower seeds in lieu of rennet as the use of herbs and spices in cheese proliferated.

Although exact dates are obscure, Table 15–1 lists dates that researchers believe to be approximate cheese-making benchmarks.

VALUE TO CHEESE MAKERS AND CONSUMERS

Among the various attributes of cheese, those in the religious orders prized its high nutritional value. As with many culinary traditions, including the making of bread and wine, the men of the church did more to preserve and advance the craftsmanship of making cheese than any others. While kings came and went in the Middle Ages, the monks continued to improve the production methods and flavors of their cheeses. Motivated by their strict adherence to fasting—100 days per year—members of religious orders were ardent about meeting their own dietary needs and tastes.

FIGURE 15–3
Ancient cheese making

Table 15–1   Early Cheese Making

7000 BC	Ancient Sumerian and Mesopotamian cultures of the Tigris-Euphrates basin raised cows and sheep, and engaged in dairy production.
3000 BC	The first historical reference to cheese, found in a Sumerian frieze.
800 BC	Homer mentions cheese in his *Iliad*.
329 BC	Greek historian named Xenophon wrote about a goat cheese that had already been made for centuries.
54 BC	Julius Caesar invades Britain and finds the Britons making Cheshire cheese.
50 AD	The Roman food writer, Columella, outlines the basic steps of cheese making.
702	Japan develops first written law codes and establishes regulations for the making of dairy products, including cheese.
800	Gorgonzola is first made in Italy.
1070	Roquefort cheese is discovered in France.
1200	Parmesan cheese and Pont l'Évèque are made.
1400	Ementhaler Swiss cheese is first produced in the canton of Bern's Ementhal Valley.
1680	A French document refers to Camembert as "a very good cheese, well suited to aid digestion washed down with good wines."
1722	Gruyère cheese is introduced into France.
1740	The London cheese shop Paxton & Whitfield opens, selling Cheddars, Gloucesters, Cheshires, Stiltons, and other English cheeses.
1815	The first factory for the mass production of Swiss cheese opens in Bern.
1824	Colby cheese is developed in Vermont.
1851	The first American cheese factory is established in Rome, New York.
1865	Marin French Cheese Company opens in Petaluma, California.
1876	McCadam Creamery opens in Heuvelton, New York (moved to Chateaugay, New York in 1934).

Beyond the needs of the monks, farmers had their own requirements to address. The making of cheese provided a solution to many problems that plagued the average farmer and citizen. Cheese was beneficial in many ways: it was a means of preserving milk, it was easy to store, and it provided food for carriage.

A Means of Preservation

In addition to its nutritional value, cheese was beneficial as a productive means of using excess milk. With the lack of refrigeration, milk that was not consumed immediately soon spoiled. Butter was also a popular by-product of milk, so any surplus milk was immediately used for producing cheese and butter. These products became much more shelf stable and could be preserved into the winter months when holding livestock was difficult. Cheese also preserved longer than milk during the summer when fresh milk could sour in a matter hours.

Ease of Storage

Food storage was always an issue for earlier peoples. The use of springhouses, cold cellars, smokehouses, and pottery jars were commonplace as civilizations developed. However, storing fresh food for the cold months was always a concern. And the worry was not only for feeding the people but also for feeding their livestock. It was not uncommon for many sheep, goats, and calves to be slaughtered before winter, because grain was too scarce to store during the cold months for feeding all of the animals. (In fair weather, animals were generally allowed to wander and eat the grasses and grains growing wild near the settlement.) Cheese provided a concentrated nutritional resource for milk products over the winter months, when less milk was available. It did not require much space, although it was often coated in salt, ash, or strong spices to deter maggots and bugs, even when kept in clay pots.

Food for Carriage

Another value of cheese was its portability. Cheese became a staple for those who worked in the fields and needed a quick mid-day repast. Their meals would often consist of cheese, pâté or sausage, coarse bread, fruit, and wine—not unlike a picnic meal we might enjoy today. For them, it was an early form of convenience food. And when Julius Caesar sent his Roman legion to conquer Gaul, they carried bundles of cheese for the long march. Cheese then became an early form of army K-rations.

Cheese Today

Today, there are four main categories of cheese production recognized in France, Italy, Switzerland, and other cheese-producing countries. They include fermier, artisanal, cooperative, and industrial.

- **Fermier.** Also known as Chalet d'Alpage, Buron, or Mountain Hut, the cheese is made in a farmhouse. The individual producer uses the milk of animals (e.g., cows, goats, or sheep) raised on the farm to make the cheese following traditional methods. Only raw milk from that farm may be used. The production is on a small scale.
- **Artisanal.** An individual producer uses the milk of animals raised on his or her farm, or buys milk to make the cheese. The production is on a small to medium scale.
- Cooperatives. Also known as a *Caseificio,* the cheese is made in a single dairy with milk provided by members of the cooperative. The production is on a medium to large scale.
- Industrial. The milk is bought from a number of producers, sometimes from distant regions. The production is always on a large scale.

MODERN EXPLOSION OF CHEESE VARIETIES

Interest in cheese on the world market has created an explosion of varieties. With the combination of small cottage-industry producers and large cheese houses, more than a thousand varieties of cheese exist. It is believed that General Charles de Gaulle once quipped, "How can anyone be expected to govern a country with 325 cheeses?" Forty years later, it is estimated that France alone now produces up to 700 different cheeses. Indeed, it is estimated that of the cheeses that we know today, 90 percent were invented in the last 100 years.

In the United States, cheese consumption has also grown dramatically. According to Max McCalman, maître fromager of New York's fashionable Artisanal Cheese Center, "Fifteen years ago, Americans ate only 25 pounds of cheese annually per person. Today, each person is consuming on the average of 40 pounds of cheese per year" (*Worlds of Flavor 6th Annual International Conference & Festival,* St. Helena, California, November, 2003).

Classifications of Cheese

The **classification of cheese** is one of the great culinary debates that continue today. Most agree that there are so many different variables to consider in the making of cheese that no classification system can effectively represent all varieties.

Common considerations include country of origin, region of production, milk variety, fat content, texture, aging, and ripening agent. Some authors seek to reduce the classifications to only four headings, with multiple variations stemming from these labels, whereas others insist there are 18 major categories of cheese. In a few cases, some cheeses are placed in different categories, depending on the opinion of the writer.

One approach is to classify cheeses by their ripening agent and from where the ripening process begins. This classification system has five categories:

- Unripened
- Using mold as the agent and ripened from the outside
- Using mold as the agent and ripened from the inside
- Using bacteria as the agent and ripened from the outside
- Using bacteria as the agent and ripened from the inside

For ease of discussion, we categorize cheeses in the following common manner, based on their texture and method of ripening.

- **Soft fresh (unripened) cheeses.** Soft fresh cheeses are uncooked and **unripened**, and generally have a fresh, mild, and creamy flavor and texture (Figure 15–4). They often have a slight tinge of tartness to them, but are not very acidic. Soft fresh cheeses are high-moisture cheeses that will keep under refrigeration for only a few weeks. Within this category are favorites like Feta, cottage cheese, Ricotta, cream cheese, Neufchâtel, Queso Blanco, and Mascarpone.

- **Soft ripened cheeses.** These are cheeses that are best distinguished by their soft, velvety surfaces, often referred to as the "blooming rind" (Figure 15–5). *Penicillium candidum* is sprayed or dusted on the surface of the cheese, allowing them to ripen from the outside toward the center. These cheeses ripen quickly and are at their peak for only a few days. They are available in different degrees of richness, including single, double, and triple creams. Favorites in this category include Bel Paese, Brie, and Camembert.

- **Semisoft cheeses.** Also known as Trappist cheeses, semisoft cheeses can be traced to the monasteries, having originated in the Middle Ages (Figure 15–6). They are characterized by mild and buttery tastes (Gouda), but may also be pungently strong and aromatic (Limburger). Other popular varieties in this category include Edam, Fontina, Port-Salut, Muenster, and Liederkranz.

Note: Some writers classify Edam and Gouda as firm cheeses.

- **Blue cheeses.** Also known as blue-veined cheeses, these varieties are easily characterized by the lines and spots of blue and green mold that form in and on the cheese (Figure 15–7).

FIGURE 15–4 A selection of soft fresh cheeses. ©2005 Wisconsin Milk Marketing Board, Inc.

FIGURE 15–5 A selection of soft ripened cheeses. ©2005 Wisconsin Milk Marketing Board, Inc.

FIGURE 15–6 A selection of semisoft ripened cheese. ©2005 Wisconsin Milk Marketing Board, Inc.

These cheeses are prized for their pungent tastes and creamy textures. The most famous include Danish Blue, English Stilton, French Roquefort, Italian Gorgonzola, and American Maytag.

■ **Pasta filata cheeses.** Pasta filata, or string curd cheeses, are made by mixing matured curds with hot water, and then pulling and stretching until they are firm (Figure 15–8). Popular varieties include string cheese, mozzarella, and provolone.

FIGURE 15–7 A selection of blue cheeses. ©2005 Wisconsin Milk Marketing Board, Inc.

FIGURE 15–8 A selection of pasta filata cheeses. ©2005 Wisconsin Milk Marketing Board, Inc.

■ **Firm cheeses.** These varieties have a firm, solid texture suitable for easy slicing (Figure 15–9). They range from the very mild, like Colby, to the very sharp, like aged Cheddar. Other popular varieties include Monterey Jack, Swiss Ementhaler, Gruyère, and Jarlsburg.

■ **Hard cheeses.** Also known as very hard or grating cheeses or *grana,* these cheeses have a dry, grainy texture (Figure 15–10). These cheeses are often sold already grated. Common varieties include Parmesan, Pecorino Romano, Queso Enchilada, Mimolette, aged Asiago, Grana

FIGURE 15–9 A selection of firm cheeses. ©2005 Wisconsin Milk Marketing Board, Inc.

FIGURE 15–10 A selection of hard cheeses. ©2005 Wisconsin Milk Marketing Board, Inc.

Padano, and the most celebrated grana: Parmigiano-Reggiano, which must age at least 2 years, giving it the characteristic salty, grainy texture.

In addition to the traditional varieties of cheese, **processed cheeses** are also popular on the world market (Figure 15–11). These cheeses are made in an artificial manner, unlike the naturally made cheeses just listed, which are all made by curdling milk and ripening the curds. Processed cheeses are manufactured by grinding one or more natural cheeses (sometimes aged and new), then blending with flavorings, colors, and emulsifiers. The mixture is then heated for

FIGURE 15–11 A selection of processed cheeses. ©2005 Wisconsin Milk Marketing Board, Inc.

pasteurization and hardened in molds. As such, these cheeses do not age or ripen, and have an extensive shelf life.

Similarities in Cheeses Throughout the World

What may be obvious to the reader is still worth mentioning: There is a marked similarity in cheeses produced around the world. Even though some countries are particularly noted for a certain classification of cheese, like France and the highly praised soft ripened cheeses, most countries produce a variety of cheeses that parallel other cheeses made on different continents. One need only look at the blue-veined cheeses, the Cheddars, the creams and yogurts, and the Swiss-styled cheeses to note their likenesses.

Perhaps this is by happenstance, perhaps by design. Possibly it is a function of trade and travel. What is certain is that there is a global demand for all varieties of cheese.

The Varied Uses of Kitchen-Made Cheeses and Creams

The knowledge of cheese—its variety, production, application, and care—is certainly worthy information for the modern garde manger. As has been demonstrated, the making of cheese is a celebrated art form, and its culinary acceptance is universal. The following are examples of the many applications of this dairy chemistry:

- Cooking ingredient
- Stuffing and binding agent
- Accompaniments
- Functional garniture
- Appetizers
- Sandwiches and rollups
- Soups
- As a course (Figure 15–12)

FIGURE 15–12 A plated cheese course

Cheese Production

In many ways, there are few differences between cheeses that are made commercially for retail sale and those made by chefs for use in their own food service operations. The size of the production is one obvious difference, and time available to prepare the cheese is another. In commercial operations, the cheese is the only product; it is the reason to be in business. Therefore, no amount of effort to realize a quality product is too much for the commercial operation. Conversely, cheese making is merely one of the countless activities that a garde manger chef must pursue, and it holds no greater importance than many of the other required tasks. For this important distinction, garde manger chefs tend to make cheeses that require fewer steps and less time for ripening than those made by commercial operations.

> **TIP** Cheeses made in the garde manger chef's food service operation may be improved with the use of non-homogenized milk (as is used in commercial operations) or skimmed milk with a 10% ratio of heavy cream added.

COMMERCIALLY MADE CHEESES

Although it is understood that each variety of cheese has its own unique procedure that differentiates it from the other cheeses on the market, certain steps are common in most cheese production. These are 10 basic steps to commercial cheese making:

1. The incoming milk is tested for purity and quality.
2. The milk is weighed and heat-treated or pasteurized (Figure 15–13).
3. A starter culture or enzymes are added to help curdle the milk, while also helping determine the ultimate flavor and texture.
4. The rennet (a milk-clotting enzyme) is added to coagulate the milk and to form a gel-like mass.
5. The coagulated mass is cut into small pieces to begin the separation of the **curds** (solids) from the **whey** (liquid) (Figure 15–14).

FIGURE 15–13 Milk being heated in a vat

FIGURE 15–14 Milk being stirred with a harp

6. The curds and whey are cooked and stirred until the desired temperature and firmness of curd is achieved (Figure 15–15).
7. The whey is drained for further treatment and other uses (Figure 15–16).
8. The curds are salted and manipulated according to their particular cheese variety.
9. The curds are pressed into a cheese mold to form their characteristic shape; the curds **knit** and release any additional whey (Figure 15–17).

FIGURE 15–15 Temperature being taken

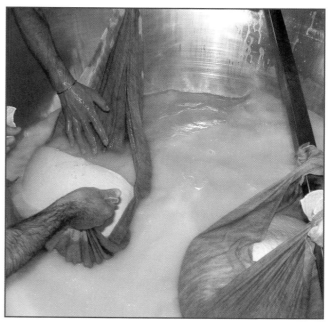

FIGURE 15–16 Cheese curds hanging in cheesecloth over whey

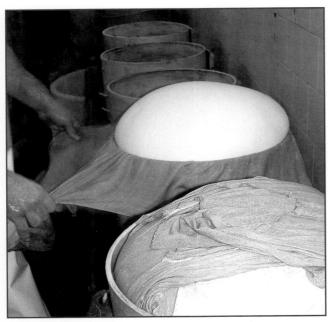

FIGURE 15–17 Cheese being placed in mold

FIGURE 15–18 Cheese floating in brine tanks

Note: Some commercially made cheeses are brined or coated with *Penicillium candidum* before ripening (Figure 15–18).

10. For cheeses that are aged, they are stored in humidity and temperature-controlled rooms to allow full development of flavor and texture, known as **ripening** (Figure 15–19).

KITCHEN-MADE CHEESES: THE EASE OF AVAILABILITY

As discussed in Chapter 2, two of the basic tenets of production management are the cross-utilization of products and the planning for leftovers. The fabrication of cheese meets these requirements, in that the equipment required to make cheese is common to kitchens, as are most of the ingredients.

FIGURE 15–19 Cheese in drying room

Kitchen-made cheeses are easily available to chefs interested in diversifying their talents, customizing their foods, and training their staffs. They become another method of creating signature dishes for their operation.

Basic Equipment Identification

The commonly used equipment for making cheeses can, for the most part, be found in any commercial kitchen (Figure 15–20). It is important to the cheese-making process that the equipment be of proper material and in sanitary condition. The following items are ordinarily used for kitchen-made cheeses and creams:

- Measuring cups and spoons
- Dairy thermometer
- Double-boiler pots
- Stainless steel slotted spoon
- Curd knife (with stainless steel blade)
- Commercial cheesecloth
- Butter muslin
- Cheese molds
- Cheese press
- Cheese boards
- Cheese mats

CARE AND SANITATION OF EQUIPMENT

Of vital importance in cheese making, as with any food preparation, is the proper selection, care, and sanitation of equipment. There are a few precautions that should be addressed in the use of the equipment.

All utensils or equipment that come in contact with milk at any stage of the cheese-making process must be of certain material. Acceptable materials include glass, stainless steel, copper, or enamel-lined vessels. During the process of cheese making, the milk becomes highly acidic. The curds may absorb the metallic salts found in aluminum or cast iron, thereby transferring an undesirable flavor and potentially dangerous substance onto the cheese.

FIGURE 15–20 Basic cheese-making equipment

In addition to making the correct selection of utensils and equipment, these must always be carefully cleaned before and after cheese making. Most failures in cheese making are caused by the use of unsanitary equipment. The process of cheese making is founded on the action of "friendly bacteria." Introducing harmful bacteria will create an unwanted variable that could produce disastrous results.

Cleaning and Sterilization of Equipment

The chef must both clean and sterilize the utensils and equipment used in cheese making. Cleaning involves the removal of fat, food particles, and other residue left on the equipment that is seen by the naked eye. It is best to start by rinsing under cold water, before washing and scrubbing in hot, soapy water. Immersing the milk-coated equipment in hot water will cook the milk products and adhere them even more strongly to the equipment. A quick rinse under cold water first eliminates this problem. Also, it is important to always wash the equipment *before* use.

Sterilizing involves the killing of harmful bacteria that is not seen by the naked eye but that is often present on cooking equipment and work surfaces. To sterilize, the chef may choose one of several methods:

- Immerse nonporous equipment and utensils in boiling water for 5 minutes.
- Steam nonporous equipment and utensils for 5 minutes in a tightly covered container.
- Boil or steam porous equipment, such as wood, cheese boards, and mats, for at least 20 minutes.
- Plastic equipment, including food grade materials, should not be boiled or steamed. These items, such as strainer baskets and spatulas, should be sterilized with a solution of bleach and water. The correct solution to use is 2 tablespoons (30 mL) of household bleach to 1 gallon (3.8 L) of water.
- Clean and sterilize all counters and work surfaces using a cleaning towel that has been rinsed in the bleach solution.

Note: All equipment sterilized with bleach must be rinsed thoroughly to remove all traces of sodium hypochlorite, which will inhibit the growth of friendly cheese-making bacteria.

Basic Ingredient Identification

One of the beneficial attributes of cheese is its simplicity of ingredients (Figure 15–21). From only a few items, many varieties of cheese can be made, which benefits the food service operator. The following ingredients are the foundation for numerous kitchen-made and commercially made cheeses. They are not all used in one single recipe, but they may be used separately depending on which cheeses are desired and what ingredients are available to the cheese maker.

- Cow's milk. Whole cow's milk is more than 87 percent water, with the balance made up of proteins, minerals, lactose (milk sugar), milkfat, vitamins, and some trace elements. There are several forms of milk produced from the natural whole milk of the cow. Whole milk contains less than 3.7 percent butterfat (or more from cows in some areas of England and France because of their breed and rich diet), whereas low-fat milks contain anywhere from 0.5 to 2 percent butterfat. Homogenating the milk fractures the fat globules and distributes them throughout, preventing them from separating and rising to the top as cream. For this reason, it is more difficult to make cheese from homogenized milk because the curd solidifies less. Skim and low-fat milk are good for making low-fat cottage cheese and hard cheeses.
- Goat's milk. Whole goat's milk contains just over 87 percent water, with the balance divided between nutrients, minerals, and vitamins. Whole goat's milk contains nearly 3.8 percent butterfat, although some breeds produce more or less. Goat's milk has 13 percent less lactose than cow's milk. The milkfat particles are small, making it superior to cow's milk for digestibility. Because all of the beta-carotene is already converted to colorless vitamin A, goat's milk and its products are whiter than cow's milk and its products.

- Buttermilk. Buttermilk is cultured milk that is produced by culturing any milk with appropriate characterizing bacteria. The addition of certain characterizing ingredients and lactic acid–producing bacteria permits the product to be labeled *cultured buttermilk.*
- Cream. Creams vary in milkfat content and are labeled accordingly. The ranges are as follows:
    - Half-and-half consists of a mixture of milk and cream containing between 10.5 and 18 percent milkfat.
    - Light cream contains between 18 and 30 percent milkfat.
    - Light whipping cream contains between 30 and 36 percent milkfat.
    - Heavy cream contains not less than 36 percent milkfat.
- Yogurt. Yogurt is the product resulting from the culturing of a mixture of milk and cream products with the lactic acid–producing bacteria *Lactobacillus bularicus* and *Streptococcus thermophilus.* Sweeteners, flavorings, and other ingredients may be added. Yogurt contains not less than 3.25 percent milkfat, unless it is a low-fat variety.
- Fresh starters. Cheese cultures are necessary to inoculate the milk with friendly bacteria. They function to aid in the coagulation of the milk and in the flavor development of the cheese. They are divided into two categories: **mesophilic starter**, which thrives around room temperature and cannot survive at higher temperatures, and **thermophilic starter**, which is used when the curd is cooked to as high as 132°F (55°C). Both cultures can be purchased from cheese-making supply companies, or they can be made by the chef. The following recipes can be used to make these starters.

Mesophilic Starter

1. Sanitize all equipment and utensils. Rinse well to remove any trace of detergent or sanitizer.
2. Use 2 cups (473 mL) of fresh cultured buttermilk.
3. Allow the buttermilk to reach room temperature, about 70°F (21°C).
4. Allow the buttermilk to ripen for about 6 to 8 hours. Store-bought buttermilk does not have a high enough concentration of bacteria to serve as a starter culture. The resulting buttermilk will be much thicker and sourer than when you started. It should have the consistency of fresh yogurt; continue to ripen if it does not.
5. Pour culture into a full-sized ice cube tray and freeze.
6. Once frozen, remove the cubes and put into a sealed container or plastic freezer bags. Label the container. The resulting ice cubes are each 1 ounce (30 mL) of mesophilic starter.
7. Add thawed cubes to recipes, as required. The cubes will keep for about 1 month.
8. To make more starter, thaw one cube and add into 2 cups (473 mL) of fresh milk. Mix thoroughly with a fork. Allow to stand at room temperature (70°F; 21°C) for 16 to 24 hours or until the consistency of fresh yogurt.

Thermophilic Starter

1. Sanitize all equipment and utensils. Rinse well to remove any trace of detergent or sanitizer.
2. Use 2 cups (473 mL) of fresh whole milk. Heat it to 185°F (85°C) on the range top or in a microwave. Be careful not to heat too high or the cream will separate.
3. Allow the milk to cool to at least 125°F (52°C).
4. Add one heaping tablespoon (15 mL) of fresh yogurt (either homemade or store bought "live and active culture," plain).
5. Mix the yogurt into the milk thoroughly with a fork or a whisk.
6. Using a double boiler on a low setting, keep the mixture at 110°F (44°C) for 8 to 10 hours until a firm yogurt has set.
7. Pour culture into a full-sized ice cube tray and freeze.
8. Once frozen, remove the cubes and put into a sealed container or plastic freezer bags. Label the container to distinguish it from the mesophilic culture. The resulting ice cubes are each 1 ounce (30 mL) of thermophilic starter.

9. Add thawed cubes to recipes, as required. The cubes will keep for about 1 month.

10. To make more starter, substitute one thawed cube as the fresh yogurt used in Step 4, and repeat steps.

■ Rennet. The active ingredient of rennet is the enzyme chymosin, also known as *rennin*. The usual source of rennet is the stomach of slaughtered newly born calves. However, vegetarian cheeses are made using rennet from either fungal or bacterial sources, such as fig leaves, melon, and safflower. Rennet is available commercially in tablet or liquid form.

■ Vinegar. The acetic acid in vinegar is a mild acid that can be used to coagulate the milk.

■ Lemon juice. The citric acid in lemon juice is a mild acid capable of "denaturing" the proteins in the milk globules.

■ Tartaric acid. This is a natural crystalline compound found in plants, especially those with tart characteristics such as tamarind, lemon, and unripe grapes. The principal acid in wine, tartaric acid is the component that promotes graceful aging and crispness of flavor. One of the by-products of tartaric acid is cream of tartar, which is used in baking and candy making. Tartaric acid is available from wine- and cheese-making supply stores.

■ Salt. Kosher or coarse salt are used for mixing with the curds to add flavor and inhibit growth of undesirable microbes. It is also used for making brines, in which some types of cheeses are soaked to form thick, tough outer rinds.

Basic Steps in Cheese Making

The central process involved in making all natural cheeses is to curdle the milk so that it forms curds and whey. Milk left unrefrigerated for a period will curdle quite naturally as the milk sours and forms an acid curd.

Production methods used today help the curdling process by adding starters and rennet. The starter is a bacterial culture that produces lactic acid, and the rennet is a coagulating enzyme that speeds the separation of solids (curds) and liquids (whey). The two basic categories of starter cultures are mesophilic, used to make cheeses like Gouda and Cheddar, and thermophilic, used to make cheeses like Swiss and Parmigiano-Reggiano. The least sophisticated cheeses are the fresh, unripened varieties such as cottage cheese and Mascarpone. These are made by heating the milk

FIGURE 15–21 Basic cheese-making ingredients

and letting it stand, treating it with a lactic starter to help the acid development and then cutting and draining the whey from the cheese. The cheese can then be salted and eaten fresh. This is the simplest, most basic form of cheese, which may also be used as a foundation for making other cheese products.

In addition to the basic fresh kitchen-made cheeses, the garde manger department can produce other cheeses. It is up to the chef to determine the cost/benefit to the effort involved. The steps include the following: milk preparation, acidification and coagulation, cutting and pressing the curd, whey separation, finishing and forming, ripening, and forming rinds.

MILK PREPARATION

In most American commercial food service operations, chefs prefer to use homogenized milk because it has been pasteurized. The process of **pasteurization** kills the harmful bacteria by heating the milk, but it also eliminates the friendly bacteria. In other parts of the world, chefs have more access to milk that has not yet been pasteurized. And sometimes an adjustment to the milkfat or casein content is made.

The milk is poured into the pot to be used for making the cheese. Milk should be poured down the side of the vessel at first, rather than directly into the bottom of the pot, until there are a few inches of milk in the container. This process prevents any undesired scorching of the milk. Some gentle stirring is required to prevent the cream from separating.

ACIDIFICATION AND COAGULATION

The first step in cheese making is **acidification**. This is the lowering of the pH (increasing acid content) of the milk to make it more acidic. Typically bacteria perform this process. Bacteria feed on the lactose in the milk, producing lactic acid as a by-product. With time, increasing amounts of lactic acid lower the pH of the milk. Acid is essential to the production of good cheese. However, if there is too much acid in the milk, the cheese will be crumbly; if not enough acid is present, the curd will be pasty.

After acidification, coagulation begins. **Coagulation** is the converting of milk into curds (solidified mass) and whey (sour liquid). Several methods are used, either singularly or in combination, causing the curd to form: (1) acidifying the milk by bacterial action, which produces lactic acid, (2) coagulating the milk with rennet or a similar coagulant, (3) direct addition of an organic acid, such as acetic acid, citric acid, lactic acid, or tartaric acid.

As the pH of the milk changes, the structural nature of the casein proteins changes, leading to curd formation. Essentially, the casein proteins in the milk form a curd that entraps fat and water. Although acid alone is capable of causing coagulation, the most common method is enzyme coagulation. The physical properties of enzyme-coagulated milk are better than those of milk coagulated purely with acid. Curds produced by enzyme coagulation achieve lower moisture content without excessive hardening.

CUTTING AND PRESSING THE CURD

After the coagulation sets the curd, the curd is cut. This step is usually accompanied with the process of heating the curd. Cutting the curd allows the whey to escape, while heating it increases the rate at which the curd contracts and squeezes out the whey. The purpose is to make a hard curd, although cheese at this stage is still quite pliable. The main difference between a soft curd and a hard curd is the amount of water remaining in the curd.

Once the curds have sufficiently hardened, salting and shaping begins. In this part of the process, salt is added to the cheese. Salt is added for flavor and to inhibit the growth of undesirable microbes. Large curds are formed as smaller curds are pressed and knit together. This often involves the use of a cheese press.

WHEY SEPARATION

In order to separate the curds from the whey, draining is necessary. This is done by a variety of means. Simple unsalted cheeses (e.g., Queso Blanco) are traditionally made by scooping the uncut curd into a cloth bag and allowing the whey to drain out, sometimes hastened by kneading or squeezing the bag of curd. When the draining is finished, taking anywhere from 4 hours to 1 day, the cheese is ready for consumption.

In commercial cheese making, the whey is simply drained from the bottom of the cheese-making vessel through a gate equipped with a strainer. When making cheese in a garde manger kitchen, the curds and whey are generally poured through a large, fine-meshed China cap or a colander that has been lined with cheesecloth. The curd may be allowed to sit undisturbed for a time after draining, depending upon the particular recipe, to allow the curd to knit and become firm. Often, particularly when making large quantities, the curd is cut into large blocks to allow further whey draining.

FINISHING AND FORMING

Some cheeses are salted before the final forming. In these cases, the curd is cut into small pieces with a cheese harp or long knife, and coarse salt is added with continuous stirring of the curd to ensure even distribution of the salt. A few cheeses are made without this step, and some of them are salted later by immersion in a brine solution. The addition of salt does several things: (1) it aids in the further removal of whey, (2) it acts as a preservative, and (3) it enhances flavor. Some cheeses, such as Feta, are heavily salted and stored in brine. A few cheeses are not salted at all.

The salted curd pieces are then put into forms or molds that have numerous small holes to allow the remaining whey to escape. The cheese is either pressed to expel more whey, or the cheese is settled and allowed to drain without pressure. When allowed to settle, the cheese must be turned top-for-bottom several times over a period of several days, to obtain uniform draining. The cheese is then removed from the mold and is often wrapped or waxed at that time. The hard cheeses will then be ripening, while fresh cheeses are ready for consumption.

RIPENING

The shaped cheese is then allowed to age for various durations. During this time, bacteria will continue to grow in the cheese and change its chemical composition, resulting in flavor and texture changes in the cheese. The variety and quality of cheese being made is determined by the bacteria that is active at this stage in the cheese-making process and by the length of time the cheese is aged. Max McCalman, maître fromager, says about cheese, "Every cheese is in a constant state of change, and a given cheese may pass through as many as six or seven stages of ripeness."

Sometimes an additional microbe is added to a cheese. Blue-veined cheeses are inoculated with a *Penicillium* spore that creates their aroma, flavor, and colorful veining. Such blue- and green-veined cheeses are internally molded and ripen from the inside out.

On the other hand, **surface-ripened cheeses** (e.g., Camembert and Brie) have their outside coatings treated with a different type of *Penicillium* spore, which creates a feathery white mold, referred to as a *blooming* or *flowery rind*. Other surface-ripened cheeses have their surfaces smeared with a bacterial broth. These cheeses are called **washed rind** varieties, because they must be washed regularly during their ripening period to prevent their interiors from drying out. The washings also help promote an even bacterial growth across the surfaces of the cheeses. Because this washing can be done with liquids as diverse as salt water, port, and brandy, it also plays a part in the final flavor of the cheese.

FORMING RINDS

The cover, or **rind** of the cheese, is formed during the ripening process. Many rinds are naturally formed, while some are created artificially. Rinds may be brushed, washed, oiled, treated with a covering of paraffin wax, or simply left untouched. Traditional Cheddars are wrapped around with a cotton cloth, called a *bandage,* and Parmigiano-Reggiano cheese is marked with a special stamp that is inserted between the cheese and the inside of the mold when it is formed.

The rind's basic function is to protect the interior of the cheese and to allow it to ripen harmoniously. Its presence thus affects the final flavor of the cheese. Salting plays an important role in rind formation. Heavily salted cheeses develop a thick, tough outer rind, typified by the Swiss and Parma range of cheeses. Cheddar, another natural rind cheese, is less salted than the Swiss and Parma varieties, and consequently has a thinner and softer rind.

Making Basic Cheese

The simplest cheeses require the least number of steps and the least equipment, and have been made the same basic way for centuries. These cheeses have neither bacterial starters nor rennet. The curd is formed by the direct addition of an acid, such as vinegar and lemon juice, to the milk. Today, these are classified as **acid-set cheeses** and are made both commercially and at home using modern practices and only slight variations of ancient recipes. (See Figure 15–22 for the steps in cheese production.)

One unusual characteristic of acid-set cheeses is that they do not melt the way mozzarella and Colby do, if at all. More commonly, they are eaten as a snack or used in cooking. Quite often they are fried; in fact, Queso Blanco (literally "white cheese" in Spanish) is sometimes called "queso para freir" ("cheese for frying"). Queso Blanco is a pressed cheese, which has a firm texture that is good for slicing. This same cheese is known as *Panir* in India. (See Figure 15–23.)

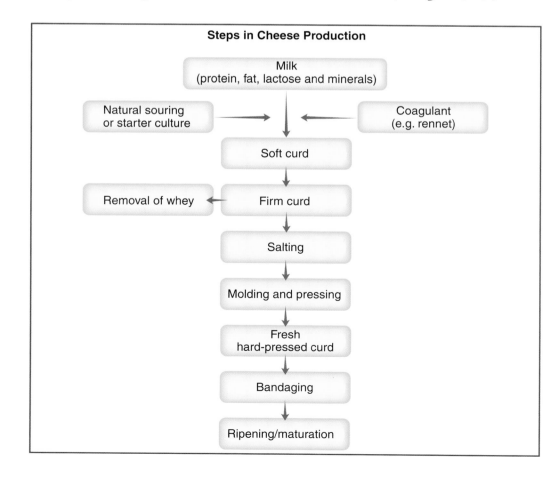

Steps in Cheese Production

Milk
(protein, fat, lactose and minerals)

Natural souring or starter culture → ← Coagulant (e.g. rennet)

Soft curd

Removal of whey ← Firm curd

Salting

Molding and pressing

Fresh hard-pressed curd

Bandaging

Ripening/maturation

FIGURE 15–22 Steps in cheese production

FIGURE 15–23 A selection of kitchen-made cheeses, including cottage cheese, mozzarella, and Queso Blanco

Note: The following recipes may be improved with the use of unhomogenized milk, or skimmed milk with a 10% ratio of heavy cream added.

RECIPE 15–1

YOGURT CHEESE

Recipe Yield: 6 to 8 ounces (177 to 237 mL)

MEASUREMENTS		INGREDIENTS
U.S.	METRIC	
1 quart	946 mL	Yogurt

PREPARATION STEPS:

1. Bring yogurt to room temperature.

2. Pour into a double piece of very fine-weave cheesecloth sitting in a colander and tie the four corners.

3. Allow the yogurt to hang until the yogurt has stopped draining, approximately 12 to 14 hours.

RECIPE 15–2

RICOTTA CHEESE

Recipe Yield: 1½ cups (355 mL)

MEASUREMENTS		INGREDIENTS
U.S.	METRIC	
1 quart	946 mL	Cow's milk
2 quarts	1.9 L	Fresh whey, no more than 2 hours old
¼ cup	59 mL	Apple cider vinegar
2 fluid ounces	59 mL	Heavy cream
2 teaspoons	10 mL	Salt

PREPARATION STEPS:

1. Bring the whey and the milk to 200°F (93.5°C).

2. Remove the pan from the stove and stir in the vinegar.

3. Pour into a very fine muslin cloth–lined colander and allow to drain until no more liquid leaves the bag.

4. Remove the cheese and add salt and herbs to taste, adding a small amount of heavy cream for a richer consistency. It will keep for 1 week.

RECIPE 15-3

QUESO BLANCO/PANIR (ALSO KNOWN AS VINEGAR CHEESE)

Recipe Yield: 6 to 8 ounces (177 to 237 mL)

TIP	This cheese can then be cut and cooked by frying or even by deep-frying without melting or loosing its shape. It can even be used in a stir-fry.

MEASUREMENTS		INGREDIENTS
U.S.	**METRIC**	
1 gallon	3.8 L	Cow's milk
¼ cup	59 mL	Vinegar
2½ ounces	71 g	Kosher salt

PREPARATION STEPS:

1. Bring the cow's milk to 180°F (82.2°C) over a direct heat source and hold it there for 4 minutes (Figure 15–24A).

2. Stir in the vinegar until the whey and curds separate (Figure 15–24B).

3. Pour the whole mixture into a cheesecloth-lined colander and tie the four ends of the cheesecloth together, or pour into a cheese mold (Figure 15–24C).

4. Allow the cheese to sit for at least 3 hours until no whey flows from the bag or mold. If a harder cheese is desired, the curds may be placed in a cheese press to expel more whey while allowing the curds to knit.

5. Unwrap and store in plastic wrap; it will last up to a week.

FIGURE 15–24A Bringing the cow's milk to a temperature of 180°F (82.2°C) over direct heat

FIGURE 15–24B Stirring in the vinegar

FIGURE 15–24C Draining through cheese mold set in colander

CREAMED LEMON GOAT'S CHEESE

Recipe Yield: 6 to 8 ounces (177 to 237 mL)

MEASUREMENTS		INGREDIENTS
U.S.	**METRIC**	
4 cups	946 mL	Goat's milk
2 to 3 each	2 to 3 each	Lemons, squeezed and the juice strained
1 ounce	28 g	Heavy cream, lightly whipped heavy cream
¼ teaspoon	1 mL	Salt

TIP Herbs and spices can be added and then used for sauces, spreads, fillings, and dips. It can also be used in cooking.

PREPARATION STEPS:

1. Heat the goat's milk to 170°F (76.7°C) over indirect heat in a double boiler (Figure 15–25A).

2. Add the juice of two of the lemons and let stand for 15 minutes (Figure 15–25B).

3. If it does not set, add the juice of the other lemon.

4. Pour the curds into cheesecloth and hang for 2 hours to separate the curds and whey (Figure 15–25C).

5. Remove the curd from the bag and add salt and lightly whipped heavy cream (Figure 15–25D).

FIGURE 15–25A Heating the goat's milk to 170°F (76.6°C) over double boiler (indirect heat)

FIGURE 15–25B Adding the juice of two lemons and allow to stand

FIGURE 15–25C Pouring the curds into cheesecloth to hang

FIGURE 15–25D Adding lightly whipped heavy cream and salt to the curd

RECIPE 15–5

MASCARPONE

Recipe Yield: 8 portions

MEASUREMENTS		INGREDIENTS
U.S.	**METRIC**	
1 quart	946 mL	Light cream or single cream
¼ teaspoon	1 mL	Tartaric acid

PREPARATION STEPS:

1. Using indirect heat in a double boiler, heat the light cream to 180°F (82.2°C). Stir in the tartaric acid for several minutes.

2. The cream should thicken to what looks like a custard with curd floating in it; if it does not, add a few drops more of tartaric acid but be very careful not to add too much or else the cheese will become very grainy.

3. Drain the cheese over a cheesecloth-lined colander for about 1 hour.

4. Place the cheesecloth into a bowl and chill overnight.

5. The cheese can be put into a container and stored until needed. It will last for 2 weeks.

RECIPE 15–6

MOZZARELLA CHEESE

Recipe Yield: 8 portions

Note: The cheese may be shaped into various sized strands, balls, sheets, rolls, or stuffed as desired.

MEASUREMENTS		INGREDIENTS
U.S.	**METRIC**	
28 cups	6.6 L	Cow's milk
1½ teaspoons	7 mL	Citric acid, dissolved in ½ cup (118 mL) water
½ teaspoon	2 mL	Liquid rennet, dissolved in ¼ cup (59 mL) of water

PREPARATION STEPS:

1. Warm the milk in a double boiler to 162°F (72.2°C) and then cool to 88°F (31.1°C) (Figure 15–26A).

2. Stir in the citric acid and the rennet and rest for 15 minutes (Figure 15–26B).

3. Bring the milk to 98°F (36.7°C) in the double boiler and hold for a further 15 minutes.

4. Cut the curd with a stainless steel knife into 1-inch (25-mm) cubes (Figure 15–26C).

5. Drain through a cheesecloth-lined colander (Figure 15–26D).

6. Divide the curd into manageable amounts, about 8 ounces (237 mL). Place the curds in a stainless steel bowl, covering with 170°F (76.7°C) water. Working with two wooden spoons, until they become pliable and resemble taffy, work the curd back and forth until it loses all its whey (Figure 15–26E).

7. At this stage, just before it solidifies, you can shape and stuff or roll the cheese.

8. When the cheese starts to show blisters on its surface, you can stop working it and form it into balls and cool in iced water (Figure 15–26F).

9. Alternately divide into smaller pieces and heat in the microwave until it becomes pliable.

FIGURE 15–26A Warming (162°F [72.2°C]) the milk in double boiler

FIGURE 15–26B Stirring in the citric acid and the rennet

FIGURE 15–26C Cutting the curd into cubes with a stainless steel knife

FIGURE 15–26D Draining the curd through the cheesecloth-lined colander

FIGURE 15–26E Forming the mozzarella with two wooden spoons

FIGURE 15–26F Cooling and setting the desired shapes in iced water

Making Basic Creams

CRÈME FRAÎCHE

Recipe Yield: 1 pint

MEASUREMENTS		INGREDIENTS
U.S.	**METRIC**	
2 cups	473 mL	Heavy cream
4 tablespoons	60 mL	Buttermilk
1 tablespoon	15 mL	Sour cream

PREPARATION STEPS:

1. Mix the cream, sour cream, and buttermilk together and let stand for 12 to 16 hours at room temperature (around 75°F [24°C]) until thickened.

2. Remove and chill. Use within 1 week.

RECIPE 15–8

ACIDULATED CREAM

Recipe Yield: 1 pint

MEASUREMENTS		INGREDIENTS
U.S.	**METRIC**	
2 cups	473 mL	Heavy cream
1 each	1 each	Lemon, juiced
1 tablespoon	15 mL	Chives, fine chopped
1 teaspoon	5 mL	Salt

PREPARATION STEPS:

1. Lightly whip the cold cream to soft peaks over ice.

2. Quickly whip in the salt, chive, and lemon juice in one motion. Store until required.

RECIPE 15–9

SCOTTISH CROWDIE

Recipe Yield: 1 quart

TIP   This old Scottish cheese is best served with scones and tea in the afternoon, but it can be used to top sautéed fish and shellfish.

MEASUREMENTS		INGREDIENTS
U.S.	**METRIC**	
1 quart	946 mL	Buttermilk
1 each	1 each	Lemon, juiced
2 ounces	57 g	Heavy cream
1 teaspoon	5 mL	Salt

PREPARATION STEPS:

1. Bring the buttermilk to 170°F (76.7°C) and add the lemon juice.

2. Pour into cheesecloth-lined colander and tie, allowing the cheese to hang until the bag has stopped draining.

3. Refrigerate overnight and remove the cheese. Add the cream and salt.

ADDING FLAVORING AGENTS

The flavors and aromas of cheese are quite extensive. They can challenge the senses and alarm the palette, or lull the tongue to boredom. One of the pleasures of working with cheese is its adaptability. A talented chef can select the proper cheese for its inherent texture and flavor qualities, or use a less flavorful variety as a foundation upon which to build.

Many kitchen-made cheeses, particularly those classified as fresh cheeses, lack the depth of taste that is characteristic of well-ripened cheeses. They have not acquired pungent flavors that are naturally developed in cheese over time, because they are not designed for that purpose. Rather, fresh cheeses are appreciated for their subtlety of flavor and creaminess of texture. And they are well suited to accepting flavoring agents.

The most common method of adding flavoring agents to cheese, is to blend it with the freshly cut curd. Salt is frequently added to the curd after it has been cut, and it is well distributed by folding it into the curd. Its purpose is for flavor and to inhibit the microbial growth of bacteria. It is at this time that other seasonings may be added. These may include chopped chiles and jalapeños, chopped olives, minced onions, diced bacon, port wine, and other varieties of cheese. The curds and flavoring agents are then pressed together for several hours, allowing the curds to knit, the whey to drain, and the resultant cheese to harden.

ADDING COATINGS

Kitchen-made cheeses are frequently left to the chef to determine their final appearance, as well as taste. The chef, who ultimately decides what is best suited for the purpose the cheese is to serve, can affect the final color, flavor, texture, and form. The cheese can be fabricated into balls, logs, or quenelles. It can be coated with fresh herbs, chopped nuts, cracked peppercorns, seeds, or spices, or merely wrapped in a shroud of pickled grape leaves. It is an empty canvass, upon which the culinary artist can express limitless creativity (Figure 15–27).

Caring for Cheese

Not unlike wine, which is considered a living beverage, cheese is a living food. And like other fresh foods, cheeses have varying degrees of shelf life. The safest practice is to order only the

FIGURE 15–27 Coated fresh cheeses. Courtesy of Jacqui Hurst/CORBIS

quantity that will be used within a reasonable amount of time. Generally speaking, the harder and more aged the cheese, the longer it will safely hold. Cheeses that are higher in moisture tend to be fresher and spoil quicker. For example, cottage cheese safely stores for no more than 2 weeks, whereas Parmigiano-Reggiano may keep an additional year after its 2-year ripening process. Soft ripened cheeses reach their peak of quality after only a few weeks and then quickly deteriorate in a few short days.

Storing cheese can be problematic and difficult, so it is best to reduce the amount and length of time that storage occurs. With its dynamic organic attributes, it is vital to the alimentary health and well-being of the consumer that proper management of cheese occurs. The use of approved food-handling gloves by the chef and of sterilized utensils will extend the storage life of the cheese. Unprotected handling of the cheese can easily transfer unhealthy bacteria. In addition to the proper sterilization of the utensils, the chef must be sure that all work surfaces such as cutting boards and display containers such as marble slabs, have been properly cleaned and sanitized.

Cheese needs to be appropriately handled, even when stored for shorter periods of time. The following are proper methods of storage:

- Wrapping cheese correctly is crucially important. The cheese needs to breathe slightly, and it holds well when wrapped in waxed paper, butcher's paper, dampened cheesecloth, or aluminum foil. Young cheeses, as well as goat's milk cheese, are better stored in tightly covered plastic containers.
- Cheese needs to be stored in a cool and humid environment. The use of refrigerators is better than storing at room temperature; however, a cold, damp fruit cellar is preferred.
- Light is damaging to cheese, and too much light will cause it to oxidize rapidly and spoil.

Serving Cheese

Even though cheese should be stored tightly wrapped in a cool, dark, humid location, ripened cheese needs to breathe slightly and come to room temperature before service. Once out of the refrigerator, the cheese should be kept wrapped until it is ready to be served. Too much oxygen will dramatically alter the goodness (desired flavor) of the cheese; slicing the cheese right before service will prevent it from drying out and losing flavor. The best serving temperatures for ripened cheese is between 69° and 75°F (20.5° and 23.8°C). Unripened cheese can be served slightly chilled.

Cheese may be enjoyed as its own entity, but it often tastes better with other foods, or with a progression of other cheeses. Because the creamy nature of cheese can coat the tongue, the palette can be cleansed with wines and sparkling beverages or acidic foods such as apples, grapes, or cornichons.

When building a cheese course or platter, it is wise to consider the adage "Less is more." Because there are limitless choices, some logical order or progression can be helpful. Using three or four cheeses in a cheese course, and four to six cheeses in a platter, is sufficient for the guest's tasting experience. Too many cheeses can lead to confusion, because the cheeses can become indistinguishable. Examples of tasting progressions can include the following:

- Milder to stronger
- Lighter to heavier
- Younger to older
- Simpler to complex
- Local and regional

The intent is to demonstrate to the customer the range and styles of cheese available. Depending on the clientele and market, it is nice to introduce lesser known examples of cheeses that also meet the needs of the cheese course and food service operation.

Professional Profile

BIOGRAPHICAL INFORMATION

Name: Howard Bunce

Place of Birth: Petaluma, California

Recipe Provided: Wild Mushroom and Goat Cheese Fondue

CULINARY EDUCATION AND TRAINING HIGHLIGHTS

After graduating from Petaluma High School in 1984, Howard Bunce went to work for the Marin French Cheese Company, which is located on a 700-acre farm in the pristine West Marin County coastal valley of California. Founded in 1865, the Marin French Cheese Company is still a family-owned and -operated business, and is the oldest continually operating cheese factory in the United States (Figure 15–28). The company started with three styles of cheese, including Schloss (castle), Camembert, and breakfast cheese, which were taken to old San Francisco on paddle wheel boats. Howard worked in many phases of the cheese business, including pasteurizing, adding cultures, hand pressing and forming, and brining. In 1995, Howard was promoted to master cheesemaker and became responsible for all facets of cheese production, including formulation and production. Under his vision and determination to expand the company's line of artisan cheeses, the company now produces more than 30 styles and varieties. The American Cheese Society has honored his work with state and national honors. His Garlic Brie, Peppercorn Brie, and Jalepeño Brie have each won Best Flavored Cheese Awards, while his Triple Cream Brie won the Best Soft Ripened Award. Howard's Triple Cream Brie, Pacific Bleu, and Plain Brie won silver and bronze medals in World Cheese Championship, held in England in 2004. When creating formulations for new cheeses, his philosophy is to try to reach the conceptual goal he holds for a cheese, and then he hopes the public likes it as well. To Howard, cheese is food; it's meant to be artisan.

ADVICE TO A JOURNEYMAN CHEF

Quality and consistency are necessary to make a great cheese; you must have both to be successful. My own company struggled in years past with that, but now we emphasize the importance of having continuous quality. Decades ago, before it was a global economy, the older generation questioned the need for change. But today, we have to be innovative in this world marketplace. If you're caught standing still, you're left behind. You must go into the marketplace to discover your niche. As an artisan company, we can't compete with everybody. We must make something unique to get our foothold in the dairy case. It's a very competitive market.

Howard Bunce

FIGURE 15–28 Marin French cheeses

RECIPE 15-10

HOWARD BUNCE'S WILD MUSHROOM AND GOAT CHEESE FONDUE

Recipe Yield: 8 portions

MEASUREMENTS		INGREDIENTS
U.S.	**METRIC**	
½ cup	118 mL	Olive oil
4 each	4 each	Garlic cloves, minced
¼ cup	59 mL	Fresh herb mixture, including parsley, sage, and thyme
To taste	To taste	Salt
To taste	To taste	Black pepper, fresh ground
1 pound	450 g	Chanterelle mushrooms, sliced
½ pound	225 g	Shiitake mushrooms, sliced
½ pound	225 g	Button mushrooms, sliced
½ cup	118 mL	White wine, dry
2 teaspoons	10 mL	Cornstarch
4 cups	1 L	Goat cheese, hard, grated
1 pound	450 g	Bread, Italian-style, cubed
1 pound	450 g	Eggplant, sliced and grilled, wedged

PREPARATION STEPS:

1. In a large saucepan, heat oil over medium heat. Add garlic, mushrooms, herbs, salt, and pepper. Cook about 8 minutes or until mushrooms are tender. Set aside.

2. In a stainless steel pan, combine wine and cornstarch and bring to a simmer. Reduce heat. Add cheese and stir until smooth.

3. Remove from heat and add mushroom mixture, stirring until well incorporated. Serve in a fondue pot accompanied by cut-up vegetables, such as grilled eggplant slices, or day-old bread cut in cubes.

REVIEW QUESTIONS

1. Classify cheeses into their categories.

2. What types of cheeses would a garde manger chef typically make?

3. What does ripening mean?

4. What are soft ripened cheeses?

5. How are curds made?

6. What is whey?

7. What does the word *artisanal* mean?

8. What is pasteurization?

9. What is acidification?

10. Explain what surface-ripened cheeses are.

ACTIVITIES AND APPLICATIONS

A. **Group Discussion**

   Discuss the use and the practicality of making cheeses in a modern kitchen.

B. **Research Project**

   In small groups, research the cheese course on a menu and report on the resurgence of the course on modern menus.

C. **Group Activity**

   Organize a blind tasting of four cheeses, paying particular attention to sharpness, acidity, creaminess, mouth feel, and texture.

D. **Individual Activity**

   Find a locally made cheese and introduce it to the other members of your class.

PART V

Displayed Arts of the Garde Manger

CHAPTER 16

Sculpting, Carving, and Modeling

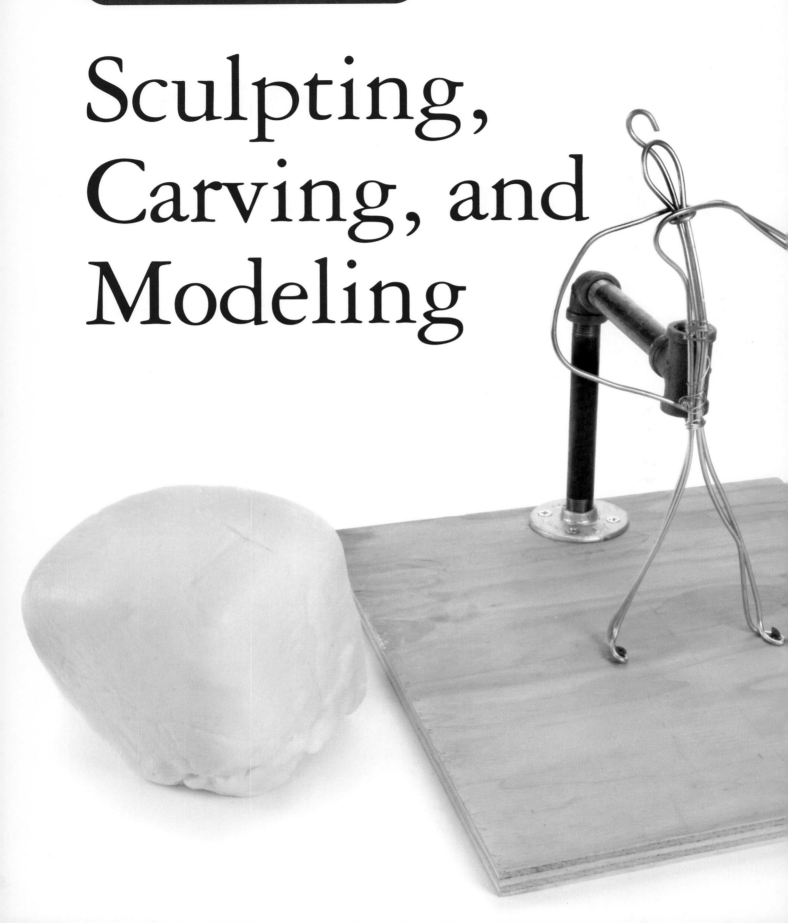

After reading this chapter, you should be able to

- discuss the breadth of edible sculpting that is done by garde manger chefs.

- explain how ice is manufactured.

- identify the tools used by carvers to sculpt ice, and explain how to care for them.

- describe how to hold and temper ice for sculpting.

- understand the various forms and uses for fruit and vegetable carvings.

- discuss various recipes and methods for making bread dough and salt dough sculptures.

- discuss fat carvings, including the origins of butter sculpting, and its limitations.

- understand the origins of tallow for sculpture and the market forms available.

- describe where styrofoam sculpture is commonly used.

The expression of beauty is in direct ratio to the power of conception the artist has acquired.

UNKNOWN

A Timeless Art Form

Sculpture is the art of producing an object, in three dimensions, that is a representation of natural or imagined forms. It includes sculpture in the round, which can be viewed from any direction, as well as incised relief in which the lines are cut into a flat surface. Sculpture, as an art form, has been a means of human expression since prehistoric times, no doubt inspired by nature's erosion of land and rock into works of beauty. The earliest human beings used stone axes and bone knives to chisel works and later refined them with the strength of bronze and sharpness of steel.

Most Stone Age statuettes were made of ivory or soft stone, although some clay human and animal figures have also been found. Small female statues, known as *Venus figurines* have been found mainly in central Europe. The Venus of Willendorf (circa 30,000 to 25,000 BC), discovered near Willendorf, Austria, is a well-known example. Later, in the Near East in the area between the Tigris and Euphrates rivers, the Sumerian and other kingdoms flourished. Materials used for sculpture during this time included basalt, diorite, sandstone, and alabaster. Copper, gold, silver, shells, and a variety of precious stones were used for high-quality sculpture and inlays. Clay was used for pottery and terra cotta sculpture. Stone was generally rare and had to be imported from other locations.

The ancient cultures of Egypt and Mesopotamia produced an enormous number of sculptural masterpieces, frequently monolithic, that had sacramental significance beyond artistic considerations. The sculptors of the ancient Americas developed superb, sophisticated techniques and styles to enhance their works, which were also symbolic in nature. In Asia, sculpture has been a highly developed art form since antiquity.

For more than 450 years, chefs and craftsmen have been producing edible sculpted displays from ice, snow, butter, tallow, bread, salt, fruit, and vegetables. Historically, these works were performed for the pleasure of the noble and the affluent. Today, anyone who dines at a hotel, country club, or cruise ship buffet will likely be treated to the skills of a talented food artisan.

Many excellent books have been written on the techniques of sculpting various and wide-ranging media; this chapter provides an overview of each of the primary media that garde manger chefs use in sculpting. The recipes and lists of materials contained herein are sufficient to begin practicing the art of food sculpture. More serious study should be made with the assistance of a single-subject reference book on this subject.

Ice Sculpting

Ice covers up to 35 million square meters of our world in the form of glaciers. The physical properties of water are unique, in that it is the only natural substance that exists in all three states of temperature normally found on earth: gas (steam), liquid (water), and solid (ice). When left absolutely still, pure water will not freeze until it reaches 15°F (−9.4°C). However, most water has some form of contaminants, such as dust and debris, or is moving or disturbed to some extent that allows it to freeze at higher temperatures, normally around 32°F (0°C). When water freezes, ice crystallizes in a hexagonal system that refracts light in eye-appealing ways, similar to a prism.

Water has an interesting and unique characteristic in that it is denser in its liquid form than in its solid form. When cakes of ice are formed, they expand to approximately 9 percent greater volume than their liquid state. This phenomenon is witnessed easily by filling an ice cube tray to its capacity with water, then placing the tray into the freezer. Upon later inspection of the tray, the frozen cubes will have grown to expand beyond their initial liquid dimensions. This expansion occurs when water molecules move farther away from each other as each crystal is formed. Air is captured between the crystals, giving the ice a lower weight-to-mass ratio. Icebergs and ice cubes float for this reason.

The actual molecular structure of the ice changes as it moves through various temperature zones. At 16°F (−8.9°C), the ice has the same density and volume as it has at 32°F (0°C). It expands with heat from 16° to 24°F (−8.9° to −4.4°C), when it then begins to contract at 24°F (−8.9°C). At 32°F (0°C), it begins melting and continues to contract until it almost reaches 40°F (4.4°C), when it again begins to expand. This information is important to the sculptor because the block will react in changing ways when being cut at different temperatures. The relative texture of the block will change from soft to hard, and back to soft again with only a few degrees of temperature fluctuation.

THE STORY OF ICE

Before the commercial manufacture of ice, **natural ice** was harvested from lakes, rivers and ponds. It is not known when man first began cooling and preserving his food with ice, or when he initially used ice for lodging or decoration. The *Book of Songs,* written in China in 600 BC, mentions the use of harvested ice that was stored in ravines covered with straw. Historians mention Roman references to ice and snow being used to cool beverages and conserve foods as far back as 52 BC. The Inuit of the Arctic Bay have made igloos for centuries; while a grand Palladian palace of ice was constructed on the banks of the Neva River in the mid 1700s.

Around 1910, the first commercially **manufactured ice**, known as *plate ice,* was made in sheets weighing close to 14,000 pounds (6,350 kg). The ice was then cut in smaller blocks, called *cakes,* depending on the need of the end user. Today, there are three forms of commercially made ice used for sculpture and display: molded, brine tank, and circulating tank.

- Molded sculptures are formed either in plastic or rubber molds that are filled with water and frozen, or directly in an ice mold machine filled with glycol.
- Placing pure chilled water in galvanized containers that are partially submerged in a brine tank filled with sodium chloride makes brine tank ice, weighing from 10 to 400 pounds (4.5 to 187 kg). During the freezing process, small contaminate particles and dust are forced to the center of the block, producing an undesirable soft feather of opaque slush.
- The circulating tank method creates clear blocks, weighing approximately 300 pounds (136 kg) by suspending a small water pump near the top of the tank that gently moves the water around the tank. As the tap water freezes upward from the bottom of the tank, oxygen bubbles and impurities are forced to the top of the water tank, thereby creating a clear and

denser block of ice. The water pump is removed near the end of the freezing process to prevent its becoming imbedded in the ice block.

This difference in the quality and clarity of the ice can be observed by comparing ice that is harvested from a lake to that which has been harvested from a river. Lake ice is cloudy and softer, because air is captured between the ice crystals as the ice forms. River ice is more clear and dense because the moving water removes air during the freezing process.

Selecting the Ice Block

Sculptors generally prefer the use of **clear ice**, because the dense ice cuts well and the final product captures and refracts the light more favorably (Figure 16–1). Ice blocks may be purchased from local icehouses, if available, or from large national distributors. Or, the sculptors can make their own ice blocks using machines such as the Clinebell CB 300X2 Block Maker (Figure 16–2). These machines have a two-block–making capacity. In addition, there are portable hoist and lift carts and other tools used to harvest and finish the ice blocks.

Making and Harvesting the Ice

To create ice for sculpting, an ice block maker is lined with a plastic liner to securely hold water. After filling the liner with water, a small pump is placed into the liquid near the top (Figure 16–3). The pump circulates the water to keep impurities from freezing into the block of ice. After 2 to 3 days (depending on the ambient temperature around the machine), the ice block has

FIGURE 16–1 A clear ice block next to an opaque block

FIGURE 16–2 Ice block maker

FIGURE 16–3 The water pump inside the ice block maker

been formed in the machine, and excess water and impurities are removed from the top of the block before harvesting by use of a common wet and dry vacuum. A stainless steel lifter plate is frozen approximately $\frac{1}{4}$ inch (6.9 mm) into each end of the block to help lift the finished block from the chamber. It is then harvested using a portable block and tackle, such as the Clinebell Hoist and Lift Cart. The tackle chains are attached to the lifter plates, and one person can easily remove the ice.

Trimming and Packaging the Ice

Once the ice has been lifted from the block maker, it is lowered onto a tilt cart in a horizontal position (Figures 16–4A and B). The clear plastic liner will still be surrounding the ice block on all sides except the topside where the water was filled. The block will frequently have higher outside edges on the topside because the sides freeze faster than the center of the block. If the block is to be slid into a storage box, then the block will need to be trimmed to a standard dimension of 40 × 20 × 10 inches (102 × 51 × 25.5 cm). This can be done using a manual handsaw, chainsaw, or a commercial ice shaver designed for this purpose while the block rests on the tilt cart. The cart is used to roll the block over to the walk-in freezer, then is tilted to stand the block upright (Figure 16–4C). If the block is to be stored in the freezer under cover of plastic, such as a garbage bag, then it does not require trimming. The excess ice can be trimmed during the sculpting process.

Holding the Ice

Care must be taken when storing the ice to prevent **sublimation** caused by the refrigeration process. In sublimation, ice changes directly to water vapor without melting. Refrigeration involves removing warm circulating air and removing moisture along with it. Therefore, blocks and sculptures will literally dehydrate if not properly covered and protected from circulating air.

Plastic bags under cardboard boxes work well to cover and insulate new ice blocks and prevent sublimation. Also, it is better to store blocks on plastic, rubber, polyethylene foam sheets, or other nonporous material to make it easier to slide the blocks in and out of the freezer. Cardboard is porous and absorbent; therefore, it is not the best material on which to store blocks of ice. Ice, when covered and insulated properly and kept below a constant state of freezing, will store for several months before decaying (Figure 16–5).

It is best to have a dedicated freezer for holding the block ice and sculptures. However, dividing one freezer can accomplish separation with the use of chain fencing and having separate doors to access both ends of the freezer. When using one freezer, without a dividing fence to hold

FIGURE 16–4A Ice being harvested

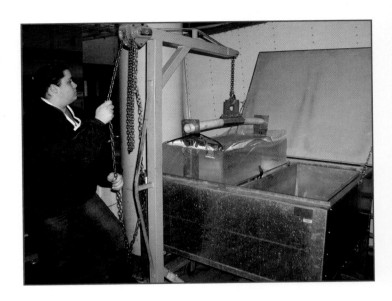

FIGURE 16–4B Setting the ice onto a cart

FIGURE 16–4C Putting the ice into the freezer

ice along with frozen foodstuffs, it is best to keep the ice out of the way of foot traffic. Ice can be damaged by accident when careless employees are in a hurry.

EQUIPMENT

Ice carvers borrowed their technology from other carving disciplines: chisels from the stone carver and chainsaws from the woodcutter. And for many years these tools satisfied the needs of

FIGURE 16-5 The covered ice in a dedicated ice freezer

the ice carver. However, it was difficult for carvers to achieve all of the soft design elements and intricate details using these basic tools.

Through the 1970s, Japanese handsaws, chippers, and chisels were applied, although less expensive wood chisels from the local hardware store were often substituted. At the end of the 1970s and into the 1980s, gas and electric chainsaws were frequently used, but their application was limited. Later improvements to the lighter electric saws made them the saws of choice because of their reduced vibration and fume-free operation.

The 1980s became the decade of the power tool and modernist ice sculpting was born. **Modernist sculptors** expanded their tool repertoire to include die grinders, rotary tools, power drills, clothes irons, hot water bags, and other devices to soften the appearance of their sculptures. Their application resulted in sculptures being created without the use of the long-standing ice chisel. **Traditionalist sculptors** still favor the pure form of carving using only handsaws, chippers, and chisels. However, few can argue the efficiency and artistic results gained by the use of power tools.

Additionally, the cost of the power tools is actually less than the cost of a quality set of ice chisels and saws. For example, a rotary tool bit is much easier to purchase and to replace than a V-chisel that costs more than 50 times as much. This economic reality is also a contributing factor to the widespread acceptance and use of power tools.

Twenty-First Century Ice Sculpting

At present, there are two, sometimes opposing, philosophies over the use of power tools in ice sculpting. A number of carvers, considered traditionalists, have chosen to avoid the use of hand-held power tools. The traditionalists even shun those power table saws that are not computerized. However, of those, a few will still use chainsaws. But the strictly classical or purist carvers believe the art of sculpting is compromised when any machines are used. Some further believe the use of CAD (computer-aided design) equipment, such as CNC (computerized numeric controlled) routers, corrupts the art by its automated production capabilities. They believe CAD equipment requires less skill to produce the ice works. In support of that position, there are even a few ice competitions that disallow the use of any and all power tools.

However, a growing number of ice artisans—particularly those who sculpt full-time for their livelihood—have chosen to embrace the use of computerization and automation with ice sculpting. Their ability to profitably create consistently high-quality products for their clientele has swayed them to blend the modern with the traditional.

Traditional Tools Used for Ice Carving

Before the current practice of using power tools, the sculptor applied traditional tools to carve his ice pieces. Traditionalists, who seek to preserve the art of ice carving as it was first practiced, use these tools to create their displays. These types of tools have been successfully used for many years, and their proper application can yield beautiful ice figures.

Most culinary arts programs use these traditional tools when teaching beginning ice sculpting. They are useful in allowing the student to become familiar with sculpting principles. As the student progresses, power tools are often introduced and the results of the two methods are compared. The following ice tools are basic to the traditional sculptor's toolbox (Figure 16–6):

- Assorted handsaws (1). Long before ice carvers used chain saws to cut and split large blocks of ice, they used the large-toothed handsaws for this work. Handsaws vary in length and tooth dimension, with the larger tooth saws being used for rough cuts and smaller tooth saws for finer cuts.

- Assorted chisels (2). Originally, ice carvers used carpenter and stonecutter chisels for shaping and shaving. However, the wood chisels had at least two faults: they usually had shorter, more impractical handles and their bevel made them virtually unsuitable. Today, an assortment of high-quality ice chisels exists, each with various blade shapes, including the narrow flat, wide flat, round (gouge), and V- (wedge) chisels.

FIGURE 16–6 Modern and traditional ice carving tools

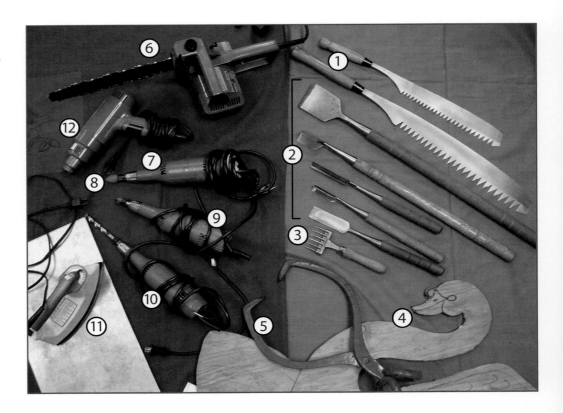

- Single prong chipper/pick. A very common old-fashioned household implement, the ice pick is used for chipping smaller sections, for cutting and punching small holes, and for scoring the ice. It is very useful for tracing and transferring template designs onto a block. The single pick must be kept sharp in order for it to be effective.

- Multiprong chipper (3). Usually designed with five or six prongs, the chipper is considered the most fundamental of the traditional ice tools. It is primarily used for quickly shaving off large sections of ice, for scratching texture onto the sculpture's surface, and for transferring patterns. The chipper should always be kept sharp; a dull chipper tends to spread out the force of the impact, which increases the risk of fracturing the block rather than cutting it as desired.

- Templates (4). Permanent templates, such as those made from wood, initially cost more than disposable templates, but repeated use of these wooden templates reduces their expense and design time. Paper templates are also commonly used.

- Ice tongs (5). As much a tool for safe handling as for lifting, tongs are practical tools for securely grabbing the large and heavy blocks. There are two styles: the *Cincinnati* or *compression*, which requires using only one hand once the tips are set in the ice, and the *Boston*, which requires the sculptor to use both hands.

    However, tool selection is very personal. The individual sculptor will choose whichever tools he or she is most comfortable using.

Modern Tools Used for Ice Sculpting

- Chainsaw (6). Although both gas and electric saws exist, sculptors primarily use electric saws because of their lighter weight and ability to carve indoors without giving off exhaust fumes. Electric saws are also favored because noncombustion engines emit less vibration and are easier to hold steady. Originally used only for making larger cuts and to remove big portions of negative space, the chainsaw also works for making straight and curved lines, rounding columns, gouging the ice, and sanding. Removing the standard bar and replacing it with a specialty bar (which is manufactured without a guard at the end of the guide bar) is necessary for plunging into or scoring the ice.

- Die grinder with converted spindle (7). A converted spindle has a threaded shaft for screwing on specialized or customized bits (i.e., rubberizers). When converting to the threaded shaft, the sculptor should use a licensed electrician or tool repairman to ensure that the integrity of the die grinder is not compromised.

- Coarse rubberizer (8). A 16-grit cone rasp is used to round, sand, and reach difficult areas of the sculpture. The rubberizer is ideal for shaping areas without removing too much ice by mistake. This action will result in a machined appearance, which may or may not be desirable. The machined marks will quickly smooth when displaying the sculpture in room temperature conditions.

- Die grinder with normal shaft (9). The die grinder has many attachments to vary the cut. This is primarily used instead of the classic V-chisel to make clean cuts, lettering, lines, and other detailing. The collet accepts up to $\frac{1}{4}$-inch (.635-cm) shanks.

- Rotary tool. Rotary tools are used with a variety of bit styles for finer detail and delicate finish work, because the collet only accepts bits with $\frac{1}{8}$-inch (.3175-cm) shanks.

- $\frac{1}{4}$-Inch straight router bit. This bit is used for lettering, drawing lines, detailing, and shaping smaller objects.

- V-Shaped router bit. This bit is used for making V-lines or grooves. The width of the line or groove is determined by how deep the bit penetrates the ice.

- Hand saw. Hand saws vary in length and tooth dimension, with the larger tooth saws being used for rough cuts and smaller tooth saws for finer cuts. The ice saw is used for splitting blocks, fusing, and marking the base. Laying the saw sideways on a piece of wood or cinder block can be used to guide the saw evenly at the base of the sculpture.

- ■ Cordless drill. The cordless drill is battery operated and requires no electrical power cord. These cordless tools tend to lose their charges faster in colder temperatures.

- ■ Aluminum. Flat clothing irons heat flat, $\frac{1}{2}$-inch (6.4-mm) or thicker, aluminum sheets or strips. They are used to evenly melt and smooth surfaces of ice and are an excellent means of fusing two flat surfaces of ice together.

- ■ Drill with $\frac{5}{8}$-inch spade bit (10). This drill is used to drill holes to drain bowls and vodka sockets. Various sized bits and hole cutters may be used to drill into the ice.

- ■ Ice tongs. Long-standing in use and design, the ice tongs have pincers to seize ice by pressure. They are used to lift, hold, or move large blocks. The two main styles are known as *Cincinnati* and *Boston.*

- ■ Clothing iron (11). A cheap, versatile heating element, the iron is used for smoothing and fusing. The iron is used to heat metal plates for flattening and fusing ice surfaces. However, steam holes in the iron do affect the even application of the heat and often leave uneven ice when applied directly to the ice.

- ■ Extension cord with rubber casing. Long enough to give the sculptor room to move about without obstruction, the cords should be rubber coated, brightly colored, insulated, heavy duty, and grounded to an outlet with a GFCI circuit breaker. Several would be desirable when using multiple power tools.

- ■ Heat gun (12). The heat gun is used in place of the propane torch to round, clean, and gloss the ice.

Caring for the Sculptor's Tools

In this world of high- and low-tech equipment, the well-worn expression "you get what you pay for" generally applies, but not always. Certainly there are various tiers of excellence among most types of products, and price can often be considered a reliable barometer to quality, but our "disposable society" mentality has even influenced the quality of machinery and their parts.

Professional sculptors often classify their tools into two distinctive categories: permanent and disposable. Derek Maxfield of Ice Sculptures Ltd. explains that some items wear out rapidly, no matter the quality; while other items hold an extended use based on their superior construction. Because Ice Sculptures Ltd. uses a lot of power tools in daily production, they also use a lot of bits and rubberizers to score and etch their sculptures. They use a computerized numeric controlled router and several electric chainsaws to fashion their works, and then a clothes iron to smooth their seams for fusing.

They classify most of their power tools as permanent items, including things such as the CNC router, electric chainsaws, and rotary tools. For these objects, they are willing to spend top dollar for extended and quality use. Most of the bits and saw chains are considered disposable, and they seek to save money on their purchase. However, rules are made to be broken, and there are a few exceptions to this rule. Some power tools, like the everyday clothes iron, are considered disposable, because there is no discernible difference in how well they function for smoothing ice and heating aluminum plates. And, there are pricey carbide-tipped chainsaw blades that get 10 times the use of standard chains.

With hand chisels, there are many levels of quality that are easily identified by price. Chisels are considered permanent tools, and the range of quality is vast. Novice ice sculptors can practice their craft using wood chisels, which are readily available and relatively inexpensive. A decent set of four wood chisels costs less than $50. As the quality of the metal (and Rockwell Hardness Scale Rating) and wood handles improves, so does the price increase. Most sculptors prefer to use Italian- or Japanese-made ice chisels, and a modest set of three starts near $200. The best quality chisels are very expensive. For example, one Misono 30-inch straight chisel sells for more than $400.

TOOL MAINTENANCE

Proper maintenance of tools is vital to the success of the sculptor. Tools must be sharp, rust-free, and in working order. Maintenance is not only essential to the longevity of expensive tools, but can also directly affect the quality of the finished sculpture. The Japanese, among many nationalities who sculpt ice, have long believed that as much time should be spent caring for their tools as using them.

Proper care begins with the sculptor's **mise en place** (everything in its place). Orderly arrangement of tools, before beginning to sculpt, allows for more organized work and less chance for accidents or misuse of the tools. Tools should be placed on cloth towels or foam to prevent them from coming in contact with the table's surface and to prevent them from falling off the table.

Bits and blades of power tools should be inspected for sharpness before use and after the sculpture has been completed. This allows time for repair or replacement. However, most of the tips are relatively inexpensive and are considered disposable.

CLEANING PROCEDURES FOR BITS, BLADES, AND CHISELS

The following procedure for cleaning bits, blades, and chisels should be carried out immediately after each sculpting session:

1. If rust is present, rub rust remover on the surface, then warm the bit, blade, or chisel with boiling water. (Cold metal will collect condensation quickly.)
2. Wipe the item completely dry with a soft cloth.
3. Coat the metal item with a light film of oil.
4. Wrap each piece individually, or place the pieces in separate slots in a lined storage tray so they do not damage each other.

SAW AND POWER TOOLS

Saws and other power tools need to be oiled regularly, and their power cords need to be inspected frequently for cuts, cracks, and exposed wires. A damaged power cord can cause arcing of electricity and shocks to the sculptor.

Attention to the condition of the chain is also vital. A standard chain can be used for approximately 10 sculptures before needing sharpening. Carbide-tipped chains can be used 90 to 100 times, but are quite expensive. It is highly recommended that a skilled professional sharpen the chain blades, because poor sharpening can permanently damage a chain. It can also cause a safety hazard and negatively affect the quality of the sculpture.

TEMPERING ICE FOR SCULPTING

In order for the sculptor to safely remove pieces of ice during the sculpting process without fear of cracking or splitting the block, the ice must be consistent with the temperature of the air around it. Ice can be sculpted inside a freezer or cooler, or outdoors in an area such as a loading dock. What is important is to prevent **thermo-shock** from happening. Failure to balance internal temperatures of the ice with the surrounding temperatures can cause thermo-shock, which results in severely damaged or weakened ice. Often the ice is **tempered**, or slowly warmed, if it is to be carved outside of the freezer. Using a warm chainsaw on a frozen block can also cause cracking due to thermo-shock.

The two most common methods of tempering ice take place either on the loading dock or in the walk-in cooler. Most chefs temper ice by transferring a block from the freezer and placing it in an acceptable area for sculpting, often the loading dock. Ice must be protected as much as possible from the direct rays of the sun, because the ultraviolet rays of the sun cause a "**greenhouse effect**" with the ice, thereby allowing it to melt from the inside out. Covering the block with a

FIGURE 16–7 Ice block tempering in a cooler

dark plastic garbage bag while it is tempering is an effective way to block the sun's rays and reduce sublimation.

When using the cooler method of tempering ice, the block is stored in a walk-in refrigerator, often overnight, and allowed to slowly warm to just above the freezing temperature. It is also best to cover the ice with a bag to prevent sublimation in the cooler (Figure 16–7). This process takes hours longer to temper the ice, but it is easier on the ice and less risky than abruptly placing frozen ice in the warmth of the outdoors.

Using a sheet of paper, the sculptor can test the ice to determine whether it has been properly tempered. If the paper can be easily removed or slid around on the raw ice, the ice has warmed sufficiently (Figure 16–8). If the paper sticks and tears when pulled, then the ice is still too frozen.

THE SCULPTOR'S TEMPLATE

Design templates are used for assisting the sculptor in defining the proportion and shape of a sculpture within a limited dimension or space. This method of design is commonly used with all media that are sculpted, including stone, wood, soap, and ice, in which the medium is carved rather than modeled.

Design templates are an accurate way of transferring a design onto a block (Figure 16–9). They are almost always used in ice competitions and when accuracy and uniformity is crucial. Templates can be made of several different materials such as wood, plastic, cardboard, or paper, depending on what one will be using them for and if they are to be reused. However, the major-

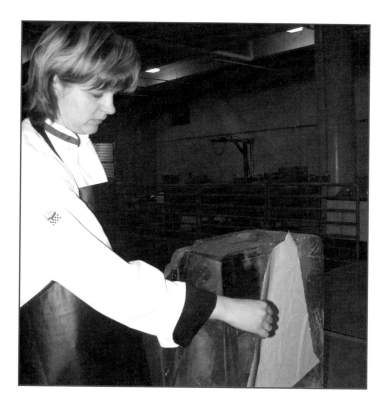

FIGURE 16–8 Paper being pulled from a block of ice

FIGURE 16–9 A sample of design templates

ity of sculptors use paper templates, while wooden or reusable templates are used for routinely made designs.

When making a template, the sculptor needs to know the dimensions of the ice block. Additionally, the sculptor must have a sketch of the design, such as a picture or drawing. Drawing out the template can be done freehand or by using a grid pattern like graph paper, although the most efficient way is to use a projector to capture a design on template paper (Figure 16–10). The projector displays the design onto tracing paper that has been taped to a wall. The paper is the same size as the block of ice being sculpted. The sculptor then traces the design onto the paper, creating the template.

After the artist has created the paper template, it needs to be applied to the ice so the sculptor can scribe the design into the ice. The transfer of designs from the template to the ice, as demonstrated in Figure 16–11, can only be achieved when the ice block temperature remains below freezing.

FIGURE 16–10 Template being transferred to butcher paper from an overhead projector

FIGURE 16–11 Using the paper template to trace the design onto the ice

The ideal conditions for attaching a paper template, etching the pattern, and then sculpting the piece would be under freezing temperatures from start to finish. This could be accomplished either in the great outdoors of the North Country on a cold winter's day, or within the confines of a spacious walk-in freezer. Thermo-shock only occurs when the temperature outside of the block varies from the inner block.

However, we realize it is likely that many people will not be sculpting in such subfreezing temperatures. For them, a few adaptations can be made. The templates can be applied to the raw block before it leaves the freezer, or immediately upon removing it from the freezer. Either way, the object is to apply the template quickly to the frozen block.

To set the paper template, the artist squirts a few sprays of cold water onto the ice. The paper template is then immediately applied and slid into position. Care needs to be taken to square the edges of the paper evenly to ensure a proper alignment.

Templates tend to tear less when using a chainsaw than when cutting with a die grinder. Once the template lines are etched into the block, approximately $\frac{1}{4}$-inch (6.9-mm) deep, the template can be removed and the block allowed to temper (when sculpting above the freezing temperature) to the ideal stage for sculpting.

From Passive to Interactive Sculptures

With a few exceptions as noted earlier, most sculptures have previously been used as ornamental centerpieces, added to an event for their aesthetic value (Figure 16–12). As Tim Ryan, President of The Culinary Institute of America wrote in the Foreword of *Ice Sculpting the Modern Way* (2004). "Properly done, ice carvings convey an aura of elegance, attention to detail, artistry, and mastery of the craft." The use of ice sculptures raises the perceived quality level of an event; they are both elegant and magical to behold.

Many sculptures today have transcended their formerly passive role as decorative centerpieces or frozen vases for flower arrangements. Now, sculptures are commonly part of the event's entertainment. Rather than being enjoyed from a distance, their modern use has been greatly expanded to involve the guests. Restaurant, banquet, and hotel customers have the opportunity to touch and interact with sculptures more than ever before. This contact by the guest has been received with great enthusiasm, elevating their dining and entertainment experience to higher levels. The following are some of the more interesting uses:

■ Punch bowls. Often lined with plastic to prevent deterioration by the liquid being held, punch bowls also work well for holding fruit and seafood. While being formed, they can be embedded with flowers and other inserts to support the theme of the event.

■ Clam shells/shrimp boats. Probably the most traditionally used ice serving vessels for holding foods, after the punch bowl, these frozen containers are popular with the guests and func-

FIGURE 16–12 Setting an ice sculpture on a buffet

tional for holding chilled shrimp, crab, and other assorted seafood (Figure 16–13). They work well in supporting tropical or nautical theme events.

- Two-dimensional (level) platters. Shaped like their silver platter counterparts, these trays are effective for maintaining proper serving temperatures of food left on display for several hours.
- Three-dimensional (tiered) platters. Adding height to a display always increases its dramatic appeal. Multitiered platters allow the chef flexibility in displaying similar or complementary categories of foods.
- Tiered wedding cakes. One of the newer uses of ice, a wedding cake from ice may be used as a replacement display in lieu of a traditional cake, as a container for ice cream to be served with the cake, or as a pedestal on which wedding cake may be displayed.
- Sorbet dishes. Small and delicate, these serving dishes are an impressive way to present an intermezzo course.
- Luges and shooter blocks. Once only used at college bars and Superbowl parties; these clever liquid dispensers have gone "highbrow." Everything from tequila to champagne can be poured from these sculptures. Using a small funnel that is attached to a clear plastic tube running through the sculpture helps prevent deterioration by the liquid being dispensed. Use of these festive sculptures has increased tremendously in the last year.
- Bars and fireplaces. These multiblock sculptures require extra care and skill in assembling, but are unique "attention grabbers" for guests. Their designs vary and may include uses as sushi bars, beverage stations, or raw bars for caviar, oysters, and other seafood.
- Message boards. Inserting clear plastic transparencies, with messages or company logos, between two sheets of ice is an effective way of promoting an activity. For example, fundraisers might present "here are the cold, hard facts..." in ice.
- Computer screens. Ice sculptures shaped like large computer screens can be used as projection screens to display static or changing information.
- Pyrotechnic displays. Cold-flame pyrotechnic explosions and heatless showers of sparks may be remotely activated to discharge from ice sculptures by using radio-controlled sparkler

FIGURE 16–13 Ice sculpture used as a service vessel. Courtesy of imagebroker/Alamy

charges, pagers, and transmitters. These somewhat pricey effects are terrific for punctuating special announcements and awards.

■ Faux-flame displays. For less money and a longer-lasting effect, fake flames made from flame-colored silk, blown about by small fans, can be used to accent fireplaces. From a distance, and in limited-light situations, these faux-flames can be very effective.

■ Lighted table centerpieces/serving pieces. Using display trays with battery-powered bottom lights, chefs can set banquet tables with individual small sculptures rather than use flower arrangements. The use of glow-sticks or glow-discs inserted into sculptures can also be an effective way of lighting smaller sculptures that function as individual serving dishes, without the need for electronics.

■ Tourist hotels. Today, tourists from all over the world can opt to stay in several commercially operated facilities constructed from ice, including the Hotel Igloo Village in Kangerlussuaq, Greenland, and the Icehotel in Jukkasjarvi, Sweden. These facilities feature rooms and furniture constructed from ice in which guests stay overnight.

Most of the uses for ice sculptures, as described herein, are additional ways a garde manger chef may "up sell" his or her services. Money budgeted by the client for one purpose, such as table decorations or china rental, may be redirected toward ice products provided by the chef. To increase the sales and services to customers, chefs should consider the many applications that ice sculptures can meet. Whether artfully presenting food, providing an elegant display for an event, emphasizing a message, or supporting a theme, the form taken by modern ice sculptures can match the function required.

Vegetable and Fruit Carvings

Among many peoples, vegetable and fruit carving has been a specialty of the Asian cultures for centuries. The Royal Court of Siam, now known as Thailand, was a huge community that included the residences of the King and the Royal Family, and also government offices, judicial offices, the military, and the Royal Chapel. The Royal Court has always been the center of Thai culture and religion, and it was there that Thai cooking developed into its own unique cuisine.

Thai cuisine was designed in harmony with the Buddhist belief of balance, and it demands that well-rounded meals stimulate and excite all five taste senses. Because of the variety of taste, texture, and temperature that the Thai people lent to their dishes, vegetables were a natural accompaniment to all meals. But serving a meal that appealed to taste alone was not enough; food had to be inviting to the eye of royalty as well.

Because they were served with all meals, and because of their wide range of colors and shapes, vegetables were a natural choice for garnishes. By perfecting the art of carving, Thai cooks could create elaborately garnished dishes with vegetables that had been transformed into something else: a carrot could become a fish, a radish could become a rose.

During the last 100 years, the Royal Court has become less distant from its citizens, and the royal art of fruit and vegetable carving has filtered into the general public. King Chulalongkorn (1868–1910), who was responsible for modernizing Thailand by providing schools, hospitals, roads, railways, and an updated military force, often gave parties and fancy dress balls for which he did the cooking himself. Royalty has always set the fashion in Thailand, and the public imitated the Royal Court's style of cooking and garnishing with vegetable sculptures.

In Thailand today, many study vegetable carving at art schools, and even women from upper class families who perform no other duties are expected to know how to make garnishes. In fact, there are some Thai princesses who teach the art of vegetable sculpture in the schools. The artistic quality of these carvings, of course, depends on individual talent. In Thailand no meal is considered complete without some form of garnish.

FIGURE 16–14 Pumpkins ready for carving. Courtesy of Cindy Haggerty/ Shutterstock

Vegetables and fruits are much more available to the apprentice carver than ice or tallow. Although the tools for sculpting vegetables and fruits are different from sculpting ice, the process of analyzing the dimension of the design is similar. Like ice sculpting, vegetable and fruit sculpting may be both additive and subtractive in nature. The chef can choose to work within the natural framework of the produce or add appendages with well-placed skewers and toothpicks.

The most common uses for fruit and vegetable carvings are as holiday decorations, vessels, garnishes, and centerpieces (Figure 16–14).

- Decorations. All countries have their own traditions, and America traditionally celebrates Halloween (All Hallow's Eve) by decorating their porches and homes with jack-o-lanterns made from carved pumpkins. Larger melons are often carved as birds and animals.
- Vessels. Hollowed out melons, gourds, and citrus fruits are functional as containers for dips, cold sauces, and dressings. They are commonly used for salad bowls of fruit and vegetables. Melons and gourds also work well as a socle in which skewered meats can be held for display and easy access. Fruits and vegetables can be etched and shaped to complement the theme of an event (Figure 16–15).

FIGURE 16–15 An attractively carved watermelon. Courtesy of Patsy A. Jacks/ Shutterstock

FIGURE 16–16 Butternut squash carved as a centerpiece

- Garnishes. Many cultures, like the Thai, use vegetable carvings to create interest on their food plates and platters. Horse carrots and cucumbers cut into shapes such as crawfish, flowers, birds, and dragons, add color and craft to a plate.
- Centerpieces. Floral arrangements made from radishes, cucumbers, turnips, rutabagas, leeks, scallions, tomatoes, potatoes, carrots, and other vegetables make stunning centerpieces for banquets (Figure 16–16). Their practicality for use is less than desirable, however, because the vegetables lose their luster and wilt after a few hours on display. The garde manger chef must selectively use these centerpieces when length of time on display will not be a problem.

Hand Tools Used for Vegetable Carving

Vegetable and fruit sculptures can be carved with a few common kitchen utensils (Figures 16–17 and 16–18). The following hand tools are basic to the vegetable carver's toolbox:

- Razor blade. Single-sided razorblades or razor knives work well to make fine cuts into the vegetables.
- Paring knife. A sharp, small blade is required for much of the work in cutting, trimming, and skinning vegetables and fruit.
- Flexible slicer. Often a flexible, long bladed knife is needed to cut melons in half or to trim the skin from melons.
- Parisiene scoops. These melon ball utensils come in different sizes; they can be shaped as oval, scalloped, and traditional round.
- Channel knife. Channel knives are helpful in creating even-sized lines and grooves in the skin of melons and gourds.
- Daisy cutters. There are a number of daisy shaped cookie cutters on the market that are useful in creating flower patterns from sliced turnips and rutabagas.

FIGURE 16–17 Vegetable and fruit carving tools

FIGURE 16–18 Garde manger chef carving vegetables

■ Round and scalloped circle cutters. Round cutters of various diameters are useful, like daisy cutters, to fashion leaves and flowers.

■ Ruler. At times, the sculptor may need to measure circumferences or distances.

■ Wooden skewers/toothpicks. Although inedible, these are the accepted means of creating stems for flowers, as long as they are hidden within scallion or leek greens.

■ Food coloring. Careful use of food coloring can be helpful in dyeing vegetables to create different colored flower petals.

Dough Modeling

Most chefs do not as commonly practice the art of dough modeling. Although a few artisans are accomplished with this edible art, it is something that is not commonly seen or taught in kitchens or culinary schools. Dough modeling is seen on a global level and is often displayed in international culinary competitions.

Most dough models are constructed by applying layers of the edible dough upon an **armature**. This process provides stability and dimension to the sculpture. The armature is commonly made from nonedible materials, so it is not to be seen; the dough material must completely hide the armature underneath its layers (Figure 16–19).

SALT DOUGH MODELING

Salt dough modeling appeals to chefs, and customers alike, because of its similarity in appearance to stone sculpture. However, salt dough modeling is an interesting blend of **additive sculpture** and **subtractive sculpture**, unlike stone sculpting, which is completely subtractive. Plus, with stone materials like marble and granite, there are various grains running within the mineral. Salt sculptures can be as large or small as the chef desires, they have an excellent shelf life, and they are practical in a warm or cold climate.

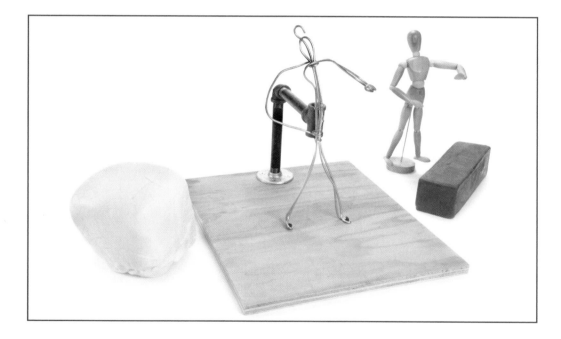

FIGURE 16–19 Armature next to modeling dough

RECIPE 16–1

SALT DOUGH

Recipe Yield: 4 pounds
(1.8 kg)

MEASUREMENTS   INGREDIENTS

U.S.	METRIC	
1 pound	450 g	Water
1 pound	450 g	Cornstarch
2 pounds	900 g	Popcorn salt
As needed	As needed	Various colored spices

PREPARATION STEPS:

1. In a medium saucepan, combine cornstarch and water until the cornstarch is in suspension.

2. Heat mixture until it thickens, stirring as needed to prevent lumps.

3. Heat salt in 400°F (204°C) oven until it is warmed.

4. Combine thickened water and salt in mixer, and blend with paddle until smooth (Figure 16–20A).

5. Cover mixture; it must be airtight to prevent hardening.

6. Sculpt and model into desired shape (Figure 16–20B and C).

7. Apply color by rubbing various spices onto the sculpture (Figure 16–20D).

FIGURE 16–20A Mixing the salt dough

FIGURE 16–20B Applying dough to armature

FIGURE 16–20C Sculpting and modeling the salt dough

FIGURE 16–20D Applying spice to sculpture for color

BREAD DOUGH MODELING

Dough modeling has its roots in ancient culture. When grand banquets were prepared for sultans, kings, and other people of noble birth, food was presented in artistic ways. Tarts and pies were decorated with edible ornaments of crude dough, and containers of bread were used to present various roasted fowl and game.

As this additive form of sculpting developed, recipes for dough that would model and shape more easily were researched and created. Still edible, this new dough lacks in flavor and tenderness what it makes up in pliability and function. Dough sculptures have graced many elaborate banquet tables, but they are not commonly used for everyday buffets because of the time and talent required to create them.

RECIPE 16–2

BREAD DOUGH I

Recipe Yield: 5 pounds
(2.3 kg)

MEASUREMENTS		INGREDIENTS
U.S.	**METRIC**	
1 pint	480 mL	Water
1 quart	960 mL	Popcorn salt
1 ounce	28 g	Granulated gelatin
1 quart	960 mL	All-purpose flour
As needed	As needed	Yellow food coloring
As needed	As needed	Granulated sugar, carmelized

PREPARATION STEPS:

1. In a medium saucepan, combine water and salt. Heat until dissolved.

2. Strain excess liquid, and cool mixture slightly.

3. Using a mixer, combine gelatin, flour, and a little food coloring with salt mixture. Mix until smooth and medium firm.

4. Adjust color as needed, but do not overcolor.

5. Place in an airtight plastic bag. Use immediately.

6. Roll out and cut into desired shapes.

7. Place pieces on flat surface, in a dry room, to harden over 24 to 48 hours.

8. Paint with carmelized sugar to give baked appearance.

RECIPE 16–3

BREAD DOUGH II

Recipe Yield: 1½ pounds
(0.68 kg)

MEASUREMENTS		INGREDIENTS
U.S.	**METRIC**	
1 pint	480 mL	All-purpose flour
½ cup	118 mL	Table salt
¾ cup	178 mL	Water
As needed	As needed	Evaporated milk

PREPARATION STEPS:

1. Place flour and salt in large mixing bowl. Blend thoroughly.

2. Add water slowly, mixing continuously. Dough will seem stiff and crumbly at first, but do not add more water.

3. Knead until dough is pliable, up to 15 minutes. If dough is not pliable after 15 minutes, water may be added a few drops at a time.

4. When dough is pliable, place in a plastic bag and remove all excess air.

5. Only work with what is needed, keeping the remaining dough covered.

6. Dough may be rolled like pie dough and cut using pattern; or, it may be formed by modeling.

7. Brush with evaporated canned milk to bring out accents of indentations.

8. Bake at 300°F (149°C) until hard and ridged, about 1 hour.

RECIPE 16–4

CORNUCOPIA DOUGH

Recipe Yield: 8½ pounds
(3.8 kg)

MEASUREMENTS		INGREDIENTS
U.S.	**METRIC**	
3 pounds	1.35 kg	Water
5 pounds	2.25 kg	High-gluten bread flour
2 ounces	56 g	Salt
2 ounces	56 g	Granulated sugar
2 ounces	56 g	Shortening
2 ounces	56 g	Egg whites
		Glaze
½ ounce	14 g	Granulated gelatin
1 pound	450 g	Water

PREPARATION STEPS:

1. Combine water, flour, salt, sugar, shortening, and egg whites in a mixing bowl.

2. Mix for 8 to 10 minutes, until dough is smooth and elastic.

3. Squeeze dough through sausage horn, or roll out on dough sheeter and cut into strips to form strands of dough.

4. Wrap around horn form and bake in 350°F (177°C) oven until hardened.

5. Meanwhile, combine gelatin and water. Allow to bloom. Heat slightly to dissolve.

6. Paint warm dough with gelatin mixture to produce a shiny glaze.

DEAD DOUGH

Recipe Yield: 6¾ pounds
(3.1 kg)

Note: The name "dead dough" is given to this dough because it does not rise, due to the large amount of salt and lack of leavening agents.

 TIP If desired, one may darken the finish before baking by coating with a mixture of kitchen bouquet and vodka.

MEASUREMENTS		INGREDIENTS
U.S.	**METRIC**	
4 each	4 each	Egg yolks
1 pint	480 mL	Water
60 ounces	1.7 kg	All-purpose flour
30 ounces	850 g	Popcorn salt
1 to 2 tablespoons	15 to 30 mL	Glycerin
As needed	As needed	Kitchen bouquet
As needed	As needed	Vodka

PREPARATION STEPS:

1. Combine yolks with water in a 30-quart mixing bowl.

2. Using paddle, blend in flour and salt.

3. Add glycerin and blend until consistency is similar to that of pie dough.

4. Divide dough into multiple pieces and wrap tightly in airtight plastic bags. Store in the refrigerator.

5. Roll or mold into desired shapes.

6. Dry in 180°F (82°C) oven until pale brown and hardened.

RECIPE 16–6

LEAF DOUGH

Recipe Yield: 2¾ pounds
(1.2 kg)

TIP	Food coloring, spices, and fine herbs may be mixed into the leaf dough to add color, depth, contrast, and texture.

MEASUREMENTS / INGREDIENTS

U.S.	METRIC	INGREDIENTS
1 pint	480 mL	Whole wheat flour
1 pint	480 mL	All-purpose flour
¼ teaspoon	1 mL	Salt
½ cup	118 mL	Vegetable oil
1 cup	237 mL	Water, cold
2 each	2 each	Egg yolks

PREPARATION STEPS:

1. In a medium-sized mixing bowl, sift together flours and salt.

2. In a separate bowl, blend remaining ingredients, mixing well.

3. Combine dry and wet ingredients, and blend until a smooth dough is formed.

4. Place dough in an airtight plastic bag and refrigerate for 1 hour.

5. Roll out dough using hand-cranked pasta machine or rolling pin and cut the dough into desired shapes (Figure 16–21A).

6. To imprint a design (in this instance, leaf veins) use the blunt edge of a knife (Figure 16–21B).

7. Bake pieces at 350°F (177°C) for 5 to 10 minutes, depending on the thickness of the dough.

FIGURE 16–21A Leaf dough being rolled out and cut into leaf shapes

FIGURE 16–21B Pressing back of knife into dough to simulate leaf veins

RECIPE 16–7

GINGERBREAD HOUSE DOUGH

Recipe Yield: 6 ¼ pounds
(2.8 kg)

MEASUREMENTS		INGREDIENTS
U.S.	**METRIC**	
		Gingerbread Dough
14 ounces	395 g	Brown sugar
22 ounces	625 g	Granulated sugar
12 ounces	337 g	Shortening
½ ounces	14 g	Baking soda
½ ounces	14 g	Salt
3 ounces	85 g	Molasses
12 each	12 ea	Eggs, large, whole
¼ ounce	7 g	Ground ginger
¼ ounce	7 g	Ground cinnamon
35½ ounces	1.2 kg	Cake flour
		Royal Icing
2 each	2 each	Egg whites
10 ounces	283 g	Powdered sugar
1 teaspoon	5 mL	Cream of tartar

PREPARATION STEPS:

1. In a stainless steel mixing bowl, cream together sugars, shortening, soda, salt, and molasses until smooth.

2. Slowly add the eggs while mixer is still blending.

3. Add sifted spices and flour, and mix until smooth.

4. Roll out dough on pastry cloth to thickness of ⅛ inch (3 mm).

5. Using patterns, cut dough into desired shapes.

6. Bake on a flat pan in level oven at 375°F (190°C.)

7. Meanwhile, make Royal Icing by mixing egg whites, powdered sugar, and cream of tartar, and heating over double boiler until icing reaches 120°F (49°C). Whip icing until stiff as possible. Keep covered with a damp cloth while using.

8. After dough pieces cool and harden, glue together and decorate with Royal Icing.

Fat Carvings

Fat carving, as the Europeans refer to it, is an art form that encompasses several media, including pastry margarine, butter, doughnut fryer shortening, and tallow. Each varies slightly in color and hardness. The harder products, such as tallow, which can be modeled or carved, are more practical for warm weather climates. While the softer media, like pastry margarine or fryer shortening, although easier to model, can sag and lose their sharp design features when displayed in warm ambient temperatures.

PASTRY MARGARINE AND BUTTER

Butter sculpture has its pedigree from an ancient Buddhist tradition. Tibetan monks would create artistic displays from butter only to have them be ruined by the warming forces of nature. To them, these temporary creations symbolized the impermanence of man—a basic tenant of Buddhism. The American form of this art has more to do with buffet presentations, sideshows, and agricultural fairs, yet serious and talented artists have worked in butter.

The Tibetans learned to periodically dip their butter creations in cold water while working on them to harden the sculpture. They often work around a stone bowl filled with cold water and use the chilled edge of the bowl as a ledge on which to place their components after dipping.

Today, chefs use cold marble slabs on which to work and often even sculpt inside a walk-in cooler. Because it is difficult to work with butter, chefs commonly blend margarine and shortening into butter to create a harder medium. Butter sculpting is not commonly done in warmer climates, for practical purposes (Figure 16–22A, B, C, and D).

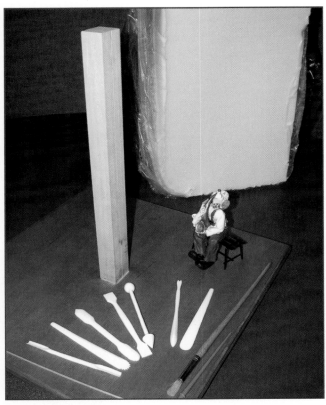

FIGURE 16–22A Elements assembled for fat carving (base with armature, sculpting tools, model, doughnut fryer shortening)

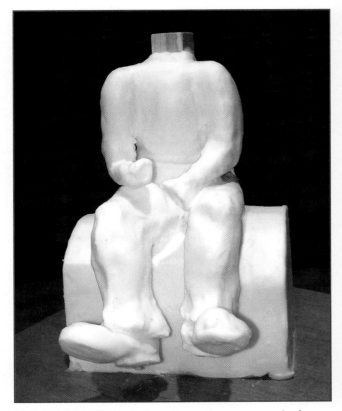

FIGURE 16–22B Building the sculpture onto the base

FIGURE 16–22C Adding decorative elements to the sculpture

FIGURE 16–22D The finished sculpture

TALLOW

Garde manger chefs have been sculpting tallow for more than 300 years. In the early seventeenth century, chefs would sculpt butter into interesting shapes for the enjoyment of their customers. To reduce the problems of heat acting upon the pure butter, the chefs of Europe began to combine butter with animal fats and later with wax to create a stabilized, creamy sculpting medium. The sculpting tallow, as it came to be known from its mixture with animal fats, was easier to work with and had a much greater shelf life. Because patrons reacted just as enthusiastically to these alternative sculptures, chefs all over the world embraced this new medium.

Chefs can make their own mixture of tallow using a simple blend of baker's shortening, bee's wax, and paraffin (see Recipe 16–8). However, the product has a tint of yellow from the bee's wax. Some chefs prefer to use multiple colors and often purchase the product in white, butter yellow, cheddar, or chocolate brown (Figure 16–23). Tallow generally is available in two forms: *sculpting* and *casting*.

Modern sculpting tallow, when worked by hand, quickly softens to the consistency of artist's clay. It can be manipulated into any shape or built up on any support structure to create sculptures of amazing size and complexity. Sculpting tallow can be used as an additive sculpture, or as a subtractive design. It can be trimmed with tallow tools and knives until its design is achieved (Figure 16–24).

FIGURE 16–23 Commercially available tallow in various colors. Courtesy of Culinart, Inc.—Global supplier of tallow

FIGURE 16–24 Garde manger chef carving tallow

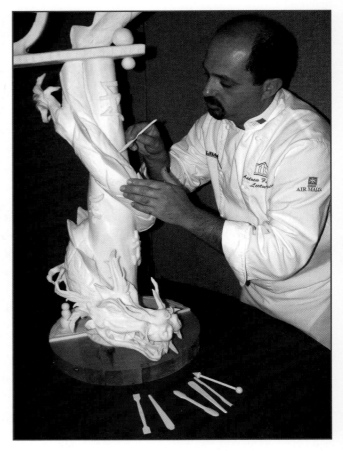

Casting tallow is extremely firm tallow that cannot be softened by hand; it requires great effort to carve on any large scale. It is designed to melt and be poured into molds to produce the final sculpture.

Both types of tallow may be melted after use, and reformed into a solid block by being poured into a large, commercial quality, plastic bucket while still warm and liquid.

RECIPE 16-8

TALLOW

Recipe Yield: 30 pounds
(13.5 kg)

MEASUREMENTS		INGREDIENTS
U.S.	**METRIC**	
10 pounds	4.5 kg	Bee's wax
10 pounds	4.5 kg	Paraffin
10 pounds	4.5 kg	Shortening

TIP This tallow will have a yellow tint compared to the commercially available product, but it is good for practice and can be melted and reused many times.

PREPARATION STEPS:

1. Melt all ingredients together in a clean stockpot. Blend while warm.

2. Allow mixture to cool.

3. Store loosely covered. No refrigeration is needed.

TIP The ratio of ingredients may be adjusted to affect desired hardness. Increase paraffin to amplify hardness or increase bee's wax to enhance the pliability of the medium.

Styrofoam

Styrofoam sculpting is most popular in the warmer climates of the Caribbean Islands, Bahamas, Polynesia, Thailand, Japan, and the rest of southern Asia. Often used in large banquets of resort hotels, these massive sculptures have a long shelf life and are frequently reused on buffets. The sculpting is done with high-density styrofoam using box cutters, heated wires, handsaws, razor blades, and knives. After being cut, the styrofoam is smoothed with sandpaper, then decorated with common household paint or royal icing (Figure 16–25).

FIGURE 16–25 Styrofoam sculpture. Courtesy of Culinart, Inc.—Global supplier of tallow

Professional Profile

Name: Derek Maxfield

Place of Birth: Ludington, Michigan

Recipe Provided: Maxfield Color Method

CULINARY EDUCATION AND TRAINING HIGHLIGHTS

Chef Maxfield began his culinary career in the late 1980s at the highly regarded Amway Grand Plaza Hotel and soon enrolled in the Culinary Arts program at Grand Rapids Community College (GRCC), where he graduated in 1992 with an associate's degree in Applied Arts and Sciences. As a student at GRCC, Derek participated in an International Culinary Work-Study Tour to Scotland with the authors and was placed to work at the Turnberry Hotel for a short period of time. While at the Amway Hotel, Derek had the good fortune to work under several talented ice sculptors and two certified master chefs, Dan Huglier and James Hanyzeski, who were members of the 1988 and 1992 Culinary Olympic Teams USA. His interest in ice sculpting was piqued at the Amway Grand Plaza Hotel, where he met and worked with his future business partner in the garde manger department. After taking ice sculpting classes at Grand Rapids Community College, Chef Maxfield competed in collegiate-level ice carving competitions, earning first place (Individual Student) at the 1990 Plymouth International Ice Spectacular. While there, he also assisted the professional-level team of Mac Winker, Dan Huglier, and Randy Finch, who went on to win the professional-level team event. After working in the garde manger department and in several of their fine dining restaurants, Derek accepted the position of sous chef at Kent Country Club. While at the country club, he started his ice business, Ice Sculptures Ltd. with Randy Finch in 1994 and worked part-time in ice until he left the club in 1996. Since that time, Chef Maxfield has gone on to fame and fortune in the ice sculpting world, being featured on many affiliates of NBC, ABC, CBS, and Fox networks. His ice sculpting work was the focus of an episode of *Ripley's Believe It or Not,* and he has been written up in countless trade magazines. Clients of Ice Sculptures Ltd. include numerous rock/pop musicians, such as Aerosmith, Elton John, AC/DC, and "Weird Al" Yankovic. He has produced sculptures for several professional football teams and the 2004 Ryder Cup, as well as numerous presidential campaigns nationally. Chef Maxfield has continued his education by taking art classes, sculpting classes, and traveling internationally, and he is a coauthor of *Ice Sculpting the Modern Way,* along with his partner Randy Finch and this author, Robert Garlough.

ADVICE TO A JOURNEYMAN CHEF

Organization is probably the most important quality in being a successful chef. Proper time management allows for more creativity, and enhanced efficiency increases profitability; simply stated, "Time is money." Organizing past notes, recipes, plate presentations, or procedures can prove invaluable. By developing a searchable database, you can easily reference this information and adjust it to meet your present needs. Creativity is another perpetual challenge for culinarians. To nurture this inventiveness, a chef must stay in constant pursuit of knowledge. Stay on the cutting edge by taking educational trips as well as classes and seminars. As a practical consideration, creativity must go hand in hand with profitability. Few and far between are the positions that encourage artistry without fiscal responsibility. Standardizing recipes, developing job descriptions, and having "mise en place" enforced in your kitchen are important tools to help keep this balance. A chef must be great at more than just cooking to prevail in the modern culinary world.

Derek Maxfield

RECIPE 16–9

CHEF DEREK MAXFIELD'S COLOR METHOD

Recipe Yield: 2 quarts
(1.9 L)

Note: Chef Maxfield, as a means of coloring ice, developed this method to prevent bleeding, shrinkage, or deterioration from the colorants. This process is considered a breakthrough in the coloring of ice (Figure 16–26).

MEASUREMENTS		INGREDIENTS
U.S.	METRIC	
2 quarts	1.9 L	Water
1¾ ounces	49 g	Powdered gelatin, unflavored
As needed	As needed	Coloring agents (nontoxic watercolor, such as powdered tempera preferred)

PREPARATION STEPS

1. Heat water to 160°F (71°C) on the stove or in the microwave.

2. Once heated, completely dissolve the gelatin in the warm water. Allow the gelatin to bloom and dissolve.

3. If creating multiple colors, divide the gelatin into several containers as needed.

4. Depending on the color intensity desired, add one or more teaspoons of the colorant to the gelatin, and carefully stir to blend the color.

5. When the colored gelatin mixture reaches 70°F (21°C), it can be safely poured into the design of a sculpture using a pitcher. It is best to apply the warm gelatin to different areas, moving around to avoid overheating an area too quickly with warm aspic. This should be done on a level table to prevent the gelatin from running.

6. Allow the gelatin to freeze solid. Pack the surface of gelatin with clean, dry snow and gently apply cold water over the snow to encase the color mixture. Allow the water to freeze solid before moving the ice.

FIGURE 16–26 Ice sculpture with assorted colors

REVIEW QUESTIONS

1. Where was ice first used for food?

2. How do you select a block of ice for carving?

3. How is ice made for ice carving?

4. What is thermo-shock?

5. What safety precautions should be observed when carving ice?

6. How is ice best held in storage for carving?

7. What is tempering?

8. Describe the use of a template.

9. What is tallow?

10. What is salt dough?

A. Group Discussion

Discuss how best to approach learning all the carving techniques so as to become a more versatile chef.

B. Research Project

In small groups, research where fruit and vegetable carving began and report your findings to your peers.

C. Group Activity

Carve a small melon into a face using a template.

D. Individual Activity

Practice modeling a small animal using clay instead of any food medium. Use a small model of an animal as a guide.

CHAPTER 17

Food Decoration, Platter Presentation, and Culinary Competition

After reading this chapter, you should be able to

- discuss the purpose of decoration and presentation.

- explain how to use gelatin.

- demonstrate how to use aspic.

- discuss the use of chaud-froid.

- discuss and demonstrate how to serve a cheese platter.

- discuss and demonstrate how to serve a gravad lox platter.

- discuss and demonstrate how to serve a caviar platter.

- discuss and demonstrate how to serve a fruit platter.

- discuss and demonstrate how to serve a vegetable platter.

- discuss and demonstrate how to serve a charcuterie platter.

- discuss and demonstrate how to serve a deli tray.

- discuss culinary competitions.

- describe how to prepare for culinary competition.

Having once decided to achieve a certain task, achieve it at all costs of tedium and distaste. The gain in self-confidence of having accomplished a tiresome labor is immense.

THOMAS A. BENNETT

The Purpose of Decoration and Presentation

The decoration of cold food for presentation, whether on a plate or on a platter, is a very specialized skill of the garde manger. One of the most important aspects of food preparation and presentation is its eye appeal. Great attention must be paid to the individual pieces of food to ensure that their shape, color, size, and shine are of the highest quality. They must be placed with the greatest of care, intricately combined together to form the overall picture that is being designed on the plate or platter. This achieved, they will give the professional appearance that should be associated with all food served by professional garde manger chefs.

It is common for these platters or plates to be part of a larger presentation, such as a buffet or competition presentation (Figure 17–1). Consideration should therefore be given to the careful planning of the individual plates or platters as to how they fit into the overall structure of the presentation. When all of these elements come together, the buffet architecture will make sense to the chefs that create it and to the customers that it is meant to serve. Careful planning also eliminates repetition of color, ingredients, shape, method of cooking, or platter layout—creating

FIGURE 17–1 Garde manger chef in a culinary salon competition. **Courtesy of Chuck Clark Photography**

flow and interest to the eye. By displaying the foods attractively on the dishes, using risers for elevation, and surrounding them with simple yet appropriate props and linens, the food will become an attractive and interesting focal point of the display.

Coating Agents

Historically, there were established practices for the presentation of most cold foods. Traditionally, they were presented cold, with a glaze or shiny liquid as a protective coating. The purpose of using this liquid was to preserve the food, improve its flavor, enhance its appearance, and act as an accompanying sauce or moistening agent. The usefulness of these coatings has been dramatically reduced in recent years due to the changes in modern equipment, refrigerated serving areas, and state-of-the-art facilities. Food coatings are also not as common and popular as they once were because many of the classical dishes they traditionally accompanied are no longer in vogue. Chefs do, however, use them a great deal in culinary competitions and minimally in presenting buffet platters and for upscale deli presentations, all helping to keep the traditions and techniques alive.

GELATIN

The product known as **gelatin** is extracted commercially from the skins, connective tissue of meat, and the bones of animals—particularly the younger ones. Gelatin, when mixed with water, is transparent and almost colorless, and is sold in a dehydrated form as a granule or as a fragile sheet (Figure 17–2). It can be purchased as clear or in a brown color. It is used to "set" such items as jellies, mousses, and savory aspics. The following are some important points to consider when using gelatin:

■ When using gelatin, it is important to avoid fresh fruits containing the enzymes *bromelain, ficin, papain,* and *actinidin,* because they can break down the gelatin and cause it to lose its thickening or setting properties. These enzymes are broken down during cooking, so it is acceptable to use the fruit in their cooked from. Some of the fruits high in these different enzymes include pineapples (bromelain), figs (ficin), papaya (papain), and kiwifruits (actinidin).

■ The infusion of large quantities of sugar can also inhibit the setting properties (gelatinization) of gelatin, and the recipes should be altered accordingly to accommodate the correct setting of the food product.

FIGURE 17–2 Sheet gelatin

- Dry unflavored gelatin should be mixed with a little cold water first for 3 to 5 minutes to moisten before adding to the hot liquid for melting. This is known as "blooming" the gelatin. This process prevents the mixture from clumping or lumping, and allows the gelatin to be evenly distributed.

Note: Prolonged exposure to heat will weaken the gelatin's thickening process and cause it to darken.

- Gelatin can be melted down and rechilled several times before the mixture loses its thickening ability.
- Bringing gelatin mixtures to a full boil will dilute its thickening properties. Gelatin is a protein that is weakened by heat.
- Gelatin takes twice as long to dissolve when used with cream or milk.
- When using sugar with unflavored gelatin, mix the sugar and gelatin first before dissolving.
- Always soak gelatin leaves (sheet gelatin) in cold water to soften, before adding it to the hot liquid for melting.
- Four sheets of leaf gelatin are equal to $2\frac{1}{2}$ teaspoons (7 g) of powdered unflavored gelatin.
- One envelope of powdered unflavored gelatin, which is normally a $\frac{1}{4}$ ounce and equal to $2\frac{1}{2}$ teaspoons (7 g), can be added to 2 cups (473 mL) of liquid to establish a standard firmness.
- Gelatin in its dry form has an indefinite shelf life, as long as it is wrapped airtight and stored in a cool, dry place.

There are other forms of gelatin or setting agents available to meet the needs of those wishing alternatives to meat products. Other forms of gelatin or setting substances can include the following:

- *Isinglass* is a type of gelatin extracted from the air bladders of fish, particularly sturgeon. It is rarely used anymore because of the lack of fish stocks available for production.
- *Carrageen* is a gelatinous thickening agent derived from seaweed that grows off the coast of Scotland and Ireland. It is sometimes called *Irish moss,* and is used in making home-brewed beer, mead, and many puddings for dessert.
- *Agar* or *agar-agar* is the Malay name for gum and is native to Japan. It is dried seaweed that is sold in blocks, powder, or strands. It is used as a setting agent and is common in Japanese cooking where it is also called *kanten* or *grass jelly.*

ASPIC-GELÉE

The term **aspic**, or *aspic jelly,* can be confusing because it can have three distinct meanings depending in what era the chef was trained. The true aspic jelly, as made by Antonin Carême, was a clarified stock that was made with the knuckle and feet of young veal. It depended purely on the natural gelatin being extracted from the collagen contained within the ingredients. This method of extracting aspic jelly has become very costly and is rarely done anymore except when the quality and expense of the terrine or item being made warrants the expense.

The modern version of aspic jelly uses a combination of well-clarified stock with the addition of commercially purchased leaf or dried gelatin. This process creates an acceptable aspic for use on pâtés and terrines, sliced meats, brushing on cold fish and shellfish, and most other presentations in the cold kitchen.

Adding pure gelatin products to clear water is the third version for making aspic and has become a popular method in recent years. Its use has become necessary because of time, labor and skill constraints, and its ability to coat and protect without interfering with the natural flavor of the food it coats. When used for enhancing the food's appearance, it is generally brushed over the finished product, creating a shine. Aspic has another major area of use in culinary competition, wherein the food that is being displayed is seldom tasted for judging. This allows the competitor to add the aspic in layers and to completely surround the food item. This heightened use not only extends the product's time and ability to be displayed, but also creates fantastic aesthetic characteristics.

The strength of the aspic jelly has to be correct, no matter what method is used to produce it. When food is going to be consumed, the aspic should be light and flavorful, with no resistance to the bite. For use in competitions where it is only viewed and not tasted, it can be as strong as is needed, allowing the glazing to last for the duration of the display (Recipe 17–1 and Recipe 17–2).

Uses for Aspic Jelly

There are many uses for aspic jelly in the garde manger department, and it is important to be consistent with the quality of the aspic used. The method of production can make the difference between a very good dish and a mediocre dish, so it is important to have good quality aspic on hand at all times. The flavor of the jelly can be changed by making it with different stocks, and by the addition of lightly flavored liquors such as Madeira or port. Aspic jelly can be used in the following ways:

- As a stabilizer in salad dressings
- To fill a pastry-encrusted pâté, to allow the slice to appear whole
- As a binding agent for mousses, parfaits, pâtés, and purées
- To brush on any sliced roasted meat, poultry, or game for enhanced presentation
- To brush on sliced terrines, pâtés, or galantines
- For brushing on individual pieces of food that will be the focal point of platters or plates
- For coating the bottom of plates as a background for food
- For coating the bottom of a platter as a base for presenting food
- Cut into shapes or chopped and spread around an accompanying dish or platter
- Layered into a vegetable terrine to allow it to set and carve easily
- For setting any other cold sauce, other than chaud-froid, that would be served on a plate or platter, including any puréed vegetable or fruit sauce, as well as any sauce made with mayonnaise, sour cream, or emulsified vinaigrette
- As a finishing shine on large food items for display on a buffet, such as whole poached salmon or trout, whole roasted racks of lamb or beef, and whole roasted geese or ducks
- To coat the bottom of molds before being filled with a cold purée to trap a design inside
- At differing strengths to coat a host of ingredients for use in competitions—almost anything that can go onto a hot plate of food
- To assist with the application of rubs and crusting on the outside of food items before being carved or served, which helps to imitate the rubbing or rolling of foods in herbs and spices
- To strengthen delicate food items, for example, terrines and pâtés submerged in aspic and then set for care during transportation

Techniques When Using Aspic

Certain techniques are useful to know and can improve the chef's skill with aspic. One of the most important things to understand about aspic is that it hardens as it cools, so time and temperature become very important. In order to be efficient while using aspic, the chef must try to keep the aspic at the optimal condition for use for as long as possible. It is not only a difficult thing to control, but aspic work can also become very frustrating and break the confidence of inexperienced chefs unfamiliar with its use. The repetitive use of aspic will aid the chef in becoming more familiar with the medium. There are some basic rules, which, if followed, will result in a better product:

- When melting aspic, it does not have to be heated; it only needs to be melted to slightly warm, never hot.
- Cool a portion of the warmed aspic over cold water. When it reaches optimum consistency, begin to use it.

■ Use the excess liquid aspic to warm up the hardening aspic to return it to the correct consistency for coating.

■ Do not leave the aspic sitting in ice water while performing a different task. It will set very quickly at the bottom of the bowl.

■ Stir the aspic gently to avoid setting; aggressive stirring will introduce unwanted bubbles into the aspic.

■ Gently transfer aspic to new bowls often to avoid lumping. Be careful not to create air bubbles when pouring.

■ Do not boil the aspic jelly; this weakens its setting qualities.

■ Avoid debris from accumulating in the aspic.

The techniques that are shown in Figures 17–3 through 17–6 are used for general kitchen use as well as for competitions, and have many small variations depending on where they are learned. Chefs often produce excellent terrines and roasted meats, only to fail in the proper coating of their showpieces with aspic. According to Metz and Ryan (1989):

> Of corollary importance to aspic guidelines is proper aspic temperature. The product is usually dipped three times (refrigerated in between dippings) before the food is ready for display. If the aspic temperature is too hot, it can remove the first coat. Too-hot aspic can drain or change the color in a product, particularly if the color is a pastel. . . . Too-cold aspic can result in coagulation of aspic which sets on product unevenly, resulting in streaks and lumps. (p. 243)

After coating the food item, it is gently placed against something warm (such as a sauce pot filled with hot water) to melt the excess aspic that has set irregularly (Figure 17–7A and B). This is done to give the pieces a clean appearance and to eliminate any drips or rough edges that detract from the food's appearance. Again this is an arduous task, and takes time and patience to do a good job. It is, however, one of the most important elements of the technique and should be done accurately and with painstaking effort. The clean and precise final appearance that is achieved will immensely enhance the overall display of the food (Figure 17–8).

FIGURE 17–3 Gelatin slowly melted over a double boiler

FIGURE 17–4 Fish terrine slice being coated by dipping

FIGURE 17–5 Pièce montée (focal piece) being coated by pouring

FIGURE 17–6 Multiple vegetables being dipped simultaneously in aspic (note the chopped herbs adhering to the vegetables)

FIGURE 17–7A Cleaning aspic by melting against a steaming pot of water

FIGURE 17–7B Melting an aspic burr

FIGURE 17–8 Dipped slices being placed back in proper sequence

RECIPE 17–1

CLEAR ASPIC FOR DISPLAY (TRIPLE STRENGTH)

Recipe Yield: 1 gallon (3.8 L)

MEASUREMENTS		INGREDIENTS
U.S.	**METRIC**	
1 gallon	3.8 L	Water, cold
8 ounces	227 g	Gelatin, clear, powdered

> **TIP** Use to coat vegetable, fish, light-colored veal, pork, chicken, and poultry (light colored) display pieces.

PREPARATION STEPS:

1. Combine the cold water and powdered gelatin in a large stainless steel bowl. Stir slowly to evenly distribute the gelatin in the water. Caution: Aggressive stirring will cause unwanted air bubbles.

2. Allow gelatin to bloom for 10 minutes.

3. Place the bowl of bloomed gelatin over a hot water bath and allow mixture to slowly dissolve only to 100°F (37°C) and become smooth.

4. Cool to approximately 75°F (24°C) before coating cooler display pieces.

RECIPE 17–2

BROWN ASPIC FOR DISPLAY (TRIPLE STRENGTH)

Recipe Yield: 1 gallon (3.8 L)

MEASUREMENTS		INGREDIENTS
U.S.	**METRIC**	
1 gallon	3.8 L	Water, cold
8 ounces	227 g	Gelatin, brown, powdered
8 ounces	227 g	Glace de viande (optional)

> **TIP** Use to coat beef, lamb, game meats, and dark-fleshed game bird (dark colored) display pieces.

PREPARATION STEPS:

1. Combine the cold water and powdered gelatin in a large stainless steel bowl. Stir slowly to evenly distribute the gelatin in the water. Caution: Aggressive stirring will cause unwanted air bubbles.

2. Allow gelatin to bloom for 10 minutes.

3. Place the bowl of bloomed gelatin over a hot water bath and allow mixture to slowly dissolve. Heat to 110°F (43°C) until smooth.

4. (Optional): Add the glace de viande to gelatin, mix well until melted.

5. Cool to approximately 75°F (38°C) before coating cooler display pieces.

RECIPE 17–3

WHITE WINE ASPIC JELLY (EDIBLE)

Recipe Yield: 5 cups (1.25 L)

MEASUREMENTS		INGREDIENTS
U.S.	**METRIC**	
10 ounces	283 g	Fatless shin of beef, ground through the large die
8 ounces	227 g	Mirepoix, fine chopped
6 each	6 each	Large egg whites
7 cups	1.7 L	Rich brown or veal stock
14 tablespoons	207 mL	Madeira wine
3 ounces	85 g	Gelatin, unflavored powdered
2 each	2 each	Tomatoes, chopped
2 tablespoons	30 mL	Black peppercorns
2 each	2 each	Bay leaves
1 tablespoon	15 mL	Salt

PREPARATION STEPS:

1. Combine the meat, mirepoix, tomatoes, peppercorns, bay leaves, salt, and egg white, beating vigorously, as if to bind.

2. Pour in the stock and bring to the boil.

3. Break the raft that forms on the top and allow to boil in this spot for 45 minutes.

4. Gently add the Madeira while still hot and strain through fine cheesecloth, being careful not to damage the raft.

5. Dissolve the gelatin in cold water, then dissolve in the warm stock.

6. Use for the coating of pâté and dark meat slices for presentation.

RECIPE 17–4

WHITE FISH ASPIC (EDIBLE)

Recipe Yield: 9 cups (2.1 L)

MEASUREMENTS		INGREDIENTS
U.S.	**METRIC**	
8 cups	1.9 L	Fish stock
2 each	2 each	Egg whites
1 cup	237 mL	White wine
4 ounces	113 g	Leaf gelatin soaked in cold water to soften

PREPARATION STEPS:

1. Whip the whites into the stock and bring to a boil, stirring continuously.

2. Simmer for 5 minutes.

3. Strain through a fine sieve.

4. Add wine to gelatin and bloom.

5. Add the wine and gelatin mixture to hot stock and cool for further use.

RECIPE 17–5

FRESH APPLE GINGER ASPIC (EDIBLE)

Recipe Yield: 1 pint (473 mL)

MEASUREMENTS		INGREDIENTS
U.S.	METRIC	
2 cups	473 mL	Golden delicious apples, freshly juiced
1 tablespoon	30 mL	Ginger, grated
1 each	1 each	Orange, zest
1 each	1 each	Lemon, zest
1 ounce	28 g	Gelatin leaves soaked in cold water to soften

PREPARATION STEPS:

1. Heat the apple juice and the ginger until it begins to steam.

2. Skim well. Remove from the pot and add the zests of lemon and orange.

3. Cool overnight; strain through a coffee filter until completely clear.

4. Add softened gelatin leaves to apple juice mixture.

5. Heat mixture to 110°F (43°C) to dissolve the gelatin.

6. Cool and store gelatinized juice until further use.

TIP This aspic jelly goes well with appetizers made with wild game, smoked fish and shellfish, and smoked duck breasts. It can also be used as a condiment on a cold buffet.

RECIPE 17–6

WATERMELON ASPIC JELLY (EDIBLE)

Recipe Yield: 2½ cups (591 mL)

MEASUREMENTS		INGREDIENTS
U.S.	METRIC	
2½ cups	591 mL	Watermelon juice
1 ounce	28 g	Sugar
1 ounce	28 g	Gelatin leaves soaked in cold water to soften

PREPARATION STEPS:

1. Pass the watermelon juice through a coffee filter several times until the juice runs clear.

2. Warm the juice in a pan and dissolve the sugar and the gelatin.

3. Strain through a fine sieve and cool for further use.

TIP This jelly can be used to coat fruits for display, as a base for plates and platters, and as an accompaniment on a buffet or an appetizer.

RECIPE 17-7

JELLIED GLACE (EDIBLE)

Recipe Yield: 1 cup (237 mL)

TIP Although glace already contains enough gelatin to set completely on its own, adding a little more allows more control at the piping or pouring stages of decoration. When using jellied glace for a sauce or for a plated presentation, the addition of the gelatin will suspend the garnish in the sauce much better, as in the case of peppercorn sauce.

MEASUREMENTS   INGREDIENTS

U.S.	METRIC	
1 cup	237 mL	Meat glace
2 each	2 each	Leaves of gelatin soaked in cold water

PREPARATION STEPS:

1. Melt the glace and while warm add the gelatin leaves to dissolve.

2. Pass through a fine sieve and cool for use.

CHAUD-FROID

The stock produced from the bones of the animals, used for presentation, can also be made into a sauce called **chaud-froid**. It can be made to coat the foods before they are carved at the buffet, which gives a very nice finish and shape to the slice. It is said that chaud-froid was invented by Louis-Alexandre Berthier, Marshal of France (under Napoleon). After eating cold fricassee of chicken and enjoying it so much, he wanted to forever adorn his table with fricassee. Chaud-froid means "first hot then cold" and was an expression that meant that cooked foods should be as appetizing when cooled down as when they are hot. This was even so when the dish was to be surrounded by all or part of its sauce.

Chaud-froid is classically made with any mother sauce that has had some form of aspic or gelatin product added to it. This makes it react to the heating and cooling process in the same way as aspic does. The same rules that apply to the use and care of aspic also apply to chaud-froid. The only exception being the rate at which it cools and sets. This is because the sauce has a thickening agent within it, so it tends to set unpredictably. The chill does not penetrate it as obviously as aspic, therefore greater attention has to be paid to the mixing of the sauce during cooling in order to avoid lumping and render a good consistency.

Uses for Chaud-Froid

Chaud-froid, as it was originally used, has in recent years tended to go out of fashion; however, it has expanded in the way it is used and from what it is made. Colored chaud-froid can be created by adding natural colors to a basic mother sauce, such as spinach and béchamel for green shades, and turmeric for yellow shades. In modern times, almost any other classification of sauce has become acceptable as chaud-froid. The chaud-froid is also presented as an under-sauce, as if it were on a platter of hot food. Another popular technique is the addition of herb and spice flecks, or delicate inlays of vegetables, to give the sauce a more interesting and aesthetic appearance. There are a lot of other uses for chaud-froid sauce, and it has become necessary to adapt the sauce to new and creative uses, including the following:

- When plating a large quantity of cold appetizer that requires a plate sauce, the sauce can have a measured amount of aspic added to it. This will allow the sauce to set on the plate but still be soft to taste when served. The advantage gained is that the sauce will not move on the plate while being stored or moved before service.
- The same principle can be used for the saucing of plates and platters for culinary competitions. This also allows the sauce to carry garnish that will enhance the appearance even more.
- The sauces can be used for flooding plates or platters to create a different colored background for the food to be presented.
- These flooded plates or platters can have shapes carefully cut from within them, and contrasting sauces inlayed in their place.
- Chaud-froid can be imitated in the cold sauce section of the garde manger using the principle of the classic mayonnaise collée. This sauce is made with two parts mayonnaise and one part any flavor of aspic, depending on the sauces used. This method can be applied to any mayonnaise-based sauce, sour cream–based sauce, emulsified vinaigrette sauce, or vegetable or fruit purées.
- These sauces can be used to bind vegetables, fruits, salad materials, or fish and shellfish for timbales, socles, or compound salad presentations.
- It can be used to coat or semicoat individual pieces of food, although the food does not have to poached as it often was in the past.
- It can be allowed to set on a flat tray, cut into attractive shapes, and used to garnish plates or platters of food.

■ It can be used for the coating of large hams, turkeys, and large fish such as salmon for decorative centerpieces.

■ It makes a very pure white canvas for the chef to decorate for a buffet.

Techniques When Using Chaud-Froid

The techniques that are associated with the use of chaud-froid are similar to that of aspic, and can be learned easily. Generally, the pieces coated with chaud-froid tend to be larger than those coated with aspic. Chaud-froid was commonly used to coat a whole item before being portioned. The techniques that have been developed for the successful coating are dipping and flooding (Figures 17–9 and 17–10.)

FIGURE 17–09 Chicken ballontine being dipped into a bucket of chaud-froid

FIGURE 17–10 Creating a flood of green chaud-froid to coat galantine

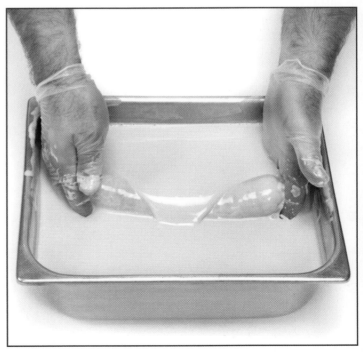

RECIPE 17–8

CLASSICAL WHITE CHAUD-FROID

Recipe Yield: 4 cups (946 mL)

Note: This will render sauces that are white and/or off-white in color.

MEASUREMENTS		INGREDIENTS
U.S.	**METRIC**	
2 cups	473 mL	Béchamel or velouté
2 cups	473 mL	Aspic jelly
1 cup	473 mL	Heavy cream

PREPARATION STEPS:

1. Boil the béchamel or velouté in a thick-bottomed pan.

2. Add the aspic jelly and reduce the sauce by one third.

3. Add the cream and bring back to a boil.

4. Strain through a fine sieve and cool.

RECIPE 17–9

CLASSICAL BROWN OR RED CHAUD-FROID

Recipe Yield: 4 cups (946 mL)

TIP Using espagnole or demi-glace will render a brown-colored sauce while using the tomato sauce will render a red sauce.

MEASUREMENTS		INGREDIENTS
U.S.	**METRIC**	
2 cups	473 mL	Espagnole, demi-glace, or tomato sauce
2 cups	473 mL	Aspic jelly

PREPARATION STEPS:

1. Boil the sauce in a thick-bottomed pan

2. Add the aspic jelly and reduce the sauce by one third.

3. Strain through a fine sieve and cool.

RECIPE 17–10

DARK RED CHAUD-FROID

Recipe Yield: 4 cups (946 mL)

MEASUREMENTS		INGREDIENTS
U.S.	**METRIC**	
3 cups	710 mL	White stock (chicken, veal, or fish)
5 ounces	142 g	Tomato paste
1 tablespoon	15 mL	Sweet paprika
1 ounce	28 g	Cornstarch
1 tablespoon	15 mL	Water
As needed	As needed	Salt and pepper
1 ounce	28 g	Leaf gelatin soaked in cold water to soften

PREPARATION STEPS:

1. Bring the stock to boil and whip in the tomato paste and the paprika.

2. Dissolve the cornstarch in the water and add to the boiling liquid.

3. Cook for 2 minutes and season well with salt and pepper.

4. Strain through a fine sieve.

5. Add the soaked gelatin leaves and cool.

RECIPE 17–11

GREEN CHAUD-FROID

Recipe Yield: 4 cups (946 mL)

Note: Watercress may be substituted for spinach.

MEASUREMENTS		INGREDIENTS
U.S.	**METRIC**	
12 ounces	340 g	Spinach
2 cups	473 mL	Light veal stock
1 cup	237 mL	Heavy cream
1½ ounces	43 mL	Cornstarch
1 tablespoon	15 mL	Water
2 ounces	57 g	Leaf gelatin, soaked in cold water
As needed	As needed	Salt and pepper

PREPARATION STEPS:

1. Mince the spinach and squeeze of all its juice. Heat the juice slowly in a saucepan. Skimming any spinach particles from the surface

2. Pass through a fine sieve.

3. Boil the stock and the cream, and thicken with the cornstarch diluted in the water. Season with salt and pepper.

4. Dissolve the gelatin in the spinach juice and add to thickened stock cream.

5. Cool and reserve for use.

RECIPE 17–12

CLASSICAL MAYONNAISE COLLÉE

Recipe Yield: 3 cups (700 mL)

MEASUREMENTS		INGREDIENTS
U.S.	**METRIC**	
1½ cups	355 mL	Strong aspic jelly
1½ cups	355 mL	Mayonnaise

TIP Sour cream may be substituted for the mayonnaise, yielding a different flavor and whiter color.

PREPARATION STEPS:

1. Warm the aspic jelly until it melts.

2. Cool the aspic until it is cool to touch but not set.

3. Whip the cooled aspic into the mayonnaise.

4. Cool further until the correct consistency is achieved for coating or piping

RECIPE 17–13

WHITE, PINK, OR GREEN MAYONNAISE COLLÉE

Recipe Yield: 4 cups (946 mL)

MEASUREMENTS		INGREDIENTS
U.S.	**METRIC**	
2 cups	473 mL	Aspic jelly
1 cup	237 mL	Sour cream (divided, if needed)
1 cup	237 mL	Mayonnaise
½ cup	118 mL	Tomato ketchup
½ cup	118 mL	Parsley coulis

PREPARATION STEPS:

1. Melt the aspic jelly and cool until almost set.

2. Combine the mayonnaise with the sour cream and mix in the aspic.

3. Cool until the correct consistency has been achieved.

   Option 1: May substitute the tomato ketchup for ½ cup (118 mL) of the sour cream for a pink collée.

   Option 2: May substitute the parsley coulis for ½ cup (118 mL) of the sour cream for a green collée.

Displaying Food for Consumption

When displaying food on a platter or a plate, the presentation should be done using the principles of general plate and platter presentation and always have the customer's ease of service in mind. The food must be laid out in a functional manner, so that the customer can immediately understand what the chef has intended as an edible portion. All the food on the platter should be equally portioned, including the accompaniments and the garnishes. It would not be appropriate, nor good business, to deny customers who are at the end of the line the same food and accompaniments available to those who go before them. The food should also be easy to remove and serve for the customers; they should not have to do any difficult cutting or digging.

When presenting food cold on a plate, the rules are the same as would be used to plate hot food (hot food on hot plates, cold food on cold plates). The sauces for cold plates are sometimes served aside of the plate in a sauceboat, but they can have aspic added to them and be presented as part of the plate.

It is very important for the platter to fit into the pattern and design of the overall buffet. The ingredients used, the predominate color of the platter, and the height of the food can all be important considerations before starting the platter design. Having established the theme of the meal, or the subject of the competition, the planning can begin for the platter itself. There are several simple rules that can help to create a successful platter:

■ When producing, portioning, and coating foods for competition, chefs often use specialty tools to create unique and attractive display pieces. Figure 17–11 shows various tools used in cutting, sculpting, and glazing display pieces with great accuracy.

■ Identify the main components or protein items of the platter and choose complementary ingredients to go along with them. This can be as simple as focusing on one food item. For example, serving duck two or three different ways on the same platter is an interesting approach to platter design.

FIGURE 17–11 Common tools used in food show competition

- If the methods of preparation for the meats are complicated, then it is appropriate to select easier accompaniments. It is best to have balance and choose simple items along with complex items of the same theme.
- The cooking methods should vary throughout the presentation and reflect the diversity of skill that the chef possesses. They should be bold and obvious; if something has been grilled, the grill marks should be evident. The food should also have the flavor of the grill.

Note: This is not so important if the foods are not to be tasted.

- The textures used should vary throughout the platter, using all available textures to the chef. They should include smooth, coarse, solid, soft or liquid, crisp and crunchy. This gives the food a more interesting mouthfeel and improves the enjoyment of the experience.
- The colors should reinforce the perception of freshness, quality, and well-executed methods of cooking. There should be no smudging or bleeding within the presentation, because this promotes the image of spoiled food.
- The shapes and sizes of the garnishes should suit the size of the platter and demonstrate well-executed knife skills. The shapes should be as natural as possible and show ingenuity by the chef when fitting them together.
- The flavors and seasonings should be well distributed around the platter. Examples include spicy with bland, rich with lean, smoky and salty with sweet, sweet with sour, and sweet with spicy.
- Any garnishing should add color, texture, taste, and interest to the plate. It should have a function and always be edible.
- The carving should be accurately executed and the shingling between slices should be exact. The carved food should create interestingly shaped lines of food. These lines should have perfect form and create flow and interest to the overall design of the platter.
- It is very important that equal attention is given to all components on the platter. This will assure the customer or judge that the chef applied the same amount of effort to all components of the platter throughout the planning.
- The platter should have a point or area on it where the eye is drawn automatically when looking at it. This is the focal point and is influenced by the placement and relationship of the various food components. The method used to layer the food is an essential part of focal point.
- The flow of the food on the platter is the result of well-balanced food in unison, providing a pleasing pathway to the focal point. This is not a skill that is achieved overnight; it should be practiced using drawings and cardboard templates (Figures 17–12 and 17–13).

When planning the design of the platter, the chef should always draw out the design on paper first to establish the focal point of the platter. Having established the furthest focal point of the food, the garde manger chef should then work backward to find the starting point for each line of food. This allows the lines of food to be clearly defined and ensure that they are bold and unbroken. It also allows the lines to always arrive at the focal point, as if they were flowing out of and into it (Figure 17–14).

The individual slices must be glazed prior to being arranged on the display platter (Figures 17–15 and 17–16). After being arranged on the display platter, the pieces should be further brushed with aspic to touch up any blemishes (Figure 17–17).

SPECIALTY PRESENTATIONS

Each **specialty presentation** has its own unique characteristics that will affect its planning and layout. If the food being presented is of a larger category compared to food that is served as an individual item, the style of presentation will alter slightly. For example, a cheese platter would certainly contain a selection of different cheeses, whereas a whole smoked salmon would generally be the singular item featured, along with a few of its accompaniments. The accompaniments

FIGURE 17–12 Sketches of platter designs showing lines flowing to the focal point

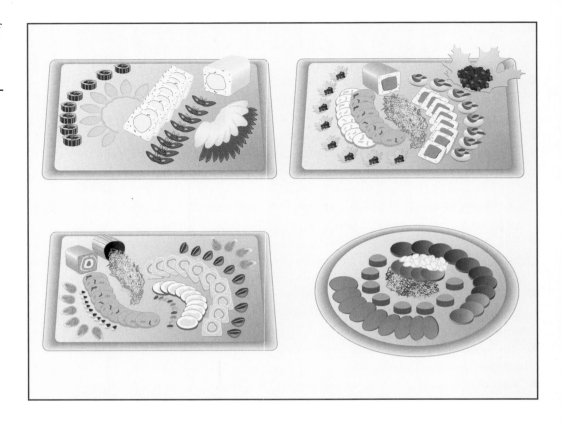

FIGURE 17–13 Cutouts of food items placed on plastic wrap–covered display platter to practice the arrangements before using the product

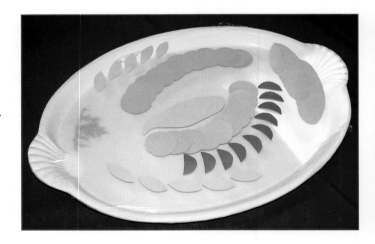

FIGURE 17–14 Slices arranged on a holding tray in the same order as they fell off the whole piece while being sliced

FIGURE 17–15 Garnishes being assembled

FIGURE 17–16 Items being arranged on show platter

FIGURE 17–17 Finished platter being touched up with fresh aspic

for any platter can be included as part of the platter design or as a part of the garnish. They can also be served to the side in small dishes that have their own serving spoons. This is more appropriate when the food has an accompanying sauce.

It is important to identify buffet food for the customer with some form of signage. Kitchen staff are always more knowledgeable about their food than their customers, and it should never be assumed that the guest can correctly identify what is presented. This is particularly important for people who have serious food allergies, such as with shellfish, nuts, and blue cheese. There should also be signs to identify the accompaniments, not only their names but also what foods they are to accompany. It is very disappointing to the chef when customers completely ignore foods that he considers the perfect accompaniment to a dish because people did not understand its purpose.

Cheese Presentations

As with any good cheese tasting, there should be a reasonable selection from different classifications of cheeses available on the platter. A range of 6 to 12 cheeses would be appropriate. These selections should be tasted in some order, or as a progression from the milder, softer, younger cheeses to the stronger, firmer, riper ones. The cheeses should therefore be laid out in such a way as to accommodate the correct tasting of the cheese. If the cheese platter is being served for a cheese tasting, then there should be little served along with the cheese. However, it is customary to serve crackers or bread, water, and fruits with cheese to aid in their tasting and digestion.

If the cheese platter is being served as part of a buffet, the cheeses should be selected from a range of styles according to the customer demographics. The cheeses should also be cut appropriately for the number of people attending, making it easier for the customers to serve themselves. However, it is recommended to limit the amount of precutting, because the cheese will oxidize and dry out, changing both the taste and texture. Cheese cutting should be done just before its

FIGURE 17–18 A summertime cheese display at a garden party

FIGURE 17–19 Simple caviar service with accompaniments. Courtesy of Foodpix

being brought to the buffet. Again, accompaniments such as crackers, butter, breads, fruits, and relishes may be served (Figure 17–18).

Caviar Presentations

Caviar can be an expensive menu item to serve buffet style, but it does give the table a grand appearance. When serving caviar, it becomes necessary to create a presentation that clearly defines the portion that the customer should take. This is accomplished by selecting a large platter that displays the container of caviar raised on a pedestal of ice. (Caviar is traditionally served in its own packing canister or in a silver dish.) The can is then surrounded with small spoons made of bone or ivory, or mini croutons that contain the portion of caviar that is intended for each guest. Caviar buffs would frown upon serving many items along with the precious egg; however, accompaniments such as grated hard-boiled egg yolks and whites, minced red onion, and lemon wedges can be served aside. Blinis, saltwater crackers, and delicate slices of brown bread are often available for guests to serve themselves (Figure 17–19).

Gravad Lox Presentations

There are not many presentations that have the brilliance of color, flavor, and aroma of lox. Gravad lox, also spelled gravid lax or gravlax, is a traditional Scandinavian means of preparing lox (salmon). Gravad lox is not smoked, but it can be served in a similar fashion. The salmon is coated with a spice mixture, which often includes sugars, dill, and spices like juniper berry. It is then weighted down to force the moisture from the fish and impart the flavorings. It is also very easy to present, because it carves into a very pretty line.

However, the great difficulty comes for the customers when they try to struggle through untangling the thinly sliced delicacies. This problem can be alleviated, to some degree, by placing a garnish of two or three slices in between each portion. A dry crouton, lemon slice, or pickle slice all work well. Alternately, the salmon could be rolled up with one of its accompaniments, such as cream cheese, and then presented. Accompaniments of cream cheese, pickles, capers, sliced sweet onions, and lemon wedges would be served aside, or on the platter, again with delicate croutons and crackers (Figure 17–20).

Fruit Platter Presentations

It is important when presenting fruit to always select varieties that are in season and sufficiently ripened. It is also important to choose fruit of differing textures, colors, and flavors for the platter presentation. The fruit should be peeled or partially peeled, portioned into bite size pieces or slices, and then arranged on the platter attractively. Some of the whole fruit can be cut attractively into designs, or carved into interesting shapes for a centerpiece and focal point of the platter (Figure 17–21).

FIGURE 17–20 A lox platter. Courtesy of Getty Images/Foodpix

FIGURE 17–21 Cut fruit display. Courtesy of Tonis Valing/Shutterstock

Vegetable Platter Presentations

Vegetables for a platter should be as fresh and crisp as possible, using as wide a variety as season and market availability allows. The vegetables should be peeled and cut into shapes that are easy for the customer to eat, as well as interesting to observe. Shapes and colors should be mixed to form attractive patterns on the platter. Gourds and squashes work well, either whole and uncut or carved, as the **pièce montée** of the platter. Appropriate dips and salad dressings are typically served to accompany the vegetables (Figure 17–22).

Charcuterie Platter Presentations

When presenting a platter of charcuterie items, they should closely represent the whole of the classification. Dry and hard cured salamis, as well as fresh and smoked sausages, should be accompanied with sliced hams and pastramis. The variety gives the platter many interesting and complementary shapes, textures, and flavors. Some of the smaller sausages can be served whole or just simply cut in half, giving another dimension to the presentation. The focal point can be an arrangement of partial sausages and salamis positioned in the center back of the display. Slices can be shingled forward into lines that appear to have originated from their larger solid part. Accompaniments of chutneys, relishes, pickles, and crusty bread should be served aside (Figure 17–23).

Deli Tray Presentations

Deli trays are extremely popular presentations that can be served to any size group, at any time of the day, depending on the type of event and customer budget. They contain a combination of sliced meats and cheeses, and are accompanied by relishes, salad items, condiments, and breads. The customers then have the option of eating the food as a sandwich, or as a salad, or a combination of both. They are very useful because they can be prepared in advance and dropped off at the location without the need for service staff (Figure 17–24).

FIGURE 17–22 Vegetable display with dip. Courtesy of Getty Images, Inc.

FIGURE 17–23 Charcuterie display. Courtesy of CuboImages srl/Alamy

FIGURE 17–24 An elegant deli presentation. Courtesy of Dragon Trifunovic/ Shutterstock

Culinary Competitions

Culinary competitions have stirred the interest of chefs from around the world for a very long time. They give the dedicated chef the opportunity to compete with national or international contemporaries in an area of professional interest and on a level playing field. Taking part in a culinary competition can be a nerve racking, yet highly exhilarating experience that develops many excellent skills and qualities in a garde manger chef. **Culinary salons** are highly charged atmospheres, and have all of the drama and excitement of sporting events (Figure 17–25). Currently, the following categories are available with American Culinary Federation sanctioned competitions:

- Cooking Professional/Student Cold Platters
- Cooking Professional/Student Cold Plated
- Patisserie/Confectionery
- Showpieces
- Team Buffet
- Hot Food Competitions

The benefits are everlasting to the individual chef who competes. His or her abilities in organization, eye for color and detail, and cooking and presentation skills are forever heightened, especially when performing under pressure. There are a host of other reasons to get involved in culinary competitions:

- Promotes camaraderie among chefs within the culinary field
- Provides inspiration to young professional chefs
- Provides a great way for chefs to network with each other
- Provides an arena for the chef to showcase skills and techniques to the public
- Offers educational rewards to the chef
- Promotes growth and research and development within the industry
- Sharpens the skills and techniques of the chef
- Allows for high levels of creativity within the industry

FIGURE 17–25 A large culinary salon. Courtesy of Chuck Clark Photography

- Encourages the use of good workmanship and nutritionally sound cooking
- Teaches economy and judicious use of products

Taking part in a culinary competition is another stepping-stone to becoming a skilled and accomplished chef. It is a worthwhile experience that also helps keep chefs abreast of the concepts, skills, and techniques that are current within the industry.

PLANNING FOR THE COMPETITION

Planning to compete in a culinary competition requires significant time and effort. It is therefore very important to start the organization process as early as possible, long before the competition date. Some obvious elements can be easily overlooked, leaving little time at the end to execute the food creation. Following a plan is the only way to be continuously successful. The following points are worth considering when planning for a competition:

- Confirm the dates and location of the competition.
- Read the current rules for the competition and make sure that they are fully understood by all those competing. Do not hesitate to seek clarification on rules and make sure the rules are the most current available.
- Some rules are specific to some areas and nations. Do not take anything for granted.
- Always ensure that the food products that are going to be used are available to you where you are and where you are going.

- Research what is current and what, if anything, the judges could be looking for.
- Concentrate on showing the judges the skills and techniques you have mastered.
- Plan a schedule of practices that is realistic and acceptable to your employer.
- Plan a progression chart of where the dishes should be by a specific time in to order ensure progress happens as meant.
- Stick to the practice schedule and always stay focused on the end goal.
- Seek out professional advice from colleagues or other seasoned competitors so as to eliminate any possible unforeseen pitfalls.
- Draw and write everything down from its conception to the final plate presentation.
- Keep the chosen dishes or presentations within the range of your skill and technical ability. Do not try to do too much.
- As competition time approaches, follow a strict regimen of nutrition and physical exercise. The competitions can be grueling and the competitors need to be fit enough to do well.
- Have checklists for all food, materials, equipment, uniforms, and all other personal items.
- Leave plenty of time to get to your destination and set up.
- Follow the entire rules specific to that competition and be prepared for any changes or equipment failures that may occur.

According to Chefs Ferdinand Metz and Tim Ryan, as identified in their *Taste of Gold,* the documented story of the 1988 U.S. Culinary Olympic Team, there are 10 common mistakes made by chefs in cold food competitions. They are listed in order of importance:

1. Poor execution of basic fundamentals
2. Making "food show" food instead of "customer food"
3. Creativity supercedes sensibility
4. Sloppy workmanship
5. Poor composition
6. Inconsistent sizes
7. Unappetizing food
8 Incorrect garnish count, portion count, or portion size
9. Sloppy aspic work
10. Poor layout

Although this list was identified nearly 20 years ago, it still accurately reflects the common shortcomings of many cold food entries in today's culinary salons.

The Devil Is in the Details

The competition hinges on the organizational abilities of the competitor, and more importantly, the attention to every detail. During practices and the lead-up to the event, a lot of time is invested in becoming skilled in several areas. But all the hard work and planning does not produce awards if details are forgotten on the day of competition. The final push to perfect the details must all be present to be successful. The devil is in the details: failure to follow an organized plan will undoubtedly result in a poorer performance, and ultimately lost points. The impor-

FIGURE 17–26 Focused
garde manger chef

tance is in achieving the personal goals as established by the competitor. In the final analysis, as a garde manger chef, one must be the strongest critic of one's own plan and of one's own work (Figure 17–26).

> Nothing in the world can take the place of persistence. Talent will not; nothing is more common than unsuccessful men with talent. Genius will not; unrewarded genius is almost a proverb. Education will not; the world is full of educated derelicts. Persistence and determination alone are omnipotent.
>
> John Quincy Adams

Professional Profile

BIOGRAPHICAL INFORMATION

Name: Michael Sullivan
Place of Birth: London, England

CULINARY EDUCATION AND TRAINING HIGHLIGHTS

Born in London's East End, just before World War II, being a world-class chef was not an obvious career choice to young Michael Sullivan. At 15 he left school before graduating and helped his musician uncle when he was playing gigs in various clubs and hotels. It was not unusual in those days for the entertainers to come through the kitchen to access the stage, and in doing so, young Michael gained an interest in the workings of these gigantic food labyrinths. It was this curiosity into the operation of a large commercial kitchen that inspired him to pursue a career in food service in the 1950s. Starting at the bottom of the kitchen ranks, he first worked at the Restaurant Frascati in London—he was the only English employee. After completing his Military National Service in the Army Catering Corp, he returned to find the restaurant had since closed down. This pushed him to find work in a series of top London restaurants, including the highly regarded Miribelle on Curzon Street. His skills and work ethic impressed one of his mentors at the restaurant and Chef Sullivan was sent to work at the legendary Maxim's in Paris in the early 1960s, where his career flourished and he worked for many years. In the early 1970s, he returned to London to work at the Carlton Tower and opened the kitchens at the Heathrow Penta Hotel in 1973, where he remained their executive chef until his retirement in 1994. Chef Sullivan was an avid culinary competitor, and he encouraged his staff to do the same as a means of training many of his young chefs in the skills of the garde manger. Altogether, he and his brigade won a total of 83 medals in world-class culinary competitions in locations such as London's Hotelympia, Frankfurt's Culinary Olympics, and Luxembourg's World Cup. He judged culinary salons in Scotland, England, Malta, and the United States, and lectured at the Westminster Catering College. It was Chef Sullivan's singular honor as National Chairman of the United Kingdom's Cookery and Food Association (a position once held by Georges-Auguste Escoffier) to lead a troupe of 100 chefs at the Queen Mother's 100th birthday celebration.

ADVICE TO A JOURNEYMAN CHEF

Author's Note: Sadly, Chef Michael Sullivan passed away in 2004 while this book was being written. We were unable to interview our friend for his advice to a journeyman chef nor ask for a favorite recipe. To that end, we offer the following, which we paraphrase from his comments made while addressing student culinary competitors from eight countries during a culinary salon he was judging in Malta.

Ours is a noble business. We are privileged daily to provide life-sustaining nourishment for those both rich and poor—it is a responsibility that requires serious commitment and a desire to aid your fellow man. And if done with care and pride, you will never be out of work nor lacking in friends. I've had the great fortune to have a wonderful family and to have worked for 45 years in many fine restaurants and hotel kitchens with colleagues I've regarded as family. I wish these treasures for you.

Michael Sullivan (paraphrased)

REVIEW QUESTIONS

1. What is the purpose of decoration and presentation?

2. What is the purpose of using aspic to coat foods?

3. How is aspic available to the chef?

4. Explain the methods of applying aspic to food.

5. Describe six other uses for aspic.

6. What is chaud-froid?

7. How is chaud-froid applied to galantines and terrines?

8. What are the other uses for chaud-froid in food displays?

9. What rules should be applied when designing a cheese platter?

10. What are the advantages of taking part in a culinary competitions?

ACTIVITIES AND APPLICATIONS

A. **Group Discussion**

Discuss the advantages of taking part in culinary competitions internationally.

B. **Research Project**

Research what it takes to actually take part in a competition using rule books from approved competitions and interviews of chefs who have already taken part.

C. **Group Activity**

Design a platter with colored cardboard cutout shapes in place of food items. Present them on a platter for review by your peers.

D. **Individual Activity**

Write an approximate budget for planning to compete in a food show, presuming (1) there are 10 practice sessions and (2) the competition will last 3 days and take place in another state.

Appendix A: Weights and Measurements

Weights and Measurements

This section has been written to give you instant access to a table of weights and measurements when you are entering recipes. The weights and measurements are only approximate: you can never be precise for many reasons:

1. The moisture contents of products vary constantly.
2. Sizes of individual pieces or particles will vary from container to container.
3. The exact weight of a gallon or a pound of product is seldom a convenient round number.
4. It would be impractical to say that a pint of water is 1 9/10 cups. It is simpler just to say 2 cups.
5. Products containing moisture become lighter as they dry out.
6. Wet products containing sugar become heavier when the moisture evaporates and they become thicker.
7. A cup of flour could weigh 4 ounces. If you sift it, it may weigh less.
8. Any measurement such as a "Level" teaspoon or "Level" cupful is seldom exactly accurate.

Gram Weight Conversion Table

OUNCES	GRAMS
1	28.35
2	56.70
3	85.05
4	113.40
5	141.75
6	170.10
7	198.45
8	226.80
9	255.15
10	283.50
11	311.85
12	340.20
13	368.55
14	396.90
15	425.25
16	453.60

POUNDS	GRAMS
1	453.6
2	907.2
2.5	1134.0
3	1136.8
4	1814.4
5	2268.0
6	2721.6
7	3175.2
8	3628.8
9	4082.4
10	4536.0
15	6804.0
20	9072.0
25	11340.0
30	13608.0
35	15876.0

Metric Size Fluid Equivalents

METRIC	U.S. FL OZ	3/4 OZ	1 OZ	1-1/8 OZ	1-1/4 OZ	1-1/2 OZ	CLOSEST PREVIOUS CONTAINER U.S. OZ
1.75 Liter	59.2	78.9	59.2	52.6	47.4	39.5	1/2 Gal. = 64 OZ
1.0 Liter	33.8	45.1	33.8	30.0	27.0	22.5	Qt. = 32 OZ
750 Milliliters	25.4	33.9	25.4	22.6	20.3	16.9	5th = 25.6 OZ
500 Milliliters	16.9	22.5	16.9	15.0	13.5	11.3	Pt = 16 OZ
200 Milliliters	6.8	9.1	6.8	6.0	5.4	4.5	1/2 Pt. = 8 OZ
50 Milliliters	1.7						Miniature = 1.6 OZ

Equivalent Measures for Fluids

3 teaspoons	=	1 tablespoon
16 tablespoons	=	1 cup
28.35 grams	=	1 ounce
1 cup	=	½ pint
2 cups	=	1 pint
2 pints	=	1 quart
4 quarts	=	1 gallon
768 teaspoons	=	1 gallon
1 lb. (water)*	=	16 fluid ounces
1 lb. (water)*	=	1 fluid pint
2 lb. (water)*	=	1 fluid quart

Volume Conversions for Recipe Writing

	TEASPOON	TABLESPOON	QUART	CUP
1 teaspoon	1.0	0.333333	0.0052083	0.020833
1 tablespoon	3.0	1.0	0.015625	0.062500
1 cup	48.0	16.0	.25	1.0
1 pint	96.0	32.0	.50	2.0
1 quart	192.0	64.0	1.0	4.0
1 gallon	768.0	256.0	4.0	16.0

Pound and Ounce to Gram

OUNCES	GRAMS
1	28.35
5	141.75
10	283.50
12	340.20
16	453.60

POUNDS	GRAMS
1	453.60
5	2268
10	4536
25	11340
50	22680

Teaspoons and Tablespoons

TEASPOONS	TABLESPOONS
3 = 0.5 ounce	1 = 3 teaspoons
6 = 1 ounce	2 = 1 ounce
48 = 1 cup	4 = 0.25 cup
96 = 1 pint	8 = 0.5 cup
192 = 1 quart	16 = 1 cup
960 = 5 quarts	128 = $\frac{1}{2}$ gallon
768 = 1 gallon	256 = 1 gallon

Decimal Equivalents of Common Fractions

FRACTION	ROUNDED TO 3 PLACES	ROUNDED TO 2 PLACES
$\frac{5}{6}$	0.833	0.83
$\frac{4}{5}$	0.8	0.8
$\frac{3}{4}$	0.75	0.75
$\frac{2}{3}$	0.667	0.67
$\frac{5}{8}$	0.625	0.63
$\frac{3}{5}$	0.6	0.6
$\frac{1}{2}$	0.5	0.5
$\frac{1}{3}$	0.333	0.33
$\frac{1}{4}$	0.25	0.25
$\frac{1}{5}$	0.2	0.2
$\frac{1}{6}$	0.167	0.17
$\frac{1}{8}$	0.125	0.13
$\frac{1}{10}$	0.1	0.1
$\frac{1}{12}$	0.083	0.08
$\frac{1}{16}$	0.063	0.06
$\frac{1}{25}$	0.04	0.04

Fahrenheit to Celsius Conversion (Approximate)

FAHRENHEIT	CELSIUS
32° F	0° C
50° F	10° C
68° F	20° C
86° F	30° C
100° F	40° C
115° F	45° C
120° F	50° C
130° F	55° C
140° F	60° C
160° F	70° C
170° F	75° C
180° F	80° C
185° F	85° C
195° F	90° C
200° F	95° C
212° F	100° C
230° F	110° C
250° F	120° C
265° F	130° C
285° F	140° C
300° F	150° C
325° F	165° C
350° F	175° C
360° F	180° C
375° F	190° C
400° F	200° C
425° F	220° C
450° F	230° C
485° F	250° C
500° F	260° C
575° F	300° C

Sizes and Capacities of Scoops

NUMBER ON SCOOP	LEVEL MEASURE
6	$\frac{2}{3}$ cup
8	$\frac{1}{2}$ cup
10	$\frac{3}{8}$ cup
12	$\frac{1}{3}$ cup
16	$\frac{1}{4}$ cup
20	$3\frac{1}{3}$ tablespoons
24	$2\frac{2}{3}$ tablespoons
30	2 tablespoons
40	$1\frac{2}{3}$ tablespoons
50	$3\frac{3}{4}$ teaspoons
60	$3\frac{1}{4}$ teaspoons
70	$2\frac{3}{4}$ teaspoons
100	2 teaspoons

Sizes and Capacities of Measuring/Serving Spoons

SIZE OF MEASURING/ SERVING SPOON	APPROXIMATE MEASURE
2 ounces	$\frac{1}{4}$ cup
3 ounces	$\frac{3}{8}$ cup
4 ounces	$\frac{1}{2}$ cup
6 ounces	$\frac{3}{4}$ cup
8 ounces	1 cup

Appendix B: Canned Goods

Common Can Sizes and Their Appropriate Contents

CAN SIZE	PRINCIPAL PRODUCTS
No. 5 squat	
75 oz. squat	
No. 10	Institution size—fruits, vegetables, and some other foods
No. 3 cyl	Institution size—condensed soups, some vegetables. Meat and poultry products. Economy family size—fruit and vegetable juices
No. 2 1/2	Family size—fruits, some vegetables
No. 2	Family size—juices, ready-to-serve soups, and some fruits
No. 303	Small cans—fruits, vegetables, some meat and poultry products, and ready-to-serve soups
No. 300	Small cans—some fruits and meat products
No. 2 (vacuum)	Principally for vacuum pack corn
No. 1 (picnic)	Small cans—condensed soups, some fruits, vegetables, meat, and fish
8 oz.	Small cans—ready-to-serve soups, fruits, and vegetables
6 oz.	

Common Can Sizes and Their Approximate Weights and Volumes

Volumes represent total water capacity of the can. Actual volume of the pack would depend upon the contents and the head space from the fluid level to top of can.

AVERAGE CAN SIZE	AVERAGE VOLUME	CANS PER FLUID	APPROX CUPS	CASE	WEIGHT
75 squat				6	4 lb 11 oz
No. 5	56			6	4 lb 2 oz
No. 10	105.1	99 to 117	12 to 13	6	6 lb 9 oz
No. 3 cyl	49.6	51 or 46	$5\frac{3}{4}$	12	46 fl oz
No. 2 1/2	28.55	27 to 29	$3\frac{1}{2}$	24	12 oz
No. 2	19.7	20 or 18	$2\frac{1}{2}$	24	1 lb 13 oz
No. 303	16.2	16 or 17	2	24 or 36	1 lb
No. 300	14.6	14 or 16	$1\frac{3}{4}$	24	$15\frac{1}{2}$ oz
No. 2 (vacuum)		12 fl oz	$1\frac{1}{2}$	24	
No. 1 (picnic)		$10\frac{1}{2}$ fl oz	$1\frac{1}{4}$	48	
8 oz.	8.3	8 fl oz	1	48 or 72	8 oz
6 oz.	5.8	6 fl oz	$\frac{3}{4}$	48	6 oz

Every industry has its unique vernacular that separates its business from others. To excel in the foodservice industry, one must speak the language. The following terms are basic to *kitchenspeak*—the dialect of the garde manger chef. Although this glossary reflects numerous innovations by the authors, many of the terms listed in this glossary are not the original definitions of the authors, but are a compilation derived from sources formally published.

A

acid   Any substance that releases hydrogen ions in a watery solution. Acids have a pH value of less than 7.

acidification   The first step in cheese making, it is the lowering of the pH (increasing acid content) of the milk to make it more acidic.

acid-set cheeses   The custard-like state that milk achieves with the addition of acidic agents such as lemon juice or vinegar.

additive sculpture   Sculpting by adding wanted material to the original mass, such as when sculpting clay, vegetables, or ice.

affinage   A French term for ripening of cheeses.

age   To store a product, often meat, under specific conditions of time, temperature, humidity and light for the purpose of improving flavor and texture.

aging   A slow chemical change that happens to meat as it ages. Aging improves flavor and tenderizes the meat.

amuse-bouches   Small "mouth amusements" served before dinner.

antipasti   An Italian platter of assorted cured meats, cheeses, and fresh vegetables served as an appetizer.

antojitos   A Mexican term for appetizers.

appetizers   The smaller portions of food that appear as the first course served at the table.

armature   The central core upon which sculpting material is placed and which adds strength and stability to a sculpture.

artisanal   A high-quality, usually handmade product such as cheese or bread.

aspic   A clarified, jellied stock that solidifies when cold. It may be flavored with wine, port, sherry, or champagne. Used to coat foodstuffs, such as terrines, or fill pâté en croûte.

as purchased (AP)   The weight or count of a product as delivered to the food service operator.

B

bain marie   A service table with openings to hold containers (bain marie inserts) over hot water or ice.

ballontine   A French term for a boned, stuffed, and rolled poultry leg and thigh served cold or hot.

barbecue   To slowly cook seasoned food over burning coals or aromatic woods.

bard   A thin slice of fat, salted fat, or bacon that is tied around meat, poultry or game to protect it and moisten it during cooking.

beef liver   A relatively large organ in a cow, it is used extensively in cooking. It is commonly associated with braising, frying, and sautéing.

binding agents   Ingredients used to cohere, unite, or hold together other ingredients to make a forcemeat.

blood   The liquid that circulates through animal's organs carrying nourishment, oxygen, heat, vitamins and other essential chemicals. Certain animal blood is used for flavor and the binding of sausages, such as with black pudding.

blue cheeses   A general category of cheeses that contain visibly blue-green veins or striations of mold that contribute to its sharp flavor and pungent aroma. The mold is a result of either dusting or inoculating with Roquefort penicillin.

braai   South African barbecue commonly cooked outside on an open barbecue.

brains   A large organ found in the cavity of an animal's skull; they are commonly poached in court bouillon, braised, or made into fritters.

brine curing   A liquid of salt, nitrites, spices, and water used for pickling meats or fish.

buffet concept   The predominant theme surrounding the event. Includes the distinguishing characteristics of a buffet service created to meet the needs of the kitchen staff, service staff, and customer.

bulbs   Subterranean buds consisting of both stems and leaves. They have flat, round, or elongated fleshy bodies with roots growing downward from their undersides and shoots and leaves sprouting from their tops.

butcher's yield test   Used to evaluate the difference in the cost of cutting portions (including the resultant by-products of trim, bones, and fat) or buying them preportioned.

C

canapés   Savory hors d'oeuvres made with a bread, cracker, or pastry base, so that they can be picked up with the fingers and eaten in one or two bites.

carpaccio   An Italian appetizer of thinly sliced raw (red) beef drizzled with olive oil and lemon juice, garnished with capers and chopped onions; so-named for the Renaissance painter Vittore Carpaccio, famous for the red colors he used in his paintings.

caul fat   A lacey fat membrane surrounding the paunch and intestines of an animal, it is used as a casing or wrap for meats and fish.

caviar   The salted roe (egg) of the sturgeon; it is literally a salt-cured fish egg.

ceviche (also cebiche and seviche)   A Latin American dish of raw seafood marinated in citrus juices (such as lime), onions, tomatoes, and chiles.

charcuterie   Cooked meat products, traditionally and primarily from pork.

chat   The East Indian term for appetizer-sized portions of food.

chaud-froid   A jellied sauce used as an edible coating for meat, poultry, or seafood. May be white, brown, red, or green.

chef de partie   An organized system to ensure against duplication of work and to further increase communication and efficiency among the staff. Developed by Georges-Auguste Escoffier.

chop   A cut of meat that is sliced through the rib bones, the loin, and sometimes the tenderloin to form a small individual cut.

chow-chow   A relish of pickles or other vegetables.

churrasco   Brazilian barbecue in which sausages, beef steaks, pork tenderloins, and all other red meats are seasoned with coarse salt before being spit-roasted and carved tableside. White meats are marinated in a mixture of garlic, salt, and lime juice before being spit-roasted.

chutney   A relish made from fresh fruits and spices, and suspended in a cooked sugar and vinegar solution.

classification   The arrangement of ingredients in classes or categories in order to make them easier to understand, according to a system.

classification of cheese   A means by which a cheese may be identified and understood by its characteristics that are common to other cheeses.

clear ice   Ice blocks that are preferred by sculptors because they are transparent and do not have a cloudy feather running down their centers. Also called *slick ice* or *crystal clear.*

client-based pricing   Bases the charges of a function on the client's perceived value rather than the direct food and labor costs to produce the event.

closed sandwiches   A category that includes all sandwiches enclosed between bread, from top to bottom.

coagulation   The point at which milk congeals into thickened mass.

Colbert   This technique differs for flat and round fish. For flat fish, the fish are skinned and the fillets are opened up exposing the bone. They are then pinned back and the bone is broken in several places. After frying, the bone is removed, exposing the fillets below. The fish is dressed in crumbs and deep-fried. For round fish, carefully remove the whole backbone, leaving 1 inch (2.5 cm) of the backbone and both fillets still attached to head and tail.

cold sandwiches   The most commonly prepared sandwich, consisting of cold bases, moistening agents, and/or fillings.

cold smoking   A method of curing, preserving and/or flavoring foods by exposing them to smoke at temperatures of 50° to 85°F (10° to 29°C) for a prolonged period. Time and temperature will depend on the specific food item.

cold wrapped sandwiches   Closed, cold sandwiches encased in bread or lettuce wrappers.

collagen casings   Sausage casings made from water-insoluble proteins found in connective tissues such as skin, ligaments, tendons, and cartilage.

combination salads   Also known as *composed salads*, combination salads are generally comprised of several different categories of ingredients that are seasoned separately but presented together on the same plate.

common cuts   These cuts are the smaller pieces that the primal cuts are broken into in order to separate the meat into levels of tenderness.

competitive-based pricing   Bases the charges to the client for a function on the relative cost charged by competitors in the market, rather than on the direct food and labor costs.

complex salads   Also known as *mixed salads,* complex salads are seasoned with flavorful dressings or marinades and usually contain multiple ingredients from the same categories of foods, including raw or cooked vegetables, fruits, meats, seafood, game, or poultry.

component parts of a sandwich   Sandwiches generally contain three specific component parts, which, when put together, constitute the completed dish. They include the structure or base, the moistening agent, and the filling.

compound butter   A mixture of softened whole butter and flavorings used as a sauce or spread for a canapé or sandwich.

contribution margin   The profit or margin that remains after product cost is subtracted from an item's selling price.

corn syrup solids   Used to add flavor and help hold the color of meat in sausage making. They are especially important in the semidry or dry-cured process because they help support the fermentation process.

cost-based pricing   Adds a standard markup to the direct food and labor costs to determine the final charges to the client.

cost of goods sold (COGS)   Describes the direct expenses, generally of food and beverage, incurred in producing a particular good. Monthly COGS are calculated by taking the opening inventory plus purchases, minus the closing inventory.

crapaudine   A method of splitting a bird through the rib cage to the point of the wing in order to open the bird flat for grilling.

cross-utilization   Both raw and prepared products are used in multiple fashions.

crown   Two racks, or a single long rack, is tied to represent a crown. The meat is tied facing in toward the canter of a circle with the bones creating a crown shape.

crustaceans   Crustaceans include shrimp, crabs, lobsters, crayfish, amphipods, isopods, ostrapods, and barnacles. This group may have more than 50 million species, and many have adapted to a life out of the water.

culinary salon   Another term used for *food show competition.*

curds   The thick substance that results when protein and fat are extracted from milk through the infusion of rennet or citric acid.

cure   To apply any of several methods of processing foods, particularly meats and fish, to retard spoilage. Usually accomplished by drying, salting, pickling, or smoking.

curing   See *cure.*

cutlets   A small cut taken from the rack, it is a small piece of meat attached to a long clean bone.

D

darne   A slice of fish taken from a large round fish through the bone and skin, resembling a steak.

délice   A neatly folded trimmed fillet that is tucked in at both ends.

demi-darne   A darne from a very large fish that has been cut in half for a better portion size.

design templates   The outline of a design that can be placed directly onto an ice block for tracing and guiding the sculptor's etching or cutting tools. May be made from wood, plastic, or metal, but is usually made from paper.

dim sum   The Chinese term for appetizer-sized portions served from carts, or à la carte, in restaurants.

dip   A thick creamy sauce or condiment (served hot or cold) to accompany raw vegetables, crackers, or other small food items.

do nhau   Vietnamese little bites or little portions of food.

drying   One of the oldest methods of preserving foods, it removes the moisture from foods, thereby extending their shelf life.

dry aging   A slow chemical change that happens to meat as it ages by hanging at the correct temperature. Dry aging dramatically improves flavor and tenderizes the meat.

dry curing   A method of curing meat or fish by packing it in salt and seasonings.

dry sugar curing   Involves the addition of sugar and nitrate and/or nitrite to the salt that is then applied directly to the surface of the meat.

duck livers   Duck livers are often enlarged by the force-feeding of farm raised ducks with enriched feed. The enlarged livers, known as fois gras, are prized for their rich, delicate flavor.

E

edible portion (EP)   The weight or count of a product after it has been trimmed, cooked, and portioned.

emulsion   A smooth temporary mixture of two unmixable liquids created by rapid blending, such as with oil and vinegar.

en colère   This technique uses a small round fish that has been cleaned and skinned. The tail of the fish is then pulled round and secured between the teeth of the fish.

en lorgnette   A small round fish that has been skinned and boned, leaving the fillets attached to the head. The fillets are then dressed in crumbs and rolled up to the side of the head. The two fillets represent the glasses of a pince-nez or handheld binoculars, with the head of the fish acting as the handle.

ergonomics   The science of adapting working conditions to the needs of the worker.

escalope   A thin boneless piece of tender meat that has been cut across the grain and flattened slightly to an even thickness.

evisceration   The removal of all the interior organs, stomach, and entrails contained in the cavity of an animal.

extenders   Less expensive ingredients, such as rice or bread, added to meats, seafood, or poultry to make forcemeat.

F

fabrication   Applies to the work needed to render food ready for consumption. For example, fish scaling, skinning, boning, pin boning, and trimming—all these tasks could be part of the fabrication of the fish.

farce   Ground, chopped, or puréed stuffing with which to fill pâtés, terrines, sausages, galantines, ballotines, or other meat or fish. A synonym for *forcemeat.*

farci   The term applies to a fish being boned and skinned with head and tail still attached, then stuffed and shaped into the fish's original shape.

fermier   A French term used to describe a farm-made cheese.

filled roll or bun sandwich   Similar to the closed sandwich, the bread is in the form of a roll or bun, and the fillings can be cold or warm and bound with a flavored dressing.

filleting   Removing the maximum amount of meat from the bone in the most efficient manner.

firm cheeses   A general category of cheeses that are aged and firm, and have a closed or opened texture. They generally have a flavor ranging from mild to very sharp, and moisture content of 30 to 40 percent.

fixed costs (expenses)   An expense that remains constant despite increases or decreases in sales volume.

flushing   Using water to wash away the salt from salted casings.

foie gras   The fattened liver of a duck or goose; it is prized as being one of the world's richest foods.

food cost   The cost of all food required/consumed for a specific period of time or event.

food cost percentage (FC%)   The cost of all food required/consumed for a specific period of time or event divided by the food sales for the same period of time or event.

food for carriage   Food designed to be eaten away from home while traveling.

food scatter stations   Similar to food zones but arranged as a series of separate tables, or islands, within a banquet. Each scatter station can consist of a singular food station, or it can be made up of several stations.

food zones   Separate groupings of similar food items generally arranged as minisections on a much larger buffet, but they can be stations unto themselves.

forcemeat   Ground, chopped, or puréed stuffing with which to fill pâtés, terrines, sausages, galantines, ballotines, or other meat or fish. A synonym for *farce*.

(to) French   Removing the excess fat and connective tissue from a roast, rack, or chop of meat, leaving the eye muscle intact and exposing the cleaned rib bone.

fried sandwiches   Any filling placed between two slices of bread and cooked by shallow frying or sautéing. They can be dipped in a coating for added texture and color.

fruits   The edible organs that develop from the ovary of flowering plants. They are usually sweet and eaten as is or used as ingredients.

furred game   This term is generally given to larger animals that have a fur-covered skin.

G

galantine   A term that indicates that poultry or any part of an animal has had its bones completely removed. The flesh is then reshaped, tied with cloth or foil, and cooked. It is generally served cold.

galette   Even pieces of fish fillets are towered on top of each other and separated by mousseline or another form of stuffing.

game birds   Originally, game birds were birds hunted for sport or food. Common game birds include duck, goose, pheasant, quail and pigeon, many of which are also ranch raised and available commercially.

garde manger   A French term meaning "keeping to eat." Historically, the position was adopted to put a responsible person in charge of keeping food fit to eat. In modern times, the position is responsible for cold food preparation, including salads, sandwiches, cold appetizers, sculptures of vegetables, fruit, or ice, kitchen-made cheeses, and charcuterie products.

gelatin   A colorless, flavorless, and odorless mixture of proteins made from animal bones, connective tissues, and other parts. When dissolved in warm water, it is used to bind or coat other foods that are chilled.

glaze   A gravy or stock reduction used to coat food, or an aspic coating.

goal value   Setting a target number that is neither a dollar figure nor a percentage of sales, above which items remain on the menu, below which items are removed or adjusted.

goujon   This is a technique of cutting fish flesh into 3- by ¼-inch (7½- by ½-cm) strips and is commonly used as a way of using up trim.

grains   The fruit or seed of grasses that bear edible seeds, including corn, wheat, and rice.

greenhouse effect   Extended exposure to sunlight, specifically the ultraviolet rays, that melts ice from the inside.

griddled sandwich   Prepared much in the same way as a fried sandwich, except it is cooked on a flat heated griddle or cast iron comel. These sandwiches can be coated with raw egg, flour, and/or breadcrumbs. They are also commonly dusted with spices or sugar when cooked.

grilled sandwiches   Encasing a filling between two slices of bread or placing the filling on the top of bread slices. The sandwich is then placed onto an open grill or char-grill for cooking, giving it the color and crispness desired.

guest experience   Includes the guest's impressions on the sanitation, safety, facility conditions, staff training, attention to detail, menu quality and service, and all of the other tangible and intangibles associated with the visit.

guilds   An association for mutual aid and the promotion of common interests.

H

hard cheeses   A general category of cheeses that are aged for an extended period of time. They have a hard and dense texture, a sharp and tangy flavor, and moisture content of approximately 30 percent.

heart   A very tough organ of the animal body that has many uses in cooking. It is normally used for sausage making and the making of terrines and pâtés. It can also be stuffed and cooked slowly by braising until rendered tender enough to eat.

hors d'oeuvres   Small portions of food that are served away from a meal. They may precede a large meal or may be used as a meal substitution.

hot sandwich   Sandwiches consisting of hot bases, moistening agents, and/or fillings.

hot smoking   A method of curing, preserving, and/or flavoring foods by exposing them to smoke at temperatures of 200° to 250°F (93° to 121°C). Time and temperature depend on the specific food item.

hot wrapped sandwiches   A closed, hot sandwich encased in bread wrappers.

J

jerk   A Jamaican preparation method in which meats and poultry are marinated in herbs and spices, then cooked over a wood fire. Pimento (allspice) wood is typically used.

job description   A listing of performance objectives, responsibilities, job specifications, and reporting relationships for a particular job.

K

kanto   Thai term for appetizer-sized portions, often served as street food.

kidneys   Small organs located in the lower back of an animal. They come in pairs and are encased in a creamy white fat. They are fried, grilled, or braised.

kitchen brigade system   A strict hierarchy of authority, responsibility, and function.

knit   The bonding of curds as they are pressed into a cheese mold to form their characteristic shape and to release any additional whey.

L

labor cost   All expenses, including payroll, required to maintain a workforce for a specific period of time or event.

labor cost percentage   All expenses, including payroll, required to maintain a workforce for a specific period of time or event divided by the food sales for the same period of time or event.

la grande cuisine   Classical French cookery founded by Marie-Antoine (Antonin) Carême, whereby he systematized techniques, standardized recipes, and began to streamline menus.

larding   The process of introducing thin strips of fat into meats that lack their own natural fats.

legumes   A Portuguese term for vegetables; they consist of a large group of plants that have double-seamed pods containing a single row of seeds.

liver   A relatively large organ found in all animals, the liver is used extensively in cooking and is commonly associated with braising, frying, and sautéing. It has a pleasant aromatic flavor.

loins   This common cut comes from the back of the animal and runs down the length of the saddle and the rack. Removing it from the saddle area is more common. It is a tender, even strip of meat.

loin cuts or steaks   A small cut that comes from a loin, it is a steak cut across the grain of the loin.

lox   Salmon that is brine cured and then typically cold smoked.

M

make or buy decision   The decision (relative to food costs, labor costs, ease of preparation, and quality) whether it is best to prepare foods versus purchasing foods ready-made.

manufactured ice   Ice intentionally formed by man rather than by nature.

marbling   Fine flecks of fat that are evenly distributed amongst the lean meat.

mazza (meze)   An Arabic term for hors d'oeuvres.

medusa   A small round or flat fish is boned, leaving the head attached to the fillets. The fillets are then cut into strips and the fish is fried in light batter to resemble Medusa's hair.

menu analysis   The disciplined act of recording the sales history of all menu items sold, and evaluating both the item's contribution to profit and customer appeal.

menu engineering   Involves the examination of menu items and cross-utilization of ingredients, periodic analysis of sales velocity reports to examine what items sell better or produce more profit or incremental revenue, consideration of where menu items appear on the menu and how they are priced, and evaluation of recipes to determine whether better sales or better food cost could be achieved by adjusting ingredients or prices. Additionally, menu engineering involves a close study of purchasing and receiving practices and inventory control.

mesophilic starter   A bacterial culture that triggers enzymatic activity for cheeses that require lower, or warm, temperatures to form curds.

metric system   A measurement system developed by the French and used worldwide. It is a decimal system in which the gram, liter, and meter are the basic units of weight, volume, and length, respectively.

mezze   A Greek and Turkish form of hors d'oeuvre. Used to describe both the social occasion preceding the meal and the food itself.

microgreens   Tiny sprouts and shoots, harvested with their roots still attached, while young and immature. Their varieties include such plants as alfalfa, clover, broccoli, beets, cress, and radishes.

mignon   A small filet that has been folded to resemble a cornet. A mignon is sometimes stuffed.

mise en place   A French term referring to having all the ingredients necessary for a dish prepared and ready to combine up to the point of cooking.

modernist sculptor   An individual who welcomes the use of power tools when ice sculpting, in preference to the use of traditional ice saws and chisels.

mousse   A French term for "foamed." It is a pâté, terrine, or mold made of smooth, puréed farce to which gelatin, whipped cream, and/or egg white is added.

mousseline   A smooth, puréed farce, usually lightened with cream and egg white, and delicately flavored.

multiplier-based pricing   To use a multiplier to arrive at a selling price for an operation's food.

N

natural casings   A natural membrane (usually from pig, beef, or sheep intestines) used to enclose sausage forcemeat.

natural ice   Ice formed by nature.

nitrate   Nitrate is a natural substance readily found throughout the plant world. Our early ancestors, lacking refrigeration, discovered the preservative powers of nitrate/nitrite. Nitrates are food additives, such as sodium nitrite and potassium nitrite, which are used to stabilize the pink color associated with cured meats, and also as antioxidants.

nitrite   Nitrate, found in fruits and vegetables, is converted to nitrite by our saliva and a digestive enzyme in our intestinal tract thereby producing over 90% of the nitrites in our bodies. Food additives such as sodium nitrite and potassium nitrite are used as curing agents, color stabilizers, and/or preservatives in processed foods, such as meat products. Nitrite blocks the growth of dangerous botulism-causing bacteria and prevents spoilage.

nonplanner's cycle   The manager is constantly busy dealing with emergencies caused through lack of organizational forethought.

nouvelle cuisine   A French expression meaning "new cooking." It is a culinary movement emphasizing freshness and lightness of ingredients, classical preparations, and innovative combinations and presentation.

O

offal   Those parts of a meat animal that can be consumed as food but are not skeletal muscle. The term means to "off fall," indicating the parts of the animal that fall off the carcass. The term is also referred to as *giblets* in poultry and game birds.

open-faced sandwich   A category that includes all sandwiches layered upon bread but not enclosed. The filling is exposed without a top bun or bread slice.

P

panadas   An egg, bread, flour, potato, or rice mixture used to lighten and bind ingredients together.

pasta filata cheeses   An Italian term describing the curds resulting from a cheese-making process. The fermented curds are heated until they are plastic-like and then stretched and spun until they are free of holes and whey.

pasteurization   The process of heating milk to a high temperature (around 161°F [72°C]).

pâtés   A farce, often encased in pastry, baked and usually served cold.

pâté en croûte   A farce encased in pastry, baked, filled with aspic, and then served cold.

paupiette   This preparation is applied to small fillets of fish that are rolled and stuffed with mousseline or forcemeat into a barrel shape.

pawed game   Applies to animals that have pawed feet as opposed to hoofed feet.

pellicle   A shiny, slightly sticky glaze that forms after washing the dry cure off the meat product and allowing it to air-dry before smoking. If it is smoked before the pellicle forms, the heat will force out more moisture and the smoke will be pushed away from the surface. The pellicle keeps moisture in and lets the surface take the smoke. The smoke itself consists of numerous tiny droplets of various natural chemicals, such as aldehydes, phenols, ketones, and carbolic acid. These chemicals tend to condense on the food being smoked, adding a tacky film to the pellicle.

performance standards   The minimum levels of performance expected of the employee for each performance objective. They are used to identify to what extent each job requirement is to be met.

pH value   The position on the pH scale with values ranging from 0 to 14, representing the degree of acidity. A value of 7 is neutral, 0 is most acidic, and 14 is most alkaline.

pickling   Using a solution of salt, vinegar, and spices to cure foods by marinating them in the solution.

pièce montée   A French term meaning "assembled part." The pièce montée is a centerpiece, usually edible, for a food display.

plié   A technique similar to délice, except the fish fillets are slightly beaten and then folded in two. It is done to reshape the fish to fit a plate easily.

pluck   The word can mean just the lungs on their own or any combination of the lungs, liver, spleen, and heart.

portion control   Methods to ensure that a portion of food or beverage is the correct size, according to the standardized recipe.

poultry   Any domesticated fowl.

powdered dextrose   Much heavier than sugar and only 70 percent as sweet, powdered dextrose is used as a browning agent in fresh breakfast or country sausage.

Prague powder #1   A commercially available cure that is used to cure all meats requiring cooking, smoking, and canning. A pound (453 g) of Prague powder #1 consists of 1 ounce (28 g) of sodium nitrite and 1 pound (453 g) of salt.

Prague powder #2   A commercially available cure that is used with dry-cured products that do not require cooking, smoking, or refrigeration.

preflushed casings   Casings that have been previously washed to remove the salt.

primal cuts   When cutting an animal into pieces, these are the standard cuts that are taken of the animal every time. They are removed in this particular way to yield the most meat and separate the tender parts from the tough parts.

prime costs (expenses)   Food, beverage, and labor costs, which collectively represent the greatest costs to an operation.

processed cheeses   Cheeses made from one or more cheeses of the same or different varieties. The processed cheese is made from grinding other cheese and mixing it with flavoring ingredients, emulsifying agents, and colorants; it is then heated and molded.

production management   The garde manger is chiefly responsible for recipe (product) development, cross-utilization (production efficiency), waste control (scrap rate control), and turning leftovers into profit.

production reports   The quantity of each item prepared and the amount sold, and their difference, is recorded daily.

Q

quality points   The important factors to bear in mind when assessing a food's quality.

quenelles   A shape given to pliable food products normally molded between two spoons of varying sizes.

R

racks   The longest section of the rib that has been removed from the spine with the eye of the meat attached.

rennet   An extract of the stomach lining of young mammals containing digestive juices (enzymes) that coagulate milk. Used to promote coagulation in cheese making.

rijsttafel   The Dutch expression for a "rice table," used to describe an Indonesian-inspired dish of spiced rice surrounded by small dishes of foods (appetizers).

rind   The outer rind (coating) that a cheese develops while air drying and aging.

ripening   A step in cheese making in which milk is allowed to undergo an increase in acidity as a result of the activity of cheese-starter culture bacteria.

roots   Those parts of the plant that grow downward away from the sun into the soil, to absorb nutrients and moisture.

S

salad   A single food, or a mixture of different foods, served with a dressing.

salad base   Serves as both the underliner on which the salad is built and as a canvass to frame the remainder of the ingredients.

salad body   The most important part and focal point of the salad, which is placed upon the base.

salad dressing   Used to moisten and flavor the other ingredients, thereby enhancing the body of the salad.

salad garnish   Adds contrast in taste, texture, color, height, and aroma to the body of a salad.

salsa   A relish of either raw or cooked fruits or vegetables intended to add flavor to other foods.

salt   A white granular substance (sodium chloride) used to season or preserve foods.

sandwich   A dish generally consisting of sliced bread and a variety of meats, cheeses, relishes, jellies, vegetables, lettuce, and condiments that can be picked up and eaten by hand as a meal.

sandwich buffet   An efficient method of feeding many people at the same time while allowing them the opportunity to choose their own ingredients. It requires the setting up of lines of sandwich materials, all in the correct order of the sandwich's construction.

sashimi   A Japanese appetizer or main course of raw fish that is often served with soy sauce, wasabi, vinegar, and Asian pepper sauce.

sausage horns   Attachments that are mounted onto meat or sausage grinders, on which sausage casings are fitted. The ground sausage meat is pressed through the horn, into the casing, to form the sausage.

sauté   A dry heat cooking method whereby high heat is transferred through a small amount of fat to a food item. The term is used for the way chickens are cut to prepare them for the method of cooking sauté. They are cut into 13 pieces, which are two breast pieces with drummet attached, two pieces of breast, two wings, two drumsticks, two thighs, and three sections of carcass.

scrap rate   The amount of loss measured in ratio to the overall quantity used.

sculpture   A three-dimensional work of art formed from solid material, such as ice.

semisoft cheeses   A general category of cheeses that are drained and molded, but not pressed, to form its body and outer rind. With a high moisture content, they generally age for a short period of time.

shari   The seasoned rice used for sushi. Also known as *zushi*.

Siamese twins of management   A strong system of planning and controls.

simple salads   Basic in nature and composition, simple salads include light salads made from a variety of one or more greens, fruits, pastas, or grains, but not in combination with each other.

skinning   Removing the skin from a fish while whole or filleted.

slaughter   The humane killing of an animal for food. The animal is stunned, usually with a blow or electrical discharge to the head, and then it is hung up and bled from one of the major blood vessels.

small cuts   These cuts are taken from the common cuts and are the final stage before single portion cooking

smoking   Applying a controlled amount of smoke to a food product.

sodium nitrite   A food additive used as an antibacterial agent and/or color stabilizer. Used primarily in cured and/or smoked fish and meats.

soft fresh cheeses   A general category of cheeses that are uncooked, unripened, and highly perishable. They generally have a creamy texture, a mild or tart and tangy flavor, and moisture content of 40 to 80 percent.

soft ripened cheeses   A general category of molded cheeses that are drained yet maintain a high moisture content.

soy protein concentrate   A product made from processed soybeans containing a very high protein content. Used as a nutrient supplement and meat extender.

specialty presentation   A unique presentation of commonplace ingredients, often regarded as an opportunity for the chef to demonstrate creativity. Specialty presentations can also be considered signature items for the establishment.

spreads   The butter, mayonnaise, cream cheese, or other paste used on bread, crackers, canapé bases, or sandwiches before the fillings are added.

standardized recipe   Ensures that menu items are properly prepared, of consistent quality, and of predictable yield and cost each time they are made. Standardized recipes also identify the particular portion size for the menu items.

studding   To add any flavorful ingredient under the skin of poultry and game birds.

sublimation   The evaporation of solid ice (frozen water molecules) into vapor that occurs when ice blocks are left uncovered in a frozen environment, such as a walk-in freezer.

subtractive sculpture   Sculpting by removing unwanted material from the original mass, such as when sculpting marble.

sulfuring   Using a food additive to dehydrate fruits.

sun drying   The ancient method used to dry food; it uses the warmth from the sun and the natural movement of the air to dehydrate the food.

suprême   (1) The individual breast of any bird, which has been completely removed with the cleaned wing bone attached. (2) A cut of fish taken from a large fillet. The fillet is cut at an angle to remove evenly weighed portions.

surface-ripened cheeses   A general category of cheeses that have ripened from the rind inward as a result of the application of mold, yeast, or bacteria to the surface.

sushi   Cold cooked rice dressed with vinegar that is shaped into bite-sized pieces and topped with raw or cooked fish, or formed into a roll with fish, egg, or vegetables and wrapped in seaweed. Usually based on the combination of several ingredients including fish (raw or cooked), rice, seaweed, vegetables, and condiments (including soy sauce, wasabi, or rice vinegar).

sushi-zu   Japanese term for "seasoned vinegar."

sweetbreads   The thymus glands from the neck and heart of young steers, calves, and lamb.

synthetic casings   Artificial casings made from edible products and used to enclose sausage forcemeat.

T

tails   This term refers to oxtails, which are the tails of beef cattle.

tandoor   A style of cooking using a clay oven that loads from the top. Meats and poultry are seasoned and threaded onto long skewers that are inserted into the oven from the narrow opening on top. Bread is also baked in the oven by slapping it onto the oven's walls.

tapas   Spanish term for appetizer-sized portions.

tartare (tartar)   A term (imprecisely) used to represent a dish prepared with raw meat or seafood. The dish has its origin as a culinary practice popular in medieval times among warring Mongolian and Turkic tribes, known as Tartars, who ate raw meat.

temperature danger zone   A temperature range in which bacteria multiplies rapidly. The U.S. Food and Drug Administration (FDA) has established this range as being between 41° and 135°F (5° and 57°C). (For this reason, foods must not be allowed to remain within this temperature range for more than 2 hours.)

tempered   The slow cooling or reduction of temperature of an ice block when placed in a subfreezing environment.

tenderizing   The actions that are taken to ensure that meat is tender when eaten, including aging, hanging, and marinating as well as physical and mechanical breakdown.

tenderloins   The most tender piece of meat found on any animal, tenderloins are located within the bone structure of the animal.

terrines   A crustless pâté cooked in earthenware, china, or metal mold lined with fat.

thermophilic starter   A sturdy bacterial culture that encourages reactivity in milk heated to higher temperatures.

thermo-shock   Weakening and cracking of the ice block caused by an extreme inconsistency between the internal temperature of the ice and the temperature outside the block. Occurs when the surface of the ice is subjected to a large change in temperature.

timbales   Small pâtés or terrines cooked in a dariole or timbale mold.

tinted cure mix   Another name for Prague powder #1.

tongue   A fleshy moveable organ located in the floor of the mouth. It is generally poached and served hot or cold. It is also pickled and served as a cold dish.

tourné   A cutting technique that results in a football-shaped finished product, generally a vegetable, with seven equal sides and blunt ends.

tournedos (beef)   A circular cut of beef primal short loin, smaller than a filet mignon.

tournedos (fish)   This technique imitates the shape of a beef tournedo, but it is made using fish. A circular cut of fish is surrounded by a different colored fish fillet and pinned together to create the shape of a tournedos.

traditionalist sculptor   An individual who shuns the use of power tools when ice sculpting, choosing instead to use the traditional ice saws and chisels.

trinity of business   The relationship between the needs of the customers, the owners, and the employees as it affects the supervisor.

tripe   Tripe comes from the first and the second stomachs of beef. From the *rumen,* or first stomach, we get the plain tripe that tends to be rather tough. From the second, or the *reticulum,* we get the honeycomb that is the more tender and the more attractive of the two.

tronçon   A slice taken from a large flat fish through the bone and skin, resembling a steak.

truffling   To add fresh truffle to beneath the skin of poultry and game birds.

truss   To secure portions of meat, game, or poultry, using butcher's twine, skewers, or pins, to maintain a desired shape during cooking.

tubers   The swollen tips of underground stems that store energy in the form of starch.

U

unity of command   The management principle that promotes "each employee should report to only one person/supervisor." Each job fits within a hierarchal order of the overall business.

unripened   Cheeses that are made by heating the milk and letting it stand, treating it with a lactic starter to help the acid development, and then cutting and draining the whey from the cheese.

V

variable (controllable) costs (expenses)   An expense in which the decision made by the food service manager can have the effect of increasing or reducing the expense.

veal   The male calves of dairy cows, which are fed on skim milk, whey, and fat. They are specially reared and slaughtered between 8 and 16 weeks old.

vegetables   The edible parts from plants, including the leaves, stalks, roots, tubers, and flowers. They are generally savory, eaten as is or cooked, and served with other savory ingredients.

venison   The meat of the red, fallow, or roe deer.

W

washed rind   Cheese that is washed regularly, with wine or another liquid, during its ripening period to prevent its interior from drying out. The washings also help promote an even bacterial growth across the surface of the cheese, giving a smooth, almost shiny appearance.

webbed game   A term indicating the connecting of the toes by a flap of skin.

wet aging   The aging of meat in vacuum packaging.

whey   The liquid portion of milk that develops after the milk protein has coagulated. Whey contains water, milk, sugar, albuminous proteins, and minerals.

wild game   Any animal that has to be hunted in the wild for food.

working factor   A multiplier used to increase or decrease the quantities of each ingredient, while maintaining a consistent ratio between the ingredients, in order to create the new yield desired by the chef.

wrappings   Edible food materials, such as skin, pastry dough, fat strips, bacon, or vegetables, used to encase food products.

Z

zakuskis   Russian form of appetizers.

Albala, K. (2002). *Eating right in the Renaissance.* Berkley, CA: University of California Press.

Albala, K. (2003). *Food in early modern Europe.* Westport, CT: Greenwood Press.

Ash, J., & Goldstein, S. (1991). *American game cooking.* Reading, MA: Addison-Wesley.

Bailey, A., Ortiz, E. L., & Radecka, H. (1980). *The book of ingredients.* London: Dorling Kindersley.

Barber, K., & Takemura, H. (2002). *Sushi: Taste and technique.* New York: DK Publishing.

Baskette, M. (2001). *The chef manager.* Upper Saddle River, NJ: Prentice Hall.

Bittman, M. (1994). *Fish: The complete guide to buying and cooking.* New York: MacMillan.

Brooks, J. (1988). *The art of accompaniment: Making condiments.* San Francisco: North Point Press.

Carafoli, J. F. (1992). *Food photography and styling.* New York: Amphoto Books.

Carroll, R. (2002). *Home cheese making.* Pownal, VT: Story Books.

Carroll, R., & Carroll, R. (1982). *Cheesemaking made easy.* Pownal, VT: Garden Way Publishing.

Chesser, J. W. (1992). *The art and science of culinary preparation.* St. Augustine, FL: The Educational Institute of the American Culinary Federation.

Ciletti, B. (2000). *Creative pickling at home.* New York: Lark Books.

Ciletti, B. (1999). *Making great cheese at home.* New York: Sterling.

Constable, G. (Ed.). (1981). *The good cook series: Hors d'oeuvre.* Alexandria, VA: Time-Life Books.

Costner, S. (1990). *Great sandwiches.* New York: Crown Publishers.

Culinary Institute of America (Ed.). (2002). *The professional chef* (7th ed.). New York: John Wiley & Sons.

Culinary Institute of America (Ed.). (2004). *Garde manger: The art and craft of the cold kitchen* (2nd ed.). New York: John Wiley & Sons.

Culinary Institute of America (Ed.). (2003). "Mediterranean flavors, American menus—Tasting the future" (p. 4). *CIA Worlds of Flavor Sixth International Conference & Festival.* St. Helena, CA.

Drysdale, J. A. (2001). *Profitable menu planning* (3rd ed.). Upper Saddle River, NJ: Prentice Hall.

Emery, W. (1980). *Culinary design and decoration.* Boston, MA: CBI Publishing.

Garlough, R., Finch, R., & Maxfield, D. (2004). *Ice Sculpting the Modern Way.* Clifton Park, NY: Thomson Delmar Learning.

Giacosa, I. G. (1992). *A taste of ancient Rome* (p. 49). Chicago: University of Chicago.

Gislen, W. (2002). *Professional cooking* (5th ed.). New York: John Wiley & Sons.

Hamaker, S. S. (2001). "Small plates add up to big profits." *Restaurants USA,* May.

Hertzberg, R., et al. (1976). *Putting food by* (2nd ed.). Brattleboro, VT: Bantam Books.

Hobson, P. (1980). *Making cheese, butter, & yogurt.* Pownal, VT: Story Books.

Janericco, T. (1990). *The book of great hors d'oeuvre.* New York: Van Nostrand Reinhold.

Jones, E. (1978). *The world of cheese.* New York: Alfred Knopf.

Kerr, G. (1997). *The gathering place.* Stanwood, WA: Camano Press.

Kinsella, J., & Harvey, D. T. (1996). *Professional charcuterie: Sausage making, curing. terrines, pâtés.* New York: John Wiley & Sons.

Kiple, K., & Ornelas, K. (Eds.). (2000). *The Cambridge world history of food.* Cambridge, UK: Cambridge University Press.

Kurlansky, M. (2002). *Salt: A world history.* New York: Penguin Books.

Kutas, R. (1999). *Great sausage recipes and meat curing* (3rd ed.). Buffalo, NY: The Sausage Maker, Inc.

Labensky, S., & Hause, A. (2002). *On cooking.* (3rd ed.). Upper Saddle River, NJ: Prentice Hall.

Labensky, S., Ingram, G. G., & Labensky, S. R. (2001). *Webster's New World dictionary of culinary arts* (2nd ed.). Upper Saddle River, NJ: Prentice Hall.

Larousse, D. P. (1995). *The hors d'oeuvre bible.* New York: John Wiley & Sons.

Larousse, D. P. (1996). *The professional garde manger.* New York: John Wiley & Sons.

Larousse, D. P. (2000). *More edible art.* New York: John Wiley & Sons.

Lieberman, J. S. (1991). *The complete off-premise caterer.* New York: Van Nostrand Reinhold.

Leto, M. J., & Bode, W. K. H. (1984). *The larder chef.* London, UK: William Heinemann.

Marrone, T. (2002). *Cookin' wild game.* Minnetonka, MN: Creative Publishing International.

McCalman, M., & Gibbons, D. (2002). *The cheese plate.* New York: Clarkson Potter Publishers.

Metz, F., & Ryan, T. (1989). *Taste of gold.* Des Plaines, IL: Cahners Publishing.

Miller, J. E., Hayes, D. K., & Dopson, L. R. (2004). *Basic food and beverage cost control* (3rd ed.). New York: John Wiley & Sons.

Miller, J., Porter, M., & Drummond, K. E. (2002). *Supervision in the hospitality industry* (4th ed.). New York: John Wiley & Sons.

Mueller, T. G. (1987). *The professional chef's book of charcuterie.* New York: Van Nostrand Reinhold.

Nicolas, J. E. (1990). *American fish and shellfish.* New York: Van Nostrand Reinhold.

Pépin, J. (1977). *La technique.* New York: Quadrangle.

Peterson, J. (1999). *Essentials of cooking.* New York: Artisan.

Reader, J. (1988). *Man on earth.* Austin, TX: University of Austin Press.

Reekie, J. (1988). *British charcuterie.* London,: Ward Lock.

Rogers, M. R. (1991). *Creative garnishing.* Philadelphia: Running Press.

Rosen, H. (1985). *Melon garnishing.* Elberon, NJ: International Culinary Consultants.

Rosenblum, M. (1996). *Olives: The life and lore of a noble fruit.* New York: North Point Press.

Sanders, E., Lewis, L., & Fluge, N. (2000). *Catering solutions: For the culinary student, foodservice operator, and caterer.* Upper Saddle River, NJ: Prentice Hall.

Shephard, S. (2000). *Pickled, potted, and canned: How the art and science of food preserving changed the world.* New York: Simon & Schuster.

Shiring, Sr., S. B., Jardine, R. W., & Mills, Jr., R. J. (2001). *Introduction to catering: Ingredients for success.* Clifton Park, NY: Thomson Delmar Learning.

Simonds, N. (2000). *Asian wraps.* New York: William Morrow and Company.

Sloan, E. (2004). "The flavor generation." *Flavor & The Menu.* Fall, p. 20.

Sonnenfeld, A. (2000). *Food: A culinary history.* Harmondsworth, Middlesex, UK: Penguin Books.

Sonnenschmidt, F., & Nicolas, J. F. (1997). *The professional chef's art of garde manger* (5th ed.). New York: John Wiley & Sons.

Splaver, B. R. (1982). *Successful catering* (2nd ed.). New York: Van Nostrand Reinhold.

Stein, G. M. (1981). *Caviar! Caviar! Caviar!* Secaucus, NJ: Lyle Stuart.

Strianese, A. J., & Strianese, P. P. (2001). *Math principles for foodservice* (4th ed.). Clifton Park, NY: Thomson Delmar Learning.

Symons, M. (2000). *A history of cooks and cooking.* Champaign, IL: University of Illinois Press.

Tannahill, R. (1988). *Food in history.* New York: Three Rivers Press.

Time-Life Books. (1980). *Snacks & sandwiches.* Chicago: Author.

Time-Life Books. (1981). *Preserving.* Chicago: Author.

Time-Life Books. (1987). *Variety meats* (2nd ed.). Alexandria, VA: Author.

Toussaint-Samat, M. (2001). *History of food.* Cambridge, MA: Blackwell Publishers.

Trager, J. (1995). *The food chronology.* New York: Henry Holt.

Veale, W. (1989). *Step-by-step garnishing.* Secaucus, NJ: Chartwell Books.

Visser, M. (1991). *The rituals of dinner: The origins, evolution, eccentricities, and meaning of table manners.* New York: Grove Weidenfeld.

Weinzweig, A. (2003). *Zingerman's guide to good eating.* Boston: Houghton Mifflin.

Wilson, C. A. (1991). *Food and dining in Britain* (p. 326). Chicago: Academy Chicago Publishers.

Wise. V. (1986). *American charcuterie.* New York: Viking.

Yudd, R. A. (1990). *Successful buffet management.* New York: Van Nostrand Reinhold.

The following Web sites were used as resources:
http://www.appollo4.bournemouth.ac.uk
http://bbq.about.com
http://www.craft-guild.org
http://www.culinart.net
http://www.fsis.usda.gov
http://www.hotelympia.com
http://www.hungrymonster.com/FoodFacts/
http://www.ianr.unl.edu/pubs/beef
http://muextension.missouri.edu
http://www.niaid.nih.gov/factsheets/foodbornedis.htm
http://www.rmef.org
http://www.scottishchefs.com
http://www.sea-ex.com/fish/preparat.htm
http://www.theworldwidegourmet.com
http://www.veal.org
http://vm.cfsan.fda.gov
http://whatscookingamerica.net
http://www.wheelwrights.org
http://www.woodennickelbuffalo.com

Note: As with any dynamic informational tool, Web sites will change and even disappear from time to time. Any Internet user must be aware of this fact and be prepared to investigate and discover other comparable Web sites.

DATE DUE
